Is the Style Readable?

- [ ] Each sentence understandable on *first* reading   (262)
- [ ] The most information expressed in the fewest words   (271)
- [ ] Related ideas combined for fluency   (279)
- [ ] Sentences varied in construction and length   (281)
- [ ] Each word chosen for exactness, not for camouflage   (284)
- [ ] Concrete and specific language   (291)
- [ ] No triteness, overstatements, euphemisms, or jargon   (286)
- [ ] Tone unbiased and inoffensive   (298)
- [ ] Level of formality appropriate to the situation   (294)

Are Design, Visuals, and Mechanics Appropriate?

- [ ] An inviting and accessible format: white space, fonts, and so on   (355)
- [ ] A design that accommodates audience needs and expectations   (359)
- [ ] Adequate, clear, and informative headings   (367)
- [ ] Adequate visuals, to clarify, emphasize, or organize   (307)
- [ ] Appropriate displays for specific visual purposes   (314)
- [ ] All visuals fully incorporated with the text   (347)
- [ ] All visuals free of distortion   (345)
- [ ] All pages numbered and in order   (360)
- [ ] Supplements that accommodate diverse audience needs   (376)
- [ ] Correct spelling, punctuation, and grammar   (605)

COVER: *A mere fantasy, Klee's "machine" has no practical purpose—in the real world it would produce a meaningless, chirping noise. Another type of noise is produced by information we cannot follow or understand. Skilled communicators transform that information into something we can use.*

Technical Writing
Seventh Edition

John M. Lannon

UNIVERSITY OF MASSACHUSETTS,
DARTMOUTH

An imprint of Addison Wesley Longman, Inc.

New York • Reading, Massachusetts • Menlo Park, California • Harlow, England
Don Mills, Ontario • Sydney • Mexico City • Madrid • Amsterdam

Executive Editor: Anne Elizabeth Smith
Developmental Editor: Thomas Maeglin
Text Design and Project Coordination: Ruttle, Shaw & Wetherill, Inc.
Cover Designer: Mary McDonnell
Cover Illustration: Paul Klee. *Twittering Machine* {Zwitscher-Maschine}. 1922. Watercolor, and pen
    and ink on oil transfer drawing on paper, mounted on cardboard, 25 1/4 × 19" (63.8 × 48.1 cm).
    The Museum of Modern Art, New York. Purchase. Photograph © 1996 The Museum of Modern
    Art, New York.
Art Studio: Ruttle, Shaw & Wetherill, Inc.
Photo Researcher: Nina Page
Electronic Production Manager: Angel Gonzalez Jr.
Manufacturing Manager: Willie Lane
Electronic Page Makeup: Ruttle Graphics, Inc.
Printer and Binder: R. R. Donnelley & Sons Company
Cover Printer: The Lehigh Press, Inc.

Library of Congress Cataloging-in-Publication Data

Lannon, John M.
    Technical Writing / John M. Lannon.—7th ed.
        p.   cm.
    Includes bibliographical references and index.
    ISBN 0-673-52472-8
    1. Technical Writing.   I. Title.
T11.L24   1996
808'.0666—dc20   96-19525
                                            CIP

ISBN 0-673-52472-8

345678910—DOC—999897

Brief Contents

Contents

Preface

TECHNICAL WRITING, seventh edition, is a comprehensive and flexible introduction to technical and professional communication. Designed for classes in which students from a variety of majors are enrolled, the book addresses a wide range of interests. Rhetorical principles are explained, illustrated, and applied to an array of assignments, from brief memos and summaries to formal reports and proposals. To help students develop awareness of audience and accountability, exercises embody the writing demands that arc typical throughout college and on the job.

Organization

Following a brief overview of technical writing in Chapter 1, the remaining text has five major sections:

Part I: Writing for Readers in the Workplace treats job-related writing as a problem-solving process. Students learn to think critically about the informative, persuasive, and ethical dimensions of their communications. Also, they learn about adapting to rapidly changing communication technologies, to interpersonal challenges of collaborative writing, and to the various needs and expectations of global audiences.

Part II: Information Retrieval, Analysis, and Synthesis treats research as a deliberate inquiry process. Students learn to formulate significant research questions; to explore primary and secondary sources; to record, evaluate, interpret, and document their findings; and to summarize for economy, accuracy, and emphasis.

Part III: Structural and Style Elements demonstrates strategies for organizing and expressing messages that readers can follow and understand. Students learn to control their material and to develop a style that connects with readers.

Part IV: Graphic and Design Elements treats the rhetorical implications of graphics, page design, and document supplements. Students learn to enhance a document's access, appeal, visual impact, and usability.

Part V: Specific Documents and Applications applies earlier concepts and strategies to the preparation of technical documents and oral presentations. Various letters, memos, reports, and proposals offer a balance of examples from the workplace and from student writing. Each sample document has been chosen so that students can emulate it easily.

Finally, the **appendix** offers a brief handbook of grammar, usage, and mechanics.

The Foundations of Technical Writing

- More than a value-neutral exercise in "information transfer," workplace communication—whether hand-written, electronically mediated, or face-to-face—is a complex social transaction. Each rhetorical situation places specific interpersonal, ethical, legal, and cultural demands on the writer.
- Writers with no rhetorical awareness overlook the decisions that are crucial for effective writing. Only by defining their rhetorical problem and asking the important questions can writers formulate an effective response to the problem.
- As well as being *communicators,* today's workplace professionals increasingly are *consumers* of information, who need to be skilled in the methods of inquiry, retrieval, evaluation, and interpretation that comprise the research process.
- Although it follows no single, predictable sequence, the writing process is not a collection of random activities; rather, it is a set of deliberate decisions in problem solving. Beyond emulating this or that model document, students need to understand that effective writing requires critical thinking.
- A technical writing classroom typically contains an assortment of students with varied backgrounds. The textbook, then, should offer explanations that are thorough, examples and models that are broadly intelligible, and goals that are rigorous yet collectively achievable. Moreover, the book should be flexible enough to allow for various course plans.
- As an alternative to reiterating the textbook material, classroom workshops apply textbook principles by focusing on the students' writing. These workshops call for an accessible, readable, and engaging book to serve as a comprehensive reference.

New to this Edition

- New material throughout on connecting with global audiences through careful analysis of the specific cultural context for various types of communication.
- In Chapter 4, the importance of *face saving* in any persuasive situation.
- A new Chapter 6, on communicating electronically. Coverage includes Internet resources, E-mail privacy and quality issues, E-mail communication guidelines, paperless documents, hypertext applications in writing and research, computer guidelines for writers.

- Guidelines for reviewing and editing the work of others (Chapter 7).
- A fully revised Chapter 8, on research methods for the information age. Examples of new coverage: achieving adequate depth in research, accessing government records via the Freedom of Information Act, broadening and customizing electronic searches, understanding the essentials and limitations of survey research.
- A fully revised Chapter 9, on critically evaluating and interpreting information. Examples of new coverage: copyright and fair use of printed and electronic information; criteria for evaluating sources, evidence, generalizations, and causal claims; guidelines for reaching accurate conclusions; fallacies inherent in statistical data.
- A new Chapter 10, on documenting research findings. Comprehensive coverage of MLA, APA, and CBE documentation styles; ACW style for documenting unconventional electronic sources (MOOs, FTPs, Telnet sites, and so on); a user-friendly chapter design for easy reference.
- A fully revised Chapter 12, on basic organizing strategies; partitioning and classifying, outlining, paragraphing, and sequencing. Coverage includes outlining and reorganizing on a computer and organizing for cross-cultural audiences.
- A more accessibly designed Chapter 13, on style, with new sections on avoiding excessive informality and offensive usage and on considering cultural differences in the observance of style guidelines.
- A fully revised Chapter 14, focusing on rhetorical considerations in the use of visuals. Examples of new coverage: why and when to use visuals, how to select visuals for a specific purpose and audience, how to compose an art brief, how to use photographs and color, how to explore Web sites for visual resources, how to incorporate visuals with written text.
- New sections in Chapter 15, on page design, the latest desktop publishing technology, and cultural considerations in page design.
- Greater emphasis in Chapter 19 on visual elements of instructions and on usability in online or multimedia documents.
- A fully revised Chapter 20, on workplace letters. Examples of new coverage: telephone and E-mail inquiries, commonly used job-search methods, online employment resources, résumé scanning, an example of a computer-scannable résumé, criteria for evaluating an employment offer.
- In Chapter 22, a new—and successful—proposal for a high-school building project, aimed at a wide readership.
- In Chapter 23, two fully updated analytical reports, chosen to engage a broad range of student interests.
- A fully revised Chapter 24, offering concise, practical advice on oral presentations. Examples of new coverage: avoiding presentation pitfalls, analyzing the listeners and the speaking situation, planning visuals, creating a storyboard, selecting media, preparing readable and understandable

visuals, managing the presentation, managing listener questions, addressing cross-cultural audiences.

- A revised appendix on grammar, usage, and mechanics, redesigned for easy access and cross reference.
- "In Brief" discussions of interpersonal and technology issues that are shaping workplace writing and oral communication. Examples include gender and cultural differences in collaborative work; online research, copyright, and documentation; usability in electronic documents; electronic job hunting.
- More on rhetorical, legal, ethical, and cultural considerations in word choice, definitions, product descriptions, instructions, and other forms of communication.
- New and updated examples throughout.
- A new art program and greater emphasis on visual communication and page design.
- More annotated writing samples, to highlight salient features.
- Expanded exercises in critical thinking and applications suitable for collaborative work. Examples: Chapter 2, exercise 1, Chapter 13, exercise 21; collaborative projects in Chapters 6, 8, 14 (#3), 18 (#3), and 19 (#5).
- A comprehensive educational package including an instructor's manual with test bank, chapter quizzes, and master sheets for overhead or opaque projection, and *Writer's Workshop* software (with an online handbook, writing prompts, and a citation formatting tool for documentation in MLA and APA style).

Acknowledgments

Many of the refinements in this edition were inspired by generous and insightful suggestions from the following reviewers: Mary Beth Bamforth, Wake Technical Community College; Alma Bryant, University of South Florida; Charlie Dawkins, Virginia Polytechnical Institute and State University; Robert Hogge, Weber State University; Gloria Jaffe, University of Central Florida; Jack Jobst, Michigan Technical University; Maxine Turner, Georgia Institute of Technology; JoAnn Kubala, Southwest Texas State University; Sherry Little, San Diego State University; Linda Loehr, Northeastern University; James L. McKenna, San Jacinto College; Mohsen Mirshafiei, California State University, Fullerton; Thomas Murphy, Mansfield University; Mark Rollins, Ohio University, Athens; Beverly Sauer, Carnegie Mellon University; Susan Simon, City College of the City University of New York; Anne Thomas, San Jacinto College; Jeff Wedge, Embry-Riddle University; Kristin Woolever, Northeastern University; Don Zimmerman, Colorado State University.

Dr. George Redman of Benedict College provided valuable advice on matters of content and design.

At the University of Massachusetts, Raymond Dumont was a constant source of help and ideas. For much of the new material on research, I am indebted to librarian Ross LaBaugh. Many other colleagues, graduate students, teaching assistants, and alumni offered countless suggestions. My students gave me feedback and inspiration.

For the many improvements in this edition, I am indebted to exceptional editorial guidance and support. Executive Editor Anne Elizabeth Smith provided the encouragement and the vision that inspired this project. Developmental Editor Tom Maeglin brought the energy and insight that transformed the manuscript from adequate to excellent. Development Director Patricia Rossi endured my many missed deadlines patiently and gracefully. Project Manager Janet Nuciforo gave expert attention to every detail on a virtually endless list. No collection of superlatives would be adequate to express my appreciation for your help and your friendship. Thank you all.

A special thank you to those I love: Chega, Daniel, Sarah, Patrick, and Max.

John M. Lannon

Introduction to Technical Writing

CHAPTER

1

As a technical writer, you communicate and interpret specialized information for your readers' use. Readers may need your information to perform a task, answer a question, solve a problem, or make a decision. Whether you write a memo, letter, report, or manual, the document must advance the goals of your readers and of the company or organization you represent.

Technical Writing Serves Practical Needs

Unlike poetry or fiction, which appeal mainly to our *imagination*, technical documents appeal to our *understanding*. Technical writing therefore rarely seeks to entertain, create suspense, or invite differing interpretations. Those of you who have written any type of lab or research report already know that technical writing has little room for ambiguity.

To serve practical needs in the workplace, technical documents must be reader oriented and efficient.

Technical Documents Are Reader Oriented

Instead of focusing on the writer's desire for self-expression, a technical document addresses the reader's desire for information. This doesn't mean your writing should sound like something produced by a robot, without any personality (or *voice*) at all. Your document may in fact reveal a lot about you (your competence, knowledge, integrity, and so on), but it rarely focuses on you personally. Readers are interested in *what you have done, in what you recommend*, or in *how you speak for your company;* they have only a professional interest in *who you are* (your feelings, hopes, dreams, visions). A personal essay, then, would not be technical writing. Consider this essay fragment:

Focuses on the feelings

> Computers are not a particularly forgiving breed. The wrong key struck or the wrong command typed is almost sure to avenge itself on the inattentive user by banishing the document to some electronic trash can.

This personal view communicates a good deal about the writer's resentment and anxiety but very little about computers themselves.

The following example can be called technical writing because it focuses (see italics) on the subject, on what the writer has done, and on what the reader should do:

Focuses on the subject, actions taken, and actions required

> On VR 320 terminals, *the BREAK key* is adjacent to keys used for text editing and special functions. Too often, users inadvertently strike the BREAK key, causing the program to quit prematurely. To prevent the problem, *we have* modified all database management terminals: to quit a program, *you must* now strike BREAK twice successively.

This next example also can be called technical writing because it focuses on what the writer recommends:

Focuses on the recommendation

> I recommend that our VAX 780 hardware be upgraded by a maximum addition to main memory, a disk input/output control unit, and an additional disk storage unit. This expansion will (1) increase the number of simultaneous terminal users from 60 to 80, (2) increase the system's responsiveness, and (3) provide sorely needed disk storage for word processing and company databases.

And so your document never makes you "disappear," but it does focus on that which is most important to the readers.

Technical Documents Strive for Efficiency

Professors read to *test* our knowledge; colleagues, customers, and supervisors read to *use* our knowledge. Workplace readers hate waste and demand efficiency; instead of reading a document from beginning to end, they are more likely to use it for reference, and want only as much as they need: "When it comes to memos, letters, proposals, and reports, there's no extra credit for extra words. And no praise for elegant prose. Bosses want employees to get to the point—quickly, clearly, and concisely" (Spruell 32). Efficient documents save time and energy in the workplace.

In any system, efficiency is the rate of useful output to input. For the product that comes out, how much energy goes in?

When a system is efficient, the output nearly equals the input.

Similarly, a document's efficiency can be measured by how hard readers work to understand the message. Is the product worth the reader's effort?

No reader should have to spend ten minutes deciphering a message worth only five minutes. Consider, for example, this wordy message:

An inefficient message

> At this point in time, we are presently awaiting an on-site inspection by vendor representatives relative to electrical utilization adaptations necessary for the new computer installation. Meanwhile, all staff are asked to respect the off-limits designation of said location, as requested, due to liability insurance provisions requiring the on-line status of the computer.

Inefficient documents drain the readers' energy; they are too easily misinterpreted; they waste time and money. Notice how hard we had to work with the previous message to extract information that could be expressed this efficiently:

A more efficient message

> Hardware consultants soon will inspect our new computer room to recommend appropriate wiring. Because our insurance covers only an *operational* computer, this room must remain off limits until the computer is fully installed.

When readers sense they are working too hard, they tune out the message—or they stop reading altogether.

Inefficient documents have varied origins. Even when the information is accurate, errors like the following make readers work too hard:

Causes of inefficient documents

- more (or less) information than readers need
- irrelevant or uninterpreted information
- confusing organization
- jargon or vague technical expressions readers cannot understand
- more words than readers need
- uninviting appearance or confusing layout
- no visual aids when readers need or expect them

An efficient document sorts, organizes, and interprets its information to suit the audience's needs, abilities, and interests.

Instead of merely happening, an efficient document is carefully designed to include these elements:

Elements of efficient documents

- *content* that makes the document worth reading
- *organization* that guides readers and emphasizes important material
- *style* that is economical and easy to read
- *visuals* (graphs, diagrams, pictures) that clarify concepts and relationships, and that substitute for words whenever possible
- *format* (layout, typeface) that is accessible and appealing
- *supplements* (abstracts, appendices) that enable readers with different needs to read only those sections required for their work

Reader orientation and efficiency are more than abstract rules: Writers are accountable for their documents. In questions of liability, faulty writing is no different from any other faulty product. If your inaccurate or unclear or incomplete information leads to injury or damage or loss, *you* can be held legally responsible.

Writing Is Part of Most Careers

Although you might not anticipate a career as a "writer," your writing skills will be tested routinely in situations like these:

Ways in which your career may test your writing skills

- proposing various projects to management or to clients
- writing progress reports to the boss
- contributing articles to the employee newsletter
- describing a product to employees or customers
- writing procedures and instructions for employees or customers
- justifying to management a request for funding or personnel
- editing and reviewing documents written by colleagues
- designing material that will be read on a computer screen or transformed into sound and pictures

You might write alone or as part of a collaborative team, and you will face strict deadlines.

Your value to any organization will depend on how clearly and persuasively you communicate. Many working professionals spend at least 40 percent of their time writing or dealing with someone else's writing. The higher their position, the more they write (Barnum and Fisher 9–11). Here is a corporate executive's description of some audiences you can expect:

> The technical graduate entering industry today will, in all probability, spend a portion of his or her career explaining technology to lawyers—some friendly and some not—to consumers, to legislators or judges, to bureaucrats, to environmentalists and to representatives of the press. (Florman 23)

These audiences, among countless others, expect to read efficient documents.

Here is what two top managers for an auto maker say about the effect a document can have on the organization *and* on the writer:

Writing is an indicator of job performance

> A written report is often the only record which is made of results that have come out of years of thought and effort. It is used to judge the value of the *person's* work and serves as the foundation for all future action on the project. If it is written clearly and precisely, it is accepted as the result of sound reasoning and careful observation. If it is poorly written, the results presented in it are placed in a bad light and are often dismissed as the work of a careless or incompetent worker. (Richards and Richards 6)

Good writing gives you and your ideas *visibility* and *authority* within your organization. Bad writing, on the other hand, is not only useless to readers and politically damaging to the writer, but also expensive: Written communication[1] in American business and industry costs billions of dollars yearly. More than 60 percent of the writing is inefficient: unclear, misleading, irrelevant, deceptive, or otherwise wasteful of time and money (Max 5–6).

---

1. "Written communication" includes the full range of activities involved in preparing, producing, processing, storing, and retrieving documents for re-use.

Whatever your career plans, you can expect to be a part-time technical writer. Employers first judge your writing by your application letter and résumé. In a large organization, your future may be decided by executives you've never met. One concrete measure of your job performance will be your letters, memos, and reports. As you advance, communication skill becomes more important than technical background. The higher your goals, the better you need to communicate.

The Information Age Requires Excellent Writing Skills

Desktop publishing (DTP) systems rapidly are eliminating clerical support positions in countless organizations. Gone are the days in which writers could count on secretarial help in "fixing up" a document. In today's electronic office, each writer is responsible for a document's typing, proofreading, editing, page layout, and distribution.

More importantly, information has become our prized commodity:

Information is the ultimate product

The new source of wealth is not material; it is information, knowledge applied to work to create value. The pursuit of wealth is now largely the pursuit of information, and the application of information to the means of production (Wriston 8).

In education, industry, business, and government, people create, gather, analyze, and distribute information electronically around the world. But whether the information itself finally appears on a printed page or a computer screen, it usually needs to be *written*. A computer can transmit data (or facts), but it cannot give *meaning* to the information—only the writer and reader can do so.

Computer networks have increased the speed and volume of communication, but excessive information actually can *impede* or *prevent* communication. Despite our advances in communication technology, information still needs to be processed by the human brain, as depicted in Figure 1.1. Unfortunately, it is easy for people to process information erroneously (to wrongly interpret the *meaning*).

Accidents that make headlines often result from human error in processing information—situations in which the complexity of information overwhelms the person receiving it (Wickens 2). The 1979 release of radiation and near-meltdown at Three-Mile Island; the 1984 chemical explosion in Bhopal, India; the 1989 runway collision at Los Angeles airport—these disasters occurred because vital information had been misunderstood. Similar but less publicized disasters occur routinely in the workplace. Today, more than ever, writers need to sort, organize, and interpret their material so readers can understand it.

IN BRIEF

Writing Reaches a Global Audience

Electronically linked, our global community shares social, political, and financial interests that demand cooperation as well as competition. Multinational corporations often use parts that are manufactured in one country and shipped to another for assembly into a product that will be marketed elsewhere. For example, cars may be assembled in the United States for a German automaker, or farm equipment may be manufactured in East Asia for a U.S. company. Medical, environmental, and other research crosses national boundaries, and professionals in all fields transact with colleagues from other cultures.

What kinds of documents might address global audiences? Here is a sample (Weymouth 143):

- scientific reports and articles on AIDS and other diseases
- studies of global pollution and industrial emissions
- specifications for hydroelectric dams and other engineering projects
- operating instructions for appliances and electronic equipment
- catalogs, promotional literature, and repair manuals
- contracts and business agreements

To communicate effectively across cultural and national boundaries, any document must respect not only language differences, but *cultural* differences as well. One writer offers this helpful definition of *culture:*

Our accumulated knowledge and experiences, beliefs and values, attitudes and roles—in other words, our cultures—shape us as individuals and differentiate us as a people. Our cultures, inbred through family life, religious training, and educational and work experiences. . . . manifest themselves . . . in our thoughts and feelings, our actions and reactions, and our views of the world.

Most important for communicators, our cultures manifest themselves in our information needs and our styles of communication. In other words, our cultures define our expectations as to how information should be organized, what should be included in its content, and how it should be expressed. (Hein 125)

Cultures differ over which behaviors seem appropriate or inappropriate for social interaction, business relationships, contract negotiation, and communication practices. A communication style considered perfectly acceptable in one culture may be offensive elsewhere.

Effective communicators recognize these differences but withhold judgment or evaluation, focusing instead on *similarities.* For example, needs assessment and needs satisfaction know no international boundaries; technical solutions are technical solutions without regard to nationality, creed, or language; courtesy and goodwill are universal values. In the diverse global context, the writer's challenge is to establish trust and enhance human relationships.

✯ ✯ ✯

Figure 1.1
The Brain Processes
Information, to Find
Meaning

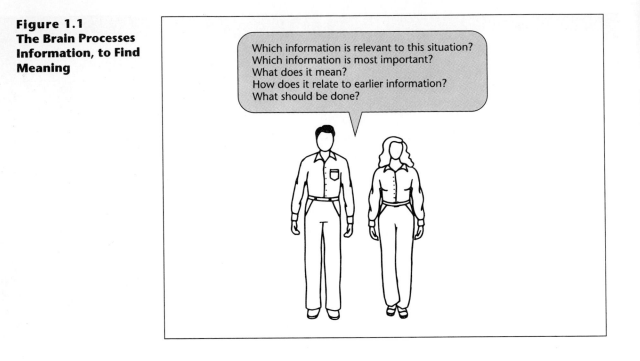

Today's readers often lack technical expertise, but they do *use* the technology. Some nontechnical workers who rely on technology include the manager using a teleconferencing network, the data-entry clerk using a computer terminal, or the bank teller using a check-verification system. Writers tell laypeople how to use software, or how to operate hardware, medical equipment, telephone systems, precision instruments, and all kinds of automated products, from computer games to microwave ovens. With so much information required, and so much available, no writer can afford to "let the facts speak for themselves."

✅ EXERCISES

1. Locate a brief example of a technical document (or a section of one). Make a photocopy, bring it to class, and explain why your selection can be called technical writing.

2. Research the kinds of writing you will do in your future career. (Begin with the *Dictionary of Occupational Titles* in your library.) You might interview a member of your chosen profession. Why will you write on the job? For whom will you write? Explain in a memo to your instructor. (See pages 506, 507 for memo elements and format.)

3. In a memo to your instructor, describe the skills you seek to develop in your technical writing course. How exactly will you apply these skills to your career?

4. Assume a friend in your major thinks that writing skills are obsolete, and that secretaries or word processors can fix any writing. Write your friend a letter, explaining why you think these assumptions are mistaken. Use examples to support your position. (See pages 467–74 for letter elements and format.)

5. Write a memo to your boss, justifying reimbursement for this course. Explain how the course will help you become more effective on the job.

✅ COLLABORATIVE PROJECT
Introducing a Classmate

Class members often will work together this semester. So that everyone can become acquainted, your task is to introduce to the class the person seated next to you. (That person, in turn, will introduce you.) To prepare your introduction, follow this procedure:

a. Exchange with your neighbor whatever personal information you think the class needs: background, major, computer experience (E-mail, Internet, other), career plans, communication needs of your intended profession, and so on. Each person gets five minutes to tell her or his story.

b. Take careful notes; ask questions if you need to.

c. Take your notes home and select only what you think will be useful to the class.

d. Prepare a one-page memo telling your classmates *who this person is.* (See pages 506, 507 for memo elements and format.)

e. Ask your neighbor to review the memo for accuracy; revise as needed.

f. Present the class with a two-minute oral paraphrase of your memo, and submit a copy of the memo to your instructor.

PART

I

Writing for Readers in the Workplace

Problem Solving in Workplace Writing

Technical Writers Face Interrelated Problems

Problem Solving Requires Critical Thinking

Critical Thinking Is Enhanced by Collaboration

Guidelines for Writing Collaboratively

V IRTUALLY all professionals specialize in solving problems (how to build this bridge, how to repair that equipment, how to improve this product, how to diagnose that ailment). Central to all specialized activity is the problem of *communicating* effectively—a complex problem we might examine more easily by considering several subordinate problems.

Technical Writers Face Interrelated Problems

No matter how sophisticated our communication technology, computers cannot *think* for us. More specifically, computers cannot offer solutions to the problems encountered by all people who write in the technical professions:

- the **information problem,** because different readers in different situations have different information needs
- the **persuasion problem,** because people often disagree about what the information means and about what should be done
- the **ethics problem,** because the interests of your company may conflict with the interests of your readers
- the **global context,** because diverse people work together on information for a diverse audience

Chapter 1 explained that information has to have meaning for its audience. But people differ in their interpretations of the facts, and so they may need persuading that one viewpoint is preferable to another. Persuasion, however, is a powerful and sometimes unethical strategy. Even the most useful and efficient document could deceive or harm. Therefore, solving the persuasion problem doesn't mean manipulating readers by using "whatever works," but rather building a case from honest and reasonable interpretation of the facts. Figure 2.1 offers one way of visualizing how these problems relate.

The scenarios that follow illustrate how a typical professional confronts the problems of communicating in the workplace.

The Information Problem

"Can I provide accurate and useful information?"

Sarah Burnes was hired two months ago as a chemical engineer for Millisun, a leading maker of cameras, multipurpose film, and photographic equipment. Sarah's first major assignment is to evaluate the plant's incoming and outgoing water. (Waterborne contaminants can taint film during production, and the production process itself can pollute outgoing water.) Management wants an answer to this question: How often should we change water filters? The filters are very expensive and difficult to change, halting production for up to a day at a time. The company wants as much "mileage" as possible from these filters, without incurring government fines or tainting its film production.

Sarah will study endless printouts of chemical analysis, review current research, do some testing of her own, and consult with her colleagues. When she

**Figure 2.1
Writers Face Four
Related Problems**

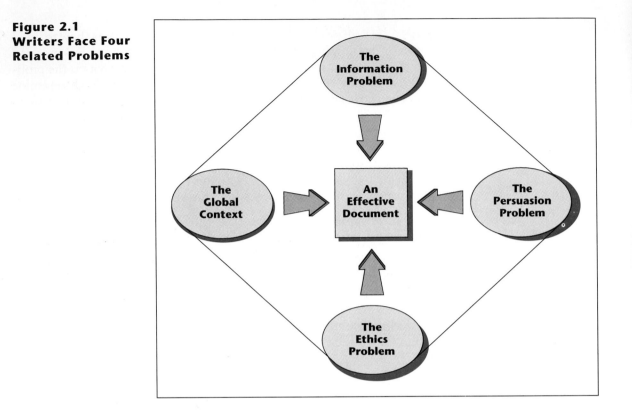

finally decides on what all the data mean, Sarah will prepare a recommendation report for her bosses.

Later, she will collaborate with the company training manager and the
maintenance supervisor to prepare a manual, instructing employees how to
check and change the filters. Trying to cut secretarial and printing costs, the
company has asked Sarah to design this manual using its new desktop publishing system. ■

Sarah's report, above all, needs to be accurate; otherwise, the company gets
fined or lowers production. Once she has processed all the information, she
has to solve the problem of giving readers what they need: *How much
explaining should I do? How will I organize? Do I need visuals?* And so on.

In other situations, Sarah will face a persuasion problem as well: for
example, when decisions must be made or actions taken on the basis of
incomplete or inconclusive facts or conflicting interpretations (Hauser 72).
In these instances, Sarah will seek reader acceptance for *her* view. Her writing
will have to be persuasive as well as informative.

The Persuasion Problem

"Can I influence readers to see things my way?"

Millisun and other electronics producers are located on the shores of a small harbor, the port for a major fishing fleet. For twenty years prior to 1980, these companies discharged directly into the harbor effluents containing metal compounds, PCBs, and other toxins. Sarah is on a multi-company team, assigned to work with the Environmental Protection Agency to clean up the harbor. Much of the team's collaboration occurs via electronic mail.

Enraged local citizens are demanding immediate action, and the companies themselves are anxious to end this public-relations nightmare. But the team's analysis reveals that any type of cleanup would stir up harbor sediment, possibly dispersing the solution into surrounding waters and the atmosphere. (Many of the contaminants are airborne.) Premature action actually might *increase* danger, but team members disagree on the degree of risk and on how to proceed.

Sarah's communication here takes on a persuasive dimension: She and her team members first have to resolve their own conflicts and produce an environmental-impact report that reflects the team's consensus. If the report recommends further study, Sarah will have to justify the delays to her bosses and the public relations office. She will have to make readers understand the dangers as well as she understands them. ■

In the above situation, the facts are neither complete nor conclusive, and views differ about what these facts mean. Sarah will have to balance the various political pressures and make a case for *her* interpretation. Moreover, as company *spokesperson*, Sarah will be expected to take a position that protects her company's interests. Some elements of Sarah's persuasion problem: *Are other interpretations possible? Is there a better way? Can I expect political fallout?*

Whenever she writes, Sarah will have to reckon with the ethical implications of her writing, with the question of "doing the right thing." Some situations will present hard choices. For instance, Sarah might feel pressured to overlook or sugarcoat or suppress facts that would be costly or embarrassing to her company. Or sometimes the best technical solution to a problem might be a poor solution in human terms (as when a heavy industry decreases local pollution by building a smokestack to disperse the emissions over hundreds of miles).

The Ethics Problem

"Can I be honest and still keep my job?"

To ensure compliance with OSHA[1] standards for worker safety, Sarah is assigned to test the air purification system in Millisun's chemical division. After finding the filters hopelessly clogged, she decides to test the air quality and discovers dangerous levels of benzene (a potent carcinogen). She reports these findings in a memo to the production manager, with an urgent recommendation that all employees be tested for benzene poisoning. The manager phones and tells Sarah to "have the filters replaced, and forget about it." Now Sarah

---

1. Occupational Safety and Health Administration.

has to decide what to do next: bury the memo in some file cabinet or defy her boss and send copies to other readers who might take action. ■

Situations that jeopardize truth and fairness present the hardest choices of all: remain silent and look the other way or speak out and risk being fired. Some elements of Sarah's ethics problem: *Is this fair? Who might benefit or suffer? What other consequences could this have?*

In addition to solving these various problems, Sarah has to reckon with the implications of communication technology: Much of her writing, done at the computer, will be produced in collaboration with others (editors, managers, graphic artists), and her audience will extend beyond her own culture.

The Global Context

"Can I connect with all these different audiences?"

Recent mergers have transformed Millisun into a multinational corporation with branches in eleven countries, all connected by computer network. Sarah can expect to collaborate with coworkers from diverse cultures on research and development, with government agencies of the host countries on safety issues, patents and licensing rights, product liability laws, and environmental concerns.

In order to standardize the sensitive management of the toxic, volatile, and even explosive chemicals used in film production, Millisun is developing automated procedures for quality control, troubleshooting, and emergency response to chemical leakage. Sarah has been assigned to work with a group of colleagues to prepare computer-based instructional packages for all personnel involved in Millisun's chemical management worldwide. ■

Sarah will have to develop working relationships with people she has never met, people she knows only via an electronic medium.

For Sarah Burnes or any of us, writing is a process of *discovering* what we want to say, "a way to end up thinking something [we] couldn't have started out thinking" (Elbow 15). Throughout this process in the workplace, we rarely work alone, but instead rely on others for information, help in writing, and feedback (Grice 29–30). We must satisfy not only our audience, but also our employer, whose goals and values ultimately shape the document (Selzer 46–47). Almost any writing for readers outside our organization will be *reviewed* for accuracy, appropriateness, usefulness, and legality before it is finally approved (Kleimann 521).

Problem Solving Requires Critical Thinking

Because writing is nothing less than *thinking* on paper, it never is merely a by-the-numbers exercise. But even though we cannot boil writing down to a formula, we can improve our chances for success through the strategy of *critical thinking*.

Critical thinking is a process of testing the worth of information or the strength of our ideas. Instead of accepting the idea at face value, we examine, evaluate, verify, analyze, weigh alternatives, and consider consequences—at every stage of that idea's development. We employ critical thinking to examine our evidence and our reasoning, to discover new connections and new possibilities, and to test the effectiveness and the limits of our solutions.

To solve writing problems as critical thinkers, we carry out the four stages of the *writing process*:

1. We work with the information.
2. We plan the document.
3. We draft the document.
4. We revise the document.

Each stage calls for deliberate decisions. And during this process our best ideas often occur (Dumont 14).

One engineering professional describes how critical thinking enriches every stage of the writing process:

Writing sharpens thinking

> Good writing is a process of thinking, writing, revising, thinking, and revising, until the idea is fully developed. An engineer can develop better perspectives and even new technical concepts when writing a report of a project. Many an engineer, at the completion of a laboratory project, senses a new interpretation or sees a defect in the results and goes back to the laboratory for additional data, a more thorough analysis, or a modified design. (Franke 13)

Figure 2.2 depicts critical thinking during the writing process. As the arrows indicate, no one stage of the process is complete until *all* stages are complete. Like the exposed tip of an iceberg, the finished document provides the only visible evidence of our labor. On the job, we often need to complete these stages under pressure of deadline.

Figure 2.3 lists the kinds of questions we answer in moving through the writing process. Chapter 7 will illustrate one writer's critical thinking as he prepares a sensitive report for his bosses.

Critical Thinking Is Enhanced by Collaboration

Electronic communication allows more and more writing to be done collaboratively. Workplace documents (especially long reports, proposals, or manuals) are produced by teams who share information, expertise, ideas, and responsibilities. Production of a software manual, for instance, relies on writers, programmers, software engineers, graphic artists, editors, reviewers, marketing personnel, and lawyers (Debs, "Recent Research" 477).

Successful collaboration brings together the best that each team member has to offer. It enhances critical thinking by providing feedback, new perspective, group support, and the chance to test one's ideas in group discussion.

**Figure 2.2
A Critical-Thinking
Model for
Technical Writing**

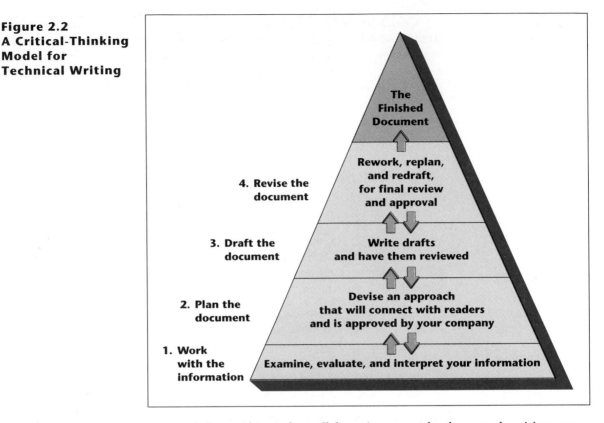

Not all members of a collaborative team do the actual writing; some might research, edit, proofread, or test the document's *usability*. But even a document with a single author can be a collaborative product, as depicted in Figure 2.4. The more important the document, the more it will be reviewed.

As in any group effort, collaborative writers can experience the types of conflict depicted in Figure 2.5. Members might fail to get along because of differences in personality, working style, commitment, standards, or ability to take criticism. Some might disagree about exactly what or how much the group should accomplish, who should do what, or who should have the final say. Some might feel intimidated or hesitant to speak out.[2] These interpersonal problems actually can worsen when the group transacts exclusively via electronic network.

In any group, members have to find ways of expressing their views persuasively, of accepting constructive criticism, or getting along and reaching agreement with others who hold different views. These skills are essential for overcoming personal differences so the group can achieve its goal.

---

2. Adapted from Bogert and Butt 51; Burnett 533–34; Hill Duin 45–46; Debs, "Collaborative Writing" 38; Nelson and Smith 61.

1. Work with the information:

- Have I defined the problem accurately?
- Is the information complete, accurate, reliable, and unbiased?
- Can it be verified?
- How much of it is useful?
- Do I need more information?
- What do these facts mean?
- What connections seem to emerge?
- Do the facts conflict?
- Are other interpretations or conclusions possible?
- Is a balance of viewpoints represented?
- What, if anything, should be done?
- Is it honest and fair?
- Is there a better way?
- What are the risks and benefits?
- What other consequences might this have?
- Should I reconsider?

2. Plan your document:

- When is it due?
- What do I want it to do?
- Who is my audience, and why will they read it?
- What do they need to know?
- What are the "political realities" (feelings, egos, cultural differences, etc.)?
- How will I organize?
- What format and visuals should I use?
- Whose help will I need?

3. Draft your document:

- How do I begin, and what comes next?
- How much is enough?
- What can I leave out?
- Am I forgetting anything?
- How will I end?
- Who needs to review my drafts?

4. Revise your document:

- How does this draft measure up?
- Does it do what I want it to do?
- Is the content worthwhile?
- Is the organization sensible?
- Is the style readable?
- Is everything easy to find?
- Does the format look good?
- Is everything accurate, complete, appropriate, and correct?
- Who needs to review and approve the final version?
- Does it advance my organization's goals?
- Does it advance my audience's goals?

Figure 2.3   Critical Thinking in the Writing Process

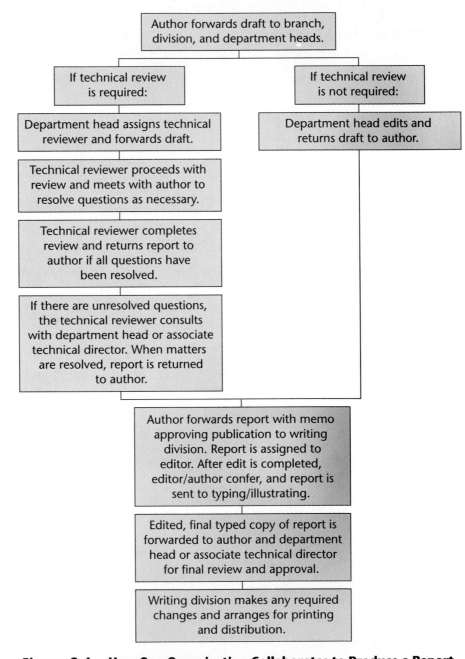

Figure 2.4   How One Organization Collaborates to Produce a Report
Source: NUSC Technical Publication Guidelines. *Naval Underwater Systems Center, Technical Information Dept.*

Figure 2.5   How Pressures of a Group Effort Can Add Up

Guidelines for Writing Collaboratively

The following guidelines[3] focus on projects in which people meet face-to-face but can apply as well to electronically mediated collaboration.

1. *Appoint a group manager.* This person will assign tasks, enforce deadlines, conduct meetings, consult with supervisors (or instructor), and generally run the show.

2. *Define a clear and definite goal.* Compose a purpose statement that spells out the project's goal and the group's plan for achieving it. Be sure each member understands the goal.

3. *Decide how the group will be organized.* Some possibilities:

   a. The group researches and plans together, but each person writes a different part of the document.

   b. Some members plan and research; one person writes a complete draft; others review, edit, revise, and produce the final version.

3. Adapted from Debs, "Collaborative Writing" 38–41; Hill-Duin 45–46; Hulbert, "Developing" 53–54; McGuire 467–68; Morgan 540–41.

Whatever the arrangement, the final revision should display a consistent style throughout—as if written by one person only.

4. *Divide the task.* Who will be responsible for which parts of the document or which phases of the project? Should one person alone do the final revision? Which jobs are hardest? Who is best at doing what (writing, editing, layout, graphics, oral presentation)? Who will make final decisions?

5. *Establish a timetable.* Specific completion dates for each phase will keep everyone focused on what is due, and when.

6. *Decide on a meeting schedule and format.* How often will the group meet, and for how long? In or out of class? Who will take notes (or minutes)? Will the instructor attend or participate?

7. *Establish a procedure for responding to the work of other members.* Will reviewing and editing be done in writing, face-to-face, as a group, one-on-one, or via computer? Will this process be supervised by the project manager or the instructor?

8. *Establish procedures for dealing with group problems.* How will gripes and disagreements be aired (to the manager, the whole group, the offending individual)? How will disputes be resolved (by vote, the manager, the instructor)? How will irrelevant discussion be avoided or curtailed? Expect some conflict, but try to use it positively, and try to identify a natural peacemaker in the group.

9. *Decide how to evaluate each member's contribution.* Will the manager assess each member's performance and, in turn, be evaluated by each member? Will members evaluate each other? What are the criteria? Figure 2.6 depicts one possible form for a manager's evaluation of members. Equivalent criteria for evaluating the manager include open mindedness, ability to organize the team, fairness in assigning tasks, ability to resolve conflicts, or ability to motivate. (Members might keep a journal of personal observations, for overall evaluation of the project.)

10. *Prepare a project management plan.* Figure 2.7 depicts a sample plan sheet. Distribute completed copies to members and the instructor.

11. *Submit progress reports regularly.* Progress reports enable everyone to track activities, problems, and rate of progress.

Beyond these guidelines, respect for other people's views and willingness to listen are essential ingredients for successful collaboration.

IN BRIEF

Gender and Cultural Differences In Collaborative Groups

Any collaborative effort stands the best chance of succeeding when each group member feels included. A big mistake is to ignore personal differences, to assume everyone shares one viewpoint, one communication style, one approach to problem solving.

Collaboration often involves working with peers—those considered of equal status, rank, and expertise. In addition to the personality differences that lead to group conflict, gender and cultural differences in collaborative groups can create perceptions of inequality.

Gender Differences

Recent research on ways men and women communicate in meetings indicates a definite gender gap. Communication specialist Kathleen Kelley-Reardon offers this assessment of gender differences in workplace communication:

> "Women and men operate according to communication rules for their gender, what experts call 'gender codes.' They learn, for example, to show gratitude, ask for help, take control, and express emotion, deference, and commitment in different ways" (88–89).

Professor Kelley-Reardon describes specific elements of a female gender code: Women are more likely than men to take as much time as needed to explore an issue, build consensus and relationship among members, use tact in expressing views, use care in choosing their words, consider the listener's feelings, speak softly, allow interruptions, make requests instead of giving commands ("Could I have the report by Friday?" versus "Have this ready by Friday."), preface assertions in ways that avoid offending ("I don't want to seem disagreeable here, but . . .").

One study of mixed-gender interaction among peers indicates that women tend to be more agreeable, solicit and admit the merits of other opinions, ask questions, and express uncertainty (e.g., with qualifiers such as *maybe, probably, it seems as if*) more often than men (Wojahn 747).

None of these traits, of course, is gender specific. Some people—regardless of gender—are more soft-spoken, contemplative, and reflective. But such traits most often are attributed to the "feminine" stereotype. Moreover, any woman who breaches the gender code, for instance, by being assertive, may be seen by peers as "too controlling" (Kelley-Reardon 6). In fact, studies suggest that women have less freedom than male peers to alter their communication strategies: less-assertive males often are still considered persuasive whereas more-assertive females often are not (Perloff 273).

Cultural Differences

International business expert David A. Victor describes cultural codes that influence inter-action in collaborative groups: Some cultures value silence more than speech, intuition and ambiguity more than hard evidence or data, politeness and personal relationships more than business relationships.

Cultures differ in their perceptions of time. Some want to get the job done immediately; others take as long as needed to weigh all the issues, engage in small talk and digressions, inquire about family, health, and other personal matters.

Cultures may differ in their willingness to express disagreement, question or be questioned, leave things unstated, touch, shake hands, kiss, hug, or backslap.

IN BRIEF

Direct eye contact is not always a good indicator of listening. In some cultures it is offensive. Other eye movements such as squinting, closing the eyes, staring away, staring at legs or other body parts are acceptable in some cultures but insulting in others.

Influence of Online Communication

Some observers argue that online communication eliminates many such problems encountered in face-to-face meetings, because every participant, in effect, has as much time as needed to contribute to the electronically mediated discussion. Also, "status cues" such as age, gender, appearance, or ethnicity virtually disappear online. Other observers disagree, arguing that the interpersonal chemistry of communication transcends specific media (Wojahn 747–48).

☆ ☆ ☆

☑ EXERCISES

(Individual or Collaborative)

1. As you respond to the following scenario,[4] consider carefully the information, persuasion, and ethical problems involved (and be prepared to discuss them in class).

   You are Manager of Product Development at High-Tech Toys, Inc. You need to send a memo to the Vice President of Information Services, explaining the following:

   a. The photocopier in your department often is out of order.

   b. The photocopier seldom is repaired satisfactorily.

   c. Either the machine is faulty or the repairperson is incompetent (but this person always appears promptly and cheerfully when summoned from Corporate Maintenance—and is a single parent raising three young children).

   d. It is difficult to get things done in your department without being able to use the photocopier.

   e. The members of your department share ideas and plans daily.

   f. You want the problem solved—but without getting the repairperson fired.

   In your memo, recommend a solution, and justify briefly your recommendation.

2. Keep a journal during a collaborative project, noting carefully what succeeded and what didn't, what interpersonal conflicts developed and how they were resolved, what other issues contributed to progress or delay, the role and effectiveness of electronic tools, and so on. In a memo report to your classmates and instructor, summarize the achievements and setbacks in your project and prepare recommendations for improving collaboration on future projects.

   In this evaluation/recommendation report, avoid attacking, blaming, or offending anyone. Offer constructive suggestions for improving collaborative work *in general*.

---

4. My thanks to Teresa Pawelczyk for the original version of this exercise.

☑ COLLABORATIVE PROJECT

Divide into small groups of mixed genders. Read "In Brief" (pages 23–24) and complete the following tasks to test the hypothesis that women and men communicate differently in the workplace.

Each group member prepares the following brief messages—without consulting with other members:

- A thank-you note to a coworker who has done you a favor.
- A note asking a coworker for help with a problem or project.
- A note asking a collaborative peer to be more cooperative or stop interrupting or complaining.
- A note expressing impatience, frustration, confusion, or satisfaction to members of your group.

- A recommendation for a friend who is applying for a position with your company.
- A note offering support to a good friend and coworker.
- A note to a new colleague, welcoming this person to the company.
- A request for a raise, based on your hard work.
- The meeting is out of hand, so you decide to take control. Write what you would say.
- Some members of your group are dragging their feet on a project. Write what you would say.

As a group, compare messages, draw conclusions about the original hypothesis, and appoint one member to present the findings to the class.

**Figure 2.6
Sample Form for
Evaluating Team
Members**

Performance Appraisal for _____
(After each item place an X in the column that applies.)

	Superior	Acceptable	Unacceptable
Dependability			
Cooperation			
Effort			
Quality of work			
Ability to meet deadlines			

Project Manager's signature

Management Plan Sheet

Project title:
Audience:
Project manager:
Team members:
Purpose statement:

Specific Assignments   **Due Dates**

Research: Research due:
Planning: Plan and outline due:
Drafting: First draft due:
Revising: Reviews due:
Preparing final document: Revision due:
Presenting oral briefing: Final document due:
                                             Progress report(s) due:

Work Schedule

Group meetings: *date*   *place*   *time*   *note-taker*
   #1
   #2
   #3
   etc.
Mtgs. w/instructor
   #1
   #2
   etc.

Miscellaneous

How will disputes and grievances be resolved?
How will performances be evaluated?
Other matters (Internet searches, E-mail routing, computer conferences, etc.)?

Figure 2.7   Sample Plan Sheet for Managing a Collaborative Project

Solving the Information Problem

A LL technical writing is for readers who will use and react to your information. You might write to *define* something—as to insurance customers who want to know what *variable annuity* means; to *describe* something—as to an architectural client who wants to know what a new addition to her home will look like; to *explain* something—as to a stereo technician who wants to know how to eliminate bass flutter in your company's new line of speakers. As depicted in Figure 3.1, you solve your information problem by enabling readers to understand exactly what you mean. An essential part of your critical thinking involves **audience analysis,** to learn all you can about your readers and how they will use your document.

Most workplace readers would prefer not to have to read your document at all. They are not interested in how smart or eloquent you are, but they *do* want to find what they need, quickly and easily.

Assess Readers' Information Needs

Good writing connects with readers by recognizing their different backgrounds, needs, and preferences. A single message may appear in several versions for several audiences. For instance, an article describing a new cancer treatment might appear in a medical journal read by doctors and nurses. A less technical version might appear in a medical textbook read by medical and nursing students. An even simpler version might appear in *Reader's Digest*. All three versions treat the same topic, but each meets the needs of a different audience.

Technical writing is intended to be *used*. You become the teacher and the reader becomes the student. Because your readers know less than you, they will have questions.

Typical Reader Questions About Workplace Documents

- What is the purpose of this document?
- Who should read the document?
- What is being described or explained?
- What does it look like?
- How do I do it?
- How did you do it?
- Why did it happen?
- When will it happen?
- Why should we do it?
- How much will it cost?
- What are the risks?

**Figure 3.1
One Problem
Confronted by
Writers**

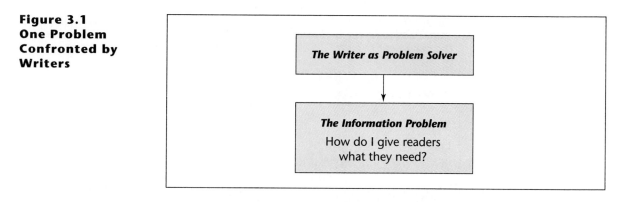

You always write to enable a specific audience to grasp the information and follow the discussion. In order to be useful, the writing must connect with the reader's level of understanding.

Identify Levels of Technicality

When you write for a close acquaintance (friend, computer crony, psychology classmate, coworker, engineering colleague, chemistry professor who reads your lab reports, or your supervisor), you know a good deal about your reader's background. You deliberately adapt your report to that reader's knowledge, interests, and needs. But sometimes you write for less defined audiences, particularly when the audience is large (when you are writing a journal article, a computer manual, a set of first-aid procedures, or a report of an accident). When you have only a general notion about your audience's background, you must decide whether your document should be *highly technical, semitechnical,* or *nontechnical,* as depicted in Figure 3.2.

The Highly Technical Document

Readers at your specialized level expect the technical facts and figures they need, without long explanations. The following report of treatment given to a heart attack victim is highly technical. The writer, an emergency room physician, is reporting to the patient's doctor. This reader needs an exact record of the patient's symptoms and treatment.

A Technical Version

Expert readers need merely the facts and figures, which they can interpret for themselves

The patient was brought to the emergency room by ambulance at 1:00 A.M., September 27, 19XX. The patient complained of severe chest pains, dyspnea, and vertigo. Auscultation and EKG revealed a massive cardiac infarction and pulmonary edema marked by pronounced cyanosis. Vital signs: blood pressure, 80/40; pulse 140/min; respiration, 35/min. Lab: wbc, 20,000; elevated serum transaminase; urea nitrogen, 60 mg%. Urinalysis showed 4+ protein and 4+ granular casts/field, indicating acute renal failure secondary to the hypotension.

**Figure 3.2
Deciding on a
Document's Level
of Technicality**

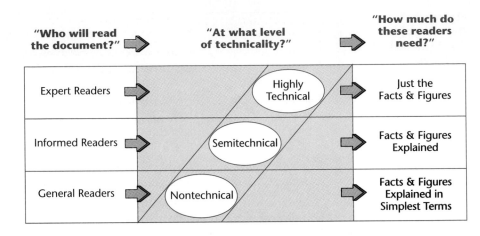

The patient received 10 mg of morphine stat, subcutaneously, followed by nasal oxygen and 5% D & W intravenously. At 1:25 A.M. the cardiac monitor recorded an irregular sinus rhythm, indicating left ventricular fibrillation. The patient was defibrillated stat and given a 50 mg bolus of Xylocaine intravenously. A Xylocaine drip was started, and sodium bicarbonate administered until a normal heartbeat was established. By 3:00 A.M., the oscilloscope was recording a normal sinus rhythm.

As the heartbeat stabilized and cyanosis diminished, the patient received 5 cc of Heparin intravenously, to be repeated every six hours. By 5:00 A.M. the BUN had fallen to 20 mg% and vital signs had stabilized: blood pressure, 110/60; pulse, 105/min; respiration, 22/min. The patient was now conscious and responsive.

This highly technical report is clear only to the medical expert. Because her reader has extensive background, this writer defines no technical terms (pulmonary edema, sinus rhythm). Nor does she interpret lab findings (4+ protein, elevated serum transaminase). She uses abbreviations her reader understands (wbc, BUN, 5% D & W). Because her reader knows the reasons for specific treatments and medications (defibrillation, Xylocaine drip), she includes no theoretical background. Her report answers concisely the main questions she can anticipate from her reader: What happened? What treatment was given? What were the results?

The Semitechnical Document

One broad class of readers may have some technical background, but less than the experts. For instance, first-year medical students have specialized knowledge, but not as much as second-, third-, and fourth-year students. Yet students in all four groups could be considered semitechnical readers. When you write for a semitechnical audience, identify the *lowest* level of understanding in the group, and write to that level. Too much explanation is better than too little.

Here is a partial version of the earlier medical report. Written at a semi-technical level, it might appear in a textbook for first-year medical or nursing students, in a report for a medical social worker, in a patient's history for the medical technology department, or in a monthly report for the hospital administration.

A Semitechnical Version

Informed but nonexpert readers need enough explanations to understand what the facts mean

Examination by stethoscope and electrocardiogram revealed a massive failure of the heart muscle along with fluid buildup in the lungs, which produced a cyanotic **discoloration of the lips and fingertips from lack of oxygen.**

The patient's blood pressure at 80 mm Hg (systolic)/40 mm Hg (diastolic) was **dangerously below its normal measure of 130/70.** A pulse rate of 140/minute was **almost twice the normal rate of 60–80.** Respiration at 35/minute was more than **twice the normal rate of 12–16.**

Laboratory blood tests yielded a white blood cell count of 20,000/cu mm (normal value: 5,000–10,000), **indicating a severe inflammatory response by the heart muscle.** The elevated serum transaminase enzymes **(produced in quantity only when the heart muscle fails)** confirmed the earlier diagnosis. A blood urea nitrogen level of 60 mg% (normal value: 12–16 mg%) **indicated that the kidneys had ceased to filter out metabolic waste products.** The 4+ protein and casts reported from the urinalysis (normal value: 0) **revealed that the kidney tubules were degenerating as a result of the lowered blood pressure.**

The patient immediately received **morphine to ease the chest pain,** followed by **oxygen to relieve strain on the cardiopulmonary system,** and an intravenous solution of **dextrose and water to prevent shock.**

The version explains (in boldface) the raw data. Exact dosages are not mentioned because the readers are not treating the patient. Normal values of lab tests and vital signs, however, make interpretation easier. (Expert readers would know these values.) Knowing what medications the patient received would be especially important to the lab technician, because some medications affect test results. For a nontechnical audience, however, the message needs further translation.

The Nontechnical Document

Readers with no specialized training expect technical data to be translated into terms they understand. Nontechnical readers are impatient with abstract theories but want enough background to help them make the right decision or take the right action. They are bored by long explanations but frustrated by bare facts not explained or interpreted. They expect a report that is clear on first reading, not one that requires review or study.

Following is a nontechnical version of our medical report. The physician might write this version for the patient's spouse who is overseas on business, or as part of a script for a documentary film about emergency room treatment.

A Nontechnical Version

General readers need everything translated into terms they understand

Heart sounds and electrical impulses both were abnormal, **indicating a massive heart attack caused by failure of a large part of the heart muscle.** The lungs were swollen with fluid and the lips and fingertips showed **a bluish discoloration from lack of oxygen.**

Blood pressure was **dangerously low, creating the risk of shock.** Pulse and respiration were **almost twice the normal rate, indicating that the heart and lungs were being overworked** in keeping oxygenated blood circulating freely.

Blood tests confirmed the heart attack diagnosis and **indicated that waste products usually filtered out by the kidneys were building up in the bloodstream. Urine tests showed that the kidneys were failing as a result of the lowered blood pressure.**

The patient was given **medication to ease the chest pain, oxygen to ease the strain on the heart and lungs, and intravenous solution to prevent the blood vessels from collapsing and causing irreversible shock.**

Nearly all interpretation (in boldface), this nontechnical version omits any mention of medications, lab tests, or normal values, because these have no meaning for the reader. The writer merely summarizes events and explains the causes of the crisis and the reasons for the particular treatment.

In some other situation, however (say, in a jury trial for malpractice), the nontechnical audience might need information about specific medication and treatment. Such a report would, of course, be much longer—a short course in emergency coronary treatment.

Each version of the medical report is useful *only* to readers at a specific level. Doctors and nurses have no need for the explanations in the two latter versions, but they do need the specialized data in the first. Beginning medical students and paramedics might be confused by the first version and bored by the third. Nontechnical readers would find both the first and second versions meaningless.

Primary and Secondary Readers

Whenever you prepare a single document for multiple readers, classify your readers as *primary* or *secondary*. Primary readers usually are those who requested the document and who will use it as a basis for decisions or actions. Secondary readers are those who will carry out the project, who will advise the primary readers about their decision, or who will somehow be affected by this decision. They will read your document (or perhaps only part of it) for information that will help them get the job done, for educated advice, or to keep up with new developments.

Often these two audiences differ in technical background. Primary readers may require highly technical messages, and secondary readers may need semitechnical or nontechnical messages—or vice versa. When you must write for audiences at different levels, follow these guidelines:

How to tailor a single document for multiple readers

1. If the document is short (a letter, memo, or anything less than two pages), rewrite it at various levels for various readers.
2. If the document exceeds two pages, address the primary readers. Then provide appendixes for secondary readers (technical appendixes when secondary readers are technical, or vice versa). Letters of transmittal, informative abstracts, and glossaries are other supplements that help nonspecialized audiences understand a highly technical report. (See pages 376–86 for how to use and prepare appendixes and other supplements.)

The next scenario shows how some documents must be tailored for both primary and secondary readers.

Tailoring a Document for Different Readers

Different readers have different information needs

You are a metallurgical engineer in a Detroit consulting firm. Your supervisor has asked that you test the fractured rear axle of a 1995 Delphi pickup truck recently involved in a fatal accident. Your job is to determine whether the fractured axle *caused* or *resulted from* the accident.

After testing the hardness and chemical composition of the metal and examining microscopic photographs of the fractured surfaces (fractographs), you conclude that the fracture resulted from stress that developed *during* the accident. Now you must report your procedure and your findings to a variety of readers.

"What do these findings mean?"

Because your report may serve as evidence in court, you must explain your findings in meticulous detail. But your primary readers (the decision makers) will be nonspecialists (the attorneys who have requested the report, insurance representatives, possibly a judge and a jury), so you will have to translate your report, explaining the principles behind the various tests, defining specialized terms such as "chevron marks," "shrinkage cavities," and "dimpled core," and showing the significance of these features as evidence.

"How did you arrive at these conclusions?"

Secondary readers will include your supervisor and outside consulting engineers who will be evaluating your test procedures and assessing the validity of your findings. Consultants will be focusing on various parts of your report, to verify that your procedure has been exact and faultless. For these readers, you will have to include appendixes spelling out the technical details of your analysis: *how* hardness testing of the axle's case and core indicated that the axle had been properly carburized; *how* chemical analysis ruled out the possibility that the manufacturer had used inferior alloys; *how* light-microscopic fractographs revealed that the origin of the fracture, its direction of propagation, and the point of final rupture indicated a ductile fast fracture, not one caused by torsional fatigue. ■

In this situation, primary readers need to know *what your findings mean,* whereas secondary readers need to know *how you arrived at your conclusions.* Unless you serve the needs of each group independently, your information will be worthless.

Develop an Audience and Use Profile

When you write for a particular reader or a small group of readers, you can focus sharply on your audience by asking the questions listed below. To answer these questions consider the suggestions that follow, and use a version of the Audience and Use Profile Sheet (Figure 3.3) for all your writing.

Reader Characteristics

Identify the primary readers by name, job title, and specialty (Martha Jones, Director of Quality Control, B.S. and M.S. in mechanical engineering). Are they superiors, colleagues, or subordinates? Are they inside or outside your organization? What is their attitude toward this topic likely to be? Are they apt to accept or reject your conclusions and recommendations? Will your report be good or bad news? For readers from other cultures, how might cultural differences affect their expectations and interpretations?

Identify also those secondary readers who might be interested in or affected by your document, or who will affect the primary reader's perception or use of your document.

Purpose of the Document

Learn why readers want the document and how they will use it. Do they merely want a record of activities or progress? Do they expect only raw data, or conclusions and recommendations as well? Will readers act immediately on the information? Do they need step-by-step instructions? Will the document be read and discarded, filed, published, distributed electronically? In your audience's view, *what* is most important? What purpose should this document achieve?

Questions About a Document's Intended Audience and Use

- Who wants the document? Who else will read it?
- Why do they want the document? How will they use it? What purpose do I want to achieve?
- What is the technical background of the primary audience? Of the secondary audience?
- How might cultural differences shape readers' expectations and interpretations?
- How much does the audience already know about the subject? What material will have informative value?
- What exactly does the audience need to know, and in what format? How much is enough?
- When is the document due?

Audience Identity and Needs

Primary reader(s): _____ *(name, title)*

Secondary reader(s): _____

Relationship: _____ *(client, employer, other)*

Intended use of document: _____ *(perform a task, solve a problem, other)*

Prior knowledge about this topic: _____ *(knows nothing, a few details, other)*

Additional information needed: _____ *(background, only bare facts, other)*

Probable questions: _____ ?

_____ ?

_____ ?

_____ ?

_____ ?

Audience's Probable Attitude and Personality

Attitude toward topic: _____ *(indifferent, skeptical, other)*

Probable objections: _____ *(cost, time, none, other)*

Probable attitude toward this writer: _____ *(intimidated, hostile, receptive, other)*

Persons most affected by this document: _____

Temperament: _____ *(cautious, impatient, other)*

Probable reaction to document: _____ *(resistance, approval, anger, guilt, other)*

Risk of alienating anyone: _____

Audience Expectations About the Document

Reason document originated: _____ *(audience request, my idea, other)*

Acceptable length: _____ *(comprehensive, concise, other)*

Material important to this audience: _____ *(interpretations, costs,*

_____ *conclusions, other)*

Most useful arrangement: _____ *(problem-causes-solutions, other)*

Tone: _____ *(businesslike, apologetic, enthusiastic, other)*

Intended effect on this audience: _____ *(win support, change behavior, other)*

Due date: _____

Figure 3.3   Audience and Use Profile Sheet

Readers' Technical Background

Colleagues who speak your technical language will understand **raw data.** Supervisors responsible for several technical areas may want **interpretations** and recommendations. Managers who have limited technical **knowledge** expect definitions and explanations. Clients with no technical **background** expect versions that spell out what the facts mean to *them* (to **their health,** pocketbook, business prospects). However, none of these **generalizations** might apply to *your* situation. When in doubt, aim for low **technicality.**

Readers' Cultural Background

Some information needs can be culturally determined. For example, **readers** in certain cultures might value thoroughness and complexity **above all: lists** of data, with every relevant detail included and explained. Readers **in other** cultures might prefer an overview of the material, with multiple **perspectives** and liberal use of graphics (Hein 125–26).

U.S. business culture generally values plain talk that spells **out the meaning** directly, but some cultures prefer indirect and somewhat ambiguous **messages,** which leave explanations and interpretation for readers **to decipher** (Leki 151; Martin and Chaney 276–77). To avoid seeming impolite, **some** readers might hesitate to request clarification or additional **information.** Even disagreement or refusal might be expressed as "We **will do our best" or** "This is very difficult," instead of "No"—to avoid offending **and to preserve** harmony (Rowland 47).

Readers' Knowledge of the Subject

Do not waste time rehashing information readers already **have. Readers** expect something *new* and *significant.* Writing has informative value[1] **when it** (1) conveys knowledge that will be new *and* worthwhile to the intended **audience;** (2) reminds the audience about something they know but ignore; **or (3)** offers fresh insight about something familiar, a new perspective.

The informative value of any document is measured by its **relevancy to** the writer's purpose and the audience's needs. As this book's audience, **for** instance, you expect to learn about technical writing, and my purpose is to **help** you do so. In this situation, which of these statements would you find **useful?**

1. Technical writing is hard work.
2. Technical writing is a process of making deliberate decisions in response to a specific situation. In this process, you discover important meanings in your topic, and give your readers the information they need to understand your meanings.[2]

---

1. Adapted from James L. Kinneavy's assertion that discourse ought to be unpredictable, in *A Theory of Discourse* (Englewood Cliffs: Prentice, 1971).

2. My thanks to Robert M. Hogge, Weber State University, for this definition.

Statement 1 offers no news to anyone who has ever picked up a pencil, and so it has no informative value for you. But 2 offers a new perspective on something familiar. No matter how much you might have struggled through decisions about punctuation, organization, and grammar, chances are you haven't viewed writing as entailing the critical thinking discussed in this book (and illustrated in Chapter 7). Because 2 provides new insight, you can say it has informative value.

The more nonessential information readers receive, the more they are likely to overlook or misinterpret the important material. Take the time to determine what your readers need, and try to give them just that.

Appropriate Details and Format

The amount of detail in your report (*How much is enough?*) will depend on what you have learned about your readers' needs. Were you asked to "keep it short" or to "be comprehensive"? Can you summarize, or does everything need spelling out? What length will they tolerate? Are the primary readers most interested in conclusions and recommendations, or do they want all the details? Have they requested a letter, a memo, a short report, or a long, formal report with supplements (title page, table of contents, appendixes, and so on)? What kinds of visuals (charts, graphs, drawings, photographs) make this material more accessible? What level of technicality will connect with primary readers?

High Technicality	The diesel engine generates 10 BTUs per gallon of fuel, as opposed to the conventional gas engine's 8 BTUs.
Low Technicality	The diesel engine yields 25 percent better fuel mileage than its gas-burning counterpart.

Every professional in the Information Age has to keep abreast of technology in order to use it effectively. What one has to know changes quickly and often.

Due Date

Does your document have a deadline? Workplace documents almost always do. Allow plenty of time to collect data, to write, and to revise. Whenever possible, ask primary readers to review an early draft and to suggest improvements.

Brainstorm for a Useful Message

When you begin working with an idea, content is raw material: ideas, insights, statistics, facts, or examples—anything that advances your message, that helps you answer this question: *How can I find something worthwhile to say, something that will convey my meaning?*

Outlines for specific writing tasks appear in later chapters, but the technique of *brainstorming* is one good way to find worthwhile content. The aim is to get all possible raw material *on paper*. Business and technical people use group brainstorming to develop ideas for new projects, marketing campaigns, problem solving, and other collaborative projects.

Some people brainstorm as a first writing step, just to get started. Others brainstorm only after writing a rough draft. Regardless of the sequence, writers usually brainstorm at some stage in the process, to ensure they discover *all* the material readers might find useful.

The brainstorming procedure is simple. Concentrate on your writing situation, and jot down *every thought*. Don't stop to judge relevance or worth, and don't worry about complete sentences or spelling. Just write everything that comes to mind. The more, the better. Trust your imagination; even the wildest idea might lead to a valuable insight.

Brainstorming invariably produces more information than you can use, and chances are you will discover more ideas while writing your document. From this broad inventory, you select only the useful material—namely, worthwhile content. (Chapter 7 shows how a working professional brainstorms while preparing an important memo to his superiors.)

☑ EXERCISES

1. Locate a short article from your field (or part of a long article or a selection from your textbook for an advanced course). Choose a piece written at the highest level of technicality you understand and then translate the piece for a layperson, as in the example on page 32. Exchange translations with a classmate from a different major. Read your neighbor's translation and write a paragraph evaluating its level of technicality. Submit to your instructor a copy of the original, your translated version, and your evaluation of your neighbor's translation.

2. Assume that a new employee is taking over your job because you have been promoted. Identify a specific problem in the job that could cause difficulty for the new employee. Write for the employee instructions for avoiding or dealing with the problem. Before writing, create an audience and use profile by answering (on paper) the questions on page 34. Then brainstorm for details. Submit to your instructor your audience and use analysis, brainstorming list, and instructions.

☑ COLLABORATIVE PROJECTS

Form teams according to major (electrical engineering, biology, etc.), and respond to the following situation.

Assume your team has received the following assignment from your major department's chairperson: An increasing number of first-year students are dropping out of the major because of low grades, stress, or inability to keep up with the work. Your task is to prepare a "Survival Guide," for distribution to incoming students. This one- or two-page memo should focus on the challenges and the pitfalls and should include a brief motivational section, along with whatever else team members think readers need.

ANALYZE YOUR AUDIENCE

Use Figure 3.3 as a guide for developing your audience and use profile.

DEVISE A PLAN FOR ACHIEVING YOUR GOAL

From the audience traits you have identified, develop a plan for communicating your information. Express your goal and plan in a statement of purpose. For example:

> The purpose of this document is to explain the challenges and pitfalls of the first year in our major. We will show how dropouts have increased, discuss what seems to go wrong, give advice on avoiding some common mistakes, and emphasize the benefits of remaining in the program.

PLAN, DRAFT, AND REVISE YOUR DOCUMENT

Brainstorm for worthwhile content (for this exercise, make up some reasons for the dropout rate if you need to), do any research that may be needed, write a workable draft, and revise until it represents your team's best work.

Appoint a team member to present the finished document (along with a complete audience and use analysis) for class evaluation, comparison, and response.

ALTERNATIVE PROJECTS

a. In a one- or two-page memo to *all* incoming students develop a "First-Year Survival Guide." Spell out the least information anyone should know in order to get through the first year.

b. In one or two pages, describe the job outlook in your field (prospects for the coming decade, salaries, subspecialties, promotional opportunities, etc.). Write for high school seniors interested in your major. Your team's description will be included in the career handbook published by your college.

c. Identify an area or situation on campus that is dangerous or inconvenient or in need of improvement (endless cafeteria lines, poorly lit intersections or parking lots, noisy library, speeding drivers, inadequate dorm security, etc.). Observe the situation as a group during a peak-use period. Spell out the problem in a letter to a specified decision maker (dean, campus police chief, head of food service) who presumably will use your information as a basis for action.

Solving the Persuasion Problem

Chapter 3 explained how writers face the *information problem: How can I make readers understand exactly what I mean?* But writers face a *persuasion problem* as well, outlined in Figure 4.1. Persuasion means trying to influence people's thinking or win their cooperation. In the workplace, persuasive efforts often are aimed at building group consensus. The size of your persuasion problem depends on who your readers are, how you want them to respond, and how firmly individuals are committed to their position.

You face a persuasion problem whenever you express a viewpoint readers might dispute. Viewpoints ordinarily are expressed in a thesis or a *claim* (a statement of the point you are trying to prove). For instance, you might want readers to *recognize* facts they've ignored:

A claim about what the facts are

> The O-rings in the space shuttle's booster rockets have a serious defect that could have disastrous consequences.

Or you might want to influence how they *evaluate* the facts:

A claim about what the facts mean

> Taking time to redesign and test the O-rings is better than taking unacceptable risks to keep the shuttle program on schedule.

Or you might want readers to *take immediate action:*

A claim about what should be done

> We should call for a delay of tomorrow's shuttle launch because the risks simply are too great.

Whenever an audience disagrees about what things mean or what is better or worse or what should be done, you face a persuasion problem.

Your own letters, memos, and reports will be asking readers to accept and act on claims like these:[1]

- We cannot meet this production deadline without sacrificing quality.
- We're doing all we can to correct your software problem.
- This hiring policy is discriminatory.
- Our software is superior to the competing brand.
- We all need to work harder.
- I deserve a raise.

Your goal might be to convert readers to a different way of thinking, to reinforce one particular way of thinking, or to create a new way of thinking. In any event, you need to make the best case for seeing things *your* way.

Assess the Political Realities

Besides their varied backgrounds, different readers have different attitudes. On one level, your readers are consumers of information; on another they are *human beings,* who react on the basis of their personality and feelings, and

---

1. This list of claims was inspired by Gilsdorf, "Executives' and Academics' Perception."

**Figure 4.1
Two Problems
Confronted by
Writers**

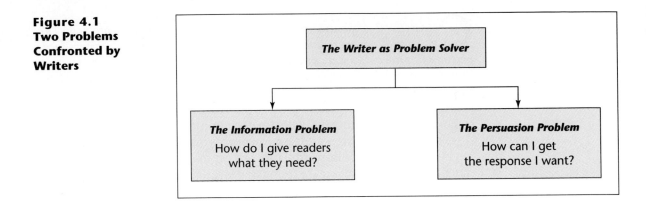

who create their own meanings for what they have just read (Littlejohn and Jabusch 5).

Any document can evoke different reactions—depending on a reader's temperament, preferences, interests, fears, biases, misconceptions, ambitions, or general attitude. Whenever readers feel their views are being challenged, they respond with questions like these:

Typical Reader Questions About a Document That Attempts to Persuade

- Says who?
- So what?
- Why should I?
- Why rock the boat?
- What's in it for me?

- What are you up to?
- What's in it for you?
- What does this *really* mean?
- Will it mean more work for me?
- Will it make me look bad?

Some readers might be impressed and pleased by your suggestions for increasing productivity; some might feel offended or threatened; others might think you are trying to make yourself look good or make them look bad. People can read much more between the lines than what actually is on the page. Such are the political realities of writing in any organization.

If you have worked with others, you already know something about office politics: how some people seek favor, influence, status, or power; how some resent, envy, or intimidate others; how some are easily threatened. Writing consultant Robert Hays sums up the writer's political situation this way:[2] "A writer must labor under political pressures from boss, peers, and

---

2. I am indebted to Hays's "Political Realities" for excellent suggestions about analyzing and addressing political realities faced by writers.

subordinates. Any conclusion affecting other people can arouse resistance" (19). Some readers might resist your suggestion for shortening lunch breaks, cutting expenses, or automating the assembly line. Or your document might be seen as an attempt to undermine your boss.

No one wants bad news; some people prefer to ignore it (as the O-ring flaw on the shuttle *Challenger's* booster rockets made all too clear). If you know something is wrong, that a project or product is unsafe, inefficient, or worthless, you have to decide whether "to try to change company plans; to keep silent; to 'blow the whistle'; or to quit" (Hays 19). Does your organization encourage or discourage outspokenness and constructive criticism? Find out—preferably before you accept the job. Ignoring political realities, you might write something that violates expectations and hurts your career.

Expect Reader Resistance

People who haven't made up their minds about what to do or think are more likely to be receptive to persuasive influence:

We rely on persuasion to help us make up our minds

> We are all consumers as well as providers of persuasion. Daily, we open OUR-SELVES to the persuasion of others. We need others' arguments and evidence. We're busy. We can't and don't want to discover and reason out everything for ourselves. We look for help, for short cuts, in making up our minds. (Gilsdorf, "Write Me" 12)

In a world overwhelmed by information, persuasion can help people "process" the information and decide on its meaning.

People who already have decided what to do or think, however, don't like to change their minds without good reason. Sometimes, even for the best reasons, people refuse to budge. The O-ring claims cited earlier were made—and convincingly supported—by engineers before the shuttle *Challenger* exploded on January 28, 1986. That such claims were ignored by decision makers illustrates how audiences can resist the most compelling arguments.

Whenever you question people's stand on an issue or try to change their behavior, expect resistance. One researcher explains why persuasion is so difficult:

Once our minds are made up, we tend to hold stubbornly to our views

> By its nature, informing "works" more often than persuading does. While most people do not mind taking in some new facts, many people do resist efforts to change their opinions, attitudes, or behaviors. (Gilsdorf, "Executives' and Academics' Perception" 61)

The bigger the readers' stake in the issue, the more personal their involvement will be, and the more resistance you can expect. Research indicates that inducing permanent change in behavior is especially difficult because people tend to revert to the familiar patterns and activities that are part of their lifestyle or work habits (Perloff 321).

**Figure 4.2
The Levels of
Response to
Persuasion**

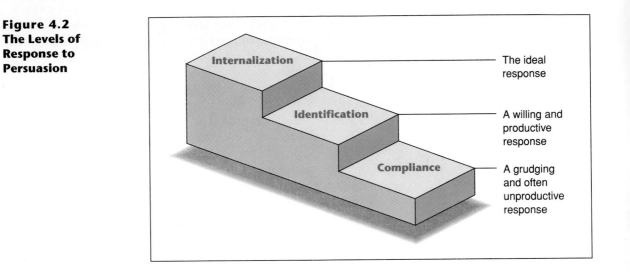

When people do yield to persuasion, they yield either grudgingly, willingly, or enthusiastically (as in Figure 4.2). Researchers categorize these responses as *compliance, identification,* or *internalization* (Kelman 51–60):

Some ways of yielding to persuasion are better than others

- *Compliance:* "I'm yielding to your demand in order to get a reward or to avoid punishment. I really don't accept it, but I feel pressured and so I'll go along to get along."
- *Identification:* "I'm yielding to your appeal because I like and believe you, I want you to like me, and I feel we have something in common."
- *Internalization:* "I'm yielding because what you're saying makes good sense and it fits my goals and values."

Although compliance is sometimes a necessary response (as in military orders or workplace safety regulations) nobody likes to be coerced. Effective persuasion relies on identification or internalization. If readers merely comply because they feel they have no choice, then you probably have lost their loyalty and goodwill—and as soon as the threat or reward disappears, you will lose their compliance as well.

Know How to Connect with Readers

Persuasive people know when to merely declare what they want, when to reach out and create a relationship, when to appeal to reason—or when to employ some combination of these strategies. These three strategies for connecting have been categorized as *hard, soft,* and *rational* (Kipnis and Schmidt 40–46). Let's call them the *power connection,* the *relationship connection,* and the *rational connection,* as shown in Figure 4.3.

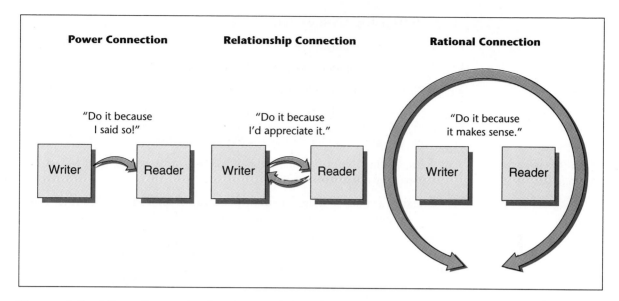

Figure 4.3   Three Strategies for Connecting with an Audience

For an illustration of these different connections, picture the following situation: Your Company, XYZ Engineering, has just developed a fitness program, based on findings that healthy employees work better, take fewer sick days, and cost less to insure. This program offers clinics for smoking, stress reduction, and weight loss, along with group exercise. In your second month on the job you receive this notice through E-mail:

Power Connection

Orders readers to show up

To:	All Employees	1/20/97
From:	G. Maximus, Human Resources Director	
Subject:	Physical Fitness	

On Monday, June 10, all employees will report to the company gymnasium at 8:00 A.M. for the purpose of choosing a walking or jogging group. Each group will meet 30 minutes three times weekly during lunch time.

How would you react to this memo—and to the person who wrote it? Here the writer seeks nothing more than compliance. Although the writer speaks of "choosing," you are given no real choice but simply ordered to show up. This kind of *power connection* is typically used by bosses and others in power. Although it may or may not achieve its goal, the power connection almost surely will alienate its audience.

Now assume instead that you received the following version of our memo. How would you react to this message and its writer?

Relationship Connection

To: All Employees   1/20/97

From: G. Maximus, Human Resources Director

Subject: An Invitation to Physical Fitness

Invites readers to participate

I realize most of you spend lunch hour playing cards, reading, or just enjoying a bit of well-earned relaxation in the middle of a hectic day. But I'd like to invite you to join our lunchtime walking/jogging club.

Leaves the choice to the reader

We're starting this club in hopes that it will be a great way for us all to feel better. Why not give it a try?

This version evokes a sense of identification, of shared feelings and goals. Instead of being commanded, readers are invited—they are given a real choice. This *relationship connection* establishes goodwill.

Often, the biggest factor in persuasion is an audience's perception of the writer. Audiences are more receptive to people they like, trust, and respect. Of course, you would be unethical in appealing to the relationship or in faking the relationship merely to hide the fact that you had no evidence to support your claim (Ross 28). Audiences need to find the claim believable ("Exercise will help me feel better") and relevant ("I personally need this kind of exercise"). The relationship connection, moreover, might strike some readers as too "chummy" to carry any real authority—and so the request might be ignored.

Here is a third version of our memo. As you read, think about the ways it makes a persuasive case.

Rational Connection

To: All Employees   1/20/97

From: G. Maximus, Human Resources Director

Subject: Invitation to Join One of Our Jogging or Walking Groups

Presents authoritative evidence

I want to share a recent study from the *New England Journal of Medicine,* which reports that adults who walk two miles a day could increase their life expectancy by three years.

Other research shows that 30 minutes of moderate aerobic exercise, at least three times weekly, has a significant and long-term effect in reducing stress, lowering blood pressure, and improving job performance.

Offers alternatives

As a first step in our exercise program, XYZ Engineering is offering a variety of daily jogging groups: The One-Milers, Three-Milers, and Five-Milers. All groups will meet at designated times on our brand new, quarter-mile, rubberized clay track.

For beginners or skeptics, we're offering daily two-mile walking groups. For the truly resistant, we offer the option of a Monday-Wednesday-Friday two-mile walk.

Offers a compromise	Coffee and lunch breaks can be rearranged to accommodate whichever group you select.
Leaves the choice to the reader	Why not take advantage of our hot new track? As small incentives, XYZ will reimburse anyone who signs up as much as $100 for running or walking
Offers incentives	shoes, and will even throw in an extra fifteen minutes for lunch breaks. And with a consistent turnout of 90 percent or better, our company insurer may be able to eliminate everyone's $200 yearly deductible in medical costs.

Here the writer shows willingness to compromise ("If you do this, I'll do that"). This *rational connection* communicates respect for the reader's intelligence *and* for the relationship by presenting good reasons, a variety of alternatives, and attractive incentives—all framed as an invitation. Whenever an audience is willing to listen to reason, the rational connection stands the best chance of succeeding.

Keep in mind that each kind of connection (or some combination) can work in particular situations. But no cookbook formula exists, and in many situations, your persuasive attempts may fail.

Ask for a Specific Decision

Unless you are giving an order, diplomacy is essential in persuasion. But don't be afraid to ask for the specific decision you want, preferably at the end of the message:

Let people know exactly what you want	Studies show that the moment of decision is made easier for people when we show them what the desired action is, rather than leaving it up to them. . . . Without this directive, people may misunderstand or lose interest in the entire message. No one likes to make decisions: there is always a risk involved. But if the writer asks for the action, and makes it look easy and urgent, the decision itself looks less risky and the entire persuasive effort has a better chance of succeeding. (Cross 3)

Let readers know what you want them to do or think.

Never Ask for Too Much

No amount of persuasion will move people to accept something they consider unreasonable. And the definition of *reasonable* depends on the individual. Employees at XYZ, for example, will differ as to which walking/jogging option they might accept.

To the jock writing the memo, a daily five-mile jog might seem perfectly reasonable, but some employees would think it outrageous. XYZ's program therefore has to offer something most of its audience (except, say, couch potatoes and those in poor health) accept as reasonable. Any request that exceeds its audience's "latitude of acceptance" (Sherif 39–59) is doomed.

Identify your audience's latitude of acceptance

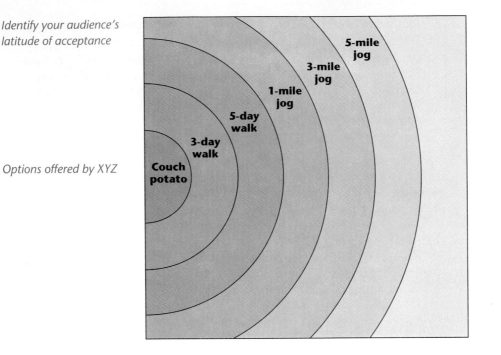

Options offered by XYZ

Recognize All Constraints

Persuasive communicators observe certain limits or restrictions imposed by their situation. These are the *constraints,* and they govern what should or should not be said, who should say it and to whom, when and how it should be said, and through which medium (printed document, computer screen, telephone, face to face, and so on). In the workplace you need to account for constraints like those in Figure 4.4.

Organizational Constraints

Organizations often have their own official constraints: for instance, schedules, deadlines, budget limitations, writing style, the way a document is organized and formatted, and its chain and medium of distribution throughout the organization. But writers also face unofficial constraints:

Decide carefully when to say what to whom

Most organizations have clear rules for interpreting and acting on (or responding to) statements made by colleagues. Even if the rules are unstated, we know who can initiate interaction, who can be approached, who can propose a delay, what topics can or cannot be discussed, who can interrupt or be interrupted, who can order or be ordered, who can terminate interaction, and how long interaction should last. (Littlejohn and Jabusch 143)

**Figure 4.4
Every Writing
Situation Poses Its
Own Constraints**

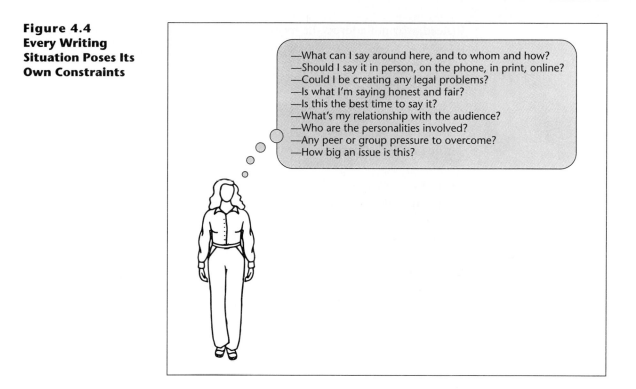

The exact rules vary among organizations, depending on whether communication channels are open and flexible or closed and rigid, on whether employee participation in decision making is encouraged or discouraged.

Although the rules of the game mostly are unspoken, anyone who ignores them (for example, going over a supervisor's head with a complaint or suggestion) invites disaster.

Airing even the most legitimate gripe in the wrong way through the wrong medium to the wrong person can be fatal to your work relationships and your career. The following memo, for instance, is likely to be interpreted by the executive officer as petty and whining behavior, and by the maintenance director as a public attack.

*Wrong way to the
wrong person*

Dear Chief Executive Officer:

Please ask the Maintenance Director to get his people to do their job for a change. I realize we're all short-staffed, but I've gotten fifty complaints this week about the filthy restrooms and overflowing wastebaskets in my department. If he wants us to empty our own wastebaskets, why doesn't he let us know?

cc: Maintenance Director

Instead, why not address the memo directly to the key person—or better yet, phone the person instead?

A better way to the right person

> Dear Maintenance Director:
>
> I wonder if we could meet to exchange some ideas about how our departments might be able to help one another during these staff shortages.

Can you identify the unspoken rules where you have worked? What happens when such rules are ignored?

Legal Constraints

Sometimes what you can say is limited by contract or by laws protecting confidentiality or customers' rights, or laws affecting product liability. For example, in a collection letter for nonpayment, you can threaten legal action but cannot threaten any kind of violence or to publicize the refusal to pay, or pretend to be an attorney (Varner and Varner 31–40). If someone requests information on one of your employees, you can "respond only to specific requests that have been approved by the employee. Further, your comments should relate only to job performance which is documented" (Harcourt 64). When writing sales literature or manuals, you and your company are liable for faulty information that leads to injury or damage.

Know how the law applies to any document you prepare. Suppose, for instance, an employee drops dead while participating in the new jogging program you've marketed so persuasively. Could you and your company be liable? Perhaps you should require physical exams and stress tests (at company expense) for participants.

Ethical Constraints

While legal constraints are defined by federal and state laws, ethical constraints are defined by good conscience, honesty, and fair play. For example, it may be perfectly legal to promote a new pesticide by emphasizing its effectiveness, while downplaying its carcinogenic effects; whether such action is *ethical*, however, is another issue entirely. To earn people's trust, you will find that "saying the right thing" involves more than legal considerations.

Persuasive skills carry tremendous potential for abuse. There is a difference between honestly presenting your best case and using deception to manipulate the reader. (Chapter 5 is devoted to various ethics problems in communication.)

Time Constraints

Persuasion often is a matter of good timing. Do you have a deadline? If not, should you delay your message, release it immediately, or what? Let's assume you're trying to "bring out the vote" among members of your professional society on some hotly debated issue: for example, whether to refuse work on any project related to biological warfare. You might want to wait until you have all the information you need or until you've analyzed the situation and planned a strategy. But you don't want to delay so long that rumors, misinformation, or paranoia cause people to harden their position *and* their resistance to your appeals. If delay might place the situation beyond your control, you might have to speak out sooner than you would like.

Social and Psychological Constraints

Too often, what we say can be misunderstood or misinterpreted. Here are just a few of the human constraints routinely encountered by communicators.

- *Relationship between communicator and audience:* Are you writing to a superior, a subordinate, or an equal? (Try not to appear dictatorial to subordinates or not to shield superiors from bad news.) How well do you and your audience know each other? Can you joke around or should you be dead serious? Do you get along or have a history of conflict? Do you trust and like one another? What you say and how you say it—and how it is interpreted—will be influenced by the relationship.

- *Audience's personality:* Researchers claim that "some people are easier to persuade than others, regardless of the topic or situation" (Littlejohn 136). Any reader's ability to be persuaded might depend on such personality traits as confidence, optimism, self-esteem, willingness to be different, desire to conform, open- or closed-mindedness, or regard for power (Stonecipher 188–89). The less your audience is open to persuasion, the harder you have to work. If you sense that your audience is totally resistant, you may want to back off—or give up altogether.

- *Audience's sense of identity and affiliation as a group:* How close-knit is the group? Does it have a strong sense of identity (as, for example, union members, conservationists, or engineering majors)? Will group loyalty or pressure to conform prevent certain appeals from working? Address the group's collective concerns.

- *Perceived size and urgency of the problem or issue:* In the audience's view, how big is this issue or problem? Has it been understated or overstated? Big problems are more likely to cause people to exaggerate their fears, anxieties, loyalties, and resistance to change—or to desperately seek some quick and easy solution. Assess the problem realistically. You don't want to downplay it, but you don't want to cause panic, either.

Writers who can assess a situation's constraints avoid serious blunders and can develop their message for greatest effectiveness.

Support Your Claims Convincingly

The persuasive argument is the one that makes the best case in the audience's view. The strength of your case depends on the *reasons* you offer to support your claims.

Persuasive claims are backed up by reasons that have meaning for the reader

When we seek a project extension, argue for a raise, interview for a job, justify our actions, advise a friend, speak out on issues of the day . . . we are involved in acts that require good reasons. Good reasons allow our audience and ourselves to find a shared basis for cooperating. . . . In speaking and writing, you can use marvelous language, tell great stories, provide exciting metaphors, speak in enthralling tones, and even use your reputation to advantage, but what it comes down to is that you must speak to your audience with reasons they understand. (Hauser 71)

To see how reasons support persuasive claims, imagine yourself in the following situation: As documentation manager for Bemis Software, a rapidly growing company, you supervise preparation and production of all user manuals. The present system for producing manuals is inefficient because three respective departments are involved in (1) assembling the required material, (2) word processing and designing, and (3) publishing the manuals. As a result, much time and energy are wasted as a manual goes back and forth among software specialists, communication specialists, and the art and printing department. After studying the problem and calling in a consultant, you decide that greater efficiency could be achieved if desktop publishing software were installed in all computer terminals. This way, all employees involved could contribute to all three phases of the process. To sell this plan to bosses and coworkers you will need good reasons, in the form of *evidence* and *appeals to readers' needs and values* (Rottenberg 104–06).

Offer Convincing Evidence

Evidence is any factual support from an outside source. Evidence is a powerful element in persuasion—as long as it measures up to readers' standards. Discerning readers evaluate evidence by using these criteria (Perloff 157–58):

- *The evidence has quality.* Instead of sheer quantity, readers expect evidence that is strong, specific, new, or different.
- *The sources are credible.* Readers want to know where the evidence comes from, how it was collected, and who collected it.
- *The evidence is considered reasonable.* It falls within a reader's "latitude of acceptance" (Sherif 39–59).

Common types of evidence include factual statements, statistics, examples, and expert testimony.

Factual Statements. A *fact* is something whose existence can be demonstrated by observation, experience, research, or measurement.

Offer the facts

| Many of our competitors already have desktop publishing networks in place.

When your space and your reader's tolerance are limited—as they usually are—be selective. Decide which facts best support your case.

Statistics. Numbers can be highly convincing. Before considering other details of your argument, many workplace readers are interested in the "bottom line": costs, savings, profits (Goodall and Waagen 57).

Give the numbers

> After a cost/benefit analysis, our accounting office estimates that an integrated desktop publishing network will save Bemis 30 percent in production costs and 25 percent in production time—savings that will enable the system to pay for itself within one year.

Numbers also can be highly misleading. Any statistics you present have to be accurate, trustworthy, and easy for readers to understand and verify. (See pages 182–86 for ways to avoid faulty statistical reasoning.) Always cite your source.

Examples. By showing specific instances of your point, examples help audiences *visualize* the idea or the concept. For example, the best way to explain what you mean by "inefficiency" in your company is to show one or more instances of it:

Show what you mean

> The figure illustrates the inefficiency of Bemis's present system for producing manuals:

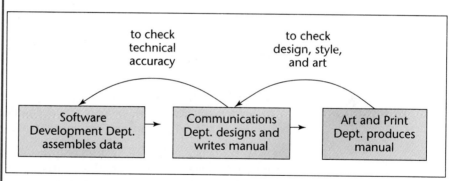

> A manual typically goes back and forth through this cycle three or four times, wasting time and effort in all three departments.

Good examples have persuasive force; they give readers something solid, a way of understanding even the most surprising or unlikely claim. Use examples that the audience can identify with and that fit the point they are designed to illustrate.

Expert Testimony. Expert testimony lends authority and credibility to any claim.

Cite the experts

> Ron Catabia, nationally recognized networking consultant, has studied our needs and strongly recommends we move ahead with the integrated network.

To be credible, however, an expert has to be unbiased and considered reliable by the audience.

Although solid evidence can be persuasive, evidence alone isn't always enough to influence the reader. At Bemis, for example, the bottom-line might be very persuasive for company executives but could mean little to some managers and employees who will be asking: Does this threaten my authority? Will I have to work harder? Will I fall behind? Is my job in danger? These readers will have to perceive some benefit beyond company profit.

Appeal to Common Goals and Values

Audiences expect a writer to share their goals and values. If you hope to create any kind of consensus, you have to identify at least one goal you and your audience have in common: "What do we all want most?"

Bemis employees, like most people, share these goals: job security, a sense of belonging, control over their jobs and destinies, a growing and fulfilling career. Any persuasive recommendation will have to take these goals into account. For example:

Appeal to shared goals

> I'd like to show how desktop publishing skills, instead of threatening anyone's job, would only increase career mobility for all of us.

Our goals are shaped by our values (qualities we believe in, ideals we stand for): friendship, loyalty, ambition, honesty, self-discipline, equality, fairness, achievement, among others (Rokeach 57–58). Beyond appealing to common goals, you can appeal to shared values.

At Bemis, you might appeal to the commitment to quality and achievement shared by the company and by individual employees:

Appeal to shared values

> None of us needs reminding of the fierce competition in the software industry. The improved collaboration among networking departments will result in better manuals, keeping us on the front line of quality and achievement.

Give your audience reasons that have real meaning for *them* personally. For example, in a recent study of teenage attitudes about the negative consequences of smoking, respondents listed these reasons for not smoking: bad breath, difficulty concentrating, loss of friends, and trouble with adults. None of them listed dying of cancer—presumably because this last reason carries little meaning for young people personally (Bauman et al. 510–30).

Consider the Cultural Context

Reaction to persuasive appeals can be influenced by a culture's customs and values.[3] Cultures might differ in their willingness to debate, criticize, or express disagreement or emotion. They might differ in their definitions of

---

3. Adapted from Beamer 293–95; Gestelend 24; Hulbert, "Overcoming" 42; Jameson 9–11; Kohl et al. 65; Martin and Chaney 271–77; Nydell 61; Thrush 276–77; Victor 159–66.

"convincing support," or they might observe special formalities in communicating. Expressions of feelings and concern for one's family might be valued more than logic, fact, statistics, research findings, or expert testimony. Some cultures consider the *source* of a message as important as its content. Establishing trust and building a relationship might weigh more heavily than proof and might be an essential prelude to getting down to business.

Cultures might differ in their attitudes toward the environment, big business, technology, or competition. They might value delayed gratification more than immediate reward, stability more than progress, time more than profit, politeness more than candor, age more than youth.

One essential element of reader expectations in all cultures is the primacy of *face saving:*

Face saving is every reader's bottom line

> *Face saving* [is] the act of preserving one's prestige or outward dignity. People of all cultures, to a greater or lesser degree, are concerned with face saving. Yet . . . [its] importance . . . varies significantly from culture to culture. . . . Indirectness in high face-saving cultures is viewed as consideration for another's sense of dignity; in low face-saving cultures, indirectness is seen as dishonesty. (Victor 159–61)

Readers lose face when they are offended by overly explicit argument, confused by a message that seems too vague, or embarrassed by the plain, unvarnished truth. They lose face when they think they are being contradicted, blatantly criticized, or addressed too informally. They lose face when their customs are ignored or their values trivialized.

Intentional or not, any perceived insult to a person's dignity ends all possibility for meaningful communication. Without violating ethical standards for honesty and fairness, effective communication in any cultural context enables recipients to save face.

CAUTION: Violating a reader's cultural frame of reference is offensive, but so is reducing individual complexity to a simplistic set of cultural stereotypes. Any generalization about any group presents a highly limited picture and in no way accurately characterizes all members of the group—or any members, for that matter. No laundry list of cultural features can replace our intuitive sensitivity toward those with whom we transact.

No particular formula for persuasion can guarantee positive results. Your approach depends on the people involved, the setting, the relationships, the issue, and the variables discussed earlier.

People rarely change their minds quickly or without good reason. A truly resistant audience will dismiss even our best arguments and may end up feeling threatened and resentful. Even when the audience is receptive, our initial attempts can fail. Often, the best we can do is avoid disaster and create an opportunity for people to appreciate the merits of our case.

IN BRIEF

Questions for Analyzing Cross-Cultural Audiences[4]

About Accepted Behavior?

- Formalities for making requests, expressing disagreement, criticism, or praise
- Preferred form for greetings or introductions (first or family names, titles)
- Willingness to criticize or request clarification
- Willingness to argue, debate, or express disagreement
- Willingness to be contradicted
- Willingness to express emotion (pleasure, gratitude, anger)
- Importance of trust and relationship building
- Importance of politeness and euphemism and leaving certain things unsaid
- Preference for casual or formal interaction
- Preference for directness and plain talk or for indirectness and ambiguity
- Preference for rapid decision making or for extensive analysis of a topic

About the Social and Legal System?

- Social/political system inflexible or open
- Class distinctions
- Democratic, egalitarian ideals or rank-conscious, authoritarian system
- Relative importance of the law versus interpersonal trust
- Formality of the contract process: a mere handshake or extensive legal documents
- Extent of attorney involvement
- Extent to which the legal system enforces contractual agreements
- Role of gift giving (viewed as bribery or as a display of respect)

About Values and Attitudes?

- Attitude toward the environment, big business, technology, competition, risk taking, status, youth versus age, rugged individualism versus group loyalty
- Preference for immediate reward or delayed gratification, progress or stability
- Importance of gender equality, interaction, and differences in the workplace
- Importance of the personal relationship in a business transaction
- Importance of time ("Time is money!" or "Never rush!")
- Importance of feelings versus logic and facts, results versus relationships
- Importance of candor versus saving face and sparing other people's feelings
- Extent of belief in fate, luck, or destiny
- View of our culture (admiration, contempt, envy, fear)

★   ★   ★

4. Adapted from Beamer 293–95; Gestelend 24; Hulbert, "Overcoming" 42; Jameson 9–11; Kohl et al. 65; Martin and Chaney 271–77; Nydell 61; Thrush 276–77; Victor 159–66.

Observe Persuasion Guidelines

Later chapters offer specific guidelines for various persuasive documents. But beyond attending to specific requirements of a particular document, remember this principle:

No matter how brilliant, any argument rejected by its audience is a failed argument.

If readers find cause to dislike you or conclude that your argument has no meaning for them personally, they usually reject *anything* you say. Connecting with an audience means being able to see things from their perspective. The following guidelines can help you make that connection.

1. *Assess the political climate.* Can you be outspoken? Who will be affected by your document? How will they react? How will your motives be interpreted? Will the document enhance your reputation or damage it? The better you assess readers' political feelings, the less likely your document will backfire. Do what you can to earn confidence and goodwill:

   • Be diplomatic; try not to make anyone look bad or lose face.
   • Be aware of your status in the organization; don't overstep.
   • Don't expect anyone to be perfect—including yourself.
   • Ask your intended readers to review early drafts.

   When reporting company negligence, dishonesty, stupidity, or incompetence, expect political fallout. Decide beforehand whether you want to keep your job or your dignity (more in Chapter 5).

2. *Learn the unspoken rules.* Know the constraints on what you can say, to whom you can say it, and how and when you can say it.

3. *Be clear about what you want.* Diplomacy is important, but people won't like having to guess about your purpose.

4. *Never make a claim or ask for something you know readers will reject outright.* Be sure readers can live with whatever you're requesting or proposing. Offer a genuine choice.

5. *Anticipate your audience's reaction.* Will they be defensive, surprised, annoyed, angry, or what? Try to address their biggest objections beforehand. Express your judgments ("We could do better") without making people defensive ("It's your fault").

6. *Decide on a connection (or combination of connections).* Does the situation call for you to merely declare your position, appeal to the relationship, or appeal to common sense and reason?

7. *Avoid an extreme persona.* **Persona** is the image or impression of the writer's personality suggested by a document's tone. Resist the urge

to "sound off," no matter how strongly you feel, because audiences tune out aggressive people, no matter how sensible the argument. Try to be likable and reasonable. Admit the imperfections in your case—a little humility never hurts. Don't hesitate to offer praise when it's deserved.

8. *Find points of agreement with your audience.* Focusing early on a shared value, goal, or concern can reduce conflict and help win agreement on later points.

9. *Never distort the opponent's position.* A sure way to alienate people is to cast the opponent as more of a villain or simpleton than the facts warrant.

10. *Try to concede something to the opponent.* Surely the opposing case is based on at least one good reason. Acknowledge the merits of that case before arguing for your own. Instead of seeming like a know-it-all, show some empathy and willingness to compromise.

11. *Use only your best material.* Not all your reasons or appeals will have equal strength or significance. Decide which material—from your *audience's* view—best advances your case.

12. *Make no claim or assertion unless you can support it with good reasons.* "Just because" does not constitute adequate support!

13. *Use your skills responsibly.* Persuasive skills are easily abused. People who feel they have been bullied, manipulated, or deceived most likely will become your enemies. Know when to back off.

14. *Seek a second opinion of your document before you release it.* Ask someone you trust and who has no stake in the issue at hand.

15. *Decide on the appropriate medium.* Given the specific issue and audience, should you communicate in person, in print, by phone, E-mail, FAX, newsletter, bulletin board, or what? (See also page 102.) Should all recipients receive your message via the same medium?

Figure 4.5 illustrates how our guidelines are employed in an actual persuasive situation. This letter is from a company that distributes systems for generating electrical power from recycled steam (cogeneration). President Tom Ewing writes a persuasive answer to a potential customer's question: "Why should I invest in the cogeneration system you are proposing for my plant?" As you read the letter, notice the kinds of evidence and appeals that support the opening claim. Notice also how the writer focuses on reasons important to the reader.

July 20, 19XX

Mr. Richard White, President
Southern Wood Products
Box 84
Memphis, TN 37162

Dear Mr. White:

The writer states his claim

In our meeting last week, you asked me to explain why we have such confidence in the project we are proposing. Let me outline what I think are excellent reasons.

Offers first reason

Gives examples

First, you and Don Smith have given us a clear idea of your needs, and our recent discussions confirm that we fully understand these needs. For instance, our proposal specifies an air-cooled condenser rather than a water-cooled condenser for your project because water in Memphis is expensive. And besides saving money, an air-cooled condenser will be easier to operate and maintain.

Offers second reason
Appeals to shared value (quality)

Gives example

Further examples

Second, we have confidence in our component suppliers and they have confidence in this project. We don't manufacture the equipment; instead, we integrate and package cogeneration systems by selecting for each application the best components from leading manufacturers. For example, Alias Engineering, the turbine manufacturer, is the world's leading producer of single-stage turbines, having built more than 40,000 turbines in 70 years. Likewise, each component manufacturer leads the field and has a proven track record. We have reviewed your project with each major component supplier, and each guarantees the equipment. This guarantee is of course transferable to you and is supplemented by our own performance guarantee.

Appeals to reader's goal (security)

Offers third reason

Third, we have confidence in the system design. We developed the CX Series specifically for applications like yours, in which there is a need for both a condensing and a backpressure turbine. In our last meeting, I pointed out the cost, maintenance, and performance benefits of the CX Series. And although the CX Series is an innovative design, all components are fully proven in many other applications, and our suppliers fully endorse this design.

Cites experts

Figure 4.5   Supporting a Claim with Good Reasons
Reprinted with permission of Thomas S. Ewing, President, Ewing Power Systems, So. Deerfield, MA 01373.

Richard White, July 20, 19XX, p. 2

Closes with best reason

Appeals to shared value (trust) and shared goal (success)

Finally, and perhaps most important, you should have confidence in this project because we will stand behind it. As you know, we are eager to establish ourselves in Memphis-area industries. If we plan to succeed, this project must succeed. We have a tremendous amount at stake in keeping you happy.

If I can answer any questions, please phone me. We look forward to working with you.

Sincerely,

EWING POWER SYSTEMS, INC.

Tom Ewing

Thomas S. Ewing
President

Figure 4.5   Supporting a Claim with Good Reasons   *Continued*

A CHECKLIST FOR CROSS-CULTURAL DOCUMENTS

Use this checklist[5] to verify that your documents respect audience diversity. (Page numbers in parentheses refer to the first page of discussion.)

❑ Does the document enable everyone to save face? (55)

❑ Is the document sensitive to the culture's customs and values? (7)

❑ Does the document conform to the safety and regulatory standards of the country? (80)

❑ Does the document provide the expected level of detail? (36)

❑ Does the document avoid possible misinterpretation? (303)

❑ Is the document organized in a way that readers will consider appropriate? (246)

❑ Does the document observe interpersonal conventions important to the culture (accepted forms of greeting or introduction, politeness requirements, first names, family names, titles, and so on)? (56)

❑ Does the document's tone reflect the appropriate level of formality or casualness one would expect? (303)

❑ Is the document's style appropriately direct or indirect? (476)

❑ Is the document's format consistent with the culture's expectations? (372)

❑ Does the document embody universal standards for ethical communication? (82)

☑ EXERCISES

1. Assume you work for a technical marketing firm proud of its reputation for honesty and fair dealing. A handbook being prepared for new personnel includes a section titled "How to Avoid Abusing Your Persuasive Skills." All employees have been asked to contribute to this section by preparing a written response to the following:

> Share a personal experience in which you or a friend were the victim of persuasive abuse in a business transaction. In a one- or two-page memo, describe the situation and explain exactly how the intimidation, manipulation, or deception occurred.

Write the memo and be prepared to discuss it in class.

2. Find an example of an effective persuasive letter. In a memo to your instructor, explain why and how the message succeeds. Base your evaluation on the persuasion guidelines, pages 57–58. Attach a copy of the letter to your evaluation memo. Be prepared to discuss your evaluation in class.

Now, evaluate a poorly written document, explaining how and why it fails.

3. Think about some change you would like to see on your campus or at your part-time job. Perhaps you would like to make something happen, such as a campus-wide policy on plagiarism, changes in course offerings or requirements, more access to computers, a policy on sexist language, or a day care center. Or perhaps you would like to improve something, such as the grading system, campus lighting, the system for student evaluation of teachers, or the promotion system at work. Or perhaps you would like to stop something from happening, such as noise in the library or sexual harassment at work.

Decide whom you want to persuade, and write a memo to that audience. Anticipate carefully your audience's implied questions, such as:

- *Do we really have a problem or need?*
- *If so, should we care enough about it to do anything?*
- *Can the problem be solved?*

5. This list was largely adapted from Caswell-Coward 265; Weymouth 144; Beamer 293–95; Martin and Chaney 271–77; Victor 159–61.

- *What are some possible solutions?*
- *What benefits can we anticipate? What liabilities?*

Can you envision additional audience questions? Do an audience and use analysis based on the profile sheet, page 65.

Don't think of this memo as the final word but as a consciousness-raising introduction that gets the reader to acknowledge that the issue deserves attention. At this early stage, highly specific recommendations would be premature and inappropriate.

4. Challenge an attitude or viewpoint that is widely held by your audience. Maybe you want to persuade your classmates that the time required to earn a bachelor's degree should be extended to five years or that grade inflation is watering down your school's education. Maybe you want to claim that the campus police should (or should not) wear guns. Or maybe you want to ask students to support a 10 percent tuition increase in order to make more computers and software available.

Do an audience and use analysis based on the profile sheet, page 65. Write specific answers to the following questions: What are the political realities? What kind of resistance could you anticipate? How would you connect with readers? What about their latitude of acceptance? Any other constraints? What reasons could you offer to support your claim?

In a memo to your instructor, submit your plan for presenting your case. Be prepared to discuss your plan in class.

☑ **COLLABORATIVE PROJECTS**

1. Assume that you work for an environmental consulting firm that is under contract with various countries for a range of projects, including these:

- A plan for rain forest regeneration in Latin America and Sub-Saharan Africa
- A plan to decrease industrial pollution in Eastern and Western Europe
- A plan for "clean" industries in developing countries
- A plan for organic agricultural development in Africa and India

- A joint American/Canadian plan to decrease acid rain
- A plan for developing alternative energy sources in Southeast Asia.

Each project will require environmental impact statements, feasibility studies, grant proposals, and a legion of other documents, often prepared in collaboration with members of the host country, and in some cases prepared by your company for audiences in the host country: from political, social, and industrial leaders to technical experts and so on.

For such projects to succeed, people from different cultures have to communicate effectively and sensitively, creating goodwill and cooperation.

Before your company begins work in earnest with a particular country, your coworkers will need to develop a degree of cultural awareness. Your assignment is to select a country and to research that culture's behaviors, attitudes, values, and social system in terms of how these variables influence the culture's communication preferences and expectations. What should your colleagues know about this culture in order to communicate effectively and diplomatically? Do the necessary research using the questions from In Brief (page 56) as a guide.

Prepare a recommendation report in memo form. Be prepared to present your findings in class.

2. Often, workplace readers need to be *persuaded* to accept recommendations that are controversial or unpopular. This project offers practice in dealing with the persuasion problems of communicating within organizations.

Divide into teams. Assume that your team agrees strongly about one of these recommendations and is seeking support from classmates and instructor (and administrators, as potential readers) for implementing the recommendations.

CHOOSE ONE GOAL

a. Your campus Writing Center always needs qualified tutors to help first-year composition students with writing problems. On the other hand, students of professional writing need to sharpen their own skill in editing, writing, motivation, and diplomacy. All students in your class, therefore, should be assigned to the Writing Center during the semester's final half, to serve as tutors for twenty hours (beyond normal course time).

b. To prepare students for communicating in an automated work environment, at least one course assignment (preferably the long report) should be composed, critiqued, and revised on disk. Students not yet skilled on a word processor will be required to develop the skill by midsemester.

c. This course should help individuals improve at their own level, instead of forcing them to compete with stronger or weaker writers. All grades, therefore, should be Pass/Fail.

d. To prepare for the world of work, students need practice in peer evaluation as well as self-evaluation. Because this textbook provides definite criteria and checklists for evaluating various documents, students should be allowed to grade each other and to grade themselves. These grades should count as heavily as the instructor's grades.

e. In preparation for writing in the workplace, no one should be allowed to limp along, just getting by with minimal performance. This course, therefore, should carry only three possible grades: A, B, or F. Those whose work would otherwise merit a C or D would instead receive an Incomplete, and be allowed to repeat the course as often as needed to achieve a B grade.

f. To ensure that all graduates have adequate communication skills for survival in a world in which information is the ultimate product, each student in the college should pass a writing proficiency examination as a graduation requirement.

ANALYZE YOUR AUDIENCE

Your audience here consists of classmates and instructor (and possibly administrators). From your recent observation of this audience, what reader characteristics can you deduce?

Follow the model in Figure 4.6 for designing a profile sheet to record your audience-and-use analysis, and to duplicate for use throughout the semester. (Feel free to improve on the design and content of our model.)

Following is one possible set of responses to questions about audience identity and needs for goal "e" from the previous list.

- *Who is my audience?* Classmates and instructor (and possibly some college administrators).

- *How will readers use my information?* Readers will decide whether to support our recommendation for limiting possible grades in this course to three: A, B or F.

- *How much is the audience likely to know already about this topic?* Everyone here is already a grade expert, and will need no explanation of the present grading system.

- *What else does the audience need to know?* The instructor should need no persuading; he or she knows all about the quality of writing expected in the workplace. But some of our classmates probably will have questions like these: Why should we have to meet such high expectations? How can this grading be fair to the marginal writers? How will I benefit from these tougher requirements? Don't we already have enough work here?

We will have to answer questions by explaining how the issue boils down to "suffering now" or "suffering later," and that one's skill in communication will determine one's career advancement.

DEVISE A PLAN FOR ACHIEVING YOUR GOAL

From the audience traits you have identified, develop a plan for justifying your recommendation. Express your goal and plan in a statement of purpose. Here is an example for Goal *e*:

> The purpose of this document is to convince classmates that our recommendation for an A/B/F grading system deserves their support. We will explain how skill in workplace writing affects career advancement, how higher standards for grading would help motivate students, and how our recommendation could be implemented realistically and fairly.

PLAN, DRAFT, AND REVISE YOUR DOCUMENT

Brainstorm for worthwhile content, do any research that may be needed, write a draft, and revise as often as needed to produce a document that stands the best chance of connecting with your audience.

Appoint a member of your team to present the finished document (along with a complete audience and use analysis) for class evaluation and response.

Audience Identity and Needs

Primary reader(s): _____ *(name, title)*

Secondary reader(s): _____

Relationship: _____ *(client, employer, other)*

Intended use of document: _____ *(perform a task, solve a problem, other)*

Prior knowledge about this topic: _____ *(knows nothing, a few details, other)*

Additional information needed: _____ *(background, only bare facts, other)*

Probable questions: _____ ?

_____ ?

_____ ?

_____ ?

Audience's Probable Attitude and Personality

Attitude toward topic: _____ *(indifferent, skeptical, other)*

Probable objections: _____ *(cost, time, none, other)*

Probable attitude toward this writer: _____ *(intimidated, hostile, receptive, other)*

Organizational climate: _____ *(receptive, repressive, creative, other)*

Persons most affected by this document: _____

Temperament: _____ *(cautious, impatient, other)*

Probable reaction to document: _____ *(resistance, approval, anger, guilt, other)*

Risk of alienating anyone: _____

Audience Expectations About the Document

Reason document originated: _____ *(audience request, my idea, other)*

Cultural context: _____ *(importance of indirectness, face saving, other)*

Acceptable length: _____ *(comprehensive, concise, other)*

Material important to this audience: _____ *(interpretations, costs,*

_____ *conclusions, other)*

Most useful arrangement: _____ *(problem-causes-solutions, other)*

Tone: _____ *(businesslike, apologetic, enthusiastic, other)*

Intended effect on this audience: _____ *(win support, change behavior, other)*

Due date: _____

Figure 4.6   Audience and Use Profile Sheet

Solving the Ethics Problem

············

CHAPTERS 3 and 4 explain how audience analysis helps us tailor informative and persuasive communication (so we can complete the project, win the contract, or the like). But an *effective* message (one that achieves its purpose) isn't necessarily an *ethical* message. Think of examples from advertising: "Our artificial sweetener is composed of proteins that occur naturally in the human body (amino acids)" or "Our potato chips contain no cholesterol." Such claims are technically accurate but misleading: amino acids in certain sweeteners can alter body chemistry to cause headaches, seizures, and possibly brain tumors; potato chips contain saturated fat—which produces cholesterol. While the advertisers' facts may be accurate, they often are incomplete and imply misleading conclusions.

Whether the miscommunication occurs deliberately or through neglect, a message is unethical when it leaves readers at a disadvantage or prevents readers from making their best decision. Therefore, writers ultimately face the threefold problem outlined in Figure 5.1. Ethical communication is measured by standards of honesty, fairness, and concern for everyone involved (Johannesen 1).

Recognize Unethical Communication

Thousands of people are injured or killed yearly in avoidable accidents—the result of faulty communication that prevented intelligent decision making. Following are descriptions of tragedies caused ultimately by unethical communication.

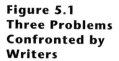

Unethical communication has consequences

- On March 3, 1974, as a Turkish Airlines DC-10 leaving Paris reached 12,000 feet, a cargo door burst open, causing a crash that killed all 346 people. Immediate cause: a poorly designed locking mechanism gave way under high pressure. Ultimate cause: the faulty design had been recognized and documented since 1969, and cargo doors had burst open during a 1970 test and a 1972 flight (in which the pilot managed to land

**Figure 5.1
Three Problems
Confronted by
Writers**

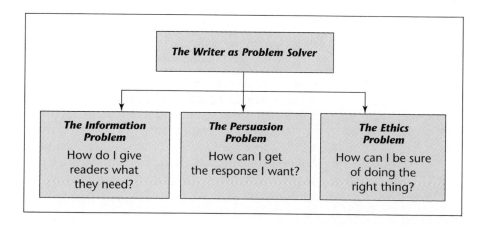

safely). On January 27, 1972, a head engineer wrote a memo to his superiors warning that "in the twenty years ahead of us, DC-10 cargo doors will come open, and I expect this to usually result in the loss of the airplane." This flaw easily could have been corrected. But, unwilling to admit design errors and under pressure from competitors to get the DC-10 flying as early as possible, manufacturers suppressed these data (Burghardt 4–5).

- In the early 1980s, thousands of workers (or their survivors) filed personal-injury suits against a leading manufacturer of asbestos products. The plaintiffs held the company responsible for respiratory afflictions ranging from lung cancer to emphysema. Immediate cause: exposure to asbestos fibers. Ultimate cause: for roughly fifty years the company hid from its workers the deadly facts about asbestos exposure—and even denied workers access to their own medical records. In 1963, the company's medical director offered this reason to justify the suppression of information: "As long as [the employee] is not disabled, it is felt that he should not be told of his condition so that he can live and work in peace, and the company can benefit from his many years of experience." When the coverup finally was revealed, the company tried to evade lawsuits by filing for bankruptcy (Mokhiber 15).

These catastrophes make for dramatic headlines, as did the revelation that a government-operated nuclear facility in Hanford, Washington, had knowingly leaked radiation for years without local residents being informed. But more routine examples of deliberate miscommunication rarely are publicized. Messages like the following succeed by *saying whatever works.*

- A person lands a great job by exaggerating his credentials, experience, or expertise.
- A marketing specialist for a chemical company negotiates a huge bulk sale of its powerful new pesticide by downplaying its carcinogenic hazards.
- To meet the production deadline on a new auto model, the test engineer suppresses, in her final report, data indicating that the fuel tank could explode upon impact.
- A manager writes a strong recommendation to get a friend promoted, while overlooking someone more deserving.

Can you recall some instances of deliberate miscommunication from your personal experience or that you learned about in the news? What were some of the effects of this miscommunication? How might the problems have been avoided?

To save face, escape blame, or get ahead, anyone might be tempted to say what people want to hear or to suppress bad news or make it seem "rosier." Some of these decisions are not simply black and white. Here is one engineer's description of the gray area in which issues of product safety and quality often are decided:

Ethical decisions are not always "black and white"

> The company must be able to produce its products at a cost low enough to be competitive. . . . To design a product that is of the highest quality and consequently has a high and uncompetitive price may mean that the company will not be able to remain profitable, and be forced out of business. (Burghardt 92)

Do you emphasize to a customer the need for extra careful maintenance of a highly sensitive computer—and risk losing the sale? Or do you downplay maintenance requirements, focusing instead on the computer's positive features? Do you tell a white lie so as not to hurt a colleague's feelings, or do you "tell it like it is" because you're convinced that lying is wrong in any circumstance? The decisions we make in these situations often are influenced by the pressures we feel.

Expect Social Pressure to Produce Unethical Communication

Pressure to get the job done can cause normally honest people to break the rules. At some point in your career you might have to choose between doing what your employer wants ("just follow orders" or "look the other way") and doing what you know is right. Maybe you will be pressured to ignore a safety hazard in order to meet a project deadline:

Pressure to "look the other way"

> Just as your automobile company is about to unveil its hot, new pickup truck, your safety engineering team discovers that the reserve gas tanks (installed beneath the truck but *outside* the frame) can explode on impact in a side collision. The company has spent a small fortune developing and producing this new model and doesn't want to hear about this problem.

Companies often face the contradictory goals of *production* (which means *making* money on the product) and *safety* (which means *spending* money to avoid accidents that may or may not happen). When productivity receives exclusive priority, safety concerns may suffer (Wickens 434–36). Thus it seems no surprise that well over 50 percent of managers studied nationwide feel "pressure to compromise personal ethics for company goals" (Golen et al. 75). These pressures come in varied forms (Lewis and Reinsch 31):

- the drive for profit
- the need to beat the competition (other organizations or coworkers)
- the need to succeed at any cost, as when superiors demand more productivity or savings without questioning the methods
- an appeal to loyalty—to the organization and to its way of doing things

Figure 5.2 depicts how such pressures can add up.

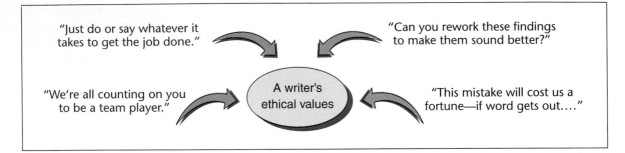

"Just do or say whatever it takes to get the job done."

"Can you rework these findings to make them sound better?"

"We're all counting on you to be a team player."

A writer's ethical values

"This mistake will cost us a fortune—if word gets out...."

Figure 5.2   How Workplace Pressures Can Influence Ethical Values

Here is a vivid reminder of how organizational pressure can result in disastrous communication: On January 28, 1986, the space shuttle *Challenger* exploded 43 seconds after launch, killing all seven crew members. Immediate cause: two rubber O-ring seals in a booster rocket permitted hot exhaust gases to escape, igniting the adjacent fuel tank. (See Figure 5.3.) Ultimate cause: the O-ring hazard had been recognized since 1977 and documented by engineers but largely ignored by management. (Managers had claimed that the O-ring system was safe because it was "redundant": each primary O-ring was backed up by a secondary O-ring.)

Moreover, in the final hours, engineers argued against launching because that day's low temperature would drastically increase the danger of both primary and secondary O-rings failing. But, under pressure to meet schedules and deadlines, managers chose to relay only a highly downplayed version of these warnings to the NASA decision makers who were to make *Challenger's* fatal launch decision.

The following analysis of key events and documents illustrates the role of miscommunication in the Challenger accident.[1]

1. More than six months before the explosion, officials at Morton Thiokol, Inc., manufacturer of the booster rockets, received a memo from engineer R. M. Boisjoly (Presidential Commission 49). Boisjoly described how exhaust gas leakage on an earlier flight had eroded O-rings in certain noncritical joints. The memo emphatically warned of possible "catastrophe" in some future shuttle flight if O-rings should fail to seal a critical joint. Boisjoly called for renewed attention to the O-ring problem. Marked COMPANY PRIVATE, Boisjoly's urgent message never was passed on to top-level decision makers at NASA.

2. During the months preceding *Challenger's* fatal launch, Boisjoly and other Morton Thiokol engineers complained, in writing, about the lack of management attention or support on the O-ring issue. These

1. Adapted from Winsor; Pace; Rowland; and Gouran et al.

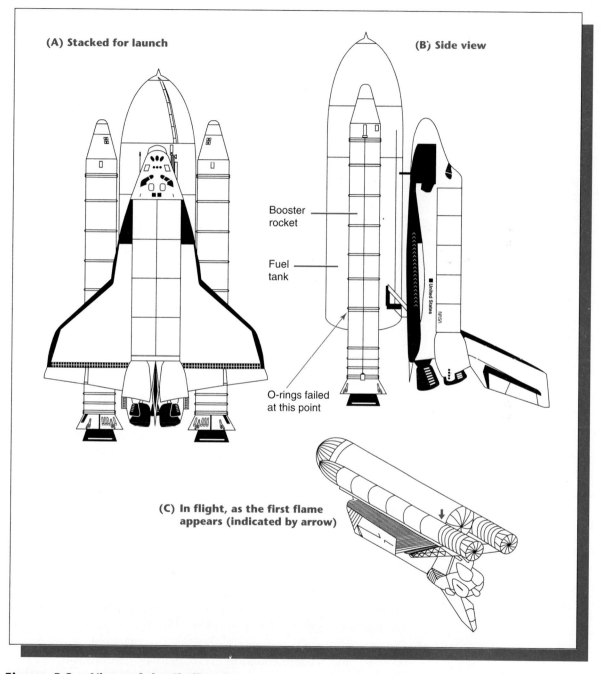

Figure 5.3   Views of the Challenger

Source: Report to the Presidential Commission on the Space Shuttle Challenger *Accident. Washington, D.C.: U.S. Government Printing Office, 1986: 3, 26.*

complaints were largely ignored within the company—and they never reached the customer, NASA decision makers.

3. On the evening of January 27, managers and engineers at Morton Thiokol debated whether to recommend the January 28 launch. In addition to the yet-unsolved problem of O-ring erosion from exhaust gases, engineers were especially concerned that predicted low temperatures could harden the rubber O-rings, preventing them from sealing the joint at all. No prior flight had launched below 53 degrees Fahrenheit. At this temperature—and even up to 75 degrees—some blow-by (exhaust leakage) had occurred. O-ring temperature on January 28 would be barely above 30 degrees.

   Arguing against the launch, Roger Boisjoly presented a chart to his Thiokol colleagues (Presidential Commission 89). The chart showed that the O-rings would take longer to seal because lower temperatures would make the rubber hard and less pliable, and that the worst exhaust leakage had occurred on a January 1985 flight, when the O-ring temperature was 53 degrees. At 30 degrees, the O-rings might not seal at all, Boisjoly warned.

   Boisjoly and his supporters made their point, and Thiokol management decided to recommend no launch until the temperature reached at least 53 degrees.

4. In a teleconference with NASA's Marshall Space Center, Thiokol managers relayed their no-launch recommendation to the next level of decision makers. But the recommendation was rejected. One Thiokol manager's testimony:

   Mr. Mulloy [a NASA official] said he did not accept that recommendation, and Mr. Hardy said he was appalled that we would make such a recommendation. (Presidential Commission 94)

   Refusing to accept the facts or the engineers' assessment of risk, NASA asked Thiokol to reconsider its recommendation.

5. The Thiokol staff met once again, the engineers and one manager continuing to oppose the launch. The managers then met *without* the engineers, and the reluctant manager was told to "take off your engineering hat and put on your management hat" (Presidential Commission 93). Despite the engineering evidence to the contrary, Thiokol managers finally concluded that the O-ring system had an acceptable margin of safety (Presidential Commission 108).

6. Despite continued engineering objections, Thiokol management reversed its recommendation (Presidential Commission 97) to Marshall and Kennedy Space Centers. Top NASA decision makers never were told about Thiokol engineers' objections to the low-temperature launch or about the level of concern about O-ring erosion in prior shuttle flights. The launch took place on schedule.

Here are some conclusions of the Presidential Commission investigating the fatal launch decision (104):

- The Commission was troubled by what appears to be a propensity of management at Marshall to contain potentially serious problems and to attempt to resolve them internally rather than communicate them forward.
- The Commission concluded that the Thiokol management reversed its position and recommended the launch . . . at the urging of Marshall and contrary to the views of its engineers in order to accommodate a major customer.

Unethical communication played a key role in the *Challenger* disaster.

Never Confuse Teamwork with *Groupthink*

Any successful organization relies on teamwork, everyone cooperating to get the job done. But teamwork is not the same as *groupthink* (Janis 9).

Groupthink occurs when group pressure prevents individuals from questioning, criticizing, or "making waves." Group members feel a greater need for acceptance and a sense of belonging than for critically examining the issues. In a conformist climate, critical thinking is impossible. Anyone who has lived through adolescent peer pressure has already experienced a version of *groupthink*.

Yielding to pressure can be especially tempting in a large company or on a complex project, in which individual responsibility is easy to camouflage in the crowd:

How some corporations evade responsibility for their actions

Lack of accountability is deeply embedded in the concept of the corporation. Shareholders' liability is limited to the amount of money they invest. Managers' liability is limited to what they choose to know about the operation of the company. The corporation's liability is limited by Congress (the Price-Anderson Act, for example, caps the liability of nuclear power companies in the aftermath of a nuclear disaster), by insurance, and by laws allowing corporations to duck liability by altering their . . . structure. (Mokhiber 16)

All kinds of people work at all levels on a major project (for instance, the production of a new passenger airplane). Countless decisions at any level have far-reaching effects on the whole project (as in the decision to ignore the DC-10's faulty locking mechanism). But with so many people collaborating, identifying those responsible for an error often is impossible—especially when the error is one of omission, that is, of *not* doing something that should have been done (Unger 137).

"I was only following orders!"

People commit unethical acts inside corporations that they never would commit as individuals representing only themselves. (Bryan 86)

Figure 5.4
***Groupthink* Can Be a Handy Hiding Place**

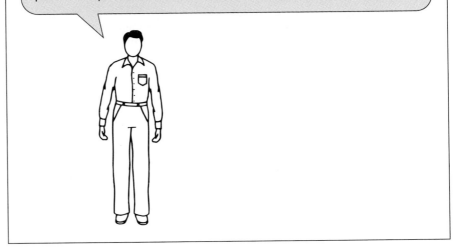

After completing their assigned task, employees too often assume their job is done. Figure 5.4 depicts the kind of thinking that enables people to deny personal responsibility for the consequences of their communication.[2]

Groupthink was exemplified by the decision to relay to top NASA officials only downplayed warnings about *Challenger's* launch. In such cases, the group suffers an "illusion of invulnerability," which creates "excessive optimism and encourages taking extreme risks" (Janis 197).

Groupthink encourages subordinates to downplay or suppress bad news in their reports to superiors, instead "stressing what they think the superior wants to hear" (Littlejohn and Jabusch 159). Morton Thiokol's final launch recommendation to NASA serves as a memorable example.

Rely on Critical Thinking for Ethical Decisions

Because of their impact on people and on your career, ethical decisions challenge your critical thinking skills:

Ethical decisions require critical thinking

- *How can I know the "right thing" in this situation?*
- *What are my obligations, and to whom, in this situation?*
- *What values or ideals do I want to stand for in this situation?*
- *What is likely to happen if I do X, or Y?*

2. My thanks to Judith Kaufman for this idea.

Can you rely on more than intuition or conscience in navigating the gray areas of ethical decisions? How will you make a convincing case against danger or folly to a roomful of people caught up in groupthink, people who ask "How do we know your way is right? Says who?" or who urge you to "take off your engineering hat and put on your management hat"?

Although ethical issues resist simple formulas, you can avoid two major fallacies that obstruct good judgment.

The Fallacy of "Doing One's Thing"

One way to oversimplify a complex ethical issue is through a misguided notion known as **ethical relativism:** "Since we have no way to agree on right or wrong, it all depends on personal preference. *Right,* then, is what I think is right!" Such denial of any reasonable criteria of course legitimizes *any* action, including the atrocities of Hitler, Charles Manson, or other fanatics who claim the "right" intentions.

The Fallacy of "One Rule Fits All"

The opposite extreme of ethical relativism is **absolutism,** the inflexible notion that one set of criteria governs all ethical decisions. If, for example, you abide absolutely by the rule "Thou shall not lie," you would violate this rule under no circumstance (even to save your family from harm).

Reasonable Criteria for Ethical Judgment

Somewhere between the extremes of relativism and absolutism are **reasonable criteria** (standards of measurement that most people would consider acceptable). These criteria for ethical judgment take the form of **obligations, ideals,** and **consequences** (Ruggiero 55–56; Christians et al. 17–18).

Obligations are the responsibilities we have to everyone involved:

- *Obligation to ourselves,* to act in our own self-interest and according to good conscience.
- *Obligation to clients and customers,* to stand by the people to whom we are bound by contract—and who pay the bills.
- *Obligation to our company,* to advance its goals, respect its policies, protect confidential information, and expose misconduct that would harm the organization.
- *Obligation to coworkers,* to promote their safety and well-being.
- *Obligation to the community,* to preserve the local economy, welfare, and quality of life.
- *Obligation to society,* to consider the national and global impact of our actions.

When the interests of these parties conflict—as they often do—we have to decide very carefully where our primary obligations lie. Can we honor all these obligations all the time?

Ideals are "notions of excellence" (Ruggiero 55), the positive values that we believe in or stand for: loyalty, friendship, courage, compassion, dignity, fairness, and whatever qualities that make us who we are.

Consequences are the beneficial or harmful results of our actions. Consequences may be immediate or delayed, intentional or unintentional, obvious or subtle (Ruggiero 56). Some consequences are easy to predict; some aren't so easy; some are impossible.

Figure 5.5 depicts the relationship among these three criteria.

The above criteria help us understand why even good intentions can produce bad judgments, as in the following situation:

What seems like the "right thing" might be the wrong thing

> Someone observes . . . that waste from the local mill is seeping into the water table and polluting the water supply. This is a serious situation and requires a remedy. But before one can be found, extremists condemn the mill for lack of conscience and for exploiting the community. People get upset and clamor for the mill to be shut down and its management tried on criminal charges. The next thing you know, the plant does close, 500 workers are without jobs, and no solution has been found for the pollution problem. (Hauser 96)

Because of their zealous dedication to the *ideal* of a pollution-free environment, the above group failed to anticipate the *consequences* of their protest or to respect their *obligation* to the community's economic welfare.

In the miscommunication surrounding the *Challenger* disaster, the group's sense of *obligation* to its client (i.e., NASA) caused it to ignore the *ideal* of honesty and the probable *consequences* of its recommendation to launch.

**Figure 5.5
Reasonable
Criteria for
Ethical Judgment**

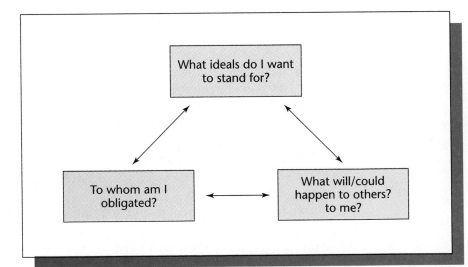

Ethical Dilemmas

Ethics decisions are especially frustrating when no single answer seems acceptable:

> [An ethical] dilemma exists whenever the conflicting obligations, ideals, and consequences are so very nearly equal in their importance that we feel we cannot choose among them, even though we must. (Ruggiero 91)

In private and public ways, such dilemmas are inescapable. For example, political candidates speak of drastic plans to eliminate the federal deficit in five years. One could argue that such a dedication to the *consequences* (or results) would violate our *obligations* (to the poor, the sick, etc.) and our *ideals* (of compassion, fairness, etc.). On the basis of our three criteria, how else might the deficit issue be considered?

Ethical dilemmas confront the medical community. For instance, in late 1992, a brain-dead, pregnant woman in Western Europe was kept alive for several weeks until her child could be delivered. Also in 1992, a California court debated the ethics of using medical findings of Nazi doctors who had experimented on prisoners in concentration camps. In terms of our three criteria, how might these dilemmas be considered?

As impossible as such dilemmas may seem, critical thinking directs our decision; without oversimplifying the issue or reaching a hasty judgment, we must consider all the criteria involved.

Anticipate Some Hard Choices

Communicators' ethical choices basically are concerned with honesty in choosing to reveal or conceal information:

- *What exactly do I report, and to whom?*
- *How much do I reveal or conceal?*
- *How do I say what I have to say?*
- *Could misplaced obligation to one party be causing me to deceive others?*

For illustration of a hard choice in workplace communication, consider the following scenario:

A Hard Choice

You are an assistant structural engineer working on the construction of a nuclear power plant near a far northern city. After years of construction delays and cost overruns, the plant finally has received its limited operating license from the Nuclear Regulatory Commission.

During your final inspection of the nuclear core containment unit on February 15, you discover a 10-foot-long, hairline crack in a section of the reinforced concrete floor, within 20 feet of the area where the cooling pipes enter the containment unit. (The especially cold and snowless winter likely has caused a frost heave under a small part of the foundation.) The crack has either just appeared or was overlooked by NRC inspectors on February 10.

The crack could be perfectly harmless, caused by normal settling of the structure; and this is, after all, a "redundant" containment system (a shell within a shell). But then again, the crack could signal some kind of serious stress on the entire containment unit, which furthermore could damage the entry and exit cooling pipes or other vital structures.

You phone your boss, who is just about to leave on a ski vacation, and who tells you, "Forget it; no problem," and hangs up.

You may have to choose between the goals of your organization and what you know is right

You know that if the crack is reported, the whole start-up process scheduled for February 16 will be delayed indefinitely. More money will be lost; excavation, reinforcement, and further testing will be required—and many people with a stake in this project (from company executives to construction officials to shareholders) will be furious—especially if your report turns out to be a false alarm. All segments of plant management are geared up for the final big moment. Media coverage will be widespread. As the bearer of bad news—and bad publicity—you suspect that, even if you turn out to be right, your own career could be hurt by what some people will see as your overreaction that has made them look bad.

On the other hand, ignoring the crack could compromise the system's safety, with unforeseeable consequences. Of course, no one would ever be able to implicate you. The NRC has already inspected and approved the containment unit, leaving you and your boss and your company in the clear. You have very little time to decide. Start-up is scheduled for tomorrow, at which time the containment system will become intensely radioactive.

What would you do? Come to class prepared to justify your decision on the basis of the obligations, ideals, and consequences involved. ■

Working professionals commonly face similar choices, the product of conflicting goals and expectations, the pressure to meet deadlines and achieve results, to be a "loyal" employee, a "team player," to consider "the bottom line." Often, these choices have to be made alone or on the spur of the moment, without the luxury of meditation or consultation.

Never Depend Only on Legal Guidelines

Can the law tell you how to communicate ethically? Sometimes. If you stay within the law, are you being ethical? Not always. Legal standards "sometimes do no more than delineate minimally acceptable behavior." In contrast, ethical standards "often attempt to describe ideal behavior, to define the best possible practices for corporations" (Porter 183). Consider these legal actions:

Deception often is legal

- The investigative TV show "20/20" exposed the common and legal practice for trucks to haul garbage or toxic substances (such as formaldehyde) one way and then haul food products (such as juice concentrates) on the return trip—without ever informing the customer.

**Figure 5.6
Some Legal Lies in
the Workplace**

"Count on us to meet your deadline!"	"This product will last you for years!"	"Trust our experts to solve your problems!"	"You're our #1 priority!"
Promises you know you can't keep	Assurances you haven't verified	Credentials you don't have	Inflated claims about your commitment

- It is perfectly legal to advertise a cereal made with oat bran (which allegedly lowers cholesterol) without mentioning that another ingredient in the cereal is coconut oil (which raises cholesterol).

Lying is rarely illegal, except in cases of lying under oath or breaking a contractual promise (Wicclair and Farkas 16). But putting aside these and other illegal lies, such as defamation of character or lying about a product so as to cause injury, we see plenty of room for the kinds of legal lies depicted in Figure 5.6. Later chapters cover other kinds of legal lying, such as page design that distorts the real emphasis or words that are deliberately unclear, misleading, or ambiguous.

What then are a communicator's legal guidelines? Besides obscenity laws (not especially relevant here), workplace writing is regulated by the types of laws described below.

- *Laws against libel* prohibit any false written statement that maliciously attacks or ridicules anyone. A statement is considered libelous when it damages someone's reputation, character, career, or livelihood or when it causes humiliation or mental suffering. Material that is damaging but *truthful* would not be considered libelous unless it were used intentionally to cause harm. In the event of a libel suit, a writer's ignorance is no defense; even when the damaging material has been obtained from a source presumed reliable, the writer (and publisher) are legally accountable.[3]

- *Copyright laws* protect the ownership rights of authors—or of their employers, in cases where the writing was done as part of one's employment (Girill 48).

- *Law protecting software* makes illegal duplication of copyrighted software a felony. Conviction for a first offense carries up to five years in

---

3. Thanks to my colleague Peter Owens for the material on libel.

prison and fines up to $250,000. The Software Publisher's Association estimates that software piracy costs the industry more than $2 billion yearly ("On Line" 29).

- *Laws against deceptive or fraudulent advertising* make it illegal, for example, to falsely claim or imply that a product or treatment will cure cancer, or to represent and sell a used product as new. Fraud can be defined as "lying that causes another person monetary damage" (Harcourt 64).

- *Liability laws* define the responsibilities of authors, editors, and publishers for damages resulting from the use of incomplete, unclear, misleading, or otherwise defective information. The misinformation might be about a product (failure to warn about the toxic fumes from a spray-on oven cleaner) or a procedure (misleading instructions in an owner's manual for using a tire jack). Even if misinformation is given out of ignorance, the writer is liable (Walter and Marsteller 164–65).

  Legal standards with which product literature must comply vary from country to country. A document must satisfy the legal standards for safety, health, accuracy, language, or other issues for any country in which it will be distributed. For example, instructions for any product requiring assembly or operation have to carry warnings as stipulated by the laws of the country in which the product will be sold. Inadequate documentation, as judged by that country's standards, can result in a lawsuit (Caswell-Coward 264–66; Weymouth 145).

Laws regulating communication practices are few because such laws traditionally have been seen as threats to our freedom of speech (Johannesen 86). Many companies, however, have legal departments you can turn to with questions about a document's legality.

Recognizing that ethical behavior is a matter of commitment, and not of legislation, most professions have developed their own ethics guidelines. If your field has its own formal code, obtain a copy.

Understand the Potential for Communication Abuse

On the job, you write in the service of your employer. Your effectiveness is judged by how well your documents speak for the company and advance its interests and agendas (Ornatowski 100–01). You walk the proverbial line between telling the truth and doing what your employer expects (Dombrowski 97).

Workplace writing influences the thinking, actions, and welfare of different people: customers, investors, coworkers, the public, policy makers—to name a few. These people are victims of communication abuse whenever we give them information that is less than the truth as we know it. Following are some examples of such abuses.

Suppressing Knowledge the Public Deserves

Except for disasters that make big news (Bhopal, *Challenger,* Chernobyl) people hear plenty about the *wonders* of technology: how fluoride eradicates tooth decay, how smart bombs and cruise missiles never miss their targets, how nuclear power will solve our energy problems—but they rarely hear about the failures or the dangers (Staudenmaier 67).

In fact, the pressure to downplay failure sometimes results in censorship. For instance, some prestigious science journals have refused to publish studies linking chlorine and fluoride in drinking water with cancer risk, and fluorescent lights with childhood leukemia. The papers allegedly were rejected as part of widespread suppression of news about dangers of technological products (Begley 63).

Exaggerating Claims About Technology

Organizations that have a stake in a particular technology are especially tempted to exaggerate its benefits, potential, or safety.

Unwarranted claims help technology sell

> An entrepreneur needs financiers. Scientists in a large corporation need advocates high enough in the hierarchy to allocate funds. And government-supported researchers at universities and national labs have an obvious incentive to overstate their progress and understate the problems that lie ahead: the better the chances for success, the more money an agency is willing to shell out. (Brody 40)

If your organization depends on outside funding (as in the defense or space industry), you might find yourself pressured to make unrealistic promises.

Stealing or Divulging Proprietary Information

Proprietary information is any document or idea that can be considered the exclusive property of the company in which it originated. Proprietary documents may include company records, test and experiment results, surveys paid for by clients, market research, minutes of meetings, plans, and specifications (Lavin 5). In theory, such information is legally protected,[4] but it remains vulnerable to sabotage or theft. Rapid developments in technology create fierce competition among rival companies for the very latest intelligence, giving rise to measures like these:

Examples of corporate espionage

> Companies have been known to use business school students to garner information on competitors under the guise of conducting "research." Even more commonplace is interviewing employees for slots that don't exist and wringing them dry about their current employer. (Gilbert 24)

---

4. Although the Uniform Trade Secrets Act is designed to curb such abuses, this federal law is subject to broad and varied interpretation by individual state courts (Gilbert 22).

Moreover, employees within a company can leak confidential information to the press or to anyone else who has no legal right to know.

Mismanaging Electronic Information

With so much information stored in databases (by schools, employers, government agencies, mail order retailers, credit bureaus, banks, credit card companies, insurance companies, pharmacies) questions of how we combine, use, and disseminate the information become increasingly important (Finkelstein 471). Moreover, a database is easier to alter than its printed equivalent; one simple command can wipe out or transform the facts.

Withholding Information People Need to Do Their Jobs

Nowhere is the adage that "information is power" more true than among coworkers. One sure way to sabotage a colleague is to withhold vital information about the task at hand.

Beyond these deliberate communication abuses is this reality: *all* information is a matter of personal or social interpretation (Dombrowski 97); therefore, "objective reporting," practically speaking, is impossible. What we say on the job and how we say it are influenced by the expectations of our employer and by our own self-interest.

Exploiting Cultural Differences

Cross-cultural documents carry great potential for communication abuse. Based on its level of business experience, technological development, or financial need, a particular culture might be especially vulnerable to manipulation or deception. Some countries, for instance, are persuaded to purchase, from U.S. companies, pesticides and other chemicals whose use is banned in the United States. Other countries witness depletion of their natural resources or exploitation of their labor force by more developed countries— all in the name of progress. All communication in all cultural contexts should embody universal standards of honesty and fairness.

Know Your Communication Guidelines

How do we balance self-interest with the interests of others—the organization, the public, our customers? How can we be practical and responsible at the same time? Here are two practical guidelines for ethical communication (Clark 194):

1. *Give the audience everything it needs to know.* To see things as clearly as you do, people need more than just a partial view. Don't bury readers in needless details, but do make sure they get all the facts and get them straight.
2. *Give the audience a clear understanding of what the information means.* Even when all the facts are known, they can be misinterpreted. Do all you can to ensure that your readers understand the facts as you do.

We have seen how the *Challenger* tragedy resulted from decision makers knowing too little or misunderstanding what they did know (Clark 194): "Low temperatures will harden the O-rings and will compromise their ability to seal the rocket joint." The *meaning* of this data depended on whether it was interpreted as a technical fact ("Exhaust leakage means an explosion") or a social fact ("Another launch delay means our company looks bad") (Ornatowski 98).

- To the engineers—the technical experts, but not the decision makers— the technical fact meant that the risk was serious enough to warrant a launch delay.
- To lower-level decision makers, the social fact meant that the risk was acceptable because of the social pressure to launch on schedule.
- To the ultimate decision makers at NASA, the data meant little, because lower-level decision makers provided information that was neither sufficient nor clear enough for the risk to be understood, or even recognized.

Because of social pressure to keep an employer happy, vital facts and truthful interpretations were suppressed.

The Ethics Checklist on page 85 incorporates additional guidelines from various chapters. Use the checklist for any document you prepare or for which you are responsible.

Decide Where and How to Draw the Line

Suppose your employer asks you to do something unethical—for example, alter data to cover up a violation of federal pollution standards. If you decide to resist, your choices seem limited: resign or go public (i.e., blow the whistle).

Walking away from a job isn't easy, however, and whistle-blowing can spell career disaster (Rubens 330). Many organizations refuse to hire anyone blacklisted as a whistle-blower (Wicclair and Farkas 19). Even if you aren't fired, expect your job to become hellish. Here is one communicator's gloomy assessment, based on personal experience:

Know what to expect

> Most of the ethical infractions [you] witness will be so small that blowing the whistle will seem fruitless and self-destructive. And leaving one company for another may prove equally fruitless, given the pervasiveness of the problem. (Bryan 86)

Moreover, laws offer little protection for whistle-blowers. In general, employers are immune to lawsuits by employees who have been dismissed unfairly but who have no contract or union agreement specifying length of employment. The *employment-at-will rule* stipulates that employees can be dismissed at any time and for any cause, or even for no cause, at the employer's discretion (Unger 94). Certain laws, however, do supersede the employment-at-will rule:

- The Federal False Claims Act allows an employee to initiate a lawsuit, in the government's name, against a contractor for defrauding the government (for example by overcharging for military parts). The aggrieved employee can receive up to 25 percent of money recovered by the government as a result of the suit. Also, this law allows employees of government contractors to sue when they are punished for whistle-blowing (Stevenson 7). Critics contend that such laws can be exploited by employees greedy for profit or by disgruntled employees who seek revenge on the company.

- Anyone who reports employer violations to a regulatory agency such as the Federal Aviation Administration (FAA), Nuclear Regulatory Commission (NRC), Occupational Health and Safety Administration (OSHA) and who is punished can request a labor department investigation. Employees whose claims are ruled valid can win reinstatement and reimbursement for back pay and legal expenses.[5]

- As of 1994, laws protecting whistle-blowers from employer retaliation were on the books in twenty-two states.

Even with such protections, an employee who takes on the company without the backing of a labor union or other powerful group can expect lengthy court battles and disruption of life and career. Exactly where you draw the line (on having your integrity or health instead of your job) will be strictly your own decision.

If you do decide to take a stand, be reasonable and cautious, and follow these suggestions (Unger 127–30):

- *Get your facts straight, and get them on paper.* Don't blow matters out of proportion, but do keep a paper trail in case of legal proceedings.

- *Appeal your case in terms of the company's interests.* Instead of being pious and judgmental ("This is a racist and sexist policy, and you'd better get your act together"), focus on what the company stands to gain or lose ("Promoting too few women and minorities makes us vulnerable to legal action").

- *Aim your appeal toward the right person.* If you have to go to the top, find someone who knows enough to appreciate the problem and who has enough clout to make something happen.

- *Get professional advice.* Contact an attorney and your professional society for advice about your legal rights.

Before accepting a job offer, do some discreet research about the company's ethical reputation. (Of course you can learn only so much about a

---

5. Although employees legally are entitled to speak confidentially with OSHA inspectors about violations of health and safety in their company, a recent survey revealed that the inspectors themselves feel such laws offer employees little actual protection against company retribution (Kraft 5).

company before actually working there.) Some companies have ombudspersons, who help employees lodge complaints. Others offer hotlines for advice on ethics problems or for reporting violations. Also, realizing that good ethics are good business, companies increasingly are developing codes for personal and organizational behavior. Without such supports, don't expect to last long as an ethical employee in an unethical organization.

Remember that very few employers tolerate any public statement, no matter how truthful, that makes the company look bad.

A Final Note: Sometimes the right choice is obvious, but often not so obvious. No one has any sure way of always knowing what to do. This chapter is only an introduction to the inevitable hard choices that, throughout your career, will be yours to make and to live with.

AN ETHICS CHECKLIST FOR COMMUNICATORS

Use this checklist[6] to help your documents reflect reasonable, ethical judgment. (Numbers in parentheses refer to the first page of discussion.)

❑ Do I avoid exaggeration, understatement, sugarcoating, or any distortion or omission that leaves readers at a disadvantage? (67)

❑ Do I make a clear distinction between "certainty" and "probability"? (79)

❑ Am I being honest and fair? (67)

❑ Have I explored all sides of the issue and all possible alternatives? (126)

❑ Are my information sources valid, reliable, and unbiased? (174)

❑ Do I actually believe what I'm saying, instead of being a mouthpiece for groupthink or advancing some hidden agenda? (73)

❑ Would I still advocate this position if I were held publicly accountable for it? (74)

❑ Do I provide enough information and interpretation for readers to understand the facts as I know them? (82)

❑ Am I reasonably sure this document will harm no innocent persons or damage their reputation? (79)

❑ Am I respecting all legitimate rights to privacy and confidentiality? (81)

❑ Do I inform readers of the consequences or risks (as I am able to predict) of what I'm advocating? (68)

❑ Do I state the case clearly, instead of hiding behind jargon and generalities? (286)

❑ Do I give candid feedback or criticism, if it is warranted? (74)

❑ Am I distributing copies of this document to every person who has the right to know about it? (471)

❑ Do I credit all contributors and sources of ideas and information? (191)

6. Adapted from Brownell and Fitzgerald 18; Bryan 87; Johannesen 21–22; Larson 39; Unger 39–46; Yoos 50–55.

☑ EXERCISES

1. Prepare a memo (one or two pages) for distribution to first-year students in which you introduce the ethical dilemmas they will face in college. For instance:

   - If you receive a final grade of A by mistake, would you inform your professor?
   - If the library loses the record of books you've signed out, would you return them anyway?
   - Would you plagiarize—and would that change in your professional life?
   - Would you support lowering standards for student athletes if the team's success was important for the school's funding and status?
   - Would you allow a friend to submit a paper you've written for some other course?

   What other ethical dilemmas can you envision? Tell your audience what to expect, and give them some *realistic* advice for coping. No sermons, please.

2. Only about two dozen companies cause one-third of toxic-waste pollution in the U.S.—and it's all perfectly legal. Who are the biggest polluters, and why do they get away with it?

3. In your workplace communications, you may end up facing hard choices concerning what to say, how much to say, how to say it, and to whom. Whatever your choice, it will have definite consequences. Be prepared to discuss the following cases in terms of the obligations, ideals, and consequences involved. Can you think of similar choices you or someone you know has already faced? What happened?

   - While traveling on assignment that is being paid for by your employer, you visit an area in which you would really like to live and work, an area in which you have lots of contacts but never can find time to visit on your own. You have five days to complete your assignment, and then you must report on your activities. You complete the assignment in three days. Should you spend the remaining two days checking out other job possibilities, without reporting this activity?
   - As a marketing specialist, you are offered a lucrative account from a cigarette manufacturer; you are expected to promote the product. Should you accept the account? Suppose instead the account were for beer, junk food, suntanning parlors, or ice cream. Would your choice be different? Why, or why not?
   - You have been authorized to hire a technical assistant, and so you are about to prepare an advertisement. This is a time of threatened cutbacks for your company. People hired as "temporary," however, have never seemed to work out well. Should your ad include the warning that this position could be only temporary?
   - You are one of three employees being considered for a yearly production bonus, which will be awarded in six weeks. You've just accepted a better job, at which you can start anytime in the next two months. Should you wait until the bonus decision is made before announcing your plans to leave?
   - You are marketing director for a major importer of coffee beans. Your testing labs report that certain African beans contain roughly twice the caffeine of South American varieties. Many of these African varieties are big sellers, from countries whose coffee bean production helps prop otherwise desperate economies. Should your advertising of these varieties inform the public about the high caffeine content? If so, how much emphasis should this fact be given?
   - You are research director for a biotechnology company working on an AIDS vaccine. At a national conference, a researcher from a competing company secretly offers to sell your company crucial data that could speed discovery of an effective vaccine. Should you accept the offer?

☑ COLLABORATIVE PROJECT

After dividing into groups, study the following scenario and complete the assignment: You belong to the Forestry Management Division in a state whose year-round economy depends almost totally on forest products (lumber, paper, etc.) but whose summer economy is greatly enriched by tourism, especially from fishing, canoeing, and other outdoor activities. The state's poorest area is also its most scenic, largely because of the virgin stands of hardwoods. Your division has been facing growing political pressure from this area to allow logging companies to harvest the trees. Logging here would have good and bad consequences: for the foreseeable future, the area's economy would benefit greatly from the jobs created; but traditional logging practices would erode the soil, pollute waterways, and decimate wildlife, including several endangered species—besides posing a serious threat to the area's tourist industry. Logging, in short, would give a desperately needed boost to the area's standard of living, but would put an end to many tourist-oriented businesses and would change the landscape forever.

Your group has been assigned to weigh the economic and environmental impacts of logging, and prepare recommendations (to log or not to log) for your bosses, who will use your report in making their final decision. To whom do you owe the most loyalty here: the unemployed or underemployed residents, the tourist businesses (mostly owned by residents), the wildlife, the land, future generations? The choices are by no means simple. In cases like this, it isn't enough to say that we should "do the right thing," because we are sometimes unable to predict the consequences of a particular action—even when it seems the best thing to do. In a memo to your supervisor, tell what action you would recommend and explain why. Be prepared to defend your group's ethical choice in class on the basis of the obligations, ideals, and consequences involved.

Communicating Electronically

No one can predict how communication technology will revolutionize the ways in which we gather, organize, write, read, and use information. Nor can we predict technology's ultimate impact on our work habits and relationships. What we do know is that more people than ever are likely to work on a document, more are likely to read the document, and more documents are likely to be paperless.[1] In this *virtual* workplace, fiber optic networks connect employees, consultants, and colleagues from anywhere, enabling all types of collaboration. This collaboration on a global scale is made possible by revolutionary developments in communication technology.

Exploring the Internet

A worldwide computer network, the Internet, has evolved from Arapanet, a network created in 1969 for university computer researchers to exchange files and share data. As the Net's potential for the open exchange of information became recognized, network sites increased and commercial online services began offering access to general subscribers. By 1989 the Internet had become a global vehicle for ideas and information of all kinds (Levy 179–80).

Today's Internet connects more than 20 million computer users (researchers, students, businesspeople, etc.). Internet users in the United States alone are expected to reach nearly 30 million by the year 2000 (Hutheesing 110).

Commercial online services such as *Prodigy, Compuserve, America Online, Delphi,* and *Microsoft Network* provide Internet access via "gateways," along with aids for navigating its many resources, which include the following:

Selected Internet resources

- *Usenet,* a global bulletin board for sharing information and getting answers to questions on topics of common interest, categorized according to particular *newsgroups.* Newsgroups offer public access to more than 10,000 topics ranging from rain-forest preservation to intellectual property laws to witchcraft.

- *Listservs,* discussion groups on more specialized topics (for example, cancer research). Listserv access is limited to subscribers.

- *Gopher,* a broad-based tool for retrieving information by searching database sites, library catalogs, and electronic publications worldwide for information under specific subject headings (for example, "Health Effects of Electromagnetic Radiation"). To help users keep track of items searched and retrieve them if needed, *Gopher* provides a bookmark for each item searched.

---

1. Keep in mind that paper documents are unlikely to disappear anytime soon and that much of our electronic communication continues to be *written.* Even for documents that are paperless, writers face essentially the same set of information, persuasion, and ethics problems.

IN BRIEF

The Evolution of Electronic Communication

Office communication has evolved dramatically over the last half of the twentieth century, as the practices and technology outlined below reveal.

The Year 1950—Communication in a U.S. Company

- The company is housed in one building or a cluster of buildings to which employees commute daily.
- Managers dictate letters and memoranda to secretaries, who type a single copy on a typewriter, using carbon paper, stencils, or ditto paper to create multiple copies for distribution by hand or by mail, usually to readers somewhere in the U.S.
- Typed manuscripts for annual reports, brochures, manuals and other external documents are sent to a printing firm that designs and typesets the material and prints it over a period of several weeks or months.
- Incoming mail is sorted, delivered, read, forwarded, or stored in file cabinets along with company forms, records, correspondence, reports, and other paper documents.
- In-house news, announcements, and company developments are distributed via printed newsletters or posted on bulletin boards.
- Employees tend to work and write in the relative isolation of an office or cubicle. Workplace discussions take place in meeting rooms, by phone, or via memos and notes.
- Manuals and reference books are stored on bookshelves or in desk drawers. Information gathering often is conducted by travel to libraries, other companies, or institutions that house the needed information.
- Printed texts tend to be read front to back.

The Year 2000—Communication in a "Virtual" Company

- The company may have branches across the state, the nation, or the world to which many employees "commute" electronically. (This includes freelance workers, employed by other companies as well)
- Managers use word processing systems to compose their own letters and memoranda for distribution via E-mail to readers across the building or across the globe.
- Via desktop publishing (DTP) networks, the composing, layout, graphics design, typesetting, and printing of external documents are done rapidly in-house on personal computers.
- Optical scanners take an electronic snapshot of any paper document produced or received by a company, including incoming mail. Stored online, this image can be retrieved and edited, printed out, faxed, or delivered via E-mail to customers or colleagues, or posted on an electronic bulletin board. Recipients can add comments or data before forwarding the imaged document to other readers (Laplante 37–41). Company forms (requisitions, accident reports, etc.) can be produced, filed, updated, distributed, filled out, and delivered electronically (Strizich 122).
- Workplace discussions and document sharing occur via E-mail, voicemail, or video conferencing networks. Electronic bulletin boards post and announce day-to-day developments for employees or readers worldwide: newsletters, price lists, changes or updates in policies or procedures, press releases (Gruman 127).
- Employees work and write collaboratively, sharing resources in a joint effort (as in devel-

IN BRIEF

oping a proposal or a marketing plan). Drafts circulated electronically enable colleagues to add comments directly on the manuscript (Gruman 128). Multimedia systems present text, graphics, sound, and animated material retrieved from a computer file. Various people networked from different locations work on the electronic document and view and comment on one another's "work in progress" using electronic pens (Weinberg 75).

- Electronic "readers" translate a printed or electronic document into speech, enabling employees to listen to memos, letters, reports, or articles as they commute to work, have lunch, or do something else (Weinberg 75). These talking documents include online manuals that provide animated demonstrations of specific procedures and letters that offer musical accompaniment (Figgins 201–06).

- Online databases store information from books, magazines, newspapers, journals, etc., and can be searched rapidly via worldwide information networks for the latest stock market quotations, trends in global weather patterns, sites of recent disease outbreaks, and so on. A single compact disk can store an entire encyclopedia, a medical dictionary, or interactive manuals and lessons.

- Electronic texts tend to be read nonsequentially, with readers navigating their own paths and choosing various routes to explore (Grice and Ridgway 37).

☆   ☆   ☆

- *Archie,* for locating and downloading free computer files or software from file-transfer sites worldwide.
- *Veronica, Jughead,* and *WAIS (Wide Area Information Service),* for specialized searches of various types.
- *World-Wide Web,* a global network of databases, documents, images, and sounds. Information from anywhere in the Web network can be accessed and explored via navigation programs such as *Mosaic* or *Netscape,* known as "browsers." Hypertext links (see page 99) among *Web* resources enable users to explore information along different paths by clicking on key words or icons that reveal additional paths for browsing and discovery.

The *Web* offers powerful business applications as a research and marketing tool: for advertising, learning about new products or new companies, updating product information, or ordering products (Teague 236, 238). Each organization advertises its services and products on the Web via its own *home page,* a type of electronic billboard that introduces the organization and provides links to additional pages that individual users can explore according to their information needs.

See pages 149–50 for the use of Internet tools in research.

**Figure 6.1
A Typical E-mail
Message**

Ihabicht@umassd.edu, powens@umassd.edu, rdumont@umassd.edu, td,8/7/96   1

```
      To: Ihabicht@umassd.edu, powens@umassd.edu, rdumont@umassd.edu,
          tdace@umassd.edu, ltravers@umassd.edu
    From: ethompson@umassd.edu (Ed Thompson)
 Subject: meeting of "program review" committee

Hello All,

          You remember the Program Review Committee, don't you?
          As far as I know, the department still needs to produce two twelve-
page review documents (one for undergrad. and one for Professional Writing) and
a five-page document of 1, 2 and 5 year plan goals for the department. The due
date, unless it's pushed back, is October 2. I guess we'd better get started
soon.
          Could everyone make an initial planning meeting before we get into
the hectic rush at the start of the semester? I'd like to meet on Wednesday,
August 30th, at 10:00 a.m. Let me know if you can't make it. By the way, the
long-awaited "data books" finally arrived in late June. I managed to get enough
copies for everyone on the committee and will leave them with John to be picked
up at your convenience.



          Best,
          Ed
```

Using Electronic Mail

Perhaps the most widely used application on the Internet is electronic mail, which provides a connection to discussion forums on Listserv and Usenet and carries routine, day-to-day communication as well. E-mail transmits an electronic document via networked computer terminals to recipients in the same building or across the globe. Specific codes direct the message to any electronic mailbox designated by the sender—or to all the mailboxes on a mailing list. Alerted by an audio signal, the recipient opens the on-screen mailbox, reads the message, and then either responds, files the message, prints it out, forwards it, or deletes it. E-mail messages can be exchanged instantaneously or at the convenience of the communicating parties. Figure 6.1 shows a typical E-mail message.

E-mail Benefits

Compared to phone, fax, or conventional mail (or even face-to-face conversation, in some cases), E-mail offers benefits:

- *E-mail is fast, convenient, efficient, and relatively unintrusive.* Unlike conventional mail, which can take days to travel, E-mail travels instantly. Although a fax network can transmit printed copy rapidly, E-mail eliminates paper shuffling, dialing, and a host of other steps. Moreover, E-mail makes for efficiency by eliminating "telephone tag." It

E-mail facilitates communication and collaboration

is less intrusive than the telephone, leaving the choice of when to read and respond to a message entirely up to the recipient.

■ *E-mail is democratic.* With few exceptions, E-mail messages appear as plain print on a screen, with no special typestyles, fancy letterheads, page design, or paper texture—enabling readers to focus on the message instead of the medium.[2] E-mail also allows for transmission of messages by anyone at any level in an organization to anyone at any other level. For instance, the mail clerk conceivably could E-mail the company president directly, whereas a conventional memo or phone call would be routed through the chain of management or screened by administrative assistants (Goodman 33–35). In addition, people who are ordinarily shy in face-to-face encounters may be more willing to express their views in an E-mail conversation.

■ *E-mail can foster creative thinking.* E-mail dialogues involve a give-and-take, much like a conversation. Writers feel encouraged to express their thoughts spontaneously, thinking as they write, without worrying about page design, paragraph structure, perfect phrasing or the like. The focus is on conveying one's meaning to the recipients who in turn will respond with thoughts of their own. This relatively free exchange of views can lead to all sorts of new insights or ideas (Bruhn 43).

■ *E-mail is excellent for collaborative work and research.* Collaborative teams keep in touch via E-mail, and researchers contact people who have the answers they need. Especially useful for collaborative work is the E-mail function that enables documents or electronic files of any length to be attached and sent for downloading by the receiver.

E-mail Privacy Issues

Although E-mail connects increasing millions of users, no specific laws protect any computer conversation from eavesdropping or snooping.[3] Gossip, personal messages, or complaints about the boss or a colleague—all might be read by unintended receivers. Employers often claim legal right to monitor *any* of their company's information, and some of these claims can be legitimate:

Monitoring of E-mail by an employer is legal

In some instances it may be proper for an employee to monitor E-mail, if it has evidence of safety violations, illegal activity, racial discrimination, or sexual improprieties, for instance. Companies may also need access to business

---

2. Recent multimedia developments allow charts, graphs, 3-D images, sound, voice, animation, or video to be added to certain E-mail messages, but many of these features would be considered inappropriate—if not frivolous—in routine correspondence.

3. Whereas the phone company and other private carriers are governed by FCC laws protecting privacy, no such legal protection has been developed for communication on the Internet (Peyser and Rhodes 82).

information, whether it is kept in an employee's drawer, file cabinet, or computer E-mail. (Bjerklie, "E-Mail" 15)

E-mail privacy can be compromised in other ways as well. Some notable examples:

E-mail offers no privacy

- Everyone on a group mailing list—intended reader or not—automatically receives a copy of the message.
- Even when deleted from the system, messages often live on for years, saved in a backup file.
- Anyone who gains access to your network and your private password can read your document, alter it, use parts of it out of context, pretend to be its author, forward it to whomever, plagiarize your ideas, or even author a document or conduct illegal activity in your name. (One partial safeguard is encryption software, which scrambles the message, enabling only those who possess the special code to unscramble it.)

E-mail Quality Issues

Free exchange via computer screen adds to the *quantity* of information exchanged, but not always to its *quality:*

A useful definition of "information"

Claude Shannon, father of communication theory . . . once said, "Information is news that makes a difference. If it doesn't make a difference, it isn't information." A radio traffic report about a car crash up ahead is information if you can still change your route. But if you are already stuck in traffic, the message is . . . useless. (Rothschild 25)

Following are specific ways in which information quality can be compromised by E-mail communication:

E-mail does not always promote quality in communication

- The ease of sending and exchanging messages can generate overload and junk mail—a party announcement sent to 300 employees on a group mailing list, or the indiscriminate mailing of a political statement to dozens of newsgroups ("spamming").
- Some electronic messages may be poorly edited and long-winded.
- Off-the-cuff messages or responses might offend certain recipients. E-mail users often seem less restrained about making rude remarks ("flaming") than they would be in a face-to-face encounter.
- Recipients might misinterpret the tone. "Emoticons" or "Smileys," punctuation cues that signify pleasure :-), displeasure :-(, sarcasm ;-), anger >:-< and other emotional states, offer some assistance but are not always an adequate or appropriate substitute for the subtle cues in spoken conversation. Also, common E-mail abbreviations (FYI, BTW, HAND—which mean "for your information," "by the way," and "have a nice day") might strike some readers as too informal.

E-mail Guidelines

Recipients who consider an E-mail message poorly written, irrelevant, offensive, or inappropriate will only end up resenting the sender. These guidelines offer suggestions for effective E-mail use.[4]

- Use E-mail to reach a lot of people quickly with a relatively brief, informal message.
- Don't use E-mail to send confidential information: employee evaluations, criticism of people, proprietary information, or anything that warrants privacy.
- Don't use E-mail to send formal correspondence to clients or customers, unless they request or approve this method beforehand.
- Don't use the company E-mail network for personal correspondence or for anything that is not work related.
- Check your distribution list before each mailing, to be sure the message reaches all intended primary and secondary readers but no unintended ones.
- Assume your E-mail correspondence is permanent and could be read by anyone anytime. Ask yourself whether you've written anything you couldn't say to another person face-to-face. Avoid spamming and flaming.
- Before you forward an incoming message to other recipients, be sure to obtain permission from the sender. (See page 176 for E-mail copyright issues.)
- Limit your message to a single topic, and keep the whole thing focused and concise. (Yours may be just one of many messages confronting the recipient.) Don't ramble.
- Use a clear subject line to identify your topic ("Subject: Request for Beta test data for Project #16"). This helps recipients decide whether to read the message immediately and makes it easier to file and retrieve for later reference.
- Refer clearly to the message to which you are responding ("Here are the Project 16 Beta test data you requested on Oct. 10").
- Try to keep the sentences and paragraphs short, for easy reading.
- Don't write in FULL CAPS—unless you want to SCREAM at the recipient!
- Close with a signature section that names your company or department, phone and fax number, and any other information the recipient might consider relevant.

4. Adapted from Bruhn 43; Goodman 33–35, 167; Kawasaki 286; Nantz and Drexel 45–51; Peyser and Rhodes 82.

Telecommuting

Nearly 10 million Americans telecommute: researching, writing, and performing other jobs on a home terminal linked to a company or worldwide network. On a *video conferencing* network, coworkers in separate locations can edit and comment on each other's work while observing each other's reactions as they proceed. Consider this example:

Computer networks are redefining the "workplace"

> For instance, as part of the effort to restore the biodiversity of Everglades National Park, a team of civil engineers, ecologists, and biologists is designing new flowways and levees to transport unpolluted water into the Everglades ecosystem. Team members might work from various sites, meeting electronically to share and refine design ideas, compare research findings, and edit reports and proposals. (Boucher 32–33)

By the early twenty-first century, additional millions will be using laptop or palmtop computers in this virtual workplace.

Some people worry about the prospect of transactions that are electronically mediated (instead of face-to-face). Even the sharpest video images are no substitute for human presence (Brittan 50), and studies indicate that as many as 80 percent of workers resist the idea of communicating exclusively via machine (Bjerklie, "Telecommuting" 21). As we search for new ways of relating to coworkers and audiences, we wonder about the computer's ultimate impact on human relations. Despite these anxieties, however, we can look forward to increasing collaboration in which people share their own knowledge while benefiting from the knowledge of others.

Source: Drawing by C. Barsotti; Copyright ©1992 The New Yorker Magazine, Inc.

Good grief, Bradbury! How long have you been working at home?

The Future of Telecommuting

You've heard the hype. We asked the experts. Here's the *real* timetable.

By the end of this year, 9.2 million Americans will call themselves telecommuters—employees who use networked computers and cellular telephones to work outside the traditional office—according to a recent study by Link Resources Corp. Telecommuting can increase productivity and lower overhead costs, but running a "virtual office" raises important questions about decentralized management, worker responsibility, and loneliness for employees who have been cut loose from familiar work environments. Yet emerging technologies will make it easier to stay connected even as we move away from our burlap cubicles. *Wired* asked five experts to look at the future of telecommuting.—*David Pescovitz*

	One-Fifth US Workers are Telecommuters	Universal Desktop Videoconferencing	Global Wireless Telephone Number	Fortune 500 "Virtual Corporation"
Franklin D. Becker	2005	2005	1998	never
Joe Carter	1999	1998	2010	1999
N. Fredric Crandall	2010	2005	1998	2000
Tom Newhouse	1999	2010	2000	unlikely
Van Romine	2000	1997	1998	1997
Bottom Line	2003	2003	2001	1999

Franklin D. Becker director, international workplace studies program, Cornell University; partner, @Work Consulting Group

Joe Carter managing director, Andersen Consulting Center for Strategic Technology

N. Fredric Crandall PhD; founding partner, The Center for Workforce Effectiveness Inc.

Tom Newhouse owner/principal, Thomas J. Newhouse-Design, an industrial-design firm working primarily in the areas of office furniture and major appliances

Van Romine director, Institute for Telework

Our experts predict the number of telecommuters will triple in the next 15 years to 20 percent of the US work force, driven by stricter air-quality regulations, improved communications networks, and rising demand for adaptability in the business world. However, Crandall warns that the work-at-home route can lead to a "virtual dead end" of employee isolation and bureaucratic ineffectiveness. "The idea is to get people into the field so they can be more responsive to customer requirements," he says, "rather than trying to adapt to individual lifestyles." Becker adds that "a variety of easily accessed telework centers are likely to function at least as well or better than a home office."

Even as remote work grows in popularity, desktop full-motion videoconferencing remains beyond the reach of most telecommuters. Thin pipes and high cost are preventing the "virtual water cooler" from being bundled, like CD-ROMs and modems, with business-use computers. "When the cost of desktop videoconferencing falls below US $500 and when ordering ISDN* is as easy as getting a burger at the local diner, things will start to shift," says Romine. But by then, Carter believes, "low bandwidth peer-to-peer networking, shared applications, and shared workspace tools that don't require video" may already fulfill most collaborative needs.

As global business increases, our experts think global wireless communication will expand as well. While most businesspeople don't need to talk on the phone while climbing the Himalayas, Newhouse thinks hand-held telephones that allow for global roaming will be useful in "the 24-hour world of finance and market analysis, where minutes mean millions. The same can be true for political organizations." But, according to Carter, international business politics may put personal satellite telephones on hold. "The complexities of crossing national borders and dealing with other countries' telecommunication monopolies are difficult to work out," he says.

Most of our experts agree that the first Fortune 500 "virtual corporation"—a company without a traditional central head-quarters—will still occupy some physical real estate. Becker predicts that large virtual corporations will maintain multi-use hubs combining meeting and communication centers, employee lounges, and classroom space for teaching new skills to a geographically dispersed work force. On the other hand, Newhouse believes "widely varied human personalities and job skill types" will keep at least half of a company's employees in a main office. "The head-quarters," he says, "will be wherever the CEO is."

---

* Integrated Services Digital Network (uses ordinary telephone lines to transmit multimedia messages). *Source: Courtesy of* Wired *Magazine, October 1995:68.*

Word Processing and Desktop Publishing

Besides eliminating much of the mechanical drudgery of drafting and revising, *word processing* provides page design and graphics options. *Groupware* (a group authoring system) enables writers in different locations to prepare a document collaboratively. Finished documents can be printed, distributed via E-mail, or filed and catalogued electronically for easy retrieval. Documents or parts of documents used repeatedly (*boilerplate*) can be retrieved, modified as needed, and inserted in some other document.

Beyond the actual writing, *desktop publishing* enables users to divide pages into columns, adjust the size and shape of graphics, insert predrawn images, and print headlines. With these and other options, trained, skilled users produce newsletters, brochures, pamphlets, manuals, and other publications that previously required graphic artists and print shops.

Creating Paperless Documents

With the exception of manuals and reference books, information in traditional paper documents (letters, memos, reports, proposals) usually is structured to be read in strict, linear sequence: front-to-back. The later information builds on the information that precedes it. E-mail documents, often printed out at their destination, usually display this linear structure. But some types of paperless documents enable users to design their own information sequence as they search for specific chunks of information or merely browse through parts of a topic in no particular order, as one might read a newspaper or encyclopedia (Grice and Ridgway 35–43). Because readers will design their own information sequence, these documents are written in small, discrete *modules,* that can stand alone in meaning.

Online Documentation

People who use computers in their jobs need instructions and training for operating the systems and understanding their many features. Online documentation is designed to support specific tasks and provide answers to specific questions.

Although computers come with printed manuals, the computer itself is becoming the preferred training medium. In computer-based training, *online documentation* on the screen itself explains how the system works and how to use it. Types of online documentation include error messages, reference guides, troubleshooting advice, tutorial lessons, interactive exercises with immediate feedback, and help and review options to accommodate different learning styles. Instead of leafing through a printed manual, users find what they need by typing a simple command.

While online documentation offers an exciting new communications medium, the preparation of electronic instruction presents unique challenges (discussed on pages 99–101).

**Figure 6.2
Topics (or Files)
in a Hypertext
Network
Are Linked
Electronically**

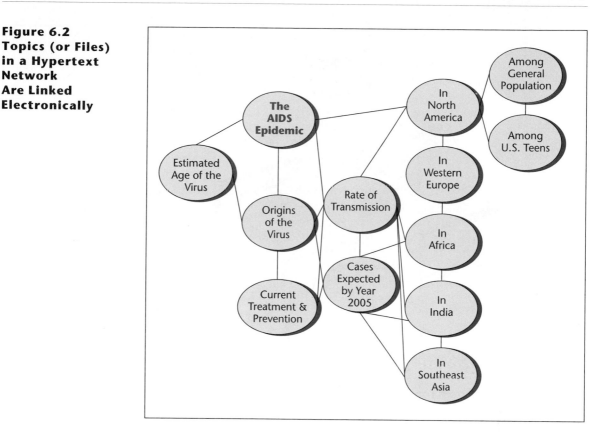

Hypertext

One extension of online documentation increasingly used in tutorial and training software is *hypertext,* information in electronic form designed for nonlinear reading. Unlike printed text, designed to be read front-to-back, a hypertext document offers various informational paths through the material, the particular path (or paths) determined by *what* and *how much* a reader wants to know. *Hypermedia* expands the applications of hypertext by adding graphics, while *multimedia* adds sound and animation.

Besides its instructional value, hypertext is a research and reference tool for navigating complex cross-references and retrieving information electronically. In a hypertext system, a topic can be explored from any angle, at any level of detail. Assume, for instance, you are researching the AIDS epidemic with the use of a hypertext database. The database contains chunks of related topics organized in a network (or web) of files linked electronically (Horton 22), as in Figure 6.2.

After accessing the initial file ("The AIDS Epidemic"), you navigate the network in any direction you wish, choosing which file to open and where to

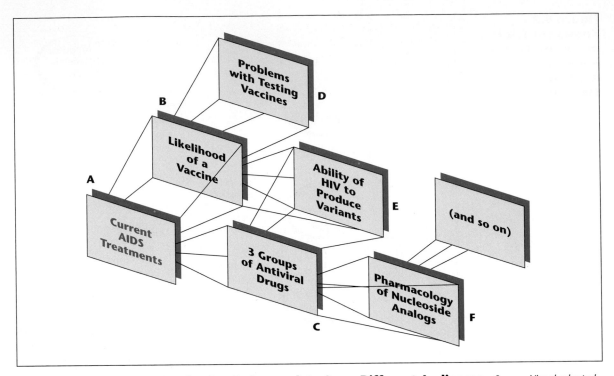

Figure 6.3   Hypertext Topics Can Be Layered, to Serve Different Audiences. *Source: Visual adapted from Horton, William. "Is Hypertext the Best Way to Document Your Product?"* Technical Communication 39.1 (1991): 25. *Used with permission from the Society for Technical Communication, Arlington, VA.*

go next. The files themselves might be printed words, graphics, sound, or animation. Freed from the fixed page sequence of a printed document, users customize their searches.

Hypertext also accommodates different levels of audience needs by offering multiple layers of information, from general to specialized, as depicted in Figure 6.3.

Communication specialist William Horton explains the instructional power achieved through hypertext layering:

Hypertext can increase a reader's depth of knowledge

Paper . . . lacks depth. All information must be on the same level or layer. With hypertext, however, the screen can have deeper reserves of information. These deeper layers do not clutter the screen, but are available if needed.

[In Figure 6.3] Topic A is the top layer. It is written at the level of detail and generality appropriate for mainstream users [seeking basic information].

If this surface explanation is sufficient, the reader need look no deeper. However, Topics B and C lie just below the surface and may be brought forth by the selection of their references within the text of Topic A. . . . Even deeper layers are in Topics D, E, and F. ("Is Hypertext" 25)

Persons navigating a hypertext document invent their own text and so can discover combinations, relationships, and chains of knowledge that could not possibly be expressed in the sequential pages of a printed document (Bernstein 42).

Following is an application of hypertext instruction in the auto industry:

A hypertext application

> A mechanic can zoom from a picture of a car engine to a video of a specific malfunctioning part, and then move directly to the text that tells how to fix it. (Morse 7)

For all their potential as research and instructional tools, hypertext documents have limitations:

- Confronting countless possible paths through the material, readers can get lost in "hyperspace." *(Where do I go next? How do I get out? How do I organize?)* Ease of navigation depends on how effectively individual chunks (or "nodes") of information are segmented and linked.
- Readers can resist a document that leaves them responsible for organizing their own learning. They generally prefer to rely on the writer for an orienting framework (Horton, "Is Hypertext" 26).
- Hypertext systems do not always enhance learning. They can in fact interfere with understanding and impede performance (Barfield, Haselkorn, and Weatbrook 22, 27). Users in one study took longer to read a hypertext document than the equivalent paper document (Rubens 36).

Despite the limitations of its early stages, hypertext remains a promising medium that challenges communicators to make decisions like these:

- How much information should be included on one screen?
- What level of detail should be presented?
- How can each chunk be written so it can be read in any order and still make sense (Horton, "Is Hypertext" 27)?
- How can the material best be linked for easiest navigation of various possible paths (Nickels-Shirk 191)?
- Which combination of media should be employed (printed words, speech, sound effects, animation, music) (Horton, "Is Hypertext" 27)?

Creating hypertext documents is no one-person job; it often requires writers, graphic artists, computer specialists, animators, and the like. This is one example of how communication technology makes collaboration both possible *and* necessary.

Computer Guidelines for Writers

The following guidelines will help you capitalize on the many writing benefits a computer can offer.

1. *Decide whether to draft by hand or by computer.* Some writers prefer to begin with pencil and paper. Others like to compose directly onscreen. Try each way before deciding which works best.

2. *Beware of computer junk.* The ease of cranking out words on a computer can result in long, windy pieces that say nothing. Edit final drafts to eliminate anything that fails to advance your meaning.

3. *Never confuse style with substance.* With laser printers, and desktop publishing and graphics software, documents can be made highly attractive, but not even the most attractive design can redeem a document whose content is worthless or inaccessible.

4. *Save and print your work often.* One way to court disaster is to write without saving or printing often enough. One wrong keystroke might cause pages of writing to disappear forever—unless you have saved them beforehand. Try to save each paragraph and print each page as you complete it.

5. *Make a backup disk.* A single electrical surge or malfunction can wipe out an entire file, floppy disk, or hard disk!

6. *Consider the benefits of revising from hard copy.* Nothing beats scribbling and scratching on the page with pen or pencil. The hard copy provides the whole text, right in front of you.

7. *Never depend only on automated "checkers."* Page 303 summarizes the limitations of many computerized aids. A synonym offered up in an electronic thesaurus may not accurately convey your intended meaning. The spell checker cannot differentiate among correct or incorrect usage of correctly spelled words such as "their," "they're," or "there" or "it's" versus "its." Although spell and grammar checkers can be a big help, they cannot evaluate style *appropriateness* (those subtle choices of phrasing that determine tone and emphasis). Even the most sophisticated writing aids are no substitute for careful proofreading.

8. *Print final copies on high-quality paper.* Inexpensive computer paper crinkles, smudges, and is hard to write on (for notes, comments, revision suggestions, and so on). Use paper with a high fiber content.

9. *Decide how to design and transmit your document.* Should the document be primarily verbal, primarily visual, or some combination? Should it travel by conventional mail, interoffice mail, E-mail? Select a design and a medium likely to have a favorable impact on your readers. Who are your readers, and what would they prefer in this situation: the solid feel of paper or the lure of the computer screen? Research indicates that younger readers prefer flashy graphics, older readers prefer traditional text, and readers in general trust text more than visual images (Horton, "Mix Media" 781).

☑ EXERCISE
(Individual or Collaborative)

Individually or in small groups, decide whether each of the following documents would be appropriate for E-mail transmission via a company E-mail network. Be prepared to explain your decisions.

Sarah Burnes' memo about benzene levels (page 15)

The "Rational Connection" memo (page 46)

The "better" memo to the maintenance director (page 50)

Tom Ewing's letter to a potential customer (page 59)

The medical report written for expert readers (page 29)

A memo reporting illegal or unethical activity in your company

A personal note to a colleague

A request for a raise or promotion

Minutes of a meeting

Announcement of a no-smoking policy

An evaluation or performance review of an employee

A reprimand to an employee

A notice of a meeting

Criticism of an employee or employer

A request for volunteers

A suggestion for change or improvement in company policy or practice

A gripe

A note of praise or thanks

A message you have received and have decided to forward to other recipients

The Problem-Solving Process Illustrated

EVERY writing situation requires deliberate decisions for *working with the information* and for *planning, drafting,* and *revising* the document. Some of these decisions are illustrated in Figure 7.1. Each writer approaches the process through a sequence of decisions that works best for *that* person. No group of decisions is complete until *all* groups are complete. This looping structure of the writing process is illustrated in Figure 7.2.

Critical Thinking in the Writing Process

The writing process is a *critical thinking* process: the writer makes a series of deliberate decisions in response to a situation. The actual "writing" (putting words on the page) is only a small part of the overall process—probably the least significant part.

In this chapter, we will follow one working writer through an important writing situation; we will see how he solves his unique information, persuasion, and ethics problems and how he collaborates to design a useful and efficient document.

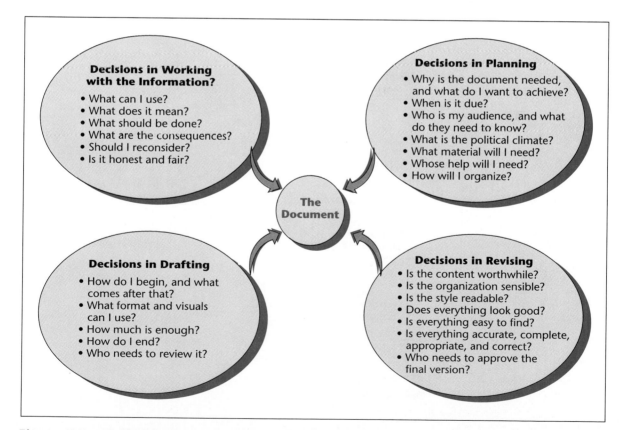

Decisions in Working with the Information?
• What can I use?
• What does it mean?
• What should be done?
• What are the consequences?
• Should I reconsider?
• Is it honest and fair?

Decisions in Planning
• Why is the document needed, and what do I want to achieve?
• When is it due?
• Who is my audience, and what do they need to know?
• What is the political climate?
• What material will I need?
• Whose help will I need?
• How will I organize?

The Document

Decisions in Drafting
• How do I begin, and what comes after that?
• What format and visuals can I use?
• How much is enough?
• How do I end?
• Who needs to review it?

Decisions in Revising
• Is the content worthwhile?
• Is the organization sensible?
• Is the style readable?
• Does everything look good?
• Is everything easy to find?
• Is everything accurate, complete, appropriate, and correct?
• Who needs to approve the final version?

Figure 7.1   Typical Decisions During the Writing Process

**Figure 7.2
A Flowchart for
the Writing
Process**

*Decisions in the writing
process are recursive: No
one stage of decisions is
complete until* all *stages
are complete.*

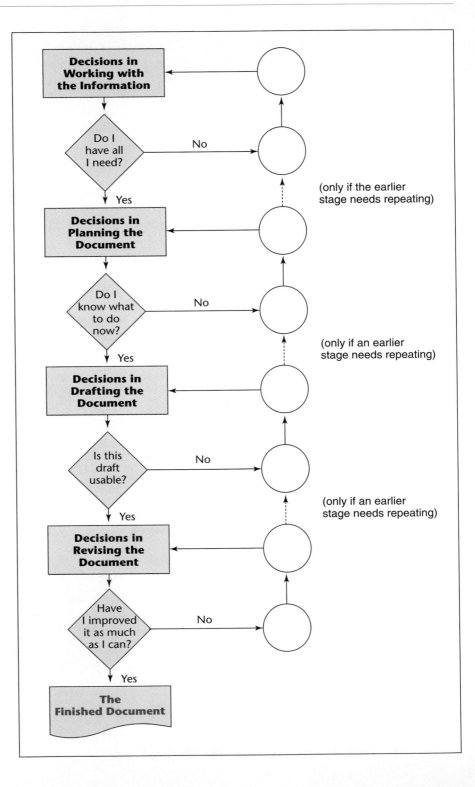

A Sample Writing Situation

The company is Microbyte, maker of portable microcomputers. The writer is Glenn Tarullo (B.S., Management; Minor: Computer Science). Glenn has been on the job three months as Assistant Training Manager for Microbyte's Marketing and Customer Service Division.

For three years, Glenn's boss, Marvin Long, periodically has offered a training program for new managers. Long's program combines an introduction to the company with instruction in management skills (time management, motivation, communication). Long seems satisfied with his two-week program but has asked Glenn to evaluate it and write a report as part of a company move to upgrade training procedures.

Glenn knows his report will be read by Long's boss, George Hopkins (Assistant Vice President, Personnel), and Charlotte Black (Vice President, Marketing, the person who devised the upgrading plan). Copies will go to other division heads, to the division's chief executive, and to Long's personnel file.

Glenn spends two weeks (Monday, October 3, to Friday, October 14) sitting in and taking notes on Long's classes. On October 14, the trainees evaluate the program. After reading these evaluations and reviewing his notes, Glenn concludes that the program was successful but could stand improvement. How can he be candid without harming or offending anyone (instructors, his boss, or guest speakers)? Figure 7.3 depicts Glenn's problem.

Glenn is scheduled to present his report in conference with Long, Black, and Hopkins on Wednesday, October 19. Right after the final class (1 P.M., Friday, the 14th), Glenn begins work on his report.

**Figure 7.3
Glenn's Threefold
Problem**

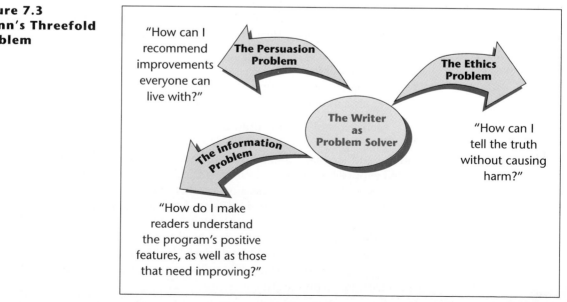

Working with the Information

Glenn spends half of Friday afternoon fretting over the details of his situation, the readers and other people involved, the political realities, constraints, and consequences. (He knows no love is lost between Long and Black, and he wants to steer clear of their ongoing conflict.) By 3 P.M., Glenn hasn't written a word. Desperate, he decides to write whatever comes to mind:

Glenn's first draft

> Although the October Management Training Session was deemed quite successful, several problems have emerged which require our immediate attention.
>
> - Too many of the instructors had poor presentation skills. A few never arrived on time. One didn't stick to the topic but rambled incessantly. Jones and Wells seemed poorly prepared. Instructors in general seemed to lack any clear objectives. Also, because too few visual aids were used, many presentations seemed colorless and apparently bored the trainees.
> - The trainees (all new people) were not at all cognizant of how the company was organized or functioned, so the majority of them often couldn't relate to what the speakers were talking about.
> - It is my impression that this was a weak session due to the fact that there were insufficient members (only five trainees). Such a small class makes the session a waste of time and money. For instance, Lester Beck, Senior Vice President of Personnel, came down to spend over one hour addressing only a handful of trainees. Another factor is that with fewer trainees in a class, less dialogue occurs, with people tending to just sit and get talked at.
> - Last but not least, executive speakers generally skirted the real issues, saying nothing about what it was really like to work here. They never really explained how to survive politically (e.g., never criticize your superior; never complain about the hard work or long hours; never tell anyone what you *really* think; never observe how few women are in executive or managerial positions, or how disorganized things seem to be). New employees shouldn't have to learn these things the hard way.
>
> In the final analysis, if these problems can be addressed immediately, it is my opinion we can look forward in the future to effectuating management training sessions of even higher quality than those we now have.

Glenn completes this draft at 5:10 P.M. Displeased with the results but not sure how to improve the piece, he asks an experienced colleague for advice and feedback. Blair Cordasco, a senior project manager, has collaborated with Glenn on several earlier projects. Cordasco agrees to study Glenn's draft over the weekend. Because of this document's sensitive nature, they decide to work on it face-to-face instead of transacting via E-mail.

At 8:05 Monday morning, Cordasco reviews the document with Glenn. First, she points out obvious style problems: wordiness ("due to the fact that"), jargon ("effectuating"), triteness ("in the final analysis"), implied bias ("weak presentation," "skirted"), among others. Can you identify other style problems in Glenn's draft?

Cordasco points out other problems. The piece is disorganized, and even though Glenn is being honest, he isn't being particularly fair. The emphasis is too critical (making Glenn's boss look bad to his superiors), and the views are too subjective (no one is interested in hearing Glenn gripe about the company's political problems). Moreover, the report lacks persuasive force because it contains little useful advice for solving the problems he identifies. The tone is bossy and judgmental. Glenn is in no position to make this kind of *power connection* (p. 45). In this form, the report will only alienate people and harm Glenn's career. He needs to be more fair, diplomatic, and reasonable.

Planning the Document

Glenn realizes he needs to begin by focusing on his writing situation. His audience and use analysis goes like this:[1]

I'd better decide *exactly* what my primary reader wants.

Long requested the report, but only because Black developed the scheme for division-wide improvements. So I really have two primary readers: my boss and the big boss.

My major question here: Am I including enough detail for all the bosses? The answer to this question will require answers to more specific questions:

Anticipated readers' questions	What are we doing right, and how can we do it better?
	What are we doing wrong, and does it cost us money?
	Have we left anything out, and does it matter?
	How, specifically, can we improve the program, and how will those improvements help the company?

Because all readers have participated in these sessions (as trainees, instructors, or guest speakers), they don't need background explanations.

I should begin with the *positive* features of the last session. Then I can discuss the problems and make recommendations. Maybe I can eliminate the bossy and judgmental tone by *suggesting improvements* instead of *criticizing weaknesses*. Also, I could be more persuasive by describing the *benefits* of my suggestions.

Glenn realizes that if he wants successful future programs, he can't afford to alienate anyone. After all, he wants to be seen as a loyal member of the company, yet preserve his self-esteem and demonstrate he is capable of making objective recommendations.

Now, I have a clear enough sense of what to do.

Statement of purpose	The purpose of my document is to provide my supervisor and interested executives with an evaluation of the workshop by describing its strengths, suggesting improvements, and explaining the benefits of these changes.

---

1. Throughout this chapter, Glenn's analysis will address *all* the areas illustrated on the audience and use profile sheet (page 65).

From this plan, I should be able to revise my first draft, but that first draft lacks important details. I should brainstorm to get *all* the details (including the *positive* ones) I want to include.

Glenn's Brainstorming List. Glenn's first draft touched on several topics; incorporating them into his brainstorming, he came up with the following list.

1. better-prepared instructors and more visuals
2. on-the-job orientation *before* the training session
3. more members in training sessions
4. executive speakers should spell out qualities needed for success
5. beneficial emphasis on interpersonal communication
6. need follow-up evaluation (in six months?)
7. four types of training evaluations:
    a. trainees' reactions
    b. testing of classroom learning
    c. transference of skills to the job
    d. impact of training on the organization (high sales, more promotions, better-written reports)
8. videotaping and critiquing of trainee speeches worked well
9. acknowledge the positive features of the session
10. ongoing improvement ensures quality training
11. division of class topics into two areas was a good idea
12. additional trainees would increase classroom dialogue
13. the more trainees in a session, the less time and money wasted
14. instructors shouldn't drift from the topic
15. on-the-job training to give a broad view of the division
16. clear course objectives, to increase audience interest and to measure the program's success
17. Marvin Long has done a great job with these sessions over the years

By 9:05 A.M., the office is hectic. Glenn puts his list aside to spend the day on work piling up. Not until 4 P.M. does he return to his report.

Now what? I should delete whatever my audience already knows or doesn't need, or whatever seems unfair or insincere: 7 can go (this audience needs no lecture in training theory); 14 is too negative and critical—besides, the same idea is stated more positively in 4; 17 is obvious brown-nosing, and I'm in no position to make such grand judgments.

Maybe I can unscramble this list by arranging items within categories (strengths, suggested changes, and benefits) from my statement of purpose.

Glenn's Brainstorming List Rearranged. Notice here how Glenn discovers additional *content* (see italic type) while he's deciding about *organization.*

Strengths of the Workshop

- division of class topics into two areas was useful
- emphasis on interpersonal communication
- videotaping of trainees' oral reports, followed by critiques

Well, that's one category done. Maybe I should combine *suggested changes* with *benefits*, since I'll want to cover them together in the report.

Suggested Changes/Benefits

- more members per session would increase dialogue and use resources more efficiently
- varied on-the-job experiences before the training sessions would give each member a broad view of the marketing division
- executive speakers should spell out qualities required for success and *future sessions should cover professional behavior, to provide trainees with a clear guide*
- follow-up evaluation in six months *by both supervisors and trainees would reveal the effectiveness of this training and suggest future improvements*
- clear course objectives and more visual aids to increase *instructor efficiency and audience interest*

Now that he has a fairly sensible arrangement, Glenn can get this list into report form, even though he probably will think of more material to add as he works. Since this is *internal* correspondence, he employs a memo format.

Drafting the Document

Glenn produces a usable draft—one containing just about everything he wants to cover. (Sentences are numbered for our later reference.)

A later draft

¹In my opinion, the Management Training Session for the month of October was somewhat successful. ²This success was evidenced when most participants rated their training as "very good." ³But improvements still are needed.

⁴First and foremost, a number of innovative aspects in this October session proved especially useful. ⁵Class topics were divided into two distinct areas. ⁶These topics created a general-to-specific focus. ⁷An emphasis on interpersonal communication skills was the most dramatic innovation. ⁸This helped class members develop a better attitude toward things in general. ⁹Videotaping of trainees' oral reports, followed by critiques, helped clarify strengths and weaknesses.

¹⁰There is a detailed summary of the trainees' evaluations attached. ¹¹Based on these and on my past observations, I have several suggestions.

- ¹²All management training sessions should have a minimum of ten to fifteen members. ¹³This would better utilize the larger number of managers involved and the time expended in the implementation of the training. ¹⁴The quality of class interaction with the speakers would also be improved with a larger group.

- [15]There should be several brief on-the-job training experiences in different sales and service areas. [16]These should be developed prior to the training session. [17]This would provide each member with a broad view of the duties and responsibilities in all areas of the marketing division.
- [18]Executive speakers should take a few minutes to spell out the personal and professional qualities essential for success with our company. [19]This would provide trainees a concrete guide to both general company and individual supervisors' expectations. [20]Additionally, by the next training session we should develop a presentation dealing with the attitudes, manners, and behavior appropriate in the business environment.
- [21]Do a six-month follow-up. [22]Get feedback from supervisors as well as trainees. [23]Ask for any new recommendations. [24]This would provide a clear assessment of the long-range impact of this training on an individual's job performance.
- [25]We need to demand clearer course objectives. [26]Instructors should be required to use more visual aids and improve their course structure based on these objectives. [27]This would increase instructor quality and audience interest.
- [28]These changes are bound to help. [29]Please contact me if you have further questions.

Although now developed and organized, this version still is some way from the finished document. Glenn has to make further decisions about his style, content, arrangement, audience, and purpose.

Blair Cordasco offers to review the piece once again and to work with Glenn on a thorough edit.

Revising the Document

At 8:15 Tuesday morning, Cordasco and Glenn begin a sentence-by-sentence revision for worthwhile content, sensible organization, and readable style. Their discussion goes something like this:

Sentence 1 begins with a needless qualifier, has a redundant phrase, and sounds insulting ("somewhat successful"). Sentence 2 should be in the passive voice, to emphasize the training—not the participants. Also, 1 and 2 are choppy and repetitious, and should be combined.

Original   In my opinion, the Management Training Session for the month of October was somewhat successful. This success was evidenced when most participants rated their training as "very good." (28 words)

Revised   The October Management Training Session was successful, with training rated "very good" by most participants. (15 words)[2]

2. Notice throughout how careful revision sharpens the writer's meaning while cutting needless words.

Sentence 3 is too blunt. An orienting sentence should forecast content diplomatically. This statement can be candid without being so negative.

Original   But improvements still are needed.

Revised   A few changes—beyond the recent innovations—should result in even greater training efficiency.

In sentence 4, "First and foremost" is trite, "aspects" is a clutter word, and word order needs changing to improve the emphasis (on innovations) and to lead into the examples.

Original   First and foremost, a number of innovative aspects in this October session proved especially useful.

Revised   Especially useful in this session were several program innovations.

Sentences 5 and 6 need combining, and content in 5 needs beefing up: *name* the two areas. If 7 is labeled the "most dramatic innovation," it ought to come last, for emphasis. In 8, "things in general" is too indefinite. And 7 and 8 need combining. Sentence 9 seems okay. All three examples would be more readable in a *list,* and with parallel phrasing (maybe starting each with an "ing" phrase to signal the "action" verb, as in 9).

Original   Class topics were divided into two distinct areas. These topics created a general-to-specific focus. An emphasis on interpersonal communication skills was the most dramatic innovation. This helped class members develop a better attitude toward things in general. Videotaping of trainees' oral reports, followed by critiques, helped clarify strengths and weaknesses.

Revised
- Dividing class topics into two areas created a general-to-specific focus: *the first week's coverage of company structure and functions created a context for the second week's coverage of management skills.*[3]
- Videotaping and critiquing trainees' oral reports clarified strengths and weaknesses.
- Emphasizing interpersonal communication skills *(listening, showing empathy, and reading nonverbal feedback) generated enthusiasm and a sense of ease about the group, their training, and the company.*

3. This revision has more words, but also much more concrete and specific detail (in italic type). Completeness of information always takes priority over word count.

As a format consideration, a couple of clear headings ("Workshop Strengths," "Suggested Changes/Benefits") would segment the text, improving readability and appearance.

In collaboration with his colleague, Glenn continues this editing and revising process throughout. (Exercise 3 asks you to identify the remaining changes.) Wednesday morning, after much revising and proofreading, Glenn prints out the final draft, shown in Figure 7.4.

Glenn's final report is both informative and persuasive. But this document did not appear magically. Glenn made deliberate decisions about purpose, audience, content, organization, and style. He sought advice and feedback on every aspect of the document. Most important, he *spent time revising*.[4]

Your Own Writing Situation

Writers work in different ways. Some begin by brainstorming. Some begin with an outline. Others hate outlining and simply write and rewrite. Some write a quick draft before thinking through their writing situation. Introductions and titles often are written last. Whether you write alone or collaborate in preparing a document, whether you are receiving feedback or providing it, no one step in the process is complete until *the whole* is complete. Notice, for instance, how Glenn sharpens his content *and* style while he organizes. Every document you write will require *all* these decisions, but rarely will you make them in the same sequence from day to day.

No matter what the sequence, *revision* is a fact of life. It is the one *constant* in the writing process. When you've finished a draft, you have in a sense only begun. Sometimes you will have more time to compose than Glenn did, sometimes much less. Whenever your deadline allows, leave time to revise.

Guidelines for Reviewing and Editing the Work of Others

Reviewing means evaluating how well a document achieves its purpose in terms of its intended audience. Is the content accurate, appropriate, useful, and legal? Is the material organized for the reader's understanding? Is the style clear, concise, fluent, exact, and engaging? Are visuals and page design effective? Whether you review the work of a subordinate, associate, or superior, you are showing the writer how you respond as a reader. This feedback helps writers envision ways of revising. Criteria for reviewing various documents appear in checklists throughout this book. (See also "Usability Testing," pages 445–47.)

---

4. A special thanks to Glenn Tarullo for his perseverance. I made his task doubly difficult by having him explain each of his decisions during this writing process.

Editing means actually fixing the piece: rephrasing or reorganizing sentences; choosing a better word; correcting spelling, usage, or punctuation; clarifying a topic sentence; improving content, organization, style, layout, or graphics. Criteria for editing are found in Chapter 13, the appendix, and listed in the Table of Correction Symbols inside the back cover.

The following guidelines should enable the reviewing and editing process to run smoothly.

1. *Read the entire piece at least twice before you comment.* Get a clear sense of the document's purpose and its intended audience. View the whole before evaluating its parts.

2. *Remember that mere correctness offers no guarantee of effectiveness.* Poor grammar, usage, punctuation, or mechanics distract readers and harm the writer's credibility. However, a correct piece of writing still might contain inappropriate rhetorical elements (inappropriate content, confusing organization, wordy style, or the like).

3. *Understand the acceptable limits of editing.* In the workplace, editing can range from cleaning up and fine tuning to an in-depth rewrite (in which case editors are cited prominently as consulting editors or co-authors). In school, however, rewriting someone's piece to the extent that it ceases to belong to the writer may constitute plagiarism (page 171).

4. *Be honest but diplomatic.* All of us benefit from honest and thoughtful feedback, but we still are sensitive to criticism—even the most constructive. Begin with something positive before moving to material needing improvement. Try to be supportive instead of judgmental.

5. *Always explain why something doesn't work.* Instead of "this paragraph is confusing," say "because this paragraph lacks a clear topic sentence, I had trouble discovering the main idea."

6. *Make specific recommendations for improvements.* Reviewing and editing is a process of diagnosing a problem and prescribing a cure. Write out your evaluation and suggestions in enough detail to give the writer a clear sense of how to proceed.

7. *Be aware that not all feedback has equal value.* Even professional reviewers and editors sometimes disagree on matters of content, organization, or style. When your own work is being reviewed or edited, you might receive conflicting opinions from different readers. In such cases, you must decide for yourself whose advice to follow (or seek your instructor's advice).

Keep in mind that a reviewer or editor helps clarify and enhance a document—without altering its original meaning.

Begins on a positive note, and cites evidence

States his claim

Gives clear examples of "innovations"

Cites the bases for his recommendations

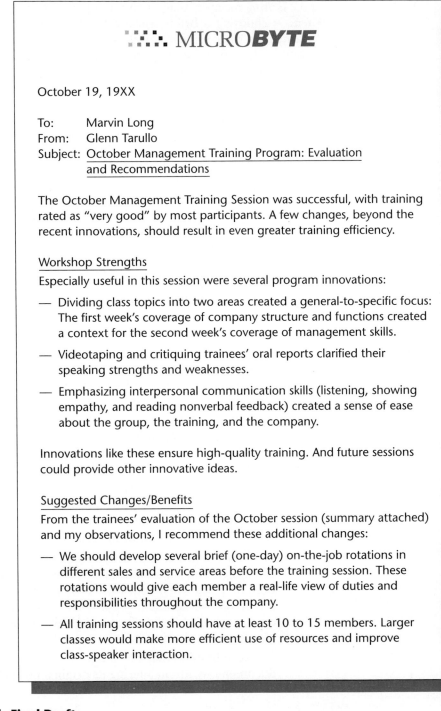

:::. MICRO*BYTE*

October 19, 19XX

To: Marvin Long
From: Glenn Tarullo
Subject: October Management Training Program: Evaluation
          and Recommendations

The October Management Training Session was successful, with training rated as "very good" by most participants. A few changes, beyond the recent innovations, should result in even greater training efficiency.

Workshop Strengths

Especially useful in this session were several program innovations:

— Dividing class topics into two areas created a general-to-specific focus: The first week's coverage of company structure and functions created a context for the second week's coverage of management skills.

— Videotaping and critiquing trainees' oral reports clarified their speaking strengths and weaknesses.

— Emphasizing interpersonal communication skills (listening, showing empathy, and reading nonverbal feedback) created a sense of ease about the group, the training, and the company.

Innovations like these ensure high-quality training. And future sessions could provide other innovative ideas.

Suggested Changes/Benefits

From the trainees' evaluation of the October session (summary attached) and my observations, I recommend these additional changes:

— We should develop several brief (one-day) on-the-job rotations in different sales and service areas before the training session. These rotations would give each member a real-life view of duties and responsibilities throughout the company.

— All training sessions should have at least 10 to 15 members. Larger classes would make more efficient use of resources and improve class-speaker interaction.

Figure 7.4   Glenn's Final Draft

Long, Oct. 19, 19XX, page 2

Supports each recommendation with convincing reasons

—We should ask instructors to follow a standard format (based on definite course objectives) for their presentation, and to use visuals liberally. These enhancements would ensure the greatest possible instructor efficiency and audience interest.

—Executive speakers should spell out personal and professional traits essential in our company. Such advice would give trainees a concrete guide to both general company and individual supervisor expectations. Also, by the next training session, we should assemble a presentation dealing with the attitudes, manners, and behavior appropriate in business.

—We should do a six-month follow-up of trainees (with feedback from supervisors as well as ex-trainees) to gain long-term insights, to measure the influence of this training on job performance, and to help design advanced training.

Closes by appealing to shared goals (efficiency and profit)

Inexpensive and easy to implement, these changes should produce more efficient training.

Copies: B. Hull, C. Black, G. Hopkins, J. Capilona, P. Maxwell, R. Sanders, L. Hunter

Figure 7.4   Glenn's Final Draft *Continued*

☑ EXERCISES

1. Think of a course you've taken that had both definite strengths and weaknesses. Assume that the instructor has asked you to evaluate the course and recommend improvements. Your secondary audience will be your instructor's chairperson and the Dean of Faculty. The instructor is up for tenure and is someone you like very much. You need to be candid in your report, but you don't want to make the instructor look bad. Write the report, being as specific as possible. (Use a fictional name for the instructor and the course.)

2. Assume that you are a training manager for XYZ Corporation. After completing this first section of the text and the course, what advice about the writing process would you have for a beginning writer who will frequently need to write reports on the job? In a one- or two-page (single-spaced) memo to new employees, explain the writing process briefly, and give a list of guidelines these beginning writers can follow.

3. Compare Glenn's second draft (page 111) with his final draft (pages 116–117). Identify all improvements in content, arrangement, and style besides those already discussed.

☑ COLLABORATIVE PROJECT

Reviewing, Editing, and Revising a Document

After several weeks in a technical writing class, you have a good sense of *what* material is covered and of *how* the material is taught. Imagine that at a recent meeting your school's Advanced Writing Committee passed this motion:

> Because of the popularity of the technical writing course, and the 200 percent enrollment increase within two years, many new sections have been added. To ensure a unified program, we suggest that all sections follow a standard syllabus and similar teaching approaches.

For help in developing a standard model, the committee has decided to survey students about to complete the technical writing course. Each student has been asked to submit a memo evaluating the section, with suggestions for improvement. The responses will form a databank for the committee's decisions about a course model that meets students' needs.

As a guide for evaluation, the committee has provided this question:

> How well do the content and teaching approach in your technical writing course fulfill your needs and expectations? How can this course best be taught, and what material should be covered? Be specific in your evaluation of strengths and weaknesses. Along with suggestions for improvement, explain how a specific change or improvement would benefit you.

Assume that Fran White, a student in some other section, has responded with this memo:

To:	The Advanced Writing Committee
From:	Fran White, Technical Writing Student
Subject:	Section 1499: Evaluation and Recommendation

This course is providing me with a great deal of useful information. Also, the teacher usually manages to hold the attention of the class very well. Learning about writing can be a pretty boring experience, but this class hardly ever is boring because the teacher does such a good job of making the material interesting. Also, the teacher is a very nice person. I'm happy to say that I've learned a number of approaches that have helped me improve my writing.

Writing skills are important in just about anyone's career, so I'm glad we have the chance to take a technical writing course before graduation. Having the course as a requirement is a good idea, because many students (myself included) probably would avoid any course that requires this much writing. It isn't until we get there that we realize how worthwhile (and difficult!) this course is for everyone. I'd like to see every section taught the way this one is: by a teacher who knows how to get a tough job done.

The only complaint I have against this teacher is the fact that he talks too much about computers. I know that computers are important, but I'd like to learn to use one for my writing instead of just being exposed to computers in general. And too many writing assignments have been saved for the latter part of the course. Everything is just stacking up, leaving me buried and confused.

Also, the class is much too large, and the layout is awful. With so many students, the teacher has no way of providing individual attention, and so too many students get lost in the crowd. We need a classroom layout that would make group editing easier.

I really enjoyed the audiovisual presentations in class and would like to see more of them. All in all, the concrete stuff was always the most useful.

In general, the material has been covered well, except for the material on oral communication.

With a few minor changes, this course would be excellent.

Before Fran can submit this memo, it will need heavy revision for content that is informative and persuasive and ethical, organization that is easy to follow, and style that is readable. Working in teams, review, edit, and revise Fran's memo.

a. As a first step, complete an audience and use profile sheet based on the following data, as well as on the details given earlier.

Primary audience: The writing committee and the department chair. Several committee members also are on the tenure committee, and they are likely to use this information for an additional purpose: to evaluate Fran's instructor for tenure. These are the decision makers. Above all, Fran wants to convey a *positive* impression of her instructor.

Secondary audience: Fran's instructor (whom Fran likes, but who deserves an honest and detailed evaluation). Fran wants to be fair to her instructor but also wants to make some realistic suggestions for improving a course so important to everyone's career.

b. Assume you helped Fran prepare the brainstorming list she should have prepared *before* writing the previous draft. To visualize what it was that specifically pleased or displeased Fran, imagine that these items would have appeared on her list:

- instructor is always willing to help students individually
- instructor always takes time in class to answer all questions thoroughly
- emphasis on planning, drafting, and revising for a specific audience is helpful
- instructor spends a lot of time encouraging us
- instructor spends too much time talking about computers and automated offices—what I need is to develop strong writing skills by *using* the computer
- now and then we should have a guest lecturer from business and industry
- we should spend more time discussing the documents *we* have prepared
- instructor spends too much time emphasizing mistakes—not enough time on the positive
- instructor should spend more time on writing, and less than the present four weeks on oral communication
- when they are used (which is not often enough), the opaque and overhead and slide projectors make things more vivid and interesting
- instructor gives a lot of feedback on our papers
- the class should have an IBM PC and a Macintosh computer (since these are the kinds we have in our campus micro labs), to illustrate how automation can affect the writing and revising process
- we should have a class with several round tables that seat 4–6 students for editing groups
- class size should be reduced from 30 to no more than 20 students
- tutoring should be available on a regularly scheduled basis for students with any type of writing problem

- we should spend at least one full week on job applications
- too few editing assignments early in the course; thus, too many at the end
- instructor should begin discussing the long report (term project) as early as the first or second week

From Fran's original draft and from the brainstorming list, select only that which is useful and appropriate for *this* audience and purpose. Compose a final draft of Fran's memo. Appoint one team member to present the document in class.

Gathering Information

R ESEARCH is done for a purpose: to answer a question, to make an evaluation, to establish a principle. Research helps you find your own answers, check your opinions against the facts, reach a conclusion that has the best chance of being valid. Major decisions in the workplace usually are based on careful research, with the findings recorded in a written report.

Depending on its sources, research is classified as *primary* or *secondary*. Primary research is an original, firsthand study of your topic or problem; primary sources include observation, interviews, questionnaires, inquiry letters, personal experiments, analysis of samples, fieldwork, or company records. Secondary research is information published by other researchers, which you then apply to your own topic or problem; secondary sources are *about* primary sources and include journal articles, encyclopedias, textbooks, reports, and handbooks. Most workplace research draws from both primary and secondary sources.

Research strategies and resources differ widely among disciplines. This chapter focuses on research for preparing a formal report.

Thinking Critically About the Research Process[1]

Research is a process of *problem solving,* in which certain procedures follow the sequence depicted in Figure 8.1. Later parts of the chapter treat these more procedural stages of the process, and Chapter 23 deals with preparing the actual research report.

**Figure 8.1
Procedural Stages
of the Research
Process**

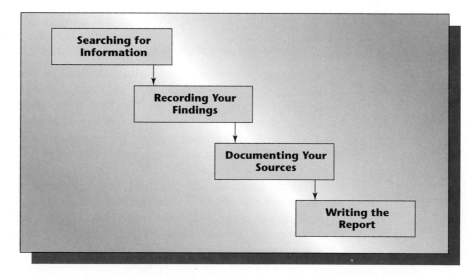

---

1. My thanks to University of Massachusetts, Dartmouth librarian Shaleen Barnes for inspiring this entire section.

**Figure 8.2
The Inquiry Stages
of the Research
Process**

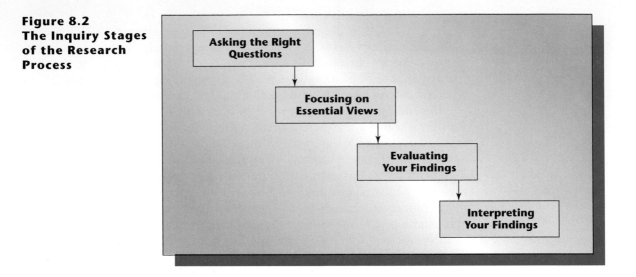

But research is not merely a by-the-numbers set of procedures ("First, do this; then, do that"). Intertwined with each of the procedural stages are the inquiry stages depicted in Figure 8.2, the many decisions that accompany any legitimate inquiry.

Asking the Right Questions

The answers you uncover will depend on the questions you ask. Assume, for instance, that you are faced with the following scenario:

Defining and Refining a Research Question

You are the public health manager for a small, New England town in which high-tension power lines run within 100 feet of the elementary school. Parents are concerned about danger from electromagnetic radiation (EMR) emitted by these power lines, in energy waves known as electromagnetic fields (EMFs). Town officials ask you to research the issue and prepare a report to be distributed at the next town meeting in six weeks. ■

Your first task is to identify the exact question or questions you want answered. Initially, the major question might be: *Do the power lines pose any real danger to our children?* After some phone calls around town and discussions at the coffee shop, you discover that townspeople actually have three major questions about electromagnetic fields: *What are they? Do they endanger our children? What can be done?*

To answer these questions, you need to consider a range of subordinate questions, like those in the Figure 8.3 tree chart. As research progresses, this chart will grow. For instance, after some preliminary reading, you learn that electromagnetic fields radiate not only from power lines but from *all* electrical

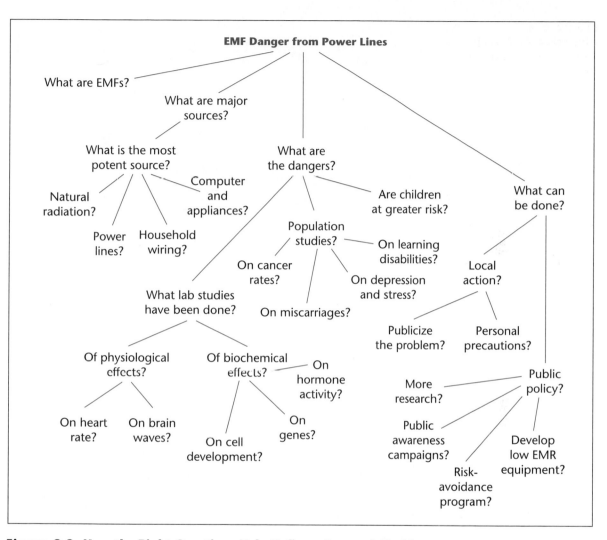

Figure 8.3 How the Right Questions Help Define a Research Problem

equipment, and even from the earth itself. So you face this additional question: *Do power lines present the greatest hazard as a source of EMFs?*

You now wonder whether the greater hazard comes from power lines or from other sources of EMF exposure. Critical thinking, in short, has enabled you to define and refine the essential questions.

Focusing on Essential Views

Assume you've settled on this research question: *Do electromagnetic fields from various sources endanger our children?* Now you can start considering information sources to consult (journals, interviews, reports, database

**Figure 8.4
Effective Research
Considers Multiple
Perspectives**

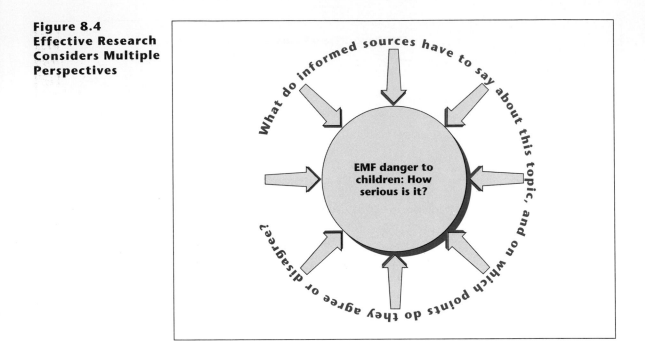

searches). For a fair and accurate picture, you need all sides of the story from up-to-date and reputable sources, as depicted in Figure 8.4. Figure 8.5 illustrates some likely sources of information.

We do research to discover the right answer—or the answer that stands the best chance of being right. Rather than settling for the first or most comforting or most convenient answer, we have an ethical obligation to consider a variety of

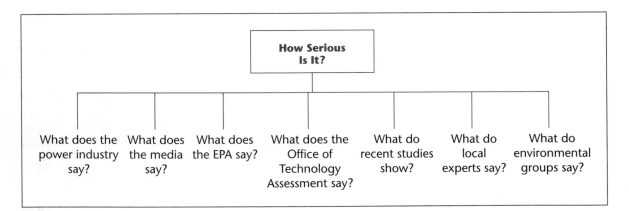

Figure 8.5 A Range of Essential Viewpoints

**Figure 8.6
Effective Research
Achieves Adequate
Depth**

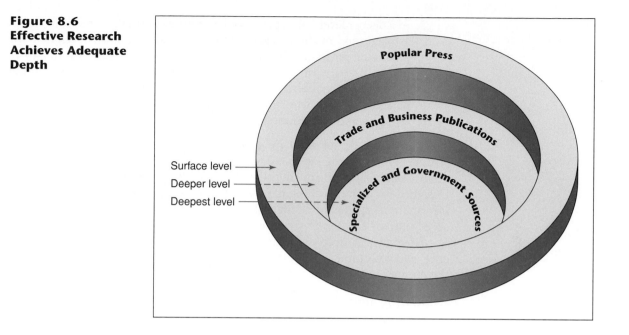

perspectives. Even expert testimony may not be the final word, because experts can disagree or they can be mistaken. To reach a balanced and informed conclusion, you need to survey the entire spectrum of significant viewpoints.

Achieving Adequate Depth in Your Search[2]

Balanced research examines a broad *range* of evidence; thorough research, however, examines that evidence at an appropriate *depth*. As depicted in Figure 8.6, different types of secondary information about any topic occupy different levels of detail and dependability.

1. At the surface level are items from the popular press (newspapers, radio, TV, general magazines). Designed for general consumption, this layer of information often offers more journalistic interpretation than factual detail.

2. At the next level are trade and business publications (*Frozen Food World, Publisher's Weekly,* and so on). Designed for readers who range from moderately informed to highly specialized, this level of information focuses more on practice than on theory, on items considered newsworthy to group members, on issues affecting the field, on public relations, on viewpoints that tend to reflect the particular biases of that field.

2. My thanks to University of Massachusetts, Dartmouth librarian Ross LaBaugh for inspiring this section.

3. At a deeper level is the specialized literature (journals from professional associations: medical, legal, engineering, and so on). Designed for practicing professionals, this level of information focuses on theory as well as practice, on descriptions of the latest studies—written by the researchers themselves and scrutinized by others for accuracy and objectivity, on debates among scholars and researchers, on reviews and critiques and refutations of prior studies and publications.

Also at this deeper level are government sources (studies and reports by NASA, EPA, FAA, the Defense Department, Congress) and corporate documents available through the Freedom of Information Act (page 141). Designed for anyone willing to investigate its complex resources, this layer of information offers hard facts and highly detailed and (in many instances) *relatively* impartial views of virtually any issue or topic in any field.

How deep is deep enough? This depends on your purpose, your audience, and your topic. But the real story and the hard facts more likely reside at the deeper levels of information.

Evaluating Your Findings

Once you have collected all the essential evidence about your topic, you need to decide how much of it is legitimate and then decide what it means.

QUESTIONS FOR EVALUATING A PARTICULAR FINDING

- Is this information accurate, reliable, and relatively unbiased?
- Can the claim be verified by the facts?
- How much of the information is useful?
- Is this the whole or the real story?
- Does something seem missing?
- Do I need more information?

Not all findings have equal value. Some information might be distorted, incomplete, or misleading. Information might be tainted by *source bias*. With such an emotional issue involving children, a source might understate or overstate certain facts, depending on whose interests that source represents (power company, government agency, parent organization, and so on).

Ethical researchers rely on evidence that represents a fair balance of views. They don't merely emphasize findings that support their own biases or assumptions.

Interpreting Your Findings

Once you have decided which of your findings seem legitimate, you need to decide what they all mean.

QUESTIONS FOR INTERPRETING YOUR FINDINGS

- What do all these facts or observations mean?
- Do any findings conflict?

- Are other interpretations possible?
- Should I reconsider the evidence?
- What are my conclusions?
- What, if anything, should be done?

The interpretation should fit the evidence and lead to an accurate conclusion—an overall judgment about what the findings mean. Perhaps you will reach a definite conclusion. (For example, "The evidence about EMF dangers seems persuasive enough for us to be concerned and to take the following actions.") Perhaps you will not.

Even the best research can produce contradictory or indefinite conclusions. For instance, some scientists question the studies linking electromagnetic radiation to health hazards. They claim these studies cannot be replicated or are flawed by statistical or procedural errors. They point out that, while some EMF studies indicate increased cancer risk, others indicate beneficial health effects. Other scientists claim that stronger EMFs are emitted by natural sources, such as earth's magnetic field, than by electrical sources (McDonald 5). An accurate conclusion would have to come from your analyzing all views and then deciding that one outweighs the others—or that only time will tell.

Never force a simplistic conclusion on a complex issue. Sometimes the best you can come up with is an indefinite conclusion: "Although controversy continues over the extent of EMF hazards, we all can take simple precautions to reduce our exposure." A wrong conclusion is far worse than no definite conclusion at all.

Figure 8.7 shows the critical-thinking decisions crucial to worthwhile research: asking the right questions about your topic, your sources, your findings, and your conclusions. Like the writing process (see Figure 7.2), the research process is recursive: stages are revisited/repeated as often as necessary. The quality of your entire research project will be determined by the quality of your *thinking* at each stage.

Searching the Literature

Where you begin your search depends on whether you seek background and basic facts or the latest information. Sources for library information appear in Figure 8.8. If you are an expert in the field, you might simply do a computerized database search or browse through specialized journals. If you have only limited knowledge or you need to focus your topic, you probably want to begin with general reference sources.[3] These can be located through the card catalog.

---

3. University of Massachusetts, Dartmouth librarian Ross LaBaugh suggests beginning with the popular, general literature, then working toward journals and other specialized sources: "The more accessible the source, the less valuable it is likely to be."

**Figure 8.7
Critical Thinking in
the Research Process**

*No single stage is
complete until all stages
are complete*

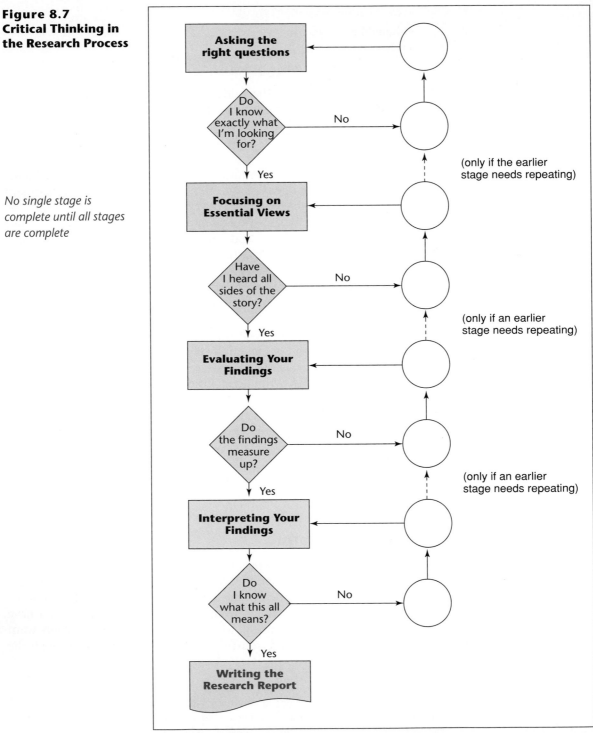

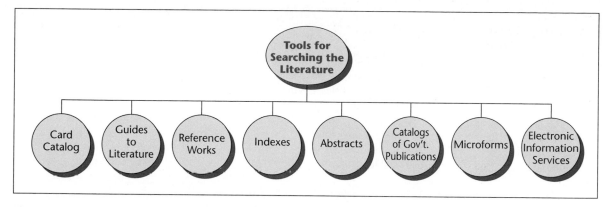

Figure 8.8 Ways You Can Search the Literature

The Card Catalog

Printed Catalog Entries. All books, reference works, indexes, periodicals, and other materials held by a library usually are listed in its card catalog under three headings: *author, title,* and *subject.* You thus have three access points for retrieving an item, as shown in Figures 8.9, 8.10, and 8.11.

First decide whether you seek a specific title, an author, or material about a given subject. Then look in the card catalog under one of those three access points (title, author, or subject).

Library of Congress Guide to Subject Headings. If you know neither authors nor titles of works on your subject, use the *subject* listing. To identify related subject headings under which you might find material on your topic, consult the *Library of Congress Subject Headings.* For material on electromagnetic radiation, for instance, you might scan the listings under *Electromagnetic energy;* among other entries, you would see these:

ELECTROMAGNETIC WAVES

Radiation

Waves

Electric waves

Electromagnetic fields

Atmospheric radiation

Microwaves

Solar radiation

ELECTROMAGNETISM IN MEDICINE

Medicine, electromagnetism in

Electrotherapeutics

Call number ———————————— RA 569.3.P54

Author ——————————————— **Pinsky, Mark A.**
Title ————————————————— The EMF book: what you should know about
electromagnetic fields, electromagnetic radiation,
and your health.

Publisher, date ——————————— New York: Warner, 1995.
Physical features ————————— 246 p., 20 cm.
ISBN number ————————————— ISBN 0–446–67004–9.

Other headings under ——————— 1. Electromagnetic fields—Health aspects.
which this work is   2. ELF magnetic fields—Health aspects. 3.
cataloged   Electric lines—Health aspects. 4. Computer
terminals—Health aspects.

Figure 8.9 Catalog Card Classified by Author

You now have an array of subjects under which to search for useful material in the card catalog.

Electronic Catalog Entries. In place of printed entries, many libraries have automated their card catalogs. Electronic card catalogs offer additional access points (beyond *author, title,* and *subject*) including:

- *Descriptor:* for retrieving works on the basis of a keyword or phrase (say, "electromagnetic" or "power lines and health") in the subject heading, in the work's title, or in the full text of its bibliographic record (its catalog entry or abstract).
- *Document type:* for retrieving works in a specific format (videotape, audiotape, compact disk, motion picture).
- *Organizations and Conferences:* for retrieving works produced under the name of an institution or professional association (Brookings Institution or American Heart Association).
- *Publisher:* for retrieving works produced by a particular publisher (Little, Brown and Co.).

RA 569.3.P54

The EMF book.
Pinsky, Mark A.

Figure 8.10 Partial Catalog Card Classified by Title

RA 569.3.P54

Electromagnetic fields—Health aspects
Pinsky, Mark A.
The EMF book: what you . . .

Figure 8.11 Partial Catalog Card Classified by Subject

- *Combination:* for retrieving works by combining any available access points (a book about a particular subject by a particular author or institution).

Figure 8.12 displays the first three screens you might encounter in an automated search using the descriptor *ELECTROMAGNETIC.* You could also narrow your search, for example, by combining the key words *ELECTROMAGNETIC and HEALTH HAZARDS.* If the computer responds to your descriptors with a "no record" screen, consult the *Library of Congress Subject Headings* for other possible key terms.

Through the *Internet,* electronic catalogs from major libraries worldwide (including the British Library, Harvard, and the Library of Congress) can be searched from home or office.

Caution: Any misspelling or typographical error in entering key terms can result in a false indication of "no record."

Guides to Literature

If you simply don't know which books, journals, indexes, and reference works are available for your topic, consult a guide to literature. For a general list of books in various disciplines, see Walford's *Guide to Reference Material* or Sheehy's *Guide to Reference Books.*

For sources in scientific and technical literature, consult Malinowsky and Richardson's *Science and Engineering Literature: A Guide to Reference Sources* or White's *Sources of Information in the Social Sciences.* For sources in specific disciplines, consult specialized guides such as *Using the Chemical Literature: A Practical Guide* or the *Encyclopedia of Business Information Sources.* Ask your librarian about literature guides for your discipline.

Reference Works

Reference works include the various resources shown in Figure 8.13. These can be a good starting point because they provide background and can lead to more specific information. One drawback to reference books is that some may be outdated. Always check the last copyright date.

All reference works will be indexed in the *Subject* card catalog, with a "Ref." designation above the call number. Many of these works can be searched electronically. Ask your librarian.

You begin by pressing any key, and the computer responds with the screen:

```
Type of searches:                          Press Help key for HELP

1  AU  = Author                     8   PU  =  Publisher
2  OC  = Organization or conference 9   SH  =  Subject heading
3  TI  = Title                      10  DT  =  Document type
4  UT  = Uniform or collective title 11       Combination
5  DE  = Descriptor                 12       ISBN
6  CN  = Call number                13       ISBN
7  SE  = Series                     14       Numeric

Enter the NUMBER of your search request and press RETURN:
```

After selecting the DE search mode, you type in your key word (ELECTRO-MAGNETIC), and then press RETURN. This next screen appears (the first of several with all 108 entries):

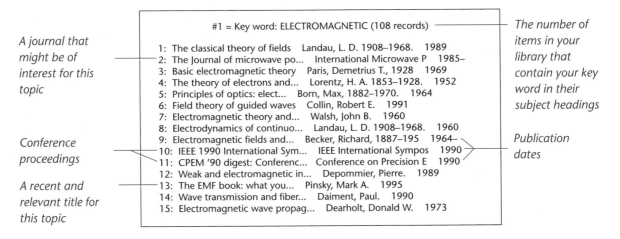

*A journal that
might be of
interest for this
topic*

*Conference
proceedings*

*A recent and
relevant title for
this topic*

*The number of
items in your
library that
contain your key
word in their
subject headings*

*Publication
dates*

```
            #1 = Key word: ELECTROMAGNETIC (108 records)

1:  The classical theory of fields   Landau, L. D. 1908–1968.   1989
2:  The Journal of microwave po...   International Microwave P   1985–
3:  Basic electromagnetic theory   Paris, Demetrius T., 1928   1969
4:  The theory of electrons and...   Lorentz, H. A. 1853–1928.   1952
5:  Principles of optics: elect...   Born, Max, 1882–1970.   1964
6:  Field theory of guided waves   Collin, Robert E.   1991
7:  Electromagnetic theory and...   Walsh, John B.   1960
8:  Electrodynamics of continuo...   Landau, L. D. 1908–1968.   1960
9:  Electromagnetic fields and...   Becker, Richard, 1887–195   1964–
10: IEEE 1990 International Sym...   IEEE International Sympos   1990
11: CPEM '90 digest: Conferenc...   Conference on Precision E   1990
12: Weak and electromagnetic in...   Depommier, Pierre.   1989
13: The EMF book: what you...   Pinsky, Mark A.   1995
14: Wave transmission and fiber...   Daiment, Paul.   1990
15: Electromagnetic wave propag...   Dearholt, Donald W.   1973
```

You select entry #13 and then press return. The computer responds with detailed bibliographic information on your selected item.

```
Selection:
01-0211132

AUTHOR       Pinsky, Mark A.
TITLE        The EMF book: what you should know about electromagnetic
             fields, electromagnetic radiation, and your health
PUBLISHER    New York: Warner
DATE         1995
PHYS. FEAT.  246 p.; 20 cm.
SUBJECTS     Electromagnetic fields—Health aspects.
             ELF electromagnetic fields—Health aspects.
             Electric lines—Health aspects.
             Computer terminals—Health aspects.
```

Figure 8.12 Searching an Electronic Card Catalog

Figure 8.13 Common Reference Works for Technical Disciplines

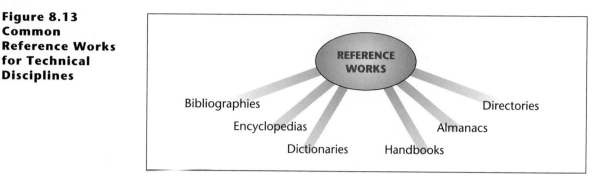

Bibliographies. Bibliographies are lists of publications about a subject, within specified dates. Although they provide a comprehensive view of major sources, bibliographies quickly become dated. Ask a librarian about recent bibliographies on your subject. (Some bibliographies are issued yearly or even weekly.)

Annotated bibliographies (which include an abstract for each entry) are most helpful because they can help you identify the most useful sources. A sample listing of bibliographies (shown with annotations):

Bibliographies

> *Bibliographic Index.* A list (by subject) of bibliographies that contain at least fifty citations; to see which bibliographies are published in your field, begin here.
>
> *A Guide to U.S. Government Scientific and Technical Resources.* A list of everything published in these broad fields by the government.
>
> *Bibliographic Guide to Business and Economics.* A list of all major business and economic publications.
>
> *Health Hazards of Video Display Terminals: An Annotated Bibliography.* One of many bibliographies focused on a highly specific subject.

Shorter, more specific bibliographies appear as parts of books and journal articles. To locate bibliographies that are whole volumes in themselves, look in the card catalog under "Bibliography" as a subject or title heading. For recent sources on electromagnetic radiation, for example, you might begin with the *Bibliographic Index*.

Encyclopedias. Use encyclopedias to quickly find basic information (which might be outdated). Sample listings:

Encyclopedias

> *Encyclopaedia Britannica*
> *Encyclopedia of Building and Construction Terms*
> *Encyclopedia of Banking and Finance*
> *Encyclopedia of Food Technology*

Journals, newsletters, and other publications from professional organizations (such as the American Medical Association or the Institute of Electrical and Electronics Engineers) are a valuable source of specialized information. The *Encyclopedia of Associations* offers a yearly listing of over 30,000 societies and organizations worldwide that range from agricultural to scientific and technical. For information on electromagnetic radiation, you might contact environmental organizations such as the Audubon Society or the Sierra Club via their Internet home page.

Dictionaries. Besides carrying general definitions, dictionaries can focus on specific disciplines or they can give biographical information. Sample listings:

Dictionaries

> *Webster's Third New International Dictionary of the English Language.* Considered the best general dictionary.
>
> *Dictionary of Engineering and Technology*
>
> *Dictionary of Telecommunications*
>
> *Dictionary of Scientific Biography*

Handbooks. Handbooks amass key facts (including formulas, tables, advice, and examples) about a field in condensed form. Often aimed at users experienced in the field, some handbooks may not be useful to newcomers. Sample listings:

Handbooks

> *Business Writer's Handbook*
>
> *Civil Engineering Handbook*
>
> *The McGraw-Hill Computer Handbook*

Almanacs. Ranging from general to specific, almanacs have factual and statistical data. Sample listings:

Almanacs

> *World Almanac and Book of Facts*
>
> *Almanac for Computers*
>
> *Almanac of Business and Industrial Financial Ratios*

Directories. Directories offer information about organizations, companies, people, products, services, statistics, or careers, often including addresses and phone numbers. This material usually is updated annually. Sample listings:

Directories

> *The Career Guide: Dun's Employment Opportunities Directory*
>
> *Directory of Computer Software*
>
> *Standard & Poor's Register of Corporations, Directors, and Executives*
>
> *Directory of New England Manufacturers*
>
> *Directory of American Firms Operating in Foreign Countries*
>
> *Directory of International Statistics*
>
> *The Internet Directory*

A growing number of directories are accessible by computer.

Reference books exist for every discipline. In researching electromagnetic radiation, you might start with titles such as the *McGraw-Hill Encyclopedia of Science and Technology* (updated yearly) or the *McGraw-Hill Directory of Scientific and Technical Terms.*

Indexes

Indexes are lists of books, newspaper articles, journal articles, or other works, as shown in Figure 8.14. They are excellent sources for current information. Because different indexes list sources in different ways, always read the introductory pages for instructions. Or ask a librarian for help.

Book Indexes. All books currently being published (up to a set date) are listed in book indexes by author, title, or subject. Sample indexes (shown with annotations):

Book indexes

> *Books in Print.* An annual listing of all books published in the United States.
>
> *Cumulative Book Index.* A monthly worldwide listing of books in English.
>
> *Forthcoming Books.* A listing every two months of U.S. books to be published.
>
> *Scientific and Technical Books and Serials in Print.* An annual listing of literature in science and technology.
>
> *New Technical Books: A Selective List with Descriptive Annotations.* Issued ten times yearly.
>
> *Technical Book Review Index.* A monthly listing (with excerpts) of book reviews.
>
> *Medical Books and Serials in Print.* An annual listing of works from medicine and psychology.

In research on electromagnetic radiation, you might check the current issue of *New Technical Books in Print* or *Scientific and Technical Books and Serials in Print.* But no book is likely to offer the very latest information because of the time required to publish a book manuscript (from several months to one year).

Figure 8.14 Useful Indexes for Technical Disciplines

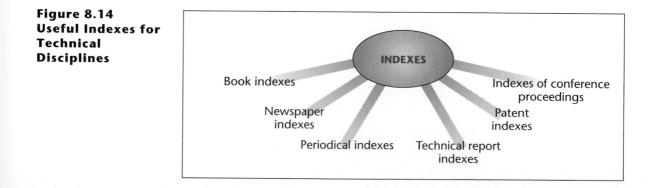

Newspaper Indexes. Most newspaper indexes list articles by subject. The *New York Times Index* is best known, but other major newspapers have their own indexes. Sample titles:

Newspaper indexes

> *Boston Globe Index*
> *Christian Science Monitor Index*
> *Wall Street Journal Index*

For research on electromagnetic radiation, you might check recent editions of the *National Newspaper Index* (in a print version or via the online catalog).

Periodical Indexes. For recent information in magazines and journals, consult periodical indexes. To find useful indexes, first decide whether you seek general or specialized information.

One most general index is the *Magazine Index,* a subject index (on microfilm) of 400 general periodicals. A popular index is the *Readers' Guide to Periodical Literature,* listing articles from 150 general magazines and journals. Because the *Readers' Guide* is updated every few weeks, you can locate current material. In research on electromagnetic radiation, you would find numerous entries under the subject heading, "Electromagnetic waves."

For specialized information, consult indexes that list journal articles in specific disciplines, such as *Ulrich's International Periodicals Directory.* Another comprehensive source of specialized information, the *Applied Science and Technology Index* carries a monthly listing, by subject, of articles in more than 200 scientific and technical journals. In research on electromagnetic radiation, you would find multiple entries under the heading "Electromagnetic fields" in recent issues of the *AS&T Index.*

For business articles, consult the *Business Periodicals Index,* with its monthly subject listing of articles and book reviews from 270 business periodicals.

Other broad indexes that cover specialized fields in general include the *Business Index* (on microfilm) and the *General Science Index.* The *Statistical Reference Index* lists statistical works not published by the government.

Along with these broad indexes, some disciplines have their own specific indexes. Sample listings:

Periodical indexes

> *Agricultural Index*
> *Education Index*
> *Energy Index*
> *F&S Index of Corporations and Industries*
> *Environment Index*
> *Index to Legal Periodicals*
> *International Nursing Index*

Ask your librarian about the best indexes for your topic and about the many indexes that can be searched by computer.

Citation Indexes. Citation indexes enable researchers to trace, through the literature, the development and refinement of a published idea, concept, or theory. Using a citation index, you can track down the specific publications in which the original material has been cited, quoted, applied, critiqued, verified, or otherwise amplified (Garfield 200). In short, you can answer this question: Who else has said what about this idea?

The *Science Citation Index,* a quarterly publication, provides a system for cross-referencing important articles on science and technology worldwide. Both the *Science Citation Index* and its counterpart, the *Social Science Citation Index,* can be searched by computer.

Technical Report Indexes. Countless government and private-sector reports written worldwide offer specialized and highly current information. (Proprietary or security restrictions, of course, restrict public access to certain corporate or government documents.) Sample indexes for these reports:

Technical report indexes

> *Scientific and Technical Aerospace Reports*
>
> *Government Reports Announcements and Index*
>
> *Monthly Catalog of United States Government Publications*

U.S. government report indexes are discussed on page 141.

Patent Indexes[4]. Over 75,000 patents yearly are issued in the United States to protect individual and company rights to new inventions, products, or processes. These patents are just a fraction of the roughly one-half million issued worldwide. As information specialists Schenk and Webster explain, patents are an excellent and often overlooked source of current information: "Since it is necessary that complete descriptions of the invention be included in patent applications, one can assume that almost everything that is new and original in technology can be found in patents." Sample indexes:

Patent indexes

> *Index of Patents Issued from the United States Patent and Trademark Office*
>
> *NASA Patent Abstracts Bibliography*
>
> *World Patents Index*

Online information about patents in fiber optics, lasers, or other technologies can be obtained through databases such as Hi Tech Patents, Data Communications, and through WPI (World Patents Index).

Indexes to Conference Proceedings. Schenk and Webster point out that many of the papers presented at the more than 10,000 yearly professional conferences are collected and then indexed in printed or computerized listings such as these:

---

4. Adapted from Schenk and Webster. Consult their work for detailed treatment of patent information and its sources, and for invaluable discussions of information sources in general.

Indexes to conference
proceedings

> *Proceedings in Print*
> *Index to Scientific and Technical Proceedings*
> *Engineering Meetings* (an *Engineering Index* database)

The very latest ideas or explorations or advances in a field often are presented during such proceedings, before appearing as journal publications.

Abstracts

Beyond indexing various works, abstracts summarize each article. The abstract can save you from going all the way to the journal in order to decide whether to read the article or to skip it.

Abstracts usually are titled by discipline. A sample list:

Collections of Abstracts

> *Biological Abstracts*
> *Computer Abstracts*
> *Engineering Index*
> *Environment Abstracts*
> *Excerpta Medica*
> *Forestry Abstracts*
> *International Aerospace Abstracts*
> *Metals Abstracts*

In researching electromagnetic radiation, you might consult *Energy Research Abstracts* under the subject heading "Electromagnetic Fields." Abstracts (such as *Energy Research Abstracts* and *Pollution Abstracts*) increasingly are searchable by computer. Check with your librarian.

For some current research, you might consult abstracts of doctoral dissertations in *Dissertation Abstracts International.*

Locating the Source. If your library does not hold the article you need, you can search an online database to identify a holding library, then request the article through interlibrary loan.

Access Tools for U.S. Government Publications

The federal government publishes maps, periodicals, books, pamphlets, manuals, monographs, annual reports, research reports, and a bewildering array of other information. Types of information available to the public include presidential proclamations, congressional bills and reports, judiciary rulings, some reports from the Central Intelligence Agency, and publications from all other government agencies (Departments of Agriculture, Commerce, Transportation, and so on). A few of the countless titles available in this gold mine of information:

Government
publications

> *Electromagnetic Fields in Your Environment*
> *Economic Report of the President*

Major Oil and Gas Fields of the Free World
Decisions of the Federal Trade Commission
Journal of Research of the National Bureau of Standards
Siting Small Wind Turbines

Much of this information can be searched online as well as in printed volumes. Your best bet for tapping this valuable but complex resource is to request assistance from the librarian in charge of government documents. If your library does not hold the publication you seek, it can be obtained through electronic access or interlibrary loan.

Here are the basic access tools for documents issued or published at government expense as well as for many privately sponsored documents.

- *The Monthly Catalog of the United States Government,* the major access to government publications and reports, is indexed by author, subject, and title.

- *Government Reports Announcements & Index* is a listing published every two weeks by the National Technical Information Service (NTIS), a federal clearinghouse for scientific and technical information—all stored in a database. The collection has summaries of over one million federally sponsored research reports published and patents issued since 1964. About 70,000 new summaries are added annually in 22 subject categories, from aeronautics to medicine and biology. Full copies of reports are available from NTIS.

- *The American Statistics Index,* a yearly guide to statistical publications by the U.S. government, is divided into two sections: *Index* and *Abstracts* (an index with summaries). The *index* volume lists material by subject and provides geographic, economic, and demographic breakdowns.

- The *Statistical Abstract of the United States,* updated yearly, offers a wide range of statistics on population, health, employment, and the like, and can be accessed via the World Wide Web. CD-ROM versions are available beginning with the 1993 edition.

In addition, the government issues *Selected Government Publications,* a monthly list of 150 titles (with descriptive abstracts). These titles range from highly general (*Questions About the Oceans*) to highly technical (*An Emission-Line Survey of the Milky Way*).

The government also publishes bibliographies on hundreds of subjects, from "Accidents and Accident Prevention" to "Home Gardening of Fruits and Vegetables." Ask your librarian for information about these subject bibliographies.

Many unpublished documents are available under the Freedom of Information Act, which grants public access to all federal agency records except for classified documents, trade secrets, certain law enforcement files, records protected by personal privacy law, and other exempted information.

Publicly accessible government records

. . . suppose you have heard that a certain toy has been recalled as a safety hazard and you want to know the details. In this case, the Consumer Product Safety Commission could help you. Perhaps you want to read the latest inspection report on conditions at a nursing home certified for Medicare. Your local Social Security office keeps such records on file. Or you might want to know if the Federal Bureau of Investigation has a file that includes you. In all these examples, you may use the FOIA to request information from the appropriate federal agency. (U.S. General Services Administration 1)

Contact the specific agency that would hold the records you seek: for workplace accident reports, the Department of Labor; for industrial pollution records, the Environmental Protection Agency, and so on.

A growing body of government information is posted to the Internet or the World-Wide Web. For example, the Food and Drug Administration's electronic bulletin board lists information on experimental drugs to fight AIDS, drug and device approvals, recalls and litigations involving drugs or devices, health fraud, and a host of related items. The Department of Energy offers a Web home page for information on human radiation experiments. Consult your reference librarian for electronic addresses of selected government agencies.

Microforms

Microform technology enables vast quantities of printed information to be reproduced and stored on rolls of microfilm or packets of microfiche. (This material is read on machines that magnify the reduced image. Ask your librarian for assistance.) Among the growing array of microform products are government documents, technical reports, newspapers, business directories, and translated documents from worldwide (Lavin 12).

A valuable business resource, for example, is *Business NewsBank,* a microfiche index to articles on business and economic development from over 450 U.S. cities. Also on microfilm are specialized indexes such as the *Business Index,* and more general indexes such as the *Magazine Index.*

Using Electronic Information Services

Compared with manual searches of printed resources, electronic searches offer greater speed. Most importantly, they offer greater access (Gibaldi 6–7, 14–15).

- *Sources can be located on the basis of limited information.* If you know only part of a full title or only an author's last name, the computer searches the database for all titles containing those words or authors sharing that last name.
- *Searches can be broadened.* A keyword search (for example, "electromagnetic fields") scans not only subject headings or titles but a work's complete bibliographic record, including its abstract. Keyword searches

therefore uncover a broad range of material that might have been overlooked in a traditional search by subject or title.

Another search expansion strategy is *truncation:* shortening a word to its root element, adding an asterisk, and searching for the word's different forms. A search using the truncated "electromag*" would retrieve bibliographic records containing the word "electromagnetism," as well as "electromagnetic field," "electromagnetic spectrum," "electromagnetic wave," "electromagnetic radiation," and so on.

■ *Searches can be customized.* The amount of information now available on most topics is overwhelming, but you can limit an electronic search to specific dates or to specific media (books, journals, newspapers, films, and so on).

You also can expand or narrow your search by using the Boolean[5] operators "AND," "OR," "NOT," "IF," "THEN," and "EXCEPT" to define relationships among various keywords. For instance, electromagnetic fields AND health" will help focus the search; "electromagnetic fields AND brain tumors" will yield even greater focus. On the other hand, "electromagnetic fields OR electromagnetic radiation" will broaden the search.

■ *A source's full text often can be accessed directly.* With a database that offers the full texts of its listings, you no longer need to search the library for a printed equivalent of the work you seek. After scanning abstracts of selected titles, you can access the full texts to be read on-screen or downloaded and printed.

In addition to electronic card catalogs, libraries increasingly offer a rapidly expanding array of electronic search services. Some libraries expect to store electronic versions of all their materials within less than two decades (Watkins 19). Moreover, major libraries across the globe are accessible by personal computer via the Internet and the World Wide Web.

Following is a sampling of the electronic search tools commonly available to researchers in virtually any discipline, working from virtually any location (Figure 8.15).

Compact Disks and Diskettes

A single CD-ROM disk stores the equivalent of an entire encyclopedia, and serves as a portable database. CD-ROM technology offers reference sources such as the *Science Citation Index, Population Statistics, Ulrich's International Guide to Periodicals,* and products ranging from corporate directories to government reports.

One useful CD-ROM disk for business information is ProQuest™, whose *ABI/INFORM* database indexes over 800 journals in management, marketing,

---

5. British mathematician and logician George Boole (1815–1864) developed the system of symbolic logic (Boolean logic) now widely used in database information retrieval.

and business since 1989, and whose *UMI* database indexes major U.S. newspapers. A keyword search of ProQuest's subject headings, titles, and abstracts yields a listing of relevant titles. You then can obtain the full bibliographic record, including the abstract, for each title of interest.

A useful CD-ROM disk for information about psychology, nursing, education, and social policy is SilverPlatter™, whose databases easily are accessed via keyword searches. Both ProQuest and SilverPlatter databases are updated frequently and subscribers (libraries and other organizations) receive revised disks on a regular basis.

Other relevant databases on CD include *Government Publications Index*™, a monthly listing of U.S. government literature, and *LegalTrac*™, a monthly listing of entries from 750 legal publications. *Lotus One Source*™ provides corporate profiles and financial summaries for over 15,000 U.S. companies, research reports on stocks, and article summaries from major business sources. Ask at your library about disk-based indexing services.

For smaller bodies of specialized information (census and stock-market data, economic profiles of U.S. cities, and so on), files on diskettes can be purchased (Lavin 19).

For personal use, the Time Almanac, Microsoft Bookshelf, and other reference CDs can be purchased.

Online Databases

Most college libraries subscribe to retrieval services that can access thousands of individual databases stored on centralized computers. From a library terminal (or in some cases, a microcomputer), you can access indexes, journals, books, monographs, dissertations, and reports. Compared with CDs, mainframe databases tend to be more specialized and more current, often updated daily (as opposed to weekly, monthly, or quarterly updating of CD databases).

Online retrieval services offer three types of databases, or some combination: *bibliographic, full text,* and *factual* (Lavin 14). Bibliographic databases list publications in a field; entries can be searched according to author, title, subject, document type (report, article, dissertation), keyword, or other access point. Some bibliographic databases include abstracts of each entry.

**Figure 8.15
Options for
Searching for
Literature
Electronically**

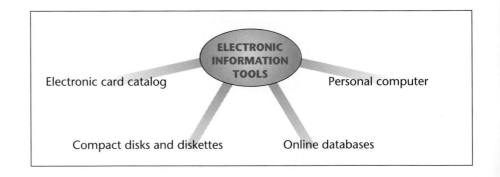

Full-text databases display the entire article or document (usually excluding graphics) directly on the computer screen, and then will print the article on command.

Factual databases provide specialized facts of all kinds: global and up-to-the-minute stock quotations, weather data, lists of new patents filed, and credit ratings of major companies, to name a few.

Until recently, searching mainframe databases has required the skills of a librarian or other trained professional. But new, user-friendly systems offer simplified menus for each step of the search process (Lavin 16).

Four popular database services are discussed in the following section. The first two services (OCLC and RLIN) help you locate titles you have already identified as useful. The next two (DIALOG and BRS) help you identify useful titles.

OCLC and RLIN. You easily can compile a comprehensive list of works on your subject at any library networked with the Online Computer Library Center or the Research Libraries Information Network. OCLC and RLIN databases store millions of records with the same information found in a printed card catalog. Using a networked terminal, you type in author or title. Within seconds, you get a listing of the publication you seek and information about where to find it. If your library doesn't have the publication, your librarian can activate the Interlibrary Loan System (ILS). The system forwards requests to libraries holding the material. Once a lender indicates (via its terminal) that it will supply the material, the system stops forwarding the request and notifies your librarian that the request has been filled. Your order will arrive at the library by mail in a week or so. (Pages 449–53 show instructions for using an OCLC terminal.)

OCLC recently has added a variety of bibliographic databases, along with a more user-friendly search system (Lavin 66). Ask if these enhancements are available in your library.

DIALOG. Many libraries subscribe to DIALOG, a comprehensive network of independent databases covering a broad range of technical subjects. You retrieve information by typing in key terms that enable the computer to scan bibliography lists for titles containing those terms. Say you need information on possible *health hazards* from *household electrical equipment*. You instruct the computer to search MEDLINE®, a medical database (one of more than 150 databases in science, technology, medicine, business, and so on), for titles including the words italicized above (or synonymous words, such as *risk, danger, appliances*). The system would provide full bibliographies and, often, abstracts of the most recent medical articles on your topic.

Besides bibliographic information on published works, dissertations, and conference papers, DIALOG also provides financial and product information about companies, names and addresses of company officers, statistical data, and patent information. Here are just a few of DIALOG's databases:

DIALOG databases

Career Placement Registry

Claims/U.S. Patents

Conference Papers Index

Electronic Yellow Pages (for Retailers, Services, Manufacturers)

Enviroline

Index Medicus

International Software Database

Oceanic Abstracts

U.S. Exports

Water Resources Abstracts

Despite its expense ($150 hourly or more for many of its databases), many college libraries subscribe to DIALOG. Companies who have full-time database researchers also subscribe. CD-ROM versions of many specific DIALOG databases (including MEDLINE) are available to paid subscribers.

BRS. Bibliographic Retrieval Services (BRS) is another popular database providing bibliographies and abstracts from life sciences, physical sciences, business, or social sciences. These are a few from the more than fifty BRS databases:

BRS databases

American Chemical Society Journals

Dissertation Abstracts International

Government Reports Announcements & Index

Harvard Business Review

International Pharmaceutical Abstracts

Military and Federal Specifications and Standards

Monthly Catalog of United States Government Publications

PATDATA (U.S. Patents)

Pollution Abstracts

Robotics Information Database

College libraries increasingly offer BRS service.

A Sample Automated Search. Assume you are continuing research on electromagnetic radiation. You have searched the print or manual indexes, and for a comprehensive view you decide on an automated search using your library's DIALOG service. You ask a librarian for help, and the two of you begin your search.

After logging onto the DIALOG system, you instruct the computer to search the MEDLINE database using the keywords *electromagnetic, health, hazard.* The computer responds with a listing of articles in that database whose titles or abstracts contain a combination of those key words. (In a full-text database, the computer would search for works containing that combination of key words anywhere in the text.)

Partial Listing of Article Titles from an Electronic Search

1. Long-term effects of a 50Hz electric field on the life-expectancy of mice
2. Possible mechanisms by which extremely low frequency magnetic fields affect opoid function
3. Variation in cancer risk estimates for exposure to powerline frequency electromagnetic fields: a meta-analysis for comparing EMF measurement methods[6]

The title that seems most relevant to your topic is number 3, so you decide to print the full bibliographic record on this article.

Full Citation for One Article from an Electronic Search

Access number

Work title

Authors and affiliation

Periodical

Document description (article, letter, report, review)

Abstract

09390402 MEDLINE Number: 95320402

Variation in cancer risk estimates for exposure to power line frequency electromagnetic fields: a meta-analysis comparing measurement methods.

Miller MA; Murphy JR; Miller TI; Ruttenber AJ
Dept. of Preventive Medicine and Biometrics, Univ. of Colorado Health Sciences Center, Denver, USA

Risk Analysis (United States) Apr 1995, 15(2), p. 281–7

Document type: JOURNAL ARTICLE; META-ANALYSIS

We used meta-analysis to synthesize the findings from eleven case-control studies on cancer risks in humans exposed to 50–60 Hertz powerline electromagnetic fields (EMFs). Pooled estimates of risk are derived from different EMF measurement methods and types of cancer. EMF measurement methods are classified as: wiring configuration codes, distance to power distribution equipment, spot measurements of magnetic fields, and calculated indices based on distance to power distribution equipment and historic load data. Pooled odds ratios depicting the risk of cancer by each measurement type are presented for all cancers combined, leukemia for all age groups and childhood leukemia. The wire code measurement technique was associated with a significantly increased risk for all three cancer types, while spot measures consistently showed non-significant odds ratios. Distance measures and the calculated indices produced risk estimates that were significant only for leukemia.

---

6. A form of secondary research, *meta-analysis*, is a comprehensive review of recent and relevant studies on a topic in an attempt to synthesize the larger meaning of these collective data. After evaluation of their methods, flawed studies are rejected. Findings from the remaining studies are compared in various ways to identify significant patterns that might suggest some overall conclusion.

You read the abstract and decide to review the whole article available from your library, from interlibrary loan, or directly from DIALOG's database. You then return to the title listing, for other promising titles.

Retrieval Services for Home and Office

Using a modem, a phone line, and an access provider such as *America Online,* you can search Internet databases from a home, office, or laptop computer. You can join various Internet newsgroups, subscribe to discussion lists, send E-mail inquiries to NASA or the White House, and gain access to publications that exist only in electronic form. Using navigational and retrieval tools such as *Netscape* and *Mosaic,* you can explore hypertext sites on the World Wide Web, locate experts in all types of specialties, and read the latest articles in journals such as *Nature* or *Science* or the latest newspaper listings of jobs in your field.

As an Internet user, you can subscribe to commercial online services such as *ORBIT Search Service* (for scientific, medical, and technical information) or *Dow-Jones News Retrieval* (for corporate news releases, new product news, business reports). For online databases in your field, consult the *Gale Directory of Databases,* volume 1 or ask your librarian.[7]

Benefits and Limitations of Automated Searches

Online searches have advantages over manual searches (that is, flipping pages by hand).

- They are rapid: you can review ten or fifteen years of an index in minutes.
- They are detailed: beyond listing titles and sources, an automated search often provides abstracts.
- They are current: the index usually comes online about six weeks before the printed copies.
- They are thorough: the system can search not only for titles but for key words (or word combinations) in the title *or* the abstract.
- They are efficient: you can extract from the database only the information you need.

Automated searches have limitations as well. Most computerized bibliographies include no entries before the mid-1960s; earlier information requires a manual search. Also, a manual search provides the whole database (the bound index or abstracts). As you browse, you often *randomly* discover something useful. This randomness is impossible with an automated search,

---

7. For guides to Internet resources, see the most recent editions of these works: Eric Brawn, *The Internet Directory,* New York: Fawcett; Harley Hahn and Rick Stout, *The Internet Complete Reference,* Berkeley: Osborne-McGraw; Edward T. L. Hardie and Vivian Neou, ed., *Internet Mailing Lists,* Englewood Cliffs: SRI-Prentice; Anne Okerson, ed., *Directory of Electronic Journals, Newsletters, and Academic Discussion Lists,* Washington: Association of Research Libraries.

IN BRIEF

A Sampling of Information Resources on the *Internet*

The Virtual Library

The virtual library has no edifice, no walls, and is open day or night. It is an electronic assemblage of library collections available worldwide to anyone with a computer, a phone line, a modem, and an *Internet* source provider or a gateway furnished by a university or local library.

In addition to the electronic resources discussed in this chapter, the virtual library offers a wide array of specialized information (James-Catalano 26–28).

- *Project Gutenberg:* an online, full-text collection of classic books that will include roughly 10,000 works.
- Electronic magazines (E-Zines): an example of information available *only* in electronic form.
- Tables of contents and article summaries from hundreds of printed magazines.
- Conference proceedings, journal articles, medical information, literature from political and social organizations.

These are a mere sampling of a rapidly growing list of resources.

E-mail Inquiries

The global E-mail network is excellent for contacting knowledgeable people in any field worldwide. E-mail addresses are becoming increasingly accessible via *Netfind,* a locator program that searches various local directories listed on the Internet (Steinberg 27). But unsolicited and indiscriminate E-mail inquiries might offend the recipient.

Special-Interest Groups

As part of the information they share via electronic bulletin board (*Usenet*) or mailing lists (*Listservs*), newsgroups and online discussion groups publish their answers to "frequently asked questions" (FAQs) about their particular topic of interest (acupuncture, AIDS research, sexual harassment, etc.). Figure 8.16 displays the announcement for one such discussion list.

FAQ lists can be a good source for the distilled wisdom of any knowledgeable group (Steinberg 26). But FAQs reflect the biases of those who contribute to them and those who edit them (Maeglin 5). A group's commitment to a particular viewpoint might politicize information and produce all sorts of inaccuracies (Snyder 90). Misinformation in any form—albeit nobly motivated and unintentional—is nonetheless misleading.

Cyberservants

How can users navigate, filter, and process the sheer volume of information available on the *Internet*? Electronic agents known as *Cyberservants* promise some relief. These software programs search, filter, sort, retrieve, analyze, and highlight information that is tailored to the needs of the particular user.

You are probably familiar already with two types of cyberservants (Dyson and Negroponte 95): Automatic Teller Machines (ATMs) that search your bank account for the information you need and guide you through various transactions; and automated phone answering ("Press One for product information," and so on).

Researchers use other types of cyberservants:

- Comprehensive retrieval systems, such as *SavvySearch,* that simultaneously employ dozens of Internet search engines (*Yahoo, InfoSeek, WebCrawler,* etc.), develop a customized search plan, and rank each source on the basis of its usefulness to the researcher.

IN BRIEF

- Supercomputers that process statistics, price lists, weather, or health data, to produce computer models that identify patterns, probabilities, and trends or provide "what if" scenarios (as in assessing nuclear meltdown risk at a power plant).
- "Bozo filters" that sift through E-mail messages, weeding out the nonessential and giving priority to others on the basis of particular names or other keywords.
- "Personalized newspaper" programs that monitor hundreds of news and information

sources, select the news most relevant to the individual subscriber (e.g., about a particular company, industry, or medical treatment), and assemble and deliver the document via fax or E-mail (Hafner 77).

Look for a coming generation of electronic "mentors [that] search databases for information useful to a particular individual, and . . . spark the user's creativity with questions and facts like those from a human consultant" ("Electronic" 56).

★   ★   ★

---

which can give you the illusion of having surveyed all that is known on your topic.[8] Finally, automated searches can be expensive, depending on how many databases you search and how long you spend online. (A BRS search costs about $30.) Some schools offer students one free search, but if your school doesn't, you pay the cost.

For any automated search, a manual (random) search of printed resources almost always is needed as well. A thorough search calls for a preliminary conference with a trained librarian or electronic research professional (page 157).

Conducting Interviews

Work-related research often requires firsthand, or primary, sources as shown in Figure 8.17. An excellent primary source for data that cannot be found in any publication is the personal interview. Much of what an expert knows may never be submitted for publication (Pugliano 6). Also, respondents can be useful sources of referral to other respondents or sources of information.

---

8. University of Massachusetts, Dartmouth librarian Charles McNeil cautions against assuming that computer access yields the best material: "The material in the computer is what is cheapest to put there." Librarian Ross LaBaugh alerts users to a built-in bias in databases: "The company that assembles the bibliographic or full-text database often includes a disproportionate number of its own publications." Like any collection of information, a database can reflect the biases of its assemblers.

**Figure 8.16
Description of a
Discussion List**

Source: Professional Ethics
Report. *Washington:
American Assn. for the
Advancement of Science,
Summer 1994: 7.*

Announcing the new, electronic discussion list . . .
Perspectives on Ethical Issues in Science and Technology

The AAAS Scientific Freedom, Responsibility and Law Program announces the creation
of an electronic discussion list, **Perspectives on Ethical Issues in Science and
Technology.** The new list will provide a forum for those working to articulate, study,
and respond to ethical issues raised by advances in science and technology, and will
likely be of interest to scientists, engineers, policymakers, students and others involved
with issues at the intersection of ethics, science and technology.

As a special feature, facilitators will be invited to make presentations based on specific
topics, case studies, or policy proposals in a virtual conference setting, with the
facilitator moderating a discussion with the other subscribers via e-mail. Presentations
will be invited on:

• scientific/research integrity or misconduct;

• ethical implications of new advances in science or technology;

• teaching ethics in the sciences and engineering;

• the role of scientific and engineering associations in promoting ethical conduct; and

• ethical issues raised by emerging social and public policy in science and technology.

Suggestions for other topics are welcome from subscribers. Additional types of
information shared on the network will include: calls for papers; announcements of
publications, courses, conferences and other meetings; reviews of articles and books;
requests for information or contacts regarding particular issues; items related to
educational, research, or advocacy opportunities; information on new regulations or
codes; and news of the professional activities of subscribers.

The discussion list is based on a listserv account and is accessible to anyone who can
receive and send e-mail. A listserv account is a service on BITNET that distributes
messages or files to a group of people with a common interest. A list of subscribers
and their e-mail addresses is stored on the server (the program/hardware that
maintains the listserv accounts at a particular site or node). When an item is posted to
the listserv address, it is automatically distributed to the e-mail address of all
subscribers. This new listserv is public, meaning that anyone can post a message to
the listserv address and it will be forwarded to all subscribers. Upon subscribing, you
will receive an e-mail message from the listserv that provides directions on how to
post messages and how to get further information about other listservs.

To subscribe to the listserv send a one-line message to:
listserv@gwuvm.gwu.edu
that reads:
subscribe AAASEST <firstname> <lastname>

For further information about the network, contact Alexander Fowler, Scientific
Freedom, Responsibility and Law Program, AAAS, 1333 H Street, NW, Washington,
DC 20005; (202) 326-7016; Fax (202) 289-4950; E-mail: afowler@aaas.org

**Figure 8.17
Sources for
Primary Research**

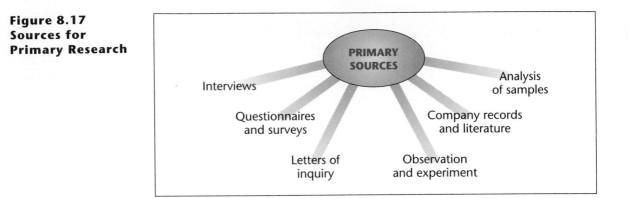

Of course, an expert opinion can be just as mistaken or biased as anyone else's. Like the wise patient who seeks a second opinion about a serious medical condition, the wise researcher seeks a balance or range of expert opinions about a complex problem or controversial issue—not only from a company engineer or environmentalist, for example, but from independent and presumably more objective third parties as well (such as a professor or a journalist who has studied the issue).

Identify Your Purpose

Know *exactly* what you are seeking. Be sure to review the literature *first*. Learn all you can about the topic before you set out to interview anyone. Suppose you decide to interview three authorities on electromagnetic radiation. Determine the information you seek from each respondent:

Purpose statement

> The purpose of my interview with Anne Hector, Chief Engineer at Northport Electric Company [a town-owned utility], is to discuss a recent company survey of community awareness and attitudes about electromagnetic radiation. Also, I will inquire about the company's plans and procedures for risk avoidance.

As soon as you can write out your purpose clearly and have determined the specific information this person might provide that is not available elsewhere, contact your respondent.

Contact the Respondent

Phone or write to request an interview at your respondent's convenience. If you plan a phone interview, call or write in advance, asking the respondent to stipulate a convenient time for the interview. Always give respondents ample notice and time to prepare.

Prepare for the Interview

Plan questions that elicit the specific information you seek. Write out each question on a separate notecard (on which you can summarize the response). Above all, know your subject. Effective interviews usually involve more than asking a set of preplanned questions and recording the responses. As in a good conversation, follow up on interesting comments with additional, unplanned questions.

Make your questions clear and specific. Vague, unspecific questions elicit unfocused answers or leave the respondent asking "What do you mean?"

Vague

| How is your company dealing with the EMR problem?

Which problem—public relations, public awareness, potential liability, danger to electrical workers, danger to the community, or what? Here are clear and specific versions:

More focused

| What specific procedures is your company developing for risk avoidance?
| Does the company have plans for promoting community awareness of EMR?

Avoid questions that can be answered with a simple "yes" or "no," because such answers provide little information:

Uninformative

| Do you think technology can find a way to decrease EMR hazards?

Instead, phrase your questions to elicit detailed information:

More informative

| Of the various technological solutions that have been proposed or considered, which do you think might be most effective?

Avoid questions that reflect a particular bias or that invite a particular stance on the issue:

Biased

| Don't you agree that EMR hazards have been overstated?

Instead, allow the respondent to express her or his own views without your influence:

More impartial

| Do you think EMR hazards have been understated or overstated?

Be ready with follow-up questions that will enable you to probe beneath the surface of an issue:

Follow-up questions

| Why or why not?
| What more needs to be done?
| What would you recommend?

The responses you obtain will only be as good as the questions you ask.

Observe Interview Guidelines

You—not your respondent—are responsible for a competent and productive interview. These guidelines should help:

1. Dress appropriately and arrive on time.
2. Begin by thanking your respondent in advance.
3. Restate the purpose of your interview.
4. Tell your respondent why you think he or she can be helpful.

5. Discuss your plans for using the information.

6. Ask (preferably before you arrive) if your respondent objects to being quoted or taped. (Although the most accurate way to record responses, taping can make people uncomfortable.)

7. Cultural differences, of course, are important in determining the appropriate technique for a particular interview. If your respondent is a member of a different culture, consider the level of formality, politeness, directness, relationship building, and other behaviors considered appropriate in that culture.

8. Ask your questions clearly, following the order in which you prepared them.

9. Be assertive but courteous. Ask pointed questions, but remember that the respondent is doing you a favor.

10. Let the respondent do most of the talking. Keep your opinions to yourself.

11. Stick to your interview plan. If your respondent wanders, politely bring the conversation back on track (unless the additional information is useful).

12. Be a good listener. Don't stare out the window, doodle, or ogle office staff while your respondent is speaking.

13. Be prepared to explore new areas of questioning. A respondent's answers often reveal new directions for the interview.

14. Keep note taking to a minimum. Record all numbers, statistics, dates, names, and other precise data, but don't transcribe responses word for word. Simply record significant words and phrases that later can refresh your memory.

15. If interviewing several people about the same issue, standardize your questions. Ask each respondent identical questions in identical order and phrasing. Don't inject random comments that could influence one person's responses.

16. Ask for clarification or explanation if necessary.

17. When all your questions have been answered, ask for any additional comments. If the interview is to be published, ask respondents to review the final draft before you quote them in print.

18. Offer to provide a copy of the document in which this information will be used.

19. Finally, thank your respondent and leave promptly.

20. As soon as you leave, write out a summary (or record one verbally) while the responses are fresh in your memory.

"In Brief" (page 157) shows the text of an interview with "Cybrarian" Reva Basch (an electronic information specialist who searches online databases for clients with various information needs). Notice how interviewer John Whalen probes, seeks clarification, and follows up on offered data.

Administering Surveys and Questionnaires

Surveys enable us to develop profiles and estimates about the concerns, preferences, attitudes, beliefs, or perceptions of a large, identifiable group (a *target population*) by studying representatives of that group (a *sample group*). Following are some types of assessments in which surveys can be useful:

Surveys help us make assessments like these

- *What percentage of workers hate their boss?*
- *What percentage of college students would cheat to get ahead?*
- *Is public confidence in technology increasing or decreasing?*
- *Do consumers prefer brand A or brand B?*

The tool for conducting surveys is the questionnaire. While interviews allow for greater clarity and depth, they cost more time and money, and direct contact with the interviewer can intimidate or inhibit respondents. Questionnaires are an inexpensive way to survey a large group. Respondents can answer privately and anonymously—and presumably more candidly, with time to reflect on their answers.

Questionnaires carry certain limitations:

Limitations of survey research

- *A low rate of response (often less than 30 percent).* People refuse to respond to a questionnaire that takes too much time, appears too complicated, or seems in some way threatening. They might be embarrassed by the topic or afraid of how their answers could be used.
- *Responses that might be nonrepresentative.* A survey will end up with responses only from the people who want to respond, but you will know nothing about the people who didn't respond. Those who responded might have extreme views, a particular stake in the outcome, or some other motive that represents inaccurately the population being surveyed (Plumb and Spyridakis 625–26).
- *Lack of follow-up.* Survey questions do not allow for the kind of follow-up and clarification possible with interview questions.

Even surveys done by professionals carry great potential for error. As consumers of survey research, we need to understand how surveys are designed and administered, how they are interpreted, and what can go wrong in the process. Following is an introduction to creating surveys and avoiding pitfalls along the way.

Define the Survey's Purpose

Why is this survey being done? What, exactly, is being measured? How much background research is needed? How will the survey findings be used?

Define the Target Population

Who is the exact population being studied(("the chronically unemployed," "part-time students," "computer users")? For example, in its research on science and technology activity, the 1994 *Statistical Abstract of the United States* defines "scientists and engineers" as follows:

One target population clearly defined

> Scientists and engineers are defined as persons engaged in scientific and engineering work at a level requiring a knowledge of sciences equivalent at least to that acquired through completion of a 4-year college course. Technicians are defined as persons engaged in technical work at a level requiring knowledge acquired through a technical institute, junior college, or other type of training less extensive than 4-year college training. Craftspersons and skilled workers are excluded. (606)

Lacking such definitions, we would be unable to determine who belongs in the target population and who does not.

Identify the Sample Group

How will the intended respondents be selected? How many respondents will there be? Generally, the larger the sample surveyed, the more dependable the results (assuming a well-chosen and representative sample). Will the sample be randomly chosen? In the statistical sense, "random" does not mean "chosen haphazardly": a *random sample* means that any member of the target population stands an equal chance of being included in the sample group.

Even a sample that is highly representative of the target population carries a measure of *sampling error.*

A Type of survey error

> The particular sample used in a survey is only one of a large number of possible samples of the same size which could have been selected using the same sampling procedures. Estimates derived from the different samples would, in general, differ from each other. (*Statistical Abstract* 937)

The larger the sampling error (usually expressed as the *margin of error,* discussed on page 184) the less dependable the survey findings.

Define the Survey Method

What type of data (opinions, ideas, facts, figures) will be collected? Is timing important? How will the survey be administered—in person, by mail, by phone? How will the data be collected, recorded, analyzed, and reported (Lavin 277)?

Phone and in-person surveys yield fast results and high response rates, but mail surveys are inexpensive and confidential. Recently developed for E-mail and other networks are computerized surveys that create the sense of a video game: as a question is answered, the program analyzes that response and automatically designs the next question. Respondents who dislike being quizzed by a human researcher seem more comfortable with this automated format (Perelman 89–90).

IN BRIEF

Interview with a Cybrarian

Wired: The major commercial databases tend to carry mainstream periodicals, not alternative publications. Is this bad news for the preservation of memetic diversity?

Reva Basch: It's funny—a lot of this stuff has been driven by economics. Until now, if the information didn't already exist in computer-readable form, or if the money wasn't there, it wasn't made into a database. But now, those smaller, alternative pubs are popping up on the Internet, big time. It's a lot cheaper to put your publication on the Net and maintain it than it is to go through one of the major database producers like DIALOG or Nexis. So I'm no longer waiting for alternative publications to come up on the standard commercial online services. They're doing an end run—they're putting themselves right on the Net. Which is in some ways as it should be. The trouble is, searching on the Internet still leaves much to be desired.

What do you need to make the Internet more navigable?

There are some pretty powerful text-analysis engines out there that are being worked on. But my real dream would be a Netwide, completely up-to-date, hierarchically arranged subject indexing capability that is as sophisticated as what you find in the commercial databases. I'm not holding my breath for that.

One thing that interests me is the way home pages on the World Wide Web are being used to link information. In my business, we've been asking for hypertext links to related documents for years, and it hasn't happened, except on a very limited level. And, of course, that's what the World Wide Web is all about.

Let's talk about intelligent agents. They've been described as digital butlers that roam the Infobahn gathering data for you—based on your needs—and learn more about your interests over time.

Apple has this really nauseating movie about an intelligent agent, a little dorky-looking guy with a bow tie. The user clicks on "him" and "he" goes out to get data from all over, using this very conversational dialog. It's by no means a new idea.

Last year, Paul Saffo made the really good point [*Wired* 2.03, page 74] that the next salable commodity in cyberspace is going to be point of view, in which you subscribe to an intelligent agent whose interests and points of view map yours. You say, "OK, Rush Limbaugh knowbot," or "OK, Ralph Nader knowbot, go out there and get me stuff from the Net that you think is important." Personifying the agent—I think we're going to see a lot of that. It's an interesting idea: Whose filter do you want to view the Net through?

It might also become a way for people to treat data more hermetically—to shun information that clashes with their particular ideologies.

Oh, yeah. The Dittohead view of the world.

It's sort of the opposite end of the classical media model. For years we've had the mainstream media force-feeding us all the news *they* deem fit to print, and now we face the prospect of having information become so fragmented and specialized that political consensus becomes even harder to achieve.

It depends on how it's implemented. I can't see there being any more polarization or skewed set of views than we're getting from media today. I've just got to believe that it's in the nature of

IN BRIEF

the Net and self-publishing to encourage a proliferation of ideas—may 10,000 intelligent agents bloom. There's going to be a lot more points of view than in today's media.

So, what will you be doing 10 years from now when we all have digital butlers and cyber-valets?

A hell of a lot less direct online searching. But there are always going to be people who, even though they know they can search for themselves, aren't going to be interested in trying. All they want is the data; they'll be willing to pay somebody else for that.

Is it addicting to have all this information under your power?

It's addicting yet self-limiting, because as long as the systems charge what they do, it's like the '80s and coke: you can work yourself into the poorhouse.

You raise a good point. Online searching is very expensive. And with public libraries cutting back on reference services, aren't we in danger of making information a commodity that only the élite can afford?

That's a question I would have answered very differently a year ago. The hype about the Internet is driving the commercial online services to change prices and offer more user-friendly interfaces. One of the second-tier online services, Data Times, just announced a $39.95-a-month flat-fee package for newspapers and trade journals—with considerable overlap of DIALOG's databases. There are no connect charges and it costs maybe a couple of bucks for each article retrieved. With this, you're getting down to the level of the average person.

How will living and interacting online warp our cognition and perhaps change the way we interact offline? You've written about how mainlining text at 9600 bits per second for a living has complicated your non-virtual life.

I notice it in particular when I read for pleasure. I just can't keep my eyes still. I have to remind myself to slow down and say, "Hey, you're reading for style, not content, stop browsing, start reading." It does have an accelerating effect on life. At parties, I'll scan the people: "not interesting, not interesting." Which is *awful*—*sort* of looking over their shoulders for the next person who might *add value.* It's a terrible, terrible thing to do.

I'll tell you something else I've discovered—I am less and less satisfied with superficial social connections. Online, especially on The Well [A California-based online bulletin board system] you really get into it, perhaps because of the conferencing software; there is one deep, deep conversation devoted to a particular issue. I find that affects my relationships with offline friends, especially people I haven't seen for a while. There's a lack of depth and context and continuity in a lot of my face-to-face relationships. I think the whole quality of human interaction is changing.

So what is it about this emerging modem society that makes it more intimate?

Well, part of it, of course, is the anonymity. And part of it is asynchronicity—you can compose a well-crafted, thoughtful posting offline. Also, you can deliberate—you don't have to respond in real time.

Reprinted with permission.

☆   ☆   ☆

Develop the Questionnaire

Survey questions should be easy to understand and hard to misinterpret.

Decide on the types of questions. (Adams and Schvaneveldt 202–12; Velotta 390). Survey questions can be *open-ended* or *closed-ended.* Open-ended questions allow respondents to express exactly what they're thinking or feeling in a word, phrase, sentence, or short essay:

Open-ended questions

> How much do you know about electromagnetic radiation at our school?
>
> What do you think should be done about EMR at our school?

Since one never knows what people will say, open-ended questions are a good way to uncover attitudes and obtain unexpected information. One disadvantage is that essay-type questions are hard to answer and tabulate, especially when responses are illegible or ambiguous.

When you know all the possible attitudes in your target group, and you want to measure where various people stand on the issue, choose closed-ended questions:

Closed-ended questions

> Are you interested in joining a group of concerned parents?
>
> YES _____   NO _____
>
> Characterize your degree of concern about the EMF issue at our school.
>
> HIGH _____   MODERATE _____   LOW _____   NO CONCERN _____
>
> Circle the number that indicates your view about the town's proposal to spend $20,000 to hire its own EMF consultant.
>
> 1 . . . . 2 . . . . 3 . . . . 4 . . . . 5 . . . . 6 7
> Strongly   No   Strongly
> Approve   Opinion   Disapprove

Depending on their design, closed-ended questions elicit various kinds of data. For instance, respondents may be asked to *rate* one item on a scale (from high to low, best to worse, and so on) or to *rank* two or more items (in order of importance, desirability, and so on). Other questions measure percentages or frequency.

> "How often do you . . . ?"
>
> ALWAYS _____   OFTEN _____   SOMETIMES _____   RARELY _____   NEVER _____

Other questions ask respondents to select one or more items from a list. The sample questionnaire in Figure 8.18 illustrates various designs for closed-ended questions.

Respondents generally find it easier and less threatening to check or circle an item rather than making more personal revelations in essay form. Also, closed-ended questions elicit data that is readily tabulated, measured, and analyzed because all responses are expressed in consistent terms.

Types of survey error

Despite these relative advantages, closed-ended questions create the potential for biased responses. Some people, for instance, automatically prefer items near the top of a list or the left side of a rating scale (Plumb and Spyridakis 633). Also, people generally are prone to agree rather than disagree with assertions in a questionnaire (Sherblom, Sullivan, and Sherblom 61). One way to avoid response bias is to present additional versions of a question, with listed or scaled items rearranged or the assertion rephrased.

Design an Engaging Introduction and Opening Questions. The clearer the rationale for a questionnaire, the more inclined readers are to respond. Explain the purpose of the survey. Persuade your respondents that the survey relates to their concerns, that their answers matter, and that their anonymity is assured. Whenever possible, explain how respondents will benefit from your findings, or offer an incentive (such as a copy of your final report). Keep the tone conversational, like a person talking to people.

A survey introduction

> Your answer to these questions about your views on proposed state handgun legislation will be appreciated. Your state representative will tabulate all responses so that she may continue to speak accurately for your views in legislative session. Thank you.

Some researchers include a cover letter with the questionnaire.

Begin with the easiest questions. Respondents are inclined to answer those questions that are general or more interesting or nonthreatening. Once they commit to these initial questions, they are more likely to complete any difficult questions that follow.

Make Each Question Unambiguous and Unbiased. All respondents should be able to interpret identical questions identically. Review your questions carefully for possible ambiguity:

An ambiguous question

> Do you favor foreign aid? YES _____ NO _____

"Foreign aid" might mean military, economic, or humanitarian aid, all three, or two out of three. Some respondents might support one or two types, but not all three—and their level of support might depend on the survey's timing (as in a period of military conflicts, widespread famine, or global recession). Consequently, responses to the above question, no matter how well analyzed, would produce a meaningless or misleading statistic, such as "Only 40 percent of Americans favor foreign aid," when the accurate conclusion might be "Over 95 percent of Americans favor some form of foreign aid." (See pages 182–86 for analyzing statistical findings.) Moreover, the limited choice of responses ("yes/no") in the above question reduces a wide range of possible opinions to a mere either/or choice. Allow for a full range of possible responses by rephrasing the above question:

A clear and incisive question

> Do you favor (check all that apply):
>
> _____ Our current foreign aid program of military, economic, and humanitarian assistance to other countries?
> _____ Greater military assistance in our foreign aid program?

_____ Less military assistance?

_____ Greater humanitarian assistance?

_____ Less humanitarian assistance?

_____ Greater economic assistance?

_____ Less economic assistance?

_____ No foreign aid program at all?

_____ Don't know

Avoid influencing respondents with *loaded questions* that invite or advocate a particular viewpoint or bias, as in these examples:

Loaded questions

Is foreign aid a waste of taxpayers' money on foreigners?
YES _____ NO _____

Is a wealthy nation morally responsible for helping the less fortunate?
YES _____ NO _____

Emotionally loaded words ("waste," "foreigners," "morally responsible," "radicals," "bureaucracy," "system") in a supposedly impartial survey are unethical because they carry built-in judgments that manipulate people's responses (Hayakawa 40).

Decide the Form and Range of Responses. Structure the questions so that responses can be tabulated easily, validly, and reliably. Decide on the best form of response: yes-no, multiple-choice, true-false, fill-in-the-blank, rating scale, order of ranking, and so on. But be sure that the form of response enables you to measure accurately what you intended to measure.

To ensure a full range of possible responses, include options such as "Other _____," "Don't know," "Not Applicable," or an "Additional Comments" section.

Keep the Questionnaire Short. The questionnaire should be as short as your information needs allow. Try to limit questions and response space to two sides of a single page. Phrase the questions in sentences that are as short as possible.

Add Touches That Encourage Response. Try to simplify the task and engage the reader. With mailed questionnaires, include a stamped, return-addressed envelope. Research experts Adams and Schvaneveldt point out that response rates improve when personal touches are included: letters and surveys individually typed, respondents addressed by name, the researcher's signature followed by a title—and even when postage stamps are used instead of a postal meter imprint (206).

Ensure Validity and Reliability

Validity and *reliability* are basic criteria by which we measure the dependability of any research. (Adams and Schvaneveldt 79–97; Burghardt 174–75; Crossen 22–24; Velotta 391.) *Valid research* produces findings that are correct.

A survey is valid when (1) it measures what you want to measure, (2) it measures accurately and precisely, and (3) its findings can be generalized to the target population. Valid survey questions enable each respondent to interpret each question exactly as the researcher intended.

Survey validity depends largely on trustworthy responses. Even clear, precise, and neutral questions can produce answers that are mistaken, inaccurate, or dishonest. People often try to appear more informed, responsible, or generous than they really are. They might suppress facts or opinions that reveal poor behavior, a bad attitude, or lack of will power: "How often do you take needless sick days?" "Would you lie to get ahead?" "How much TV do you watch?" And they might exaggerate or invent facts or opinions that suggest a more admirable picture: "How much do you give to charity?" "How many books do you read?" "How often do you hug your children?" Even when respondents don't know, don't remember, or have no opinion, they tend to guess in ways designed to win the researcher's approval.

Reliable research produces findings that can be replicated. A survey is reliable when its results are consistent, for instance when a respondent gives identical answers to the same survey given twice or to different versions of the same questions. Reliable survey questions enable all respondents to interpret the questions in the same way.

Much of your technical writing will be based on secondary research findings, so you will need to assess the validity and reliability of other people's research as well as your own. (See pages 174–78.)

A Sample Questionnaire

The questionnaire in Figure 8.18, sent to users of the *Statistical Abstract of the United States,* 1994 Edition, is typical of surveys done routinely to help improve products and address user needs.

Written reports of survey findings usually include an appendix (page 384) that contains a copy of the questionnaire as well as the tabulated responses.

Exploring Other Primary Sources

Whenever possible, explore all primary sources by writing letters, checking records, or observing and analyzing directly.

Inquiry Letters or Calls

Letters, phone calls, or E-mail inquiries are handy for obtaining specific information from government agencies, legislators, private companies, university research centers, trade associations, and research foundations such as the Brookings Institution and the Rand Corporation (Lavin 9). Keep in mind, however, that unsolicited inquiries, especially via phone or E-mail, can be intrusive and offensive. Letters and E-mail inquiries are discussed in greater detail in Chapter 20.

FORM **S-555**
(8-3-94)

U.S. DEPARTMENT OF COMMERCE
BUREAU OF THE CENSUS

1994 STATISTICAL ABSTRACT SURVEY

Please take a few minutes to answer the questions below. Your voluntary cooperation will help us continue to serve your needs as data users. When completed, please refold, **apply tape to the open edges at the top,** and drop in the mail. Thank you.

1a. **Which sections do you refer to frequently?** *Mark (X) all that apply.*

☐ Population
☐ Vital statistics
☐ Health
☐ Education
☐ Law enforcement
☐ Geography
☐ Parks and recreation
☐ Elections

☐ Federal Government
☐ State and local government
☐ National defense
☐ Social insurance
☐ Labor force
☐ Income
☐ Prices
☐ Banking

☐ Business
☐ Communications
☐ Energy
☐ Science
☐ Land transportation
☐ Air and water transportation
☐ Agriculture

☐ Forests and fisheries
☐ Mining
☐ Construction and housing
☐ Manufactures
☐ Domestic trade
☐ Foreign commerce
☐ Outlying areas
☐ International statistics

b. indicate topics for which you would like to see more coverage.

2. **To what degree do you find our current presentation of data in tables clear, meaningful, and easy to understand?** *Mark (X) one.*

☐ Very much   ☐ Somewhat   ☐ Not at all

3. **Which of the following do you feel currently interferes with the clarity of the tables presented?** *Mark (X) all that apply.*

☐ Hard to understand column headings and row indentations
☐ Too many numbers in the tables—too much data to absorb
☐ Not clear what numbers mean when they are rounded ("In thousands" or "in millions", for example)
☐ Too many notes in the tables

☐ Some concepts too difficult to understand
☐ Other — *Specify*

4. Please indicate which of these features you might like to see expanded or reduced in future editions.

Mark (X) the appropriate column for each feature.

	Expand	Reduce	OK as is
a. State rankings *(pp. xii–xxi)*			
b. Telephone contact list *(pp. xxii–xxiv)*			
c. Introductory text for sections			
d. Guide to Sources *(pp. 887–925)*			
e. Metropolitan Concepts and Components *(pp. 926–935)*			
f. Statistical Methodology (now called Limitations of the Data) *(pp. 936–950)*			
g. Index *(pp. 959–1011)*			
h. Charts and graphs			

5. **Indicate your level of satisfaction with the Abstract.** *Mark (x) one.*

☐ Very satisfied   ☐ Satisfied   ☐ Indifferent   ☐ Unsatisfied   ☐ Very unsatisfied

Figure 8.18 A Questionnaire About User Needs and Preferences

6. **How did you find out that the 1994 edition of the Abstract was available?** *Mark (X) all that apply.*

☐ Census and You
☐ Monthly Product Announcement
☐ Newspaper/magazine/journal piece — *If possible, specify the name of the newspaper/magazine/ journal* ↗

_____

☐ Census exhibit booth at a conference
☐ Mailed announcement from the Superintendent of Documents (catalog, flyer, etc.)

☐ Announcement from the National Technical Information Service (NTIS)
☐ Standing order: Government Printing Office
☐ Standing order: Book wholesaler (jobber)
   Received official distribution:
   ☐ Depository library
   ☐ State Data Center or Business and Industry Data Center
☐ Other —*Specify* ↗

_____

7. Please indicate which of the following publications you have used before. *Mark (X) all that apply.*

☐ 1993 edition of the Abstract
☐ 1992 or earlier editions of the Abstract
☐ 1991 State and Metropolitan Area Data Book

☐ 1988 County and City Data Book
☐ Historical Statistics of the United States
☐ 1993 CD-ROM version of the Abstract

8. **Are you interested in the following potential new products?**

			Mark (X) the appropriate column for each product.			
			Yes	No	No opinion	Depends on price
a. Updated	Electronic	On-line				
edition of	version	CD-ROM				
Historical		Diskette				
Statistics:	Printed version					
b. Portions of the Statistical Abstract available on-line						
c. Glossary of terms for the Abstract						
If you answered "Yes," in c. above — *Mark (X) for the one you would prefer.*	☐ Terms placed in text for each section ☐ One separate appendix for all terms ☐ No opinion					

9. **Would you like to see the current Guide to Sources reorganized so that the statistical publications listed are placed at the beginning of the appropriate subject sections?** *Mark (X) one.*

☐ Yes   ☐ No   ☐ No opinion

10. **What sources do you use to keep up-to-date with reference materials?** *Mark (X) all that apply.*

☐ Newsletters
☐ Magazines or journals
☐ Direct mail

☐ Trade shows/conventions
☐ PC user groups
☐ On-line services or bulletin boards

☐ Other — *Specify* ↗

_____

11. **What is your principal occupation?** *Mark (X) one.*

☐ Librarian
☐ Teacher (below college)

☐ Teacher (college or above)
☐ Journalist/writer

☐ Researcher/analyst
☐ Other — *Specify* _____

12. **Where are you employed?** *Mark (X) one.*

☐ State or local government
☐ College or university
☐ Market research/consulting

☐ International sector
☐ Federal Government
☐ Media (radio, TV, print)

☐ Public service
☐ Unemployed/retired
☐ Other — *Specify* _____

13. **What comments or suggestions do you have regarding the Abstract?**

Figure 8.18 A Questionnaire About User Needs and Preferences   *Continued*

Organizational Records and Publications

Company records (reports, memos, computer printouts, and so on) are a good primary source for data. Most organizations also publish pamphlets, brochures, annual reports, or prospectuses for consumers, employees, investors, or voters. But be alert for bias in company literature. If you were evaluating the safety measures at a local nuclear power plant, you would want the complete picture. Along with the company's literature, you would want studies and reports from government agencies and publications from environmental groups.

Personal Observation and Experiment

If possible, amplify and verify your findings with a firsthand look. Observation should be your final step, because you now know what to look for. Have a plan. Know how, where, and when to look, and jot down observations immediately. You might even take photos or make drawings.

Informed observations can pinpoint real problems. Here is an excerpt from a report investigating low morale at an electronics firm. This researcher's observations and interpretation are crucial in defining the problem:

Direct observation can be essential

> Our on-site communications audit revealed that employees were unaware of any major barriers to communication. Over 75 percent of employees claimed they felt free to talk to their managers, but the managers, in turn, estimated that fewer than 50 percent of employees felt free to talk to them.
>
> The problem involves misinterpretation. Because managers don't ask for complaints, employees are afraid to make them, and because employees never ask for an evaluation, they never get one. Each side has inaccurate perceptions of what the other side expects, and because of ineffective communications, each side fails to realize that its perceptions are wrong.

Keep in mind that even direct observation can lack validity: for instance, you might be biased about what you see (focusing on the wrong events or ignoring something important), or, instead of behaving normally, people who know they are being observed might behave in ways they think the researcher expects (Adams and Schvanveldt 244).

An experiment is a controlled form of observation designed to *verify an assumption* (e.g., the role of fish oil in preventing heart disease) or to *test something untried* (the relationship between background music and worker productivity). Each specialty has its own guidelines for experiment design.

Analysis of Samples

Workplace research can involve collecting and analyzing samples: water or soil or air, for contamination and pollution; foods, for nutritional value; ore, for mineral value; or plants, for medicinal value. Investigators analyze material samples to find the cause of an airline accident. Engineers analyze samples of steel, concrete, or other building materials to determine their load-bearing capacity. Medical specialists analyze tissue samples for disease.

✅ EXERCISES

1. Begin researching for the analytical report (Chapter 23) due at semester's end. Complete these steps. (Your instructor might establish a timetable.)

Phase One: Preliminary Steps

   a. Choose a topic of *immediate practical importance,* something that affects you or your community directly.

   b. Identify a specific audience and its intended use of your information. Complete an audience and use profile (page 65).

   c. Narrow your topic, and check with your instructor for approval.

   d. Make a working bibliography to ensure sufficient primary and secondary resources. Don't delay this step!

   e. List things you already know about your topic.

   f. Write a clear statement of purpose and submit it in a proposal memo (pages 526–28) to your instructor.

   g. Develop a tree chart of possible questions (as on page 125).

   h. Make a working outline.

Phase Two: Collecting Data (Read Chapter 9 in preparation for this phase.)

   a. In your research, move from general to specific; begin with general reference works for an overview.

   b. Skim your material, looking for high points.

   c. Take selective notes. Don't write everything down! Use notecards.

   d. Plan and administer questionnaires, interviews, and letters of inquiry.

   e. Whenever possible, conclude your research with direct observation.

   f. Evaluate and interpret your findings.

   g. Use the checklist on page 187 to reassess your research methods and reasoning.

Phase Three: Organizing Your Data and Writing Your Report

   a. Revise and adjust your working outline, as needed.

   b. Compose an audience and use analysis, like the sample on pages 570–71.

   c. Fully document all sources of information.

   d. Proofread carefully and add all needed supplements (title page, letter of transmittal, abstract, summary, appendix, glossary).

Due Dates: To Be Assigned by Your Instructor

   List of possible topics due:

   Final topic due:

   Proposal memo due:

   Working bibliography and working outline due:

   Notecards due:

   Copies of questionnaires, interview questions, and inquiry letters due:

   Revised outline due:

   First draft of report due:

   Final draft with supplements and documentation due:

2. Using the printed or electronic card catalog, locate and record the full bibliographic data for five books in your field or on your semester report topic, all published within the past year.

3. Consult the *Library of Congress Subject Headings* for alternative headings under which you might find information in the card catalog for your semester report topic.

4. List five major reference works in your field or on your topic by consulting Sheehy, Walford, or a more specific guide to literature.

5. List the titles of each of these specialized reference works in your field or on your topic: a bibliography, an encyclopedia, a dictionary, a handbook, an almanac (if available), and a directory.

6. Identify the major periodical index in your field or on your topic. Locate a recent article on a specific topic (e.g., use of artificial intelligence in medical diagnosis). Photocopy the article and write an informative abstract.

7. Consult the appropriate librarian and identify two databases you would search for information on the topic in Exercise 3.

8. Identify the major abstract collection in your field or on your topic. Using the abstracts, locate a recent article. Photocopy the abstract and the article.

9. Using technical report indexes, locate abstracts of three recent reports on one specific topic in your field. Provide complete bibliographic information.

10. Using patent indexes, locate and describe three recently patented inventions in your field, and provide complete bibliographic information.

11. Using indexes of conference proceedings, locate abstracts of three recent conference papers on *one* specific topic in your field. Provide complete bibliographic information.

12. Using the *Monthly Catalog* or *Government Reports Announcements and Index,* locate and photocopy (or download and print) a recent government publication in your field or on your topic.

13. Using OCLC, RLIN, ProQuest, SilverPlatter, or similar online or CD-ROM services, locate and copy the bibliographic record (including abstracts, if available) of four current books and four current articles in your field or on your topic.

14. If your library offers students a free search of commercial databases such as DIALOG, ask your librarian for help in preparing an electronic search for your semester report.

15. Explore Internet databases via Prodigy, Pathfinder, the Microsoft Network or a similar access provider or "on-ramp." Prepare a list of promising database resources for your report topic.

16. Locate an Internet newsgroup or discussion list related to your report topic. Download and print out the group's FAQ list.

17. Using Netscape, Spyglass, Mosaic, or a similar browsing program, search Web sites to locate resources for your report topic.

18. If your library belongs to a consortium of electronically networked libraries, search the holdings of other libraries on the network for topic resources not available in your library. Prepare a list of promising possibilities.

19. Students in your major want a listing of at least *two* of each of the following discipline-specific sources: the main reference books; indexes; periodicals; government publications; commercial, Internet, and CD-ROM databases; online newsgroups and discussion groups. Prepare the list (in memo form) and include a one-paragraph description of each source. Be prepared to discuss your listing in class.

20. Students in your major want a listing of one or two discipline-specific information sources from different depths of specialization:

a. the popular press (newspaper, radio, TV, magazines)

b. trade/business publications (newsletters and trade magazines)

c. professional literature (journals)

d. government sources (corporate data, technical reports, etc.)

   Prepare the list (in memo form) and include a one-paragraph description of each source.

21. Revise these questions to make them appropriate for inclusion in a questionnaire:

a. Would a female president do the job as well as a male?

b. Don't you think that euthanasia is a crime?

c. Do you oppose increased government spending?

d. Do you think welfare recipients are too lazy to support themselves?

e. Are teachers responsible for the decline in literacy among students?

f. Aren't humanities studies a waste of time?

g. Do you prefer Rocket Cola to other leading brands?

h. In meetings, do you think men are more interruptive than women?

22. Identify and illustrate at least six features that enhance the effectiveness of the questionnaire in Figure 8.18. (Review pages 155–62 for criteria.) Be prepared to discuss your evaluation in class.

23. Arrange an interview with someone in your field. Decide on general areas for questioning: job opportunities, chances for promotion, salary range, requirements, outlook for the next decade, working conditions, job satisfaction, and so on. Compose specific interview questions; conduct the interview, and summarize your findings in a memo to your instructor.

✔ COLLABORATIVE PROJECTS

1. Group yourselves according to major. For other students in your major, prepare a guide, in the form of a brochure, to your library's electronic resources (CD-ROM services and commercial database services, electronic catalogs, network

consortium, Internet gateways, World Wide Web access, and so on). Describe discipline-specific types of resources available via each electronic medium. Early in this project, arrange for a group tour and demonstration of your library's resources by a trained librarian. (In conjunction with this project, your instructor may assign Chapters 15 and 19.)

2. Divide into small groups, and decide on a campus or community issue or some other topic worthy of research. Elect a group manager to assign and coordinate tasks. At project's end, the manager will provide a performance appraisal by summarizing, in writing, the contribution of each team member. Assigned tasks will include planning, information gathering from primary and secondary sources, document preparation (including visuals) and revision, and classroom presentation. (See pages 21–22 for collaboration guidelines.)

   Do the research, write the report, and present your findings to the class. (In conjunction with this project, your instructor may assign Chapter 23.)

3. Group yourselves according to major. Assume that several major employers in your field are holding a job fair on campus next month and will be interviewing entry-level candidates. Each member of your group is assigned to develop a profile of *one* of these companies or organizations by researching its history, record of mergers and stock value, management style, financial condition, price/earnings ratio of its stock, growth prospects, products and services, multinational affiliations, ethical record, environmental record, employee relations, pension plan, employee-stock options or profit-sharing plans, commitment to affirmative action, number of women in upper-management, or any other features important to a prospective employee. The entire group then will edit each profile and assemble them in one single document to be used as a reference for students in your major.

4. Divide into small groups and prepare a comparative evaluation of literature-search media. Each group member will select *one* of the resources listed below and create an individual bibliography (listing at least twelve recent and relevant works on a specific topic of interest selected by the group):

   - conventional print media
   - electronic catalogs
   - CD-ROM services
   - a commercial database service such as DIALOG
   - the Internet and World Wide Web
   - an electronic consortium of local libraries, if applicable.

After carefully recording the findings and keeping track of the time spent in each search, compare the ease of searching and quality of results obtained from each type of search on your group's selected topic. Which medium yielded the most current sources (page 174)? Which provided abstracts and full texts as well as bibliographic data? Which consumed the most time? Which provided the most dependable sources (page 174)? The most diverse or varied sources (page 126)? Which cost the most to use? Finally, which yielded the greatest *depth* of resources (page 127).

Prepare a report and present your findings to the class. (In conjunction with this project, your instructor may assign Chapter 23.)

Recording and Reviewing Research Findings

A S YOU discover material during research, you confront questions like these: *How much is worth keeping? How should I record it? Can I trust this information? What, exactly, does it mean? How will I credit the source?* These latter stages of the research process call for the same quality of critical thinking required by the earlier stages in Chapter 8.

Recording the Findings

Findings should be recorded in ways that enable you to easily locate, organize, shuffle, and control the material as you work with it. Record primary research findings by using notebooks, photographs, drawings, tape recorder, videotape, or whichever medium suits your purpose. Record secondary research findings in the form of notes.

Taking Notes

Notecards are convenient because they are easy to organize and reorganize. In place of notecards, many researchers take notes on a laptop computer, using electronic file programs or database management software that allows notes to be filed, shuffled, and retrieved by author, title, topic, date, or other access point.

Follow these suggestions for using notecards:

1. Make a separate bibliography card for each work you plan to consult (Figure 9.1). Record the complete entry, using the identical citation format that will appear in your document. (See pages 194–207 for sample entries.) When searching an online catalog, you often can print out the full bibliographic record for each work, thereby ensuring accurate citation.

2. Skim the entire work to locate relevant material.

Record each bibliographic citation exactly as it will appear in your final report

> Pinsky, Mark A. *The EMF book: What You Should Know About Electromagnetic*
>
> *Fields, Electromagnetic Radiation, and Your Health.* New York: Warner, 1995.

Figure 9.1 Bibliography Card

3. Go back and decide what to record. (Use a separate card for each item.)

4. Decide how to record the item: as a quotation or a paraphrase. When quoting others directly, be sure to record words and punctuation accurately. When restating or adapting material in your own words, be sure to preserve the original meaning and emphasis.

Quoting the Work of Others

When you borrow exact wording, whether the words were written or spoken (as in an interview or presentation) or whether they appeared in electronic form, you must place quotation marks around all borrowed material. Even a single borrowed sentence or phrase, or a single word used in a special way, needs quotation marks, with the exact source properly cited.

If your notes fail to identify quoted material accurately, you might forget to credit the source in your report. Even when this omission is unintentional, writers face the charge of *plagiarism* (misrepresenting as one's own the words or ideas of someone else). Possible consequences of plagiarism include expulsion from school, the loss of your job, and a lawsuit.

In recording a direct quotation, copy the selection word for word (Figure 9.2) and include the page numbers. If your quotation omits parts of a sentence, use an *ellipsis* (three periods: . . .) to indicate each part that you have omitted from the original. If your quotation omits the end of a sentence, the beginning of the subsequent sentence, or whole sentences or paragraphs, show the ellipsis with four periods (. . . .).

Ellipsis within and between sentences

> If your quotation omits parts . . . use an ellipsis. . . . If your quotation omits the end. . . .

Be sure that your elliptical expression is grammatical and that the omitted material in no way distorts the original meaning.

Place quotation marks around all directly quoted material

Pinsky, Mark A. pp. 29–30.

"Neither electromagnetic fields nor electromagnetic radiation cause cancer per se, most researchers agree. What they may do is promote cancer. Cancer is a multistage process that requires an "initiator" that makes a cell or group of cells abnormal. Everyone has cancerous cells in his or her body. Cancer— the disease as we think of it—occurs when these cancerous cells grow uncontrollably."

Figure 9.2 Notecard for a Quotation

If you insert your own comments within the quotation, place them inside brackets to distinguish your words from those of your source:

Brackets setting off personal comments within quoted material

| "This profession [aircraft ground controller] requires exhaustive attention."

(For more on brackets, see Appendix.)

Sentences and paragraphs that include quotations must be clear and understandable. Read your sentences aloud to be sure they make sense and they read smoothly and grammatically. Generally, integrated quotations are introduced by phrases such as "Jones argues that," "Smith suggests that," so that readers will know who said what. More importantly, readers must see the relationship between the quoted idea and the sentence that precedes it. Use a transitional phrase that emphasizes this relationship by looking back as well as ahead:

An introduction that unifies a quotation with the discussion

| After you decide to develop a program, "the first step in the programming process . . . ."

Besides showing how each quotation helps advance the main idea you are developing, your integrated sentences should be grammatical:

Quoted material integrated grammatically with the writer's words

| "The agricultural crisis," Marx acknowledges, "resulted primarily from unchecked land speculation."
| "She has rejuvenated the industrial economy of our region," Smith writes of Berry's term as regional planner.

(For quoting long passages and for punctuating at the end of a quotation, see Appendix.)

Use a direct quotation only when precision, clarity, or emphasis requires the exact words from the original. Avoid excessively long quoted passages. Research writing is more a process of independent thinking, in which you work with the ideas of others in order to reach your own conclusions; you should therefore paraphrase, instead of quoting, much of your borrowed material.

Paraphrasing the Work of Others

We paraphrase not only to preserve the original idea, but also to express it in a clear, simple, direct, or emphatic way—without distorting the idea. Paraphrasing means more than changing or shuffling a few words; it means restating the original idea in your own words and giving full credit to the source.

To borrow or adapt someone else's ideas or reasoning without properly documenting the source is plagiarism. To offer as a paraphrase an original passage only slightly altered—even when you document the source—also is plagiarism. Equally unethical is to offer a paraphrase, although documented, that distorts the original meaning.

An effective paraphrase generally displays all or most of the following elements (Weinstein 3):

Elements of an effective paraphrase

- reference to the author early in the paraphrase, to indicate the beginning of the borrowed passage
- key words retained from the original, to preserve the meaning
- original sentences restructured and combined, for emphasis and fluency
- needless words from the original deleted, for conciseness
- your own words and phrases that help explain the author's ideas, for clarity
- a citation (in parentheses) of the exact source, to mark the end of the borrowed passage and to give full credit
- preservation of the author's original intent

Figure 9.3 shows an entry paraphrased from the passage in Figure 9.2. Paraphrased material takes no quotation marks, but it has to be documented to acknowledge your debt to the source. Failing to acknowledge ideas, findings, judgments, lines of reasoning, opinions, facts, or insights not considered *common knowledge* (page 192) is plagiarism—even when these are expressed in your own words.

Evaluating and Interpreting Information

Not all information is equal. Not all interpretations are equal. Whether you work with your own findings or the findings of other researchers, you need to decide if the information is valid and reliable. Then you need to decide what your information means. Figure 9.4 outlines this challenge.

For producers and consumers of research alike, evaluating and interpreting information is a complex process, with much room for error. The critical-thinking strategies described in this section help distinguish legitimate inquiry from mere information gathering.

Signal the beginning of the paraphrase by citing the author, and the end by citing the source

---

Pinsky, Mark A.

Pinsky explains that electromagnetic waves probably do not directly cause cancer. However, they might contribute to the uncontrollable growth of those cancer cells normally present—but controlled—in the human body (29–30).

Figure 9.3 Notecard for a Paraphrase

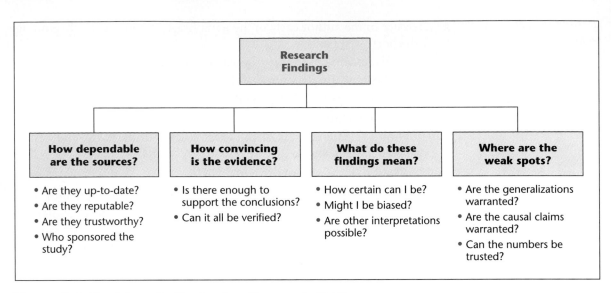

Figure 9.4 Decisions in Reviewing Research Findings

Evaluating the Sources

Not all data sources are equally dependable. A source might offer information that is out of date, inaccurate, incomplete, mistaken or biased.

Is the Source Up-to-Date? Newly published books contain information that can be more than one year old, and journal articles often undergo a lengthy process of peer review.

"How current is the information?"

Certain types of information become outdated more quickly than others. For topics that focus on *technology* (superconductivity, multimedia law, Internet censorship, alternative cancer treatments), information more than a few months old may be outdated. Except for historical or background research, sources in those areas generally should offer the most recent information available. For topics that focus on *people* (business ethics, management practices, workplace gender equality, employee motivation), information several decades old might offer valuable perspective on present situations.

Is the Source Reputable? Some sources enjoy better reputations than others. For research on alternative cancer treatments, you could depend more on reports in the *New England Journal of Medicine* or *Scientific American* than on those in scandal sheets or movie magazines. Even researchers with expert credentials, however, can disagree or be mistaken.

One way to assess a publication's reputation is to check its copyright page for information. Is the work published by a university, professional society, museum, or respected news organization? Do members of the editorial and

IN BRIEF

Copyright Protection and Fair Use of Printed Information

Copyright Law

A copyright is the exclusive legal right to reproduce, publish, and sell a literary, dramatic, musical, or artistic work. The law grants to the owner of the copyright for the work the exclusive rights to do and to authorize any of the following:

1. To reproduce the copyrighted work.
2. To prepare derivative works.
3. To distribute copies of the copyrighted work to the public by sale, rental, lease, or lending.
4. In certain cases, such as for literary and musical works, to perform the copyrighted work publicly.
5. In certain cases, such as for graphics, images, or other audiovisuals, to display the copyrighted work publicly.

Written permission must be obtained to use all copyrighted material. Works published prior to January 1, 1978 are protected for 75 years. Works published on or after January 1, 1978 are copyrighted for the life of the author plus 50 years. A work made for hire (such as an instructor's manual) is copyrighted for 75 years from publication or 100 years from creation, whichever is shorter.

Public Domain

Public domain refers to material on which copyright has expired or material that, due to its nature, is not protected by copyright. Works published in the United States 75 years prior to the current year is in the public domain. Most government publications and commonplace information, such as height and weight charts or a [metric conversion] table are considered to be in the public domain. These works occasionally contain copyrighted material used with permission and properly acknowledged. However, a new translation or version of a work in the public domain can be protected by copyright; if you are not sure of whether something is in the public domain, the safest course is to request permission.

Fair Use

Fair use is the limited use of copyrighted material without obtaining permission. The use is considered "fair" if the material is not a large or representative portion of the whole. However, the source of any material that is not your own should be acknowledged. There is no hard and fast rule or formula that can be applied to determine what or how much "use" is "fair."

Copyright law provides the following criteria to be considered in the determination of fair use:

1. The purpose and character of the use, including whether such use is of a commercial nature or is for nonprofit educational purposes.
2. The nature of the copyrighted work.
3. The amount and substantiality of the portion used in relation to the whole.
4. The effect of the use upon the potential market for or value of the copyrighted work.

When the material being quoted, though scholarly, illustrative, and/or brief, forms the core, distinguishable creative effort of the work being cited, then the use of the material cannot be considered fair.

Fair use ordinarily does not apply to use of the following: poetry, musical lyrics, dialogue of a play, entries in a diary, case studies, charts and graphs, author's notes, private letters, testing materials, or quotations for use as epigraphs.

Reprinted with permission from the HarperCollins Author's Guide. Copyright ©1995.

★ ★ ★

IN BRIEF

Copyright Protection and Fair Use of Electronic Information

The Problem

Copyright and fair use law is quite specific for printed works or works in other tangible form (paintings, photographs, music). But how do we define "fair use" (page 175) of intellectual property in electronic form? How does copyright protection apply (Dyson137)? How do fair-use restrictions apply to material used in multimedia presentations or to text or images that have been altered or reshaped to suit the user's specific needs (Steinberg, "Travels" 30)?

Information obtained via E-mail or discussion groups presents additional problems: Sources often do not wish to be quoted or named or to have early drafts made public. How do we protect source confidentiality? How do we avoid infringing on works in progress that have not yet been published? How do we quote and cite this material without violating ownership and privacy rights (Howard 40–41)?

Present Status of Electronic Copyright Law

Subscribers to commercial online databases such as DIALOG™ pay fees, and copyholders in turn receive royalties (Communication Concepts, Inc. 13). But as of this writing, few specific legal protections exist for noncommercial types of electronic information. Since April 1989, however, most works are considered copyrighted as soon as they are produced—even if they carry no copyright notice. Fair use of electronic information generally is limited to brief excerpts that serve as a basis for response—for example, in a discussion group. Except for certain government documents, no *Internet* posting is in the public domain unless it is expressly designated as such by its author (Templeton).

Until specific laws are enacted, the following examples can be considered violations of copyright (Communication Concepts, Inc. 13; Templeton):

- Downloading a work from the *Internet* and forwarding copies to other readers.
- Editing, altering, or incorporating an original work as part of your own document or multimedia presentation.
- Putting someone else's printed work online without the author's written permission.
- Copying and forwarding an E-mail message without the sender's authorization. The E-mail *text* is copyrighted, but its *content* legally may be revealed—except for proprietary information (page 81).

Certain copyright violations may exceed the boundaries of civil law and may be prosecuted as felonies (Templeton). When in doubt, assume the work is copyrighted, and obtain written permission for its use.

☆ ☆ ☆

advisory board have distinguished titles and degrees? Is the publication *refereed* (all submissions reviewed by experts prior to acceptance)?

One way to assess an author's reputation is to check citation indexes (page 139) to see what others have said about this research. Many periodicals also provide brief biographies or descriptions of authors' earlier publications and other achievements.

"Can the source be trusted?"

Is the Source Trustworthy? The *Internet* offers information that never appears in other sources, for example from listservs and newsgroups. But much of this information may reflect the bias of the special interest groups that provide it. Moreover, anyone can publish almost anything on the *Internet*—including a great deal of misinformation—without having it verified, edited, or reviewed for accuracy (Snyder 89–90).

Even in a commercial database, decisions about what to include and what to leave out depend on the biases, priorities, or interests of those who assemble that database. In general, try not to rely on any single information source.

"Who sponsored the study?"

Is the Information Biased? Much of today's research is paid for by private companies or special-interest groups, which have their own social, political, or economic agendas (Crossen 14, 19). Medical research may be sponsored by drug or tobacco companies; nutritional research, by food manufacturers; environmental research, by oil or chemical companies. Public policy research (on gun control, school prayer, seat-belt laws, endangered species) sponsored by opposing groups (environmentalists versus the logging industry) produces opposing results. Instead of a neutral and balanced inquiry, this kind of "strategic research" is designed to support one special interest or another (132–34). Those who pay for strategic research are not likely to publicize findings that contradict their original claims, opinions, or beliefs (profits lower than expected, losses or risks greater than expected). As consumers of research, we should try to determine exactly what the sponsors of a particular study stand to gain or lose from the results (234).

Evaluating the Evidence

Evidence is any finding used to support or refute a particular conclusion. While evidence can serve the truth, it also can create distortion, misinformation, and deception. For example, how much money, material, or energy does recycling really save? How good for your heart is oat bran? How well are public schools educating children? Which investments or automobiles are safest? Conclusions about such matters are based on evidence which often can be manipulated in support of one view or another. As consumers of research we have to assess for ourselves the quality of the evidence presented.

We assess the quality of evidence by examining it critically to understand its limitations, to see if findings conflict, to discover connections, similarities, trends or relationships, to determine the need for further inquiry, to envision new questions.

"Is there enough evidence?"

Is the Evidence Sufficient? Evidence is sufficient when it enables us to reach an accurate judgment or conclusion. A study of the stress-reducing benefits of low-impact aerobics, for example, would require a broad survey sample: people who have practiced aerobics for a long time, people of different genders, different ages, different occupations, and different lifestyles before they began aerobics, and so on. Even responses from hundreds of practitioners might constitute insufficient evidence unless those responses were supported by laboratory measurements of metabolism, heart rates, and blood pressure.

Personal experience usually offers insufficient evidence from which to generalize. You cannot tell whether your experience is representative, no matter how long you might have practiced aerobics. Although anecdotal evidence ("This worked great for me!") might offer a good starting point for an investigation, personal experience should be evaluated within the broader context of *all* available evidence.

"Is the evidence hard or soft?"

Can the Evidence Be Verified? Hard evidence consists of factual statements, expert opinion, or statistics that can be verified (shown to be true). Soft evidence consists of uninformed opinion or speculation, data obtained or analyzed unscientifically, and findings that have not been replicated or reviewed by experts. (Reputable news organizations employ "fact-checkers" to verify information before it appears in print.)

Evidence that seems scientific can turn out to be soft. For example, information obtained from polling often is reported in fancy charts, graphs, and impressive statistics—but it is based on public opinion, which is almost always changing (Crossen 104).

Base your conclusions on hard evidence. For example, suppose an article makes positive claims about low-impact aerobics but provides no data on measurements of pulse, blood pressure, or metabolic rates. Although these claims might coincide with your own experience, your evidence so far consists of only two opinions: yours and the author's—without scientific support (for example, tests of a broad cross-section under controlled conditions). Any conclusion at this point would rest on soft evidence. Only after carefully assessing dependable sources can you decide which conclusions are supported by the bulk of the evidence.

Interpreting the Evidence

Interpreting means trying to reach the truth of the matter: an overall judgment about what the evidence means and what conclusion or action it suggests. Unfortunately, research does not always yield answers that are conclusive or about which we can be certain. Instead of settling for the most *convenient* answer, we should pursue the most *reasonable* answer by examining critically a full range of possible meanings.

What Level of Certainty Is Warranted? As possible outcomes of research, we can identify three distinct and very different levels of certainty:

1. The definitive truth: the *conclusive answer:*

A practical definition of "truth"

> Truth is *what is so* about something, the reality of the matter, as distinguished from what people wish were so, believe to be so, or assert to be so. From another perspective, in the words of Harvard philosopher Israel Scheffler, truth is the view "which is fated to be ultimately agreed to by all who investigate." The word *ultimately* is important. Investigation may produce a

wrong answer for years, even for centuries. . . . Does the truth ever change? No. . . . One easy way to spare yourself any further confusion about truth is to reserve the word *truth* for the final answer to an issue. Get in the habit of using the words, *belief, theory,* and *present understanding* more often. (Ruggiero 21–22)

We often are mistaken in our certainty about the *truth.* For example, in the second century A.D., Ptolemy's view of the universe concluded that the earth was its center. Though untrue, this judgment was based on the best information available at that time. Ptolemy's view survived for thirteen centuries, even after new information had discredited this belief. When Copernicus and Galileo proposed more truthful views in the fifteenth century, they were labeled heretics.

Conclusive answers are the research outcome we seek, but often we have to settle for answers that are less than certain.

2. The *probable answer:* the answer that stands the best chance of being true or accurate—given the most we can know at this particular time. Probable answers are subject to revision in the light of new information.

3. The *inconclusive answer:* the realization that the truth of the matter is far more elusive, ambiguous, or complex than we expected.

"Exactly how certain are we?"

To ensure an accurate outcome, we must decide what level of certainty the findings warrant. For example, we are *highly certain* about the perils of smoking or sunburn, *reasonably certain* about the benefits of fruits and vegetables and moderate exercise, but *far less certain* about the perils of coffee drinking or electromagnetic waves or the benefits of vitamin supplements.

Do I Have a Personal Bias? To support a particular version of the truth, our own bias might cause us to overestimate (or deny) the certainty of our findings.

Personal bias is a fact of life

Expect yourself to be biased, and expect your bias to affect your efforts to construct arguments. Unless you are perfectly neutral about the issue, an unlikely circumstance, at the very outset . . . you will believe one side of the issue to be right, and that belief will incline you to . . . present more and better arguments for the side of the issue you prefer. (Ruggiero 134)

Because personal bias is hard to transcend, *rationalizing* often becomes a substitute for *reasoning:*

Reasoning versus rationalizing

You are reasoning if your belief follows the evidence—that is, if you examine the evidence first and then make up your mind. You are rationalizing if the evidence follows your belief—if you first decide what you'll believe and then select and interpret evidence to justify it. (Ruggiero 44)

Personal bias often is unconscious until we examine our own value systems, attitudes long held but never analyzed, notions we've inherited from our own backgrounds, and so on. Recognizing our own biases is a crucial first step in managing them.

"What else could this mean?"

Are Other Interpretations Possible? Perhaps other researchers would disagree with the meaning of these findings. Some controversial issues (the need for defense spending or causes of inflation) never will be resolved. Although we can get verifiable data and can reason persuasively on some subjects, no close reasoning by any expert and no supporting statistical analysis will prove anything about a controversial subject to everyone's satisfaction. For instance, one could only *argue* (more or less effectively) that federal funds will or will not alleviate poverty or unemployment.

Settling on a final meaning can be difficult. For example (Ledermen 5): What does a reported increase in violent crime on U.S. college campuses mean—especially in light of national statistics that show violent crime decreasing?

- That college students are becoming more violent?
- That some drugs and guns in high schools end up on campuses?
- That off-campus criminals see students as easy targets?

Or could these findings mean something else entirely?

- That increased law enforcement has led to more campus arrests—and thus, greater recognition of the problem?
- That crimes haven't increased, but more are being reported?

Depending on our interpretation, we might conclude that the problem is worsening—or improving!

Avoiding Errors in Reasoning

Finding the truth, especially in a complex issue or problem, often is a process of elimination, of ruling out or avoiding errors in reasoning. As we interpret, we make *inferences:* We derive conclusions about what we don't know by reasoning from what we do know (Hayakawa 37). For example, we might infer that a drug that boosts immunity in laboratory mice will boost immunity in humans, or that a rise in campus crime statistics is caused by the fact that young people have become more violent. Whether a particular inference is on target or dead wrong depends largely on our answers to one or more of these questions:

- *To what extent can these findings be generalized?*
- *Is Y really caused by X?*
- *How much can the numbers be trusted, and what do they mean?*

Following are three major reasoning errors that can distort our interpretations.

"How much can we generalize?"

Faulty Generalizations. When we accept research findings uncritically and jump to conclusions about their meaning, we commit the error of *hasty generalization*. When we overestimate the extent to which the findings reveal some larger truth, we commit the error of *overstated generalization*.

A study in Greece on the role of fruits, vegetables, and olive oil in lowering breast cancer risk was widely publicized in 1995 because of the alleged benefits of olive oil for women who consume it twice or more a day. Subsequent analysis of this study revealed that data about the women's food consumption covered only one year and were based on a single questionnaire asking respondents to estimate their previous year's diet. (Estimates of this type tend to be highly inaccurate.) Also, the study did not identify the quantities of olive oil individual users consumed. In this instance, the study's generalization about olive oil was shown to be *hasty* (based on insufficient evidence).

Further analysis revealed that only 99 respondents (of the nearly 2,500 surveyed) claimed to have consumed olive oil twice or more a day ("Olive Oil" 1). In this instance, the study's generalization was shown to be *overstated* (a limited generalization made to apply to all cases). Something true in one instance need not be true in other instances.

Although this particular study was flawed, many other studies support the generalization that fruits and vegetables do help lower the risk of cancer. Generalizing is vital and perfectly legitimate—when it is warranted.

Faulty Causal Reasoning. Causal reasoning tries to explain *why* something happened or *what* will happen, often very complex questions. Faulty causal reasoning oversimplifies or distorts the cause-effect relationship through errors like these:

Ignoring other causes	Investment builds wealth. [*Ignores the role of knowledge, wisdom, timing, and luck in successful investing.*]
Ignoring other effects	Running improves health. [*Ignores the fact that many runners get injured, and that some even drop dead while running.*]
Inventing a cause	Right after buying a rabbit's foot, Felix won the state lottery. [*Posits an unwarranted causal relationship merely because one event follows another.*]
Confusing correlation with causation	Poverty causes disease. [*Ignores the fact that disease, while highly associated with poverty, has many causes unrelated to poverty.*]
Rationalizing	My grades were poor because my exams were unfair. [*Denies the real causes of one's failures.*]

Because of bias or impatience, we can be tempted to settle for a hasty cause or to confuse possible, probable, and definite causes.

"Did X possibly, probably, or definitely cause Y?"

Sometimes a definite cause is apparent (e.g., "The engine's overheating is caused by a faulty radiator cap"), but usually much analysis is needed to isolate a specific cause. Suppose you want to answer this question: Why does our state college have no day care facilities? Brainstorming yields these possible causes:

- lack of need among students
- lack of interest among students, faculty, and staff
- high cost of liability insurance
- lack of space and facilities on campus
- lack of trained personnel
- prohibition by state law
- lack of legislative funding for such a project

Say you proceed with interviews, questionnaires, and research into state laws, insurance rates, and availability of personnel. You begin to rule out some items, and others appear as probable causes. Specifically, you find a need among students, high campus interest, an abundance of qualified people for staffing, and no state laws prohibiting such a project. Three probable causes remain: lack of funding, high insurance rates, and lack of space. Further inquiry shows that lack of funding and high insurance rates *are* issues. These obstacles, however, could be eliminated through new sources of revenue: charging a fee for each child, soliciting donations, or diverting funds from other campus organizations.

Finally, after examining available campus space and speaking with school officials, you arrive at one definite cause: lack of space and facilities. One could argue that lack of space and facilities is somehow related to funding, and the college's being unable to find funds or space may be related to student need, which is not sufficiently acute or interest sufficiently high to exert real pressure. Lack of space and facilities, however, appears to be the *immediate* cause.

When you report on your research, be sure readers can draw conclusions identical to your own on the basis of the evidence. The process might be diagrammed like this:

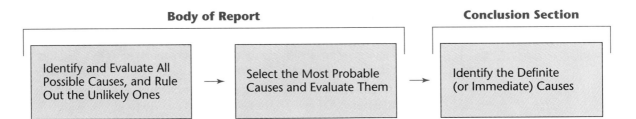

Body of Report		**Conclusion Section**
Identify and Evaluate All Possible Causes, and Rule Out the Unlikely Ones	Select the Most Probable Causes and Evaluate Them	Identify the Definite (or Immediate) Causes

Initially you might have based your conclusions hastily on soft evidence (an opinion—buttressed by a newspaper editorial—that the campus was apathetic.) Now you base your conclusions on solid, factual evidence. You have moved from a wide range of possible causes to a narrow range of probable causes, and finally to a definite cause.

Sometimes, finding a single cause is impossible, but this reasoning process can be tailored to most problem-solving analyses. Anything but the simplest effect is likely to have more than one cause. By narrowing the field, you can focus on the real issues.

Faulty Statistical Reasoning. The purpose of statistical analysis is to determine the meaning of a collected set of numbers. In primary research, our surveys and questionnaires often lead to some kind of numerical interpretation ("What percentage of respondents prefer X?" "How often does Y happen?"). In secondary research, we rely on numbers collected by primary researchers.

Numbers seem more precise, more objective, more scientific and less ambiguous than words. They are easier to summarize, measure, compare, and analyze. But numbers can be misleading. For example, radio or television phone-in surveys produce grossly distorted data: Although 90 percent of callers might express support for a particular viewpoint, callers tend to be those with the greatest anger or the strongest feelings about the issue—representing only a fraction of overall attitudes (Fineman 24). Mail-in surveys can produce similar distortion because only people with certain attitudes might choose to respond.

"How much can we trust these numbers?"

Before relying on any set of numbers, we need to know exactly where they come from, how they were collected, and how they were analyzed (Lavin 275–76). Can the numbers be trusted, and if so, what do they mean?

Faulty statistical reasoning produces conclusions that are unwarranted, inaccurate, or downright deceptive. Following are some common statistical fallacies:

- *The sanitized statistic:* Numbers are manipulated (or "cleaned up") to obscure the facts. For instance, a recently revised formula enables the government to exclude from its unemployment figures an estimated 5 million people who remain unemployed after one year—thus creating a far rosier economic picture than the facts warrant. Similar formulas allow for all sorts of sugarcoating in reports of wages, economic growth, inflation, and other statistics that affect the political climate (Morgenson 54). The College Board's recentering of SAT scores has raised the average math score from 478 to 500 and the average verbal score from 424 to 500 (a boost of almost 5 and 18 percent, respectively) although actual student performance remains unchanged (Samuelson 44).

"Exactly how well *are we doing?"*

- *The meaningless statistic:* Exact numbers are used to quantify something so inexact or vaguely defined that it should only be approximated (Huff 247; Lavin 278): "Only 38.2 percent of college graduates end up working in their specialty." "Boston has 3,247,561 rats." "Zappo detergent makes laundry 10 percent brighter." An exact number looks impressive, but certain subjects (child abuse, cheating in college, virginity, drug and alcohol abuse on the job, eating habits) cannot be quantified exactly because respondents don't always tell the truth (because of denial, embarrassment, or merely guessing). Or they respond in ways they think the researcher expects.

"How many rats was that?"

- *The undefined average:* The mean, median, and mode are confused in determining an average (Huff 244; Lavin 279). The *mean* is the result of adding up the value of each item in a set of numbers, then dividing by the number of items. The *median* is the result of ranking all the values from high to low, then choosing the middle value (or the 50th percentile, as in calculating SAT scores). The *mode* is the value that occurs most often in a set of numbers.

  Each of these three measurements represents some kind of average, but unless we know which average is being presented, we cannot interpret the figures accurately. Assume that we are computing the average salary among female vice presidents at XYZ Corporation:

Vice President	Salary
A	$90,000
B	90,000
C	80,000
D	65,000
E	60,000
F	55,000
G	50,000

"Why is everybody griping?"

In the above example, the mean salary (total salaries divided by people) equals $70,000; the median salary (middle value) equals $65,000; the mode (most frequent value) equals $90,000. Each is legitimately an average, and each could be used to support or refute a particular assertion (for example, "Women vice presidents are paid too little" or "Women vice presidents are paid too much").

Research expert Michael Lavin sums up the potential for bias in reporting averages:

> Depending on the circumstances, any one of these measurements [mean, median, or mode] may describe a group of numbers better than the other two. . . . [But] people typically choose the value which best presents their case, whether or not it is the most appropriate to use. (279)

Although the mean is the most commonly computed average, this measurement can be misleading when values on either end of the scale are extremely high or low. Suppose, for instance, that vice president A received a $200,000 salary. Because this figure deviates so far from the normal range of salary figures for B through G, it distorts the average for the whole group—increasing the mean salary by more than 20 percent (Plumb and Spyridakis 636).

■ *The distorted percentage figure:* Percentages are reported without explanation of the original numbers used in the calculation (Adams and Schvaneveldt 359; Lavin 280): "Seventy-five percent of respondents prefer our brand over the competing brand"—without mention that only four people were surveyed. Or "Sixty-six percent of employees we hired this year are women and minorities, compared to the national average of 40 percent"—without mention that only three people have been hired this year, by a company that employs 300 (mostly white males).

Another fallacy in reporting percentages occurs when the *margin of error* is ignored. This is the margin within which the true figure lies, based on estimated sampling errors in a survey. For example, a claim that most people surveyed prefer brand X might be based on the fact that 51 percent of respondents expressed this preference; but if the survey carried a 2

"Is 51 percent really a majority?"

percent margin of error, the true figure could be as low as 49 percent or as high as 53 percent. In a survey with a high margin of error, the true figure may be so uncertain that no definite conclusion can be drawn.

■ *The bogus ranking:* Items are compared on the basis of ill-defined criteria (Adams and Schvaneveldt 212; Lavin 284): "Last year, the Batmobile was the number-one selling car in America"—without mention that some competing car makers actually sold *more* cars to private individuals, and that the Batmobile figures were inflated by hefty sales to rental-car companies and corporate fleets. Unless we know how the ranked items were chosen and how they were compared (the criteria), a ranking can produce a scientific-seeming number based on a completely unscientific method.

"Which car should we buy?"

■ *The fallible computer model:* Computer models process complex assumptions to produce impressive but often inaccurate statistical estimates about costs, benefits, risks, or probable outcomes.

Assumptions are notions we take for granted, things we often accept without proof. The research process rests on assumptions like these: that a sample group accurately represents a larger target group, that survey respondents remember certain facts accurately, that mice and humans share enough biological similarities for meaningful research. For a particular study to be valid, the underlying assumptions have to be accurate.

Computer models to predict global warming levels, for instance, are based on differing assumptions about wind and weather patterns, cloud formations, ozone levels, carbon dioxide concentrations, sea levels, or airborne sediment from volcanic eruptions. Despite their seemingly scientific precision, different global warming models generate fifty-year predictions of sea-level rises that range from a few inches to several feet (Barbour 121). Other models suggest that warming effects could be offset by evaporation of ocean water and by clouds reflecting sunlight back to outer space (Monatersky 69). Still other models suggest that the 1-degree F. warming over the last 100 years may not be the result of the greenhouse effect at all, but of "random fluctuations in global temperatures" (Stone 38). The estimates produced by any model depend on the assumptions (and data) programmed in.

"Garbage in, garbage out."

Choice of assumptions might be influenced by researcher bias or the sponsors' agenda. For example, a prediction of human fatalities from a nuclear plant meltdown might rest on assumptions about availability of safe shelter, evacuation routes, time of day, season, wind direction, and structural integrity of the containment unit. But the assumptions could be manipulated to produce an overstated or understated estimate of risk (Barbour 228). For computer-modeled estimates of accident risk (oil spill, plane crash) or of the costs and benefits of a proposed project or policy (a space station, welfare reform), consumers rarely know the assumptions behind the numbers. We wonder, for example, about the assumptions underlying NASA's pre-Challenger risk assessment, in which a 1985 computer model reportedly showed an accident risk of less than 1 in 100,000 shuttle flights (Crossen 54).

- *Confusion of correlation with causation: Correlation* is the measure of association between two variables (between smoking and increased lung cancer risk, or between education and income). *Causation* is the demonstrable production of a specific effect (smoking causes lung cancer). Correlations between smoking and lung cancer or education and income signal a causal relationship that has been proven by studies of all kinds. But not every correlation implies causation. For instance, a recently discovered correlation between moderate alcohol consumption and decreased heart disease risk offers insufficient proof that moderate drinking *causes* less heart disease.

"Does a beer a day keep the doctor away?"

Many highly publicized correlations are the product of "data dredging." In this process, computers randomly compare one set of variables (various eating habits) with another set (a range of diseases). From these countless comparisons, certain relationships are revealed (say, between coffee drinking and pancreatic cancer risk). As dramatic as such isolated correlations may be, they constitute no proof of causation and often lead to hasty conclusions (Ross 135).

- *Misleading terminology:* The terms used to interpret statistics sometimes hide their real meaning. For instance, the widely publicized figure that people treated for cancer have a "50 percent survival rate" is misleading in three ways: (1) *survival* to laypersons means "staying alive," but to medical experts, staying alive for only five years after diagnosis qualifies as survival; (2) the "50 percent" survival figure covers *all* cancers, including certain skin or thyroid cancers that have extremely high cure rates, as well as other cancers (such as lung or ovarian) that rarely are curable and have extremely low survival rates; (3) more than 55 percent of all cancers are skin cancers with a nearly 100 percent survival rate, thereby greatly inflating survival statistics for other types of cancer ("Are We" 6; *Facts and Figures* 2).

"Is this good news or bad news?"

Even the most valid and reliable statistics require that we interpret the reality behind the numbers. For instance, the overall cancer rate today is higher than it was in 1910. What this may mean is that people are living longer and thus are more likely to die of cancer and that cancer today rarely is misdiagnosed—or mislabeled because of stigma ("Are We" 4). The finding that rates for certain cancers double after prolonged exposure to electromagnetic waves may really mean that cancer risk actually increases from 1 in 10,000 to 2 in 10,000.

These are only a few examples of statistics and interpretations that seem highly persuasive but that in fact cannot always be trusted. Any interpretation of statistical data carries the possibility that other, more accurate interpretations have been overlooked or deliberately excluded (Barnett 45).

Reassessing the Entire Research Process

Chapters 8 and 9 show that the research process is a minefield of potential errors, in what we do and how we reason: We might ask the wrong questions; we might rely on the wrong sources; we might collect or record data incorrectly; we might analyze or document data incorrectly. We therefore need to critically examine our methods and our reasoning before reporting findings and conclusions. The following research checklist helps guide our assessment.

Checklist for the Research Process

Use this checklist to assess your research process. (Numbers in parentheses refer to the first page of discussion.)

METHOD

❑ Did I ask the right questions? (124)

❑ Are the sources appropriately up to date? (174)

❑ Is each source reputable, trustworthy, and relatively unbiased? (174)

❑ Does the evidence clearly support all the conclusions? (177)

❑ Can all the evidence be verified? (178)

❑ Is a fair balance of viewpoints presented? (126)

❑ Has the research achieved adequate depth? (127)

❑ Has the entire research process been valid and reliable? (161)

❑ Is all quoted material clearly marked throughout the text? (171)

❑ Are direct quotations used sparingly and appropriately? (172)

❑ Are all quotations accurate and integrated grammatically? (172)

❑ Are all paraphrases accurate and clear? (172)

❑ Have I documented all sources not considered common knowledge? (192)

❑ Is the documentation consistent, complete, and correct? (191)

REASONING

❑ Am I reasonably certain about the meaning of these findings? (178)

❑ Does my final answer seem definitive, only probable, or inconclusive? (178)

❑ Am I reasoning instead of rationalizing? (179)

❑ Is this the most reasonable conclusion (or merely the most convenient)? (126)

❑ Can I rule out other possible interpretations or conclusions? (180)

❑ Have I accounted for all sources of bias, including my own? (179)

❑ Are my generalizations warranted by the evidence? (180)

❑ Am I confident that my causal reasoning is correct? (181)

❑ Can all the numbers and statistics and interpretations be trusted? (182)

❑ Have I resolved (or at least acknowledged) any conflicts among my findings? (128)

❑ Should the evidence be reconsidered? (129)

☑ EXERCISES

1. Assume you are an assistant communications manager for a new organization that prepares research reports for decision makers worldwide. (A sample topic: "What is the expected long-term impact of the North American Free Trade Agreement on the U.S. computer industry?") These clients expect answers based on the best available evidence and reasoning.

   Although your recently hired coworkers are technical specialists, few have experience in the kind of wide-ranging research required by your clients. Training programs in the research process are being developed by your communications division but will not be ready for several weeks.

   Meanwhile, your boss directs you to prepare a one- or two-page memo that introduces employees to major procedural and reasoning errors that affect validity and reliability in the research process. Your boss wants this memo to be comprehensive but not vague.

2. Assume the scenario from Exercise 1. In a memo to colleagues, offer guidelines for avoiding unintentional plagiarism in quoting, paraphrasing, and citing the work of others. Explain what to document and how, using MLA style for a parenthetical reference and a works cited entry. (Illustrate with examples, but not those from the book!)

3. From print or broadcast media or from personal experience, identify an example of each of the following sources of distortion or of interpretive error:
   - a study with questionable sponsorship or motives
   - reliance on soft evidence
   - overestimating the level of certainty
   - biased interpretation
   - rationalizing
   - faulty causal reasoning
   - hasty generalization
   - overstated generalization
   - sanitized statistic
   - meaningless statistic
   - undefined average
   - distorted percentage figure
   - bogus ranking
   - fallible computer model
   - misinterpreted statistic

Submit to your instructor your examples along with a memo explaining each error, or be prepared to discuss your material in class.

☑ COLLABORATIVE PROJECTS

1. Exercises 1, 2, and 3 from the previous section are suited for collaborative work.

2. *Evaluating sources.* Figure 10.1 (pages 204–06) lists the final sources for the research project discussed early in Chapter 8. Review pages 124–29 and then turn to Figure 10.1 and evaluate the sources on the basis of these criteria: currency, range, balance, relative objectivity, and reputability. Here are more specific questions:
   - Should the sources generally have been more current? Why, or why not?
   - Which types of sources are represented here: business, science, trade, or general-interest publications; newspapers or news magazines; scholarly journals; government reports; or primary sources? Does the range of sources seem adequate (from general to specialized)? Explain.
   - How many different viewpoints are represented here (representatives of the company, consumer advocates, independent researchers, print journalists, the media, people in the industry, investors, others)? Do the sources represent a fair balance of views on this controversial issue? Explain.
   - Which sources seem most likely to be objective or impartial? Explain.
   - Which seem most likely to be biased? Explain.
   - Which seem most expert or authoritative? Explain.
   - Which seem most comprehensive? Explain.
   - Overall, do the sources seem adequate for the topic and situation described in Chapter 8? Explain.

Make the explanations brief but informative enough to justify your evaluation.

Appoint a group manager who will lead the discussion and assign the following tasks: taking notes, reporting the group's evaluation in a memo, editing and revising the memo, orally reporting the evaluation to the class.

CHAPTER

10

Documenting Research Findings

D OCUMENTING research findings means acknowledging one's debt to each information source. Proper documentation satisfies professional requirements for ethics, efficiency, and authority.

Why You Should Document

Documentation is a matter of *ethics* in that the originator of borrowed material deserves full credit and recognition. Moreover, all published material is protected by copyright law. Failure to credit a source could make you liable to legal action, even if your omission was unintentional.

Documentation also is a matter of *efficiency*. It provides a network for organizing and locating the world's recorded knowledge. If you cite a particular source correctly, your reference will enable interested readers to locate that source themselves.

Finally, documentation is a matter of *authority*. In making any claim—for example, "A Mercedes-Benz is more reliable than a Ford Taurus"—you invite challenge: "Says who?" Data on road tests, frequency of repairs, resale value, workmanship, and owner comments can help validate your claim by showing its basis in *fact*. A claim's credibility increases in relation to the expert references supporting it. For a controversial topic, you may need to cite several authorities who hold various views, as in this next example, instead of forcing a simplistic conclusion on your material:

Citing a balance of views

> Opinion is mixed as to whether a marketable quantity of oil rests beneath Georges Bank. Cape Code geologist John Blocke feels that extensive reserves are improbable ("Geologist Dampens Hopes" 3). Oil geologist Donald Marshall is uncertain about the existence of any oil in quantity at this location ("Offshore Oil Drilling" 2). But the U.S. Interior Department reports that the Atlantic continental shelf may contain 5.5 billion barrels of oil (Kemprecos 8).

Readers of your research report expect the *complete* picture.

What You Should Document

Document any insight, assertion, fact, finding, interpretation, judgment or other "appropriated material that readers might otherwise mistake for your own" (Gibaldi and Achtert 155)—whether the material appears in published form or not. Specifically, you must document:

Sources that require documentation

- any source from which you use exact wording
- any source from which you adapt material in your own words
- any visual illustration: charts, graphs, drawings, or the like (see Chapter 14 for documenting visuals)

In some instances, you might have reason to preserve the anonymity of unpublished sources: for example, to allow people to respond candidly without fear of reprisal (as with employee criticism of the company), or to protect their privacy (as with certain material from E-mail inquiries or electronic newsgroups). You still must document the fact that you are not the originator of this material by providing a general acknowledgment in the text ("A number of employees expressed frustration with . . .") along with a general citation in your list of references or works cited ("Interviews with Polex employees, May 1996").

You don't need to document anything considered *common knowledge:* material that appears repeatedly in general sources. In medicine, for instance, it is common knowledge that foods high in fat correlate with higher incidences of cancer, so in a report on fatty diets and cancer, you probably would not need to document that well-known fact. But you would document information about how the fat/cancer connection was discovered, subsequent studies (e.g., of the role of saturated versus unsaturated fats), and any information for which some other person could claim specific credit. If the borrowed material can be found in only one specific source, and not in multiple sources, document it. When in doubt, document the source.

How You Should Document

Borrowed material has to be cited twice: at the exact place you use that material, and at the end of your document. Documentation practices vary widely, but all systems work almost identically: a brief reference in the text names the source and refers readers to the complete citation, which enables the source to be retrieved.

Many disciplines, institutions, and organizations publish their own style guides or documentation manuals. Here are a few:

Style guides from various disciplines

Geographical Research and Writing
Style Manual for Engineering Authors and Editors
IBM Style Manual
NASA Publications Manual

This chapter illustrates citations and entries for three styles widely used for documenting sources in respective disciplines:

- Modern Language Association (MLA) style, for the humanities
- American Psychological Association (APA) style, for social sciences
- Council of Biology Editors (CBE) style, for natural and applied sciences

Unless your audience has a particular preference, any of these three styles can be adapted to most research writing. Another widely used format is that of the *Chicago Manual of Style,* 14th ed., which covers documentation in the humanities, related fields, and natural sciences. Whichever style you select, use it consistently throughout the document.

MLA Documentation Style

Traditional MLA documentation of sources used superscript numbers (like this:[1]) in the text, followed by full references at page bottom (footnotes) or at document's end (endnotes) and, finally, by a bibliography. A more current form of documentation appears in the *MLA Handbook for Writers of Research Papers,* 4th ed. New York: Modern Language Association, 1995. In the new MLA style, in-text parenthetical references briefly identify the sources. Full documentation then appears in a "Works Cited" section at document's end. Footnotes or endnotes now are used only to comment on material in the text.

A parenthetical reference usually includes the author's surname and the exact page number of the borrowed material:

Parenthetical reference in the text

> A recent study indicates an elevated risk of leukemia among children exposed to certain types of electromagnetic fields (Bowman et al. 59).

Readers seeking the complete citation for Bowman can move easily to "Works Cited," listed alphabetically by author:

Full citation at document's end

> Bowman, J. D., et al. "Hypothesis: The Risk of Childhood Leukemia Is Related to Combinations of Power-Frequency and Static Magnetic Fields." Bioelectromagnetics 16.1 (1995): 48–59.

This complete citation includes page numbers for the entire article.

MLA Parenthetical References

For clear and informative parenthetical references, observe these guidelines:

- If your discussion names the author, do not repeat the name in your parenthetical reference; simply give the page number:

Citing page numbers only

> Bowman et al. explain how their recent study indicates an elevated risk of leukemia for children exposed to certain types of electromagnetic fields (59).

- If you cite two or more works in a single parenthetical reference, separate the citations with semicolons:

Three works in a single reference

> (Jones 32; Leduc 41; Gomez 293–94)

- If you cite two or more authors with the same surname, include the first initial in your parenthetical reference to each author:

Two authors with identical surnames

> (R. Jones 32)
>
> (S. Jones 14–15)

- If you cite two or more works by the same author, include the first significant word from each work's title, or a shortened version:

Two works by one author

(Lamont, Biophysics 100–01)

(Lamont, Diagnostic Tests 81)

■ If the work is by an institutional or corporate author or if it is unsigned (that is, author unknown), use only the first few words of the institutional name or the work's title in your parenthetical reference:

Institutional, corporate, or anonymous author

(American Medical Assn. 2)

("Distribution Systems" 18)

To avoid distracting your readers, keep each parenthetical reference as brief as possible. One method is to name the source in your discussion, and to place only the page number in parentheses.

Where to place a parenthetical reference

For a paraphrase, place the parenthetical reference *before* the closing punctuation mark. For a quotation that runs into the text, place the reference *between* the final quotation mark and the closing punctuation mark. For a quotation set off (indented) from the text, place the reference two spaces *after* the closing punctuation mark.

MLA Works Cited Entries

The Works Cited list includes each source you have paraphrased or quoted. In preparing the list, which you should double space, type the first line of each entry flush with the left margin. Indent second and subsequent lines five spaces. Use one character space after any period, comma, or colon.

How to space and indent entries

Following are examples of complete citations as they would appear in the Works Cited section of your document. Shown italicized below each citation is its corresponding parenthetical reference as it would appear in the text. Note capitalization, abbreviations, spacing, and punctuation in sample entries.

MLA Works Cited Entries for Books. Any citation for a book should contain the following information (found on the book's title and copyright pages): author, title, editor or translator, edition, volume number, and facts about publication (city, publisher, date).

1. Book, Single Author—MLA

Kerzin-Fontana, Jane B. Technology Management: A Handbook. 3rd ed.

Delmar, NY: American Management Assn., 1997.

Parenthetical reference: (Kerzin-Fontana 3–4)

Identify the state of publication by U.S. Postal Service abbreviations. If the city of publication is well known (Boston, Chicago, and so on), omit the state abbreviation. If several cities are listed on the title page, give only the first. For Canada, include the province abbreviation after the city. For all other countries, include an abbreviation of the country name.

Index to Sample MLA Works Cited Entries

Books

1. Book, single author
2. Book, two or three authors
3. Book, four or more authors
4. Book, anonymous author
5. Multiple books, same author
6. Book, one or more editors
7. Book, indirect source
8. Anthology selection or book chapter

Periodicals

9. Article, magazine
10. Article, journal with new pagination each issue
11. Article, journal with continuous pagination
12. Article, newspaper

Other Sources

13. Encyclopedia, dictionary, alphabetic reference
14. Report
15. Conference presentation

16. Interview, personally conducted
17. Interview, published
18. Letter, unpublished
19. Questionnaire
20. Brochure or pamphlet
21. Lecture
22. Government document
23. Document with corporate authorship
24. Map or other visual
25. Dissertation or miscellaneous items

Electronic Sources

26. Online database source
27. Computer software
28. CD-ROM source
29. Internet posting
30. E-mail
31. Web source

2. Book, Two or Three Authors—MLA

Aronson, Linda, Roger Katz, and Candide Moustafa. Toxic Waste Disposal
Methods. New Haven: Yale UP, 1996.

Parenthetical reference: (Aronson, Katz, and Moustafa 121–23)

Shorten publisher's names, as in "Simon" for Simon & Schuster or "Yale UP" for Yale University Press. For page numbers having more than two digits, give only the final two digits for the second number.

3. Book, Four or More Authors—MLA

Santos, Ruth J., et al. Environmental Crises in Developing Countries. New
York: Harper, 1994.

Parenthetical reference: (Santos et al. 9)

"Et al." is the abbreviated form of the Latin "et alia," meaning "and others."

4. Book, Anonymous Author—MLA

Structured Programming. Boston: Meredith, 1995.

Parenthetical reference: (Structured 67)

5. Multiple Books, Same Author—MLA

Chang, John W. Biophysics. Boston: Little, 1997.

---. Diagnostic Techniques. New York: Radon, 1994.

Parenthetical references: (Chang, Biophysics 123–26) (Chang, Diagnostic 87)

When citing more than one work by the same author, do not repeat the author's name; simply type three hyphens followed by a period. List the works alphabetically by title.

6. Book, One or More Editors—MLA

Morris, A. J., and Louise B. Pardin-Walker, ed. Handbook of New

Information Technology. New York: Harper, 1996.

Parenthetical reference: (Morris and Pardin-Walker 34)

For more than three editors, name only the first, followed by "et al."

7. Book, Indirect Source—MLA

Kline, Thomas. Automated Systems. Boston: Rhodes, 1992.

Stubbs, John. White-Collar Productivity. Miami: Harris, 1996.

Parenthetical reference: (qtd. in Stubbs 116)

When your source (as in Stubbs, above) has quoted or cited another source, include each source in its appropriate alphabetical place in your Works Cited list. Use the name of the original source (here, Kline) in your text and begin the parenthetical reference with "qtd. in"—or "cited in" for a paraphrase.

8. Anthology Selection or Book Chapter—MLA

Bowman, Joel P. "Electronic Conferencing." Communication and

Technology: Today and Tomorrow. Ed. Al Williams. Denton, TX: Assn.

for Business Communication, 1994. 123–42.

Parenthetical reference: (Bowman 129)

Page numbers in the entry cover the selection cited from the anthology.

MLA Works Cited Entries for Periodicals. Give all available information in this order: author, article title, periodical title, volume and issue, date (day, month, year), and page numbers for the entire article—not just pages cited.

9. Article, Magazine—MLA

DesMarteau, Kathleen. "Study Links Sewing Machine Use to Alzheimer's

Disease." Bobbin Oct. 1994: 36–38.

Parenthetical reference: (DesMarteau 36)

No punctuation separates the magazine title and date. Nor is the abbreviation "p." or "pp." used to designate page numbers. If no author is given, list all other information:

> "Distribution Systems for the New Decade." Power Technology
>
> Magazine 18 Oct. 1996: 18+.

Parenthetical reference: ("Distribution Systems" 18)

This article began on page 18 and then continued on page 21. When an article does not appear on consecutive pages, give only the number of the first page, followed immediately by a plus sign. A three-letter abbreviation denotes any month spelled with five or more letters.

10. Article, Journal with New Pagination Each Issue—MLA

> Thackman-White, Joan R. "Computer-Assisted Research." American Library
>
> Journal 51.1 (1997): 3–9.

Parenthetical reference: (Thackman-White 4–5)

Because each issue for that year will have page numbers beginning with "1," readers need the number of this issue. The "51" denotes the volume number; the "1" denotes the issue number. Omit "The" or "A" or any other introductory article from a journal or magazine title.

11. Article, Journal with Continuous Pagination—MLA

> Barnstead, Marion H. "The Writing Crisis." Journal of Writing Theory
>
> 12 (1994): 415–33.

Parenthetical reference: (Barnstead 418)

When page numbers continue from issue to issue for the full year, readers won't need the issue number, because no other issue in that year repeats these same page numbers. (Include the issue number if you think it will help readers retrieve the article more easily.) The "12" denotes the volume number.

12. Article, Newspaper—MLA

> Baranski, Vida H. "Errors in Technology Assessment." Boston Times
>
> 15 Jan. 1997, evening ed., sec. B: 3.

Parenthetical reference: (Baranski 3)

When a daily newspaper has more than one edition, cite the specific edition after the date. Omit any introductory article in the newspaper's name (not *The Boston Times*). If no author is given, list all other information. If the newspaper's name does not contain the city of publication, insert it, using brackets: "*Sippican Sentinel* [Marion, MA]."

MLA Works Cited Entries for Other Sources. Miscellaneous sources range from unsigned encyclopedia entries to conference presentations to government publications. A full citation should give this information (as available): author, title, city, publisher, date, and page numbers.

13. Encyclopedia, Dictionary, Other Alphabetic Reference—MLA

"Communication." The Business Reference Book. 1993 ed.

Parenthetical reference: ("Communication")

Begin a signed entry with the author's name. For any work arranged alphabetically, omit page numbers in the citation and the parenthetical reference. For a well-known reference book, only an edition (if stated) and a date are needed. For other reference books, give the full publication information.

14. Report—MLA

Electrical Power Research Institute (EPRI). Epidemiologic Studies of Electric

Utility Employees. (Report No. RP2964.5). Palo Alto, CA: EPRI, Nov. 1994.

Parenthetical reference: (Electrical Power Research Institute [EPRI] 27)

If no author is given, begin with the organization that sponsored the report. For any report or other document with group authorship, as above, include the group's abbreviated name in your first parenthetical reference, and then use only that abbreviation in any subsequent reference.

15. Conference Presentation—MLA

Smith, Abelard A. "Radon Concentrations in Molded Concrete." First British

Symposium in Environmental Engineering. London, 11–13 Oct. 1995.

Ed. Anne Hodkins. London: Harrison, 1996. 106–21.

Parenthetical reference: (Smith 109)

The above example shows a presentation that has been included in the published proceedings of a conference. For an unpublished presentation, include the presenter's name, the title of the presentation, and the conference title, location, and date, but do not underline or italicize the conference information.

16. Interview, Personally Conducted—MLA

Nasser, Gamel. Chief Engineer for Northern Electric. Personal Interview.

Rangeley, ME. 2 Apr. 1996.

Parenthetical reference: (Nasser)

17. Interview, Published—MLA

Lescault, James. "The Future of Graphics." Executive Views of Automation.

Ed. Karen Prell. Miami: Haber, 1997. 216–31.

Parenthetical reference: (Lescault 218)

The interviewee's name is placed in the entry's author slot.

18. Letter, Unpublished—MLA

Rogers, Leonard. Letter to the author. 15 May 1993.

Parenthetical reference: (Rogers)

19. Questionnaire—MLA

Taylor, Lynne. Questionnaire sent to 612 Massachusetts business executives. 14 Feb. 1997.

Parenthetical reference: (Taylor)

20. Brochure or Pamphlet—MLA

Investment Strategies for the 21st Century. San Francisco: Blount Economics Assn., 1997.

Parenthetical reference: (Investment)

If the work is signed, begin with its author.

21. Lecture—MLA

Dumont, R. A. "Managing Natural Gas." Lecture. University of Massachusetts at Dartmouth, 15 Jan. 1996.

Parenthetical reference: (Dumont)

If the lecture title is not known, write Address, Lecture, or Reading but do not use quotation marks. Include the sponsor and the location if available.

22. Government Document—MLA

Virginia. Highway Dept. Standards for Bridge Maintenance. Richmond: Virginia Highway Dept., 1991.

Parenthetical reference: (Virginia Highway Dept. 49)

If the author is unknown (as above), list the information in this order: name of the government, name of the issuing agency, document title, place, publisher, and date. For any congressional document, identify the house of Congress (Senate or House of Representatives) before the title, and the number and session of Congress after the title:

United States Cong. House. Armed Services Committee. Funding for the Military Academies. 103rd Congress., 2nd. sess. Washington: GPO, 1995.

Parenthetical reference: (U.S. Cong. 41)

"GPO" is the abbreviation for the U.S. Government Printing Office.

For an entry from the Congressional Record, give only date and pages:

> Cong. Rec. 10 Mar. 1994: 2178–92.

Parenthetical reference: (Cong. Rec. 2184)

23. Document with Corporate Authorship—MLA

> Hermitage Foundation. Global Warming Scenarios for the Year 2030.
>
> Washington: Natl. Res. Council, 1996.

Parenthetical reference: (Hermitage Foun. 123)

24. Map or Other Visual—MLA

> Deaths Caused by Breast Cancer, by County. Map. Scientific American
>
> Oct. 1995: 32D.

Parenthetical reference: (Deaths Caused)

If the creator of the visual is listed, list that name first. Identify the type of visual ("Map," "Graph," "Table," "Diagram") immediately following its title.

25. Unpublished Dissertation, Report, or Miscellaneous Items—MLA

> Author (if known), title (in quotes), sponsoring organization or
>
> publisher, date, page numbers.

For any work that has group authorship (corporation, committee, task force), cite the name of the group or agency in place of the author's name.

MLA Works Cited Entries for Electronic Sources. In general, citation for an electronic source with a printed equivalent should begin with that publication information (see relevant sections above). But whether or not a printed equivalent exists, any citation should enable readers to retrieve the material electronically.

26. Online Database Source—MLA

> Sahl. J. D. "Power Lines, Viruses, and Childhood Leukemia." Cancer
>
> Causes Control 6.1 (Jan. 1995): 83. MEDLINE. Online. DIALOG.
>
> 7 Nov. 1995.

Parenthetical reference: (Sahl 83)

For entries with a printed equivalent, begin with complete publication information, then the database title (underlined), the "Online" designation to indicate the medium, the service provider, and date of access. The access date is important because frequent updatings of databases can produce different versions of the material.

For entries with no printed equivalent, give the title and date of the work in quotation marks, followed by the electronic source information:

> Argent, Roger R. "An Analysis of International Exchange Rates for 1995."
>
> Accu-Data. Online. Dow Jones News Retrieval. 10 Jan. 1996.

Parenthetical reference: (Argent 4)

If the author is not known, begin with the work's title.

27. Computer Software—MLA

> Virtual Collaboration. Diskette. New York: Harper, 1994.

Parenthetical reference: (Virtual)

Begin with the author's name, if known.

28. CD-ROM Source—MLA

> Cavanaugh, Herbert A. "EMF Study: Good News and Bad News."
>
> Electrical World Feb. 1995: 8. ABI/INFORM. CD-ROM. Proquest.
>
> Sept. 1995.

Parenthetical reference: (Cavanaugh 8)

If the material also is available in print, begin with complete publication information, followed by the name of the database (underlined), "CD-ROM" designation, vendor name, and electronic publication date. If the material has no printed equivalent, list its author (if known) and its title (in quotation marks), followed by the electronic source information.

For CD-ROM reference works and other material that is not routinely updated, give the work title followed by the "CD-ROM" designation, place, electronic publisher, and date:

> Time Almanac. CD-ROM. Washington: Compact, 1994.

Parenthetical reference. (Time Almanac 74)

Begin with the author's name, if known.

29. Internet Posting (Bulletin Board, Discussion List)—MLA

> Templeton, Brad. "10 Big Myths about Copyright Explained."
>
> 29 Nov. 1994. Online posting. Listserv law/copyright-FAQ/myths/part
>
> 1. BITNET. 6 May 1995.

Parenthetical reference: (Templeton)

Begin with the author's name (if known), followed by the title of the work (in quotation marks), publication date, the "Online posting" designation, name of discussion group, name of network, and date of your access. If appropriate, you can include the online address at the end of your entry, after the word "Available." The parenthetical reference includes no page number because none is given in an online posting.

30. E-mail—MLA

> Wallin, John Luther. "Frog Reveries." E-mail to author. 12 Oct. 1996.

Cite personal E-mail as you would printed correspondence. If the document has a subject line or title, enclose it in quotation marks. For publicly posted E-mail (for a newsgroup or discussion list) include the address and the date of access.

31. Web Source—MLA

> Dumont, R. A. "An Online Course in Technical Writing." 10 Dec. 1995.
>
> Online posting. http://www.umassd.edu/englishdepartment
>
> (6 Jan. 1996).

Parenthetical reference: (Dumont 7–9)

Begin with the author's name (if known), followed by title of the work (in quotation marks), the posting date, the "Online" designation, the Web address, and the date of access. In place of (or in addition to) the Web address, include the name of the Web site (underlined), if available:

> Rogers, S. E. "Chemical Risk Assessment Guidelines." 12 Feb. 1996. OTA
>
> Online. http://www.ota.gov (10 Mar. 1996).

"OTA" stands for Office of Technology Assessment.

MLA Sample Works Cited Page

Place your Works Cited section on a separate page at the document's end. Arrange entries alphabetically by author's surname. When the author is unknown, list the title alphabetically according to its first word (excluding introductory articles). For a title that begins with a digit ("5," "6," etc.), alphabetize the entry as if the digit were spelled out.

The list of works cited in Figure 10.1 accompanies the report on electromagnetic fields, pages 560–69. In the left margin, colored numbers refer to the elements discussed on the page facing Figure 10.1. Bracketed labels identify different types of sources the first time a particular type is cited.

ACW Documentation for Unconventional Electronic Sources[1]

Unconventional electronic sources include MUDs (multi-user dungeons), MOOs (MUD Object-Oriented software), IRC (Internet Relay Chat); FTPs (File Transfer Protocols), and others listed below. Conventions for documenting these sources continue to evolve. One useful system has been developed by Professor Janice R. Walker and endorsed by The Alliance for Computers and Writing (ACW).

---

1. Discussion adapted and examples reproduced from the style sheet prepared by Janice R. Walker and its version published in Hairston, Maxine, and John J. Ruskiewicz. *The Scott, Foresman Handbook for Writers.* 4th ed. New York: Harper, 1996: 671–75.

The ACW system observes MLA conventions wherever possible, but also provides formats for unique documentation often required by unconventional sources. For example, if the author is not named, your parenthetical reference might include instead the Internet or Web site and date:

(MediaMOO 10 Mar. 1996)

If page numbers are not given, a reference might include author and date:

(Walker Apr. 1995)

The full citation for each parenthetical reference appears alphabetically on a Works Cited page at the document's end. A typical entry contains the following information (as available or appropriate):

Author's Last Name, First Name. "Title of Work." Title of Complete Work.

[protocol (e.g., ftp, telnet, gopher) and address] [search path] (date of

message or visit).

Following are examples of specific entries as they would appear in the Works Cited section of your document—along with MLA entries for conventional print and electronic sources.

FTP Site—ACW. A File Transfer Protocol site enables files to be transferred between computers via phone lines.

Bruckman, Amy. "Approaches to Managing Deviant Behavior in Virtual

Communities." ftp.media.mit.edupub/asb/papers/deviance-chi94

(4 Dec. 1995).

WWW Site—ACW. A World Wide Web site is accessed via a web browser such as *Yahoo, Netscape, InfoSeek, WebCrawler,* or *Lynx.*

Burka, Lauren P. "A Hypertext History of Multi-User Dimensions."

Mud History. http://www.ccs.neu.edu/home/1pb/mud-history.html

(5 Dec. 1995).

Telnet Site—ACW. Telnet provides users direct access to files in other computers on the Internet.

Gomes, Lee. "Xerox's On-Line Neighborhood: A Great Place to

Visit." Mercury News 3 May 1992. telnet lambda.parc.xerox.com

8888, @go#50827, press13 (5 Dec. 1995).

Synchronous Communication (MOOs, MUDs IRC)—ACW. Synchronous communication happens in "real time": The message typed in by the sender appears instantly on the screen of the recipient, as in a personal interview.

Works Cited

1 Broad, William J. "Cancer Fear Is Unfounded, Physicists Say." New York Times 14 May
2 1995, sec. A: 19.   *[newspaper article]*
3 Brodeur, Paul, "Annals of Radiation: The Cancer at Slater School." New Yorker 7 Dec.
1992: 86+.   *[magazine article]*
4 Castleman, Michael. "Electromagnetic Fields." Sierra Jan./Feb. 1992: 21–22.
5 Cavanaugh, Herbert A. "EMF Study: Good News and Bad News." Electrical World
Feb. 1995: 8. ABI/INFORM. CD-ROM. Proquest. Sept. 1995.
                              [trade-magazine article from CD-ROM database]
6 Dana, Amy, and Tom Turner. "Currents of Controversy." Amicus Journal Summer
1993: 29–32.   *[alternative press]*
de Jager, L., and L., deBruyn. "Long-Term Effects of a 50 HZ Electric Field on the Life-
Expectancy of Mice." Review of Environmental Health 10.3 (1994): 221–24.
7 MEDLINE. Online. DIALOG. 8 Mar. 1996.   *[journal article from online database]*
Des Marteau, Kathleen. "Study Links Sewing Machine Use to Alzheimer's Disease."
8 Bobbin Oct. 1994: 36-38. ABI/INFORM. CD-ROM. Proquest. Aug. 1995.
"Electrophobia: Overcoming Fears of EMFs." University of California Wellness Letter
Nov. 1994: 1.   *[newsletter]*
Goodman, E. M., B. Greenebaum, and M. T. Marron. "Effects of Electromagnetic
Fields on Molecules and Cells." International Review of Cytology 158 (1995):
279–338. MEDLINE. Online. DIALOG. 8 Mar. 1996.
9 Halloran-Barney, Marianne B. Energy Service Advisor for County Electric. E-mail to
author. 3 Apr. 1996.   *[E-mail inquiry]*
10 Jauchem, J. "Alleged Health Effects of Electromagnetic Fields: Misconceptions in the
Scientific Literature." Journal of Microwave Power and Electromagnetic Energy
26.4 (1991): 189–95.   *[journal article from print source]*
Kirkpatrick, David. "Can Power Lines Give You Cancer?" Fortune 31 Dec. 1990: 80–85.
11 Lee, J. M., Jr., et al. Electrical and Biological Effects of Transmission Lines: A Review.
U.S. Dept. of Energy. NTIS no. PC Ao6/MF A01. Washington: GPO, 1989.

Figure 10.1   A List of Works Cited (MLA Style)

Discussion of Figure 10.1

1. Center the Works Cited title at the top of the page. Use one-inch margins. Double space the entries, and order them alphabetically. For numbering works cited pages, follow numbering of text pages.

2. Indent five spaces for the second and subsequent lines of an entry.

3. Place quotation marks around article titles. Underline or italicize periodical or book titles. Capitalize the first letter of key words in all titles (also articles, prepositions, and conjunctions only if they come first or last).

4. Do not cite a magazine's volume number, even if it is given.

5. For a CD-ROM database that is updated often (such as Proquest), conclude your citation with the date of electronic publication.

6. For additional perspective beyond "establishment" viewpoints, examine "alternative" publications (such as *The Amicus Journal* and *In These Times,* in this list).

7. Conclude an online database citation with the date you accessed the source.

8. Use a period and one space to separate a citation's three major items (author, title, publication data). Skip one space after a comma or colon. Use no punctuation to separate magazine title and date.

9. Alphabetize hyphenated surnames according to the name that appears first.

10. Include the issue number for a journal with new pagination in each issue. For page numbers of more than two digits, give only the final digits in the second number.

11. For government reports, name the sponsoring agency and include all available information for retrieving the document. Use the first author's name and "et al." for works with four or more authors or editors.

Maugh, Thomas H. "Studies Link EMF Exposure to Higher Risks of Alzheimer's." Los Angeles Times 31 July 1994, sec A: 3.

12   Mevissen, M., M. Keitzmann, and W. Loscher. "In Vivo Exposure of Rats to a Weak Alternating Magnetic Field Increases Ornithine Decarboxylase Activity in the Mammary Gland by a Similar Extent as the Carcinogen DMBA." Cancer Letter 90.2 (1995): 207–14. MEDLINE. Online. DIALOG. 8 Mar. 1996.

Miller, M. A., et al. "Variation in Cancer Risk Estimates for Exposure to Powerline Frequency Electromagnetic Fields: A Meta-analysis Comparing EMF Measurement Methods." Risk Analysis 15.2 (1995): 281–87. MEDLINE. Online. DIALOG. 8 Mar. 1996.

Miltane, John. Chief Engineer for County Electric. Personal Interview. Adams, MA. 5 Apr. 1996.   *[personal interview]*

Moore, Taylor. "EMF Health Risks: The Story in Brief." EPRI Journal Mar./Apr. 1995: 7–17.

13   Moulder, John. "Power Lines and Cancer" 6 Oct. 1995. Online posting. Newsgroup powerlines.cancer.FAQ: USENET. 10 Mar. 1996.   *[World Wide Web newsgroup]*

Palfreman, Jon. "Apocalypse Not." Technology Review 24 April 1996: 24–33.

Pinsky, Mark A. The EMF Book: What You Should Know about Electromagnetic Fields, Electromagnetic Radiation, and Your Health. New York: Warner, 1995.

[book—one author]

Reiter, R. J. "Melatonin Suppression by Static and Extremely Low Frequency Electromagnetic Fields: Relationship to the Reported Increased Incidence of Cancer." Review of Environmental Health 10.3 (1994):171–86. MEDLINE. Online. DIALOG. 8 Mar. 1996.

14   Schneider, David. "High Tension: Researchers Debate EMF Experiments on Cells." Scientific American Oct. 1995: 26+.

Taubes, Gary. "Fields of Fear." Atlantic Monthly Nov. 1994: 94–108. Online. U of Virginia Electronic Text Center. Internet. 15 Mar. 1996. Available WWW:http:/etext.

15   libvirginia.edu/english.html.   *[Internet source]*

United States Environmental Protection Agency. EMF in Your Environment. Washington:

16   GPO, 1992.

White, Peter. "Bad Vibes." In These Times 28 June 1993: 14–17.

Figure 10.1   A List of Works Cited (MLA Style)   *Continued*

Discussion of Figure 10.1 *Continued* ▬▬▬▬▬▬▬▬

12. Use three-letter abbreviations for months with five or more letters.

13. Because an online conference source such as a listserv or newsgroup provides no page numbers, you can eliminate the in-text parenthetical reference by referring directly to that source in your discussion ("Dr. Jones of Harvard points out that").

14. When an article skips pages in a publication, give only the first page number followed by a plus sign.

15. When the privacy of the electronic source is not an issue (e.g., a library versus an E-mail correspondent), consider including its electronic address in your entry, after the word *Available.*

16. Shorten publisher's names (as in "Simon" for Simon & Schuster; "Knopf" for Alfred A. Knopf, Inc.; "GPO" for Government Printing Office; or "Yale UP" for Yale University Press).

Begin the citation with the name of the communicator(s) and indicate the type of communication (e.g., personal interview).

> Pine-Guest. Personal Interview. telnet world.sensemedia.net 1234
>
> (12 Dec. 1995).

Gopher Site—ACW. Gopher is a software tool for locating and searching Internet databases and retrieving and downloading files.

> Quittner, Joshua. "Far Out: Welcome to Their World Built of MUD."
>
> Published in Newsday, 7 Nov. 1993. gopher/University of
>
> Koeln/About MUDs, MOOs and MUSEs in Education/Selected
>
> Papers/newsday (5 Dec. 1995).

Listserv and Newslist—ACW. In one of these sites messages are posted about a single topic. Indicate the subject line of the posting in quotes.

> Seabrook, Richard H.C. "Community and Progress." cybermind@jefferson.
>
> village.virginia.edu (22 Jan. 1994).

E-mail—ACW. Indicate the author and the subject line of the posting. For personal E-mail entries, the address usually is omitted.

> Walker, Janice. "Electronic Documentation." Personal e-mail (1 May 1995).

APA Documentation Style

One popular alternative to MLA style appears in the *Publication Manual of the American Psychological Association,* 4th ed. Washington: American Psychological Association, 1994. APA style is useful when writers wish to emphasize the publication dates of their references. A parenthetical reference in the text briefly identifies the source, date, and page number:

Reference cited in the text

> In a recent study, mice continuously exposed to an electromagnetic field tended
>
> to die earlier than mice in the control group (de Jager & de Brun, 1994, p. 224).

The full citation then appears in the alphabetic listing of "References," at the report's end:

Full citation at document's end

> de Jager, L., & de Brun, L. (1994). Long term effects of a 50 Hz electric field
>
> on the life-expectancy of mice. Review of Environmental Health, 10
>
> (3–4), 221–224.

APA style (or some similar author-date style) is preferred in the sciences and social sciences, where information quickly becomes outdated.

APA Parenthetical References

APA's parenthetical references differ from MLA's as follows: the citation includes the publication date; a comma separates each item in the reference; and "p." or "pp." precedes the page number (which is optional in the APA system). When a subsequent reference to a work follows closely after the initial reference, the date need not be included. Here are specific guidelines:

- If your discussion names the author, do not repeat the name in your parenthetical reference; simply give the date and page number:

Author named in the text

> Researchers de Jager and de Brun explain that experimental mice
>
> exposed to an electromagnetic field tended to die earlier than
>
> mice in the control group (1994, p. 224).

When two authors of a work are named in your text, their names are connected by "and," but in a parenthetical reference their names are connected by an ampersand, "&."

- If you cite two or more works in a single reference, list the authors in alphabetical order and separate the citations with semicolons:

Two or more works in a single reference

> (Jones, 1994; Gomez, 1992; Leduc, 1996)

- If you cite a work with three to five authors, try to name them in your text, to avoid an excessively long parenthetical reference:

A work with three to five authors

> Franks, Oblesky, Ryan, Jablar, and Perkins (1993) studied the role
>
> of electromagnetic fields in tumor formation.

In any subsequent references to this work, name only the first author, followed by "et al." (Latin abbreviation for "and others").

■ If you cite two or more works by the same author published in the same year, assign a different letter to each work:

Two or more works by the same author in the same year

(Lamont 1990a, p. 135)

(Lamont 1990b, pp. 67–68)

Other examples of parenthetical references appear with their corresponding entries in the following discussion of the list of references.

APA Reference List Entries

The APA reference list includes each source that you have cited in your document. In preparing the list, which you should double space, type the first line of each entry flush with the left margin. Indent the second and subsequent lines five spaces. Use one character space after any period, comma, or colon.

How to space and indent entries

Following are examples of complete citations as they would appear in the "References" section of your document. Shown immediately below each entry is its corresponding parenthetical reference as it would appear in the text. Note the capitalization, abbreviation, spacing, and punctuation in the sample entries.

Index to Sample Entries for APA References

Books

1. Book, single author
2. Book, two to five authors
3. Book, six or more authors
4. Book, anonymous author
5. Multiple books, same author
6. Book, one or more editors
7. Book, indirect source
8. Anthology selection or book chapter

Periodicals

9. Article, magazine
10. Article, journal with new pagination each issue
11. Article, journal with continuous pagination
12. Article, newspaper

Other Sources

13. Encyclopedia, dictionary, alphabetic reference
14. Report

15. Conference presentation
16. Interview, personally conducted
17. Interview, published
18. Personal correspondence
19. Brochure or pamphlet
20. Lecture
21. Government document
22. Miscellaneous items

Electronic Sources

23. Online database abstract
24. Online database article
25. Computer software or manual
26. CD-ROM abstract
27. CD-ROM reference work
28. Electronic bulletin boards, discussion lists, E-mail

APA Entries for Books. Any citation for a book should contain all applicable information in the following order: author, date, title, editor or translator, edition, volume number, and facts about publication (city and publisher).

1. Book, Single Author—APA

Kerzin-Fontana, J. B. (1997). Technology management: A handbook (3rd

ed.). Delmar, NY: American Management Association.

Parenthetical reference: (Kerzin-Fontana, 1997, pp. 3–4)

Use only initials for an author's first and middle name. Capitalize only the first words of a book's title and subtitle and any proper names. Identify a later edition in parentheses.

2. Book, Two to Five Authors—APA

Aronson, L., Katz, R., & Moustafa, C. (1996). Toxic waste disposal methods.

New Haven: Yale University Press.

Parenthetical reference: (Aronson, Katz, & Moustafa, 1993)

Use an ampersand (&) before the name of the final author listed in an entry. As an alternative parenthetical reference, name the authors in your text and include date (and page numbers, if appropriate) in parentheses.

3. Book, Six or More Authors—APA

Fogle, S. T., et al. (1995). Hyperspace technology. Boston: Little, Brown.

Parenthetical reference: (Fogle, et al., 1995, p. 34)

"Et al." is the Latin abbreviation for "et alia," meaning "and others."

4. Book, Anonymous Author—APA

Structured programming. (1995). Boston: Meredith Press.

Parenthetical reference: (Structured Programming, 1995, p. 67)

In your list of references, place an anonymous work alphabetically by the first key word (not *The, A,* or *An*) in its title. In your parenthetical reference, capitalize all key words in a book, article, or journal title. But in your list of references, capitalize only journal titles in this way.

5. Multiple Books, Same Author—APA

Chang, J. W. (1997a). Biophysics. Boston: Little, Brown.

Chang, J. W. (1997b). MindQuest. Chicago: John Pressler.

Parenthetical references: (Chang, 1997a) (Chang, 1997b)

Two or more works by the same author not published in the same year are distinguished by their respective dates alone, without the added letter.

6. Book, One or More Editors—APA

Morris, A. J., & Pardin-Walker, L. B. (Eds.). (1996). Handbook of new
information technology. New York: HarperCollins.

Parenthetical reference: (Morris & Pardin-Walker, 1996, p. 79)

For more than five editors, name only the first, followed by "et al."

7. Book, Indirect Source

Stubbs, J. (1996). White-collar productivity. Miami: Harris.

Parenthetical reference: (cited in Stubbs, 1996, p. 47)

When your source (as in Stubbs, above) has cited another source, list your
source in the References section, but name the original source in your text:
"Kline's study (cited in Stubbs, 1996, p. 47) supports this conclusion."

8. Anthology Selection or Book Chapter—APA

Bowman, J. (1994). Electronic conferencing. In A. Williams (Ed.),
Communication and technology: Today and tomorrow. (pp.
123–142). Denton, TX: Association for Business Communication.

Parenthetical reference: (Bowman, 1994, p. 126)

The page numbers in the complete reference are for the selection cited from
the anthology.

APA Entries for Periodicals. A citation for an article should give this information (as available), in order: author, publication date, article title (without
quotation marks), volume or number (or both), and page numbers for the
entire article—not just the page cited.

9. Article, Magazine—APA

DesMarteau, K. (1994, October). Study links sewing machine use to
Alzheimer's disease. Bobbin, 36, 36–38.

Parenthetical reference: (DesMarteau, 1994, p. 36)

If no author is given, provide all other information. Capitalize the first word
in an article's title and subtitle, and any proper nouns. Capitalize all key
words in a periodical title. Underline or italicize the periodical title, volume
number, and commas (as above).

10. Article, Journal with New Pagination for Each Issue—APA

Thackman-White, J. R. (1997). Computer-assisted research. American
Library Journal, 51(1), 3–9.

Parenthetical reference: (Thackman-White, 1997, 4–5)

Because each issue for a given year has page numbers that begin at "1," readers need the issue number ("1"). The "<u>51</u>" denotes the volume number, which is underlined or italicized.

11. Article, Journal with Continuous Pagination—APA

Barnstead, M. H. (1994). The writing crisis. <u>Journal of Writing Theory</u>

<u>12</u>, 415–433.

Parenthetical reference: (Barnstead, 1994, pp. 415–416)

The "<u>12</u>" denotes the volume number. When page numbers continue from issue to issue for the full year, readers won't need the issue number, because no other issue in that year repeats these same page numbers. (You can include the issue number if you think it will help readers retrieve the article more easily.)

12. Article, Newspaper—APA

Baranski, V. H. (1997, January 15). Errors in technology assessment. <u>The</u>

<u>Boston Times</u>, p. B3.

Parenthetical reference: (Baranski, 1997, p. B3)

In addition to the year of publication, include the month and day. If the newspaper's name begins with "The," include it in your citation. Include "p." or "pp." before page numbers. For an article on nonconsecutive pages, list each page, separated by a comma.

APA Entries for Other Sources. Miscellaneous sources range from unsigned encyclopedia entries to conference presentations to government documents. A full citation should give this information (as available): author, publication date, work title (and report or series number), page numbers (if applicable), city, and publisher.

13. Encyclopedia, Dictionary, Alphabetic Reference—APA

Communication. (1993). In <u>The business reference book</u>. Boston: Business

Resources Press.

Parenthetical reference: ("Communication," 1993)

For a signed entry, begin with the author's name and publication date.

14. Report—APA

Electrical Power Research Institute. (1994). <u>Epidemiologic studies of electric</u>

<u>utility employees</u> (Report No. RP2964.5). Palo Alto, CA: Author.

Parenthetical reference: (Electrical Power Research Institute [EPRI], 1994, p. 12)

If authors are named, list them first, followed by the publication date. When citing a group author, as above, include the group's abbreviated name in your first parenthetical reference, and use only that abbreviation in any subsequent reference. When the agency (or organization) and publisher are the same, list "Author" in the publisher's slot.

15. Conference Presentation—APA

> Smith, A. A. (1996). Radon concentrations in molded concrete. In A.
>
> Hodkins (Ed.), First British Symposium on Environmental Engineering
>
> (pp. 106–121). London: Harrison Press.

Parenthetical reference: (Smith, 1995, p. 109)

The example shows a presentation included in the published proceedings of a conference. The name of the symposium is a proper name, and so is capitalized. For an unpublished presentation, include the presenter's name, year and month, title of the presentation (underlined or italicized), and all available information about the conference or meeting: "Symposium held at . . ." Do not underline or italicize this last information.

16. Interview, Personally Conducted—APA

Parenthetical reference: (G. Nasser, personal interview, April 2, 1996)

This material is considered a nonrecoverable source, and so is cited in the text only, as a parenthetical reference. If you name the interviewee in your text, do not repeat the name in your parenthetical reference.

17. Interview, Published—APA

> Jable, C. K. (1997). The future of graphics [Interview with James Lescault].
>
> In K. Prell (Ed.), Executive Views of Automation (pp. 216–231). Miami:
>
> Haber Press.

Parenthetical reference: (Jable, 1997, pp. 218–223)

Begin with the name of the interviewer, followed by the publication date, title, the designation (in brackets), and the publication information.

18. Personal Correspondence—APA

Parenthetical reference: (L. Rogers, personal correspondence, May 15, 1993)

This material is considered nonrecoverable data, and so is cited in the text only, as a parenthetical reference. If you name the correspondent in your text, do not repeat the name in your citation.

19. Brochure or Pamphlet—APA

This material follows the citation format for a book entry. After the title of the work, include the designation "Brochure" in brackets.

20. Lecture—APA

> Dumont, R. A. (1996, January 15). Managing natural gas. Lecture
>
> presented at the University of Massachusetts at Dartmouth.

Parenthetical reference: (Dumont, 1996)

If you name the lecturer in your text, do not repeat the name in your citation.

21. Government Document—APA

> Virginia Highway Department. (1991). Standards for bridge maintenance.
>
> Richmond: Author.

Parenthetical reference: (Virginia Highway Department, 1991, p. 49)

If the author is unknown, present the information in this order: name of the issuing agency, publication date, document title, place, and publisher. When the issuing agency is both author and publisher, list "Author" in the publisher's slot.

For any congressional document, identify the house of Congress (Senate or House of Representatives) and the issuing body.

> U.S. House Armed Services Committee. (1995). Funding for the military
>
> academies. Washington, DC: U.S. Government Printing Office.

Parenthetical reference: (U.S. House, 1995, p. 41)

22. Miscellaneous items (unpublished manuscripts, dissertations, and so on)—APA

> Author (if known), date of publication, title of work, sponsoring
>
> organization or publisher, page numbers.

For any work that has group authorship (corporation, committee, and so on), cite the name of the group or agency in place of the author's name.

APA Entries for Electronic Sources. APA documentation standards for electronic sources continue to be refined and defined. A sampling of currently preferred formats is presented below. Any citation for electronic media should enable readers to identify the original source (printed or electronic) and provide an electronic path for retrieving the material.

Begin with the publication information for the printed equivalent. Then in brackets name the electronic source ([On-line], [CD-ROM], [Computer software]), the protocol[2] (Bitnet, Dialog, FTP, Telnet), and any other items that define a clear path (service provider, database title, access code, retrieval number, or site address).

23. Online Database Abstract—APA

> Sahl, J. D. (1995). Power lines, viruses, and childhood leukemia [On-line].
>
> Cancer Causes Control, 6 (1), 83. Abstract from: DIALOG File:
>
> MEDLINE Item: 93–04881.

Parenthetical reference: (Sahl, 1995)

---

2. A protocol is a body of standards that ensures compatibility among the different products designed to work together on a particular network.

Note the absence of closing punctuation in the path statement. Any punctuation added to the path could interfere with retrieval.

24. Online Database Article—APA

Alley, R. A. (1995, January). Ergonomic influences on worker satisfaction

[29 paragraphs]. Industrial Psychology [On-line serial], 5(11). Available

FTP: Hostname: publisher.com Directory: pub/journals/industrial.

psychology/1995

Parenthetical reference: (Alley, 1995)

Give the length of the article [in paragraphs], after its title. Add no terminal punctuation to the availability statement.

25. Computer Software or Software Manual—APA

Virtual collaboration [Computer software]. (1994). New York: HarperCollins.

Parenthetical reference: (Virtual, 1994)

For citing a manual, replace the "Computer software" designation in brackets with "Software manual."

26. CD-ROM Abstract—APA

Cavanaugh, H. (1995). An EMF study: Good news and bad news [CD-ROM].

Electrical World, 209(2), 8. Abstract from: Proquest File: ABI/Inform

Item: 978032

Parenthetical reference: (Cavanaugh, 1995)

The "8" in the above entry denotes the page number of this one-page article.

27. CD-ROM Reference Work—APA

Time almanac. (1994). Washington: Compact, 1994.

Parenthetical reference: (Time almanac, 1994)

If the work on CD-ROM has a printed equivalent, APA currently prefers that it be cited in its printed form. As more works appear in electronic form, this convention may be revised.

28. Electronic Bulletin Boards, Discussion Lists, E-mail—APA

Parenthetical reference: Fred Flynn (personal communication, May 10, 1996)

provided these statistics.

This material is considered personal communication in APA style. Instead of being included in the list of references, it is cited directly in the text. According to APA's current standards, material from discussion lists and electronic bulletin boards has limited research value because it does not undergo the kind of review and verification process used in scholarly publications.

APA Sample List of References

APA's References section is an alphabetic listing (by author) equivalent to MLA's Works Cited section. Like Works Cited, the reference list includes only those works actually cited. (A bibliography usually would include background works or works consulted as well.) In one notable difference from MLA style, APA style calls for only "recoverable" sources to appear in the reference list. Therefore, personal interviews, E-mail messages, and other unpublished materials are cited in the text only.

The list of references in Figure 10.2 accompanies the report on technical marketing, pages 572–81. In the left margin, colored numbers denote elements discussed on the page facing Figure 10.2. Bracketed labels on the right identify different types of sources the first time a particular type is cited.

CBE Numerical Documentation Style

In the numerical system of documentation preferred by the Council of Biology Editors, each work is assigned a number the first time it is cited. This same number then is used for any subsequent reference to that work. Numerical documentation often is used in the physical sciences (astronomy, chemistry, geology, physics) and the applied sciences (mathematics, medicine, engineering, computer science).

Preferred documentation styles for particular disciplines are defined in style manuals such as these:

- American Chemical Society. *The ACS Style Guide for Authors and Editors* (1985).
- American Institute of Physics. *AIP Style Manual* (1990).
- American Mathematical Society. *A Manual for Authors of Mathematical Papers* (1990).
- American Medical Association. *Manual of Style* (1989)

One widely consulted guide for numerical documentation is *Scientific Style and Format: The CBE Manual for Authors, Editors, and Publishers*. 6th ed., 1994, from the Council of Biology Editors.

CBE Numbered Citations

In one version of CBE style, a citation in the text appears as a raised number immediately following the source to which it refers:

Numbered citations in the text

> A recent study[1] indicates an elevated leukemia risk among children exposed to certain types of electromagnetic fields. Related studies[2-3] tend to confirm the EMF/cancer hypothesis.

When referring to two or more sources in a single note (as in "[2-3]" above) separate the numbers by a hyphen if they are in sequence and by commas but no space if they are out of sequence: ("[2,6,9]").

The full citation for each source then appears in the numerical listing of references at the document's end:

References

Full citations at document's end

1. Bowman JD, et al. Hypothesis: the risk of childhood leukemia is related to combinations of power-frequency and static magnetic fields. Bioelectromagnetics 1995; 16(1): 48–59.

2. Feychting M, Ahlbom A. Electromagnetic fields and childhood cancer: meta-analysis. Cancer Causes Control 1995 May; 6(3): 275–277.

To refer again to any of these sources later in your document, use the same number.

CBE Reference List Entries

CBE's References section lists each source in the numerical order in which it was first cited. In preparing the list, which should be double spaced, begin each entry on a new line. Type the number flush with the left margin, followed by a period and a space. Align subsequent lines directly under the first word of line one.

Following are examples of complete citations as they would appear in the References section for your document.

CBE Entries for Books. Any citation for a book should contain all available information in the following order: number assigned to the entry, author or editor, work title (and edition), facts about publication (place, publisher, date), and number of pages. Note the capitalization, abbreviation, spacing, and punctuation in the sample entries.

1. Book, Single Author—CBE

1. Kerzin-Fontana JB. Technology management: a handbook. 3rd ed. Delmar, NY: American Management Assn.; 1997. 356p.

Index to Sample CBE Entries

1   **References**

2   Alderman, L. (1995, July). How you can take control of your own career.
        Money, 24(7), 38–40.   *[magazine article]*

3   Basta, N. (1988, September). Take a good look at sales engineering. Graduating
        Engineer, 32, 84–87.

4   Baxter, N. (1994). Is there another degree in your future? Washington, DC: U.S.
        Department of Labor.   *[govt. publication—author named]*

5   Campbell, M. K. (1993). Wanted: Sales reps with EE degrees. IEEE Potentials, 31(1),
        28–29.   *[journal article]*

6   College Placement Council. (1995). CPC annual (39th ed.). Bethlehem, PA: Author.
                                                      [book—author as publisher]

7   Cornelius, H., & Lewis, W. Career guide for sales and marketing (2nd ed.). New York:
        Monarch Press.   *[book with two authors]*

8   Electronic sales positions. (1996). The national job bank. Holbrook, MA: Bob Adams,
        Inc.   *[directory entry—no author named]*

        Engineering careers. (1995). The encyclopedia of careers and vocational guidance
        (9th ed.). Chicago: J. G. Ferguson.   *[encyclopedia]*

        Gradler, C., & Schrammel, K. (1994). The 1992–2005 job outlook in brief.
        Washington, DC: U.S. Department of Labor.

        Resnick, R. R. (1995, June). Business is good, NOT. Internet World, 6(6), 71–73.

        Schranke, R. W. (1985). EE and MBA: A winning combination? IEEE Potentials, 28 (1),
        13–15.

9   Solomon, S. D. (1996, January). An engineer goes to Wall Street [10 pages].
        Technology Review [On-line serial], 99(1). Available WWW:
        http://web.mit.edu/techreview/www/   *[online article]*

10   Tolland, M. (1996, April). Alternate careers in marketing. Presentation at Electro '96,
        Conference in Boston.   *[unpublished conf. presentation]*

        U.S. Department of Labor. (1994) Tomorrow's jobs. Washington, DC: Author.
                                                [govt. publication—no author named]

        Young, J. (1995, August). Can computers really boost sales? Forbes ASAP,
        84–101.   *[magazine article—no vol. or issue number]*

Figure 10.2   A List of References (APA Style)

Discussion of Figure 10.2

1. Center the References title at the top of the page. Use one-inch margins. For numbering reference pages, follow numbering of text pages. Include only recoverable data (material that readers could retrieve for themselves); cite personal interviews, unpublished lectures, electronic discussion lists, and E-mail and other personal correspondence parenthetically in the text only. See also item 10 in this list.

2. Double space entries and order them alphabetically by author's last name (excluding *A, An,* or *The*). List initials only for authors' first and middle names. Write out names of all months. In student papers, indent the second and subsequent lines of an entry five spaces. In papers submitted for publication in an APA journal, the *first* line instead is indented.

3. Do not enclose article titles in quotation marks. Underline or italicize periodical titles.

4. Capitalize the first word in article or book titles and subtitles, and any proper nouns. Capitalize all key words in magazine or journal titles.

5. Use italics or a continuous underline for a journal article's title, volume number, and the comma. Give the issue number in parentheses only if each issue begins on page 1. Do not include "p." or "pp." before journal page numbers (only before page numbers from a newspaper).

6. Identify the edition of a book in parentheses. If the author is also the publisher, use the word "Author" after the place of publication. Otherwise, write out the publisher's name in full.

7. For more than one author or editor, use ampersands instead of spelling out "and."

8. Use the first key word in the title to alphabetize works whose author is not named.

9. Omit punctuation from the end of an electronic address.

10. Treat an unpublished conference presentation as a recoverable source; include it in your list of references instead of merely citing it parenthetically in your text.

2. Book, Multiple Authors—CBE

2. Aronson L, Katz R, Moustafa C. Toxic waste disposal methods. New Haven: Yale Univ. Pr.; 1996. 316p.

3. Book, Anonymous Author—CBE

3. [Anonymous]. Structured programming. Boston: Meredith Pr.; 1995. 267p.

4. Book, One or More Editors—CBE

4. Morris AJ, Pardin-Walker LB, editors. Handbook of new information technology. New York: Harper; 1996. 345p.

5. Anthology Selection or Book Chapter—CBE

5. Bowman JP. Electronic conferencing. In: Williams A, editor. Communication and technology: today and tomorrow. Denton, TX: Assn. for Business Communication; 1994. p. 123–42.

CBE Entries for Periodicals. Any citation for an article should contain all available information in the following order: number assigned to the entry, author, article title, periodical title, date (year, month), volume and issue number, and inclusive page numbers for the article. Note the capitalization, abbreviation, spacing, and punctuation in the sample entries.

6. Article, Magazine—CBE

6. DesMarteau K. Study links sewing machine use to Alzheimer's disease. Bobbin 1994 Oct: 36–38.

7. Article, Journal with New Pagination Each Issue—CBE

7. Thackman-White JR. Computer-assisted research. American Library Jour 1997; 51(1): 3–9.

8. Article, Journal with Continuous Pagination—CBE

8. Barnstead MH. The writing crisis. Jour of Writing Theory 1994; 12: 415–433.

9. Article, Newspaper—CBE

9. Baranski VH. Errors in technology assessment. Boston Times 1997 Jan 15; Sect B: 33 (col 2).

10. Article, Online Source—CBE

10. Alley RA. Ergonomic influences on worker satisfaction. Industrial Psychology [serial online] 1995 Jan; 5(11). Available from: ftp. pub/journals/industrial psychology/1995 via the INTERNET. Accessed 1996 Feb 10.

For more detailed guidelines on CBE style, consult the CBE manual.

☑ EXERCISE

Locate the style manual for your discipline. (Ask faculty in your major or a librarian). Redesign Figure 10.1 according to the guidelines in this manual. Submit your document along with a memo outlining main differences in the two documentation styles. If your discipline stipulates no particular style, use the *APA Manual* for this assignment.

Summarizing Information

A SUMMARY is a short version of a longer document. An economical way to communicate, a summary saves time, space, and energy.

Purpose of Summaries

Chapter 8 shows how abstracts (a type of summary) aid our research by providing an encapsulated glimpse of an article or other long document. As we record our research findings, we summarize and paraphrase to capture the main ideas in a compressed form. In addition to this dual role as a research aid, summarized information is vital in day-to-day workplace transactions.

On the job, you have to write concisely about your work. You might report on meetings or conferences, describe your progress on a project, or propose a money-saving idea. A routine assignment for many new employees is to provide superiors (decision makers) with summaries of the latest developments in their field.

Given today's pace and volume of information, summaries are more vital than ever. Some reports and proposals can be hundreds of pages long. Those who must act on this information need to rapidly identify what is most important in a document. From a good summary, busy readers can get enough information to decide whether they should read the entire document, parts of it, or none of it.

Whether you summarize someone else's document or your own, your job is to communicate the *essential message* accurately and in the fewest words. The essential message in any well-written document is easy enough to identify, as in the following passage:

The original passage

> The lack of technical knowledge among owners of television sets leads to their suspicion about the honesty of television repair technicians. Although television owners might be fairly knowledgeable about most repairs made to their automobiles, they rarely understand the nature and extent of specialized electronic repairs. For instance, the function and importance of an automatic transmission in an automobile are generally well known; however, the average television owner knows nothing about the flyback transformer in a television set. The repair charge for a flyback transformer failure is roughly $150—a large amount to a consumer who lacks even a simple understanding of what the repairs accomplished. In contrast, a $450 repair charge for the transmission on the family car, though distressing, is more readily understood and accepted.

Three significant ideas comprise the essential message: (1) television owners lack technical knowledge and are suspicious of repair technicians; (2) an owner usually understands even the most expensive automobile repairs; and (3) owners do not understand or accept expenses for television repairs. A possible summary might read like this:

A summarized version

> Because television owners lack technical knowledge about their sets, they often are suspicious of repair technicians. Although consumers may understand expensive automobile repairs, they rarely understand or accept repair and parts expenses for their television sets.

This summary is almost 30 percent of the original length because the original itself is short. With a longer original, a summary might be 5 percent or less. But length is less important than informative value: an effective summary gives readers just what they need and no more. For letters, memos, or other short documents that can be read quickly, the only summary needed is usually an opening thesis or topic sentence that previews the contents.

Summaries are vital whenever people have no time to read in detail everything that crosses their desks. A recent U.S. president reportedly required all significant world news for the last twenty-four hours to be condensed into one typed page and placed on his desk, first thing each morning. Another president employed a writer who summarized articles from more than two dozen major magazines.

Elements of a Summary

All effective summaries display the elements discussed below.

- *The essential message:* The essential message is the significant material from the original: controlling ideas (thesis and topic sentences); major findings; important names, dates, statistics, and measurements; and conclusions or recommendations. Significant material does not include background; the author's personal comments or conjectures; introductions; long explanations, examples, or definitions; visuals; or data of questionable accuracy. (For these distinctions, see pages 225–28.)

- *Nontechnical style:* More people generally read the summary than any other part of a document. Write at the lowest level of technicality. Translate technical data into plain English. "The patient's serum glucose measured 240 mg%" can be translated: "The patient's blood sugar remained critically high." Of course, if you know all your readers are experts, you won't need to simplify.

- *Independent meaning:* In meaning as well as style, your summary should stand alone as a self-contained message. Readers should have to read the original only for more detail—not to make sense of the basic ideas.

- *No personal assessment:* Avoid personal comments ("This interesting report" or "The author is correct in assuming"). Add nothing to the original except for a brief clarifying definition, if needed.

- *Conciseness:* Conciseness is vital, but never at the expense of clarity and accuracy. Make the summary short enough to be economical, but long enough to be clear and comprehensive.

Critical Thinking in the Summary Process

Follow these guidelines for summarizing your own writing or another's.

1. *Read the entire original.* When summarizing another's work, get a complete picture before writing a word.
2. *Reread and underline.* Reread the original, underlining essential material. Focus on the thesis and topic sentences.
3. *Edit the underlined data.* Reread the underlined material and cross out whatever does not advance the meaning.
4. *Rewrite in your own words.* Include all essential material in the first draft, even if it's too long; you can trim later.
5. *Edit your own version.* When you have everything readers need, edit for conciseness.
   a. Cross out all needless words without harming clarity or grammar. Use complete sentences.

The summer internship in journalism gives the ~~journalism~~ student ~~first-hand~~ experience ~~at what goes~~ on ~~within~~ a ~~real~~ newspaper.

   b. Cross out needless prefaces such as "The writer argues" or "Also discussed is."
   c. Use numerals for numbers, except to begin a sentence.
   d. Combine related ideas in order to emphasize relationships (pages 279–81).
6. *Check your version against the original.* Verify that you have preserved the essential message and added no comments.
7. *Rewrite your edited version.* Add transitional expressions to reinforce the connection between related ideas.
8. *Document your source.* If summarizing another's work, cite the source immediately below the summary, and place directly quoted statements within quotation marks. (See Chapter 10 for documentation formats.)

Although the summary is written last by the writer, it is read *first* by the reader. Take the time to do a good job.

A Sample Situation

Imagine that you work in the information office of your state's Department of Environmental Management (DEM). In the coming election, citizens will vote on a referendum proposal for constructing municipal trash incinerators. Referendum supporters argue that incinerators would help solve the growing problem of waste disposal in highly populated parts of the state. Opponents argue that incinerators cause air pollution.

To clarify the issues for voters, the DEM is preparing a newsletter to be mailed to each registered voter. You have been assigned the task of researching the recent data and summarizing them for newsletter readers. Here is one of the articles marked and then summarized to show the critical thinking process outlined above.

INCINERATING TRASH: A HOT ISSUE GETTING HOTTER

Combine as orienting sentence (controlling idea)

Alarmed by the tendency of landfills to contaminate the environment, both public officials and citizens are vocally seeking alternatives. The most commonly discussed alternative is something called a *resource recovery facility*. Nearly 100 U.S. cities have built such in the last 15 years, and another 150 or so are in various stages of planning.

Include definition

These recovery facilities are a new form of an old technology. Basically, they're incinerators. But, unlike the incinerators of old, they don't just burn waste. They also recover energy. The energy is sold as steam to an industrial customer, or it is converted to electricity and sold to the local utility. (A few facilities, not many, also recover metals or other materials before using the waste as fuel.)

Include major fact

Include major statistic

A ton of trash possesses the energy content of a barrel and a half of oil. This is not a trivial amount. The United States discards 150 million tons of municipal refuse a year. If all of it were converted to energy, we could replace the equivalent of 12 percent of our oil imports.

Include major fact

At the local level, selling energy or materials not only replaces nonrenewable resources; it also provides a source of income that partly offsets the cost of operating the facility.

Include major fact

Delete explanation

The new facilities are, on average, much cleaner than the municipal incinerators of old. Many have two-stage combustion units, in which the second-stage burns exhaust gases at high temperature, converting many potential organic pollutants to less harmful emissions such as carbon dioxide. Some, especially the larger and newer facilities, also come equipped with the latest in pollution control devices.

Delete questionable point

The environmental community is uneasy with this new technology. Environmentalists have argued for many years that the best method of handling municipal trash is to recycle it—i.e., to separate the glass, metal, paper, and other materials and use them again, either without reprocessing or as raw materials in producing new products. The thought of the potential resources in municipal solid waste simply being burned, even with energy recovery, has made many environmentalists opponents of resource recovery.

Include key finding and explanation

More recently, opponents have found a stronger reason to oppose burning waste: *dioxin* in the plants' emissions. The amounts present are extremely small, measured in trillionths of a gram per cubic meter of air. But dioxin can be deadly, at least to animals, at very low levels.

What is Dioxin?

Include definition

Delete technical details

Dioxin is a generic term for any of 75 chemical compounds, the technical name for which is poly-chlorinated dibenzo-p-dioxins (PCDDs). A related group of 135 chemicals, the PCDFs or furans, are often found in association with PCDDs.

Include major fact

The most infamous of these substances, 2,3,7,8-TCDD, is often referred to as the "most toxic chemical known," This judgment is based on animal test data. In laboratory tests, 2,3,7,8-TCDD is lethal to guinea pigs at a concentration of *500 parts per trillion.* A part per trillion is roughly equivalent to the thickness of a human hair compared to the distance across the United States.

Include major point

Delete long explanation

The effects on humans are less certain, for many reasons: it is difficult to measure the amounts to which humans have been exposed; difficult to isolate the effects of dioxin from the effects of other toxic substances on the same population; and the latency period for many potential effects, such as cancer, may be as long as 20 to 30 years.

Include continuation of major point

Delete long example

Nevertheless, because of the extreme effects of this substance on animals, known releases of dioxin have generated considerable public alarm. One of the most publicized releases occurred at Seveso, Italy, in July 1976, where a pharmaceutical plant explosion resulted in the contamination of at least 700 acres of fields and affected more than 5,000 people. Dioxin was found in the soil in concentrations of 20 to 55 parts per billion.

The immediate effects on humans were nausea, headaches, dizziness, diarrhea, and an acute skin condition called chloracne, which causes burn-like sores. The effects on animals were more severe: birds, rabbits, mice, chickens, and cats died by the hundreds, within days of the explosion. In response to the explosion, the Italian provincial authorities evacuated 730 people from the zone nearest the plant, and sealed off an area containing another 5,000 people from contact with nonresidents.

Delete long example

In this country, perhaps the best known dioxin contamination incident occurred at Times Beach, Missouri, where used oil, contaminated with dioxin, was sprayed on roads as a dust suppressant. Soil samples showed dioxin at levels exceeding 100 parts per billion. While no human health effects were documented at Times Beach, a flood in December 1982 led to widespread dispersal of the contamination, as a result of which the entire town was condemned, the population evacuated, and over $30 million of Superfund money used to purchase the condemned property.

Include the most striking and familiar example

Dioxin was among the substances of concern at Love Canal. And it was the major contaminant in the chemical defoliant, Agent Orange, the subject of a lawsuit by 15,000 Vietnam veterans and dependents and an out-of-court settlement of those complaints valued at $180 million.

Include key findings

As early as 1978, trace amounts of dioxin were found in the routine emissions of a municipal incinerator. Virtually every incinerator tested since that date has shown traces of dioxin.

The Meaning of it All

Include major fact

At the request of Congress, the Agency began in 1984 a major research effort on dioxin, the National Dioxin Study. The study is intended to provide a context in which to place mounting concerns about dioxin. Research for the study was organized into seven "tiers," each tier including a group of sites at which dioxin contamination may be present.

Condense list

- Tier 1, <u>production sites</u>, includes the 10 sites at which 2,4,5-TCP, a pesticide known to have been contaminated by dioxin, was produced, <u>and</u> additional <u>sites</u> where <u>waste materials</u> from its production were <u>disposed</u>.
- Tier 2, <u>precursor sites</u>, includes 9 sites where 2,4,5-TCP was used as a precursor <u>to make other chemical products</u>, and related waste disposal sites. The chemical products included the herbicides 2,4,5-T and silvex, and hexachlorophene, a disinfectant that was widely used in soaps and deodorants, but was banned from nonprescription uses by the Food and Drug Administration in 1981.
- Tier 3 includes 60 to 70 <u>sites at which</u> 2,4,5-TCP and its derivatives were for-<u>mulated into herbicide products</u>, and associated waste disposal sites.
- Tier 4 includes a wide range of <u>combustion sources, including internal com-<u>bustion engines, wood stoves, fireplaces, forest fires, oil burners and other <u>sources</u> burning waste oil, and many others.
- Tier 5 includes 20 to 30 of the thousands of <u>sites at which dioxin-contami-<u>nated pesticides have been used</u>, for example, power line rights-of-way, forests, and rice and sugar cane fields.
- Tier 6 includes about 20 of the <u>chemical and pesticide production facilities</u> where improper quality control <u>may</u> have led to the <u>accidental production</u> <u>of dioxin</u>.
- Tier 7 includes samples from sites where the Agency least expected to find dioxin, to determine whether there are background levels of dioxin in the environment. <u>Soil samples</u> have been taken at 500 <u>randomly selected loca-<u>tions</u> across the country—200 in rural areas, 300 in urban areas—and fish have been <u>sampled</u> from over 400 locations.

Include key finding

While the study is not yet complete, data from a variety of sources have already produced disturbing—yet, perhaps, in an odd way, reassuring—<u>results</u>. Dioxin in trace amounts appears to be <u>widely present in the environment, even in</u> <u>remote locations where industrial activity and waste combustion are unlikely to</u> <u>be the source</u>.

Delete long example

Environment Canada, the Canadian EPA, also has an extensive dioxin testing program under way. One of the more startling findings of their research, conducted at a resource recovery facility on picturesque Prince Edward Island, is that the garbage delivered to the plant contained more dioxin than the plant's emissions. The source of the dioxin in this case is not known, though it

Delete speculation

could include pesticide residues or other products contaminated with dioxin during manufacturing processes.

Include key conclusion

In short, <u>dioxin is not just</u> a problem <u>created by burning municipal waste</u>. It is <u>not clear</u> at this time <u>whether municipal waste combustion is even the major</u> <u>source of dioxin in the environment</u>.

Include conclusion

Ultimately, the dioxin problem is <u>like other toxic substance issues. We know</u> less than we need to know to thoroughly evaluate the risk. The more we find

Delete personal comment

out, the more complex the issues tend to become. There is <u>no risk-free solution, since all</u> the potential <u>disposal methods may</u> result in <u>some release of toxic substances</u> to the environment. Yet those who counsel delay, to allow the collection of more data, are met with the suspicion that their real agenda is to prevent action entirely.

Include recommendations

What we do know at present <u>does not seem</u> to suggest <u>that we should stop planning</u> to build <u>resource recovery facilities</u>. What it does suggest is that we <u>proceed cautiously, inform the public</u> of both what is known and what is unknown, <u>install pollution controls</u> if the plants' uncontrolled emissions are significant or if the exposed population wants added protection, and <u>hope that continued examination</u> of all the sources and effects of dioxin <u>will eventually produce a consensus.</u>

Source: James E. McCarthy, *Congressional Research Service Review* Apr. 1986: 19–21.

Assume that in two early drafts of your summary, you rewrote and edited; for coherence and emphasis, you inserted transitions and combined related ideas. Here is your final draft.

INCINERATING TRASH: A HOT ISSUE GETTING HOTTER (A Summary)

Because landfills often contaminate the environment, trash incinerators (resource recovery facilities) are becoming a popular alternative. Nearly 100 are operating in U.S. cities, and 150 more are planned. Besides their relatively clean burning of waste, these incinerators recover energy, which can be sold to offset operating costs. One ton of trash has roughly the energy content of 1.5 barrels of oil. Converting all U.S. refuse to energy could reduce oil imports by 12 percent.

Unfortunately, incinerator emissions contain very small amounts of dioxin (a generic name for any of 75 related chemicals). Even low dioxin levels can be deadly to animals. In fact, animal tests have helped label one dioxin substance "the most toxic chemical known." Although effects on human beings are less certain, news of dioxin in the environment creates public alarm, as evidenced at Love Canal and by the successful Agent Orange lawsuit by 15,000 Vietnam veterans. Almost every municipal incinerator tested since 1978 has shown traces of dioxin.

At Congress's request, the Environmental Protection Agency in 1984 began the National Dioxin Study. Sites included herbicide and pesticide production and waste-disposal facilities, combustion sources such as woodstoves and forest fires, areas of dioxin-contaminated pesticide use such as rice and sugarcane fields, and random soil and fish samples nationwide. Findings indicate that trace amounts of dioxin are widely present in the environment, even in areas remote from industry and waste combustion. Waste incineration is by no means the only source of environmental dioxin—and may not even be the major source.

At this stage, we know too little to evaluate the risk. And no risk-free waste-disposal solution in fact exists. But no evidence so far suggests that we stop planning

> incinerators. We should, however, move cautiously, fully informing the public, installing pollution controls as needed, and searching for a better solution.
>
> *Source:* James E. McCarthy, *Congressional Research Service Review* Apr. 1986: 19–21.

The version above is trimmed, tightened, and edited: word count is reduced to roughly 20 percent of the original. A summary this long serves well in many situations, but other audiences might want a briefer and more compressed summary—say, 10 to 15 percent of the original:

> **A More Compressed Summary**
>
> Because landfills often contaminate the environment, trash incinerators (resource recovery facilities) are becoming a popular alternative across the United States. Besides their relatively clean burning of waste, these incinerators recover energy, which can be sold to offset operating costs.
>
> Unfortunately, incinerator emissions contain very small amounts of dioxin, a chemical proven so deadly to animals, even at low levels, that it has been labeled the most toxic chemical known. Although its effects on human beings are less certain, news of dioxin in the environment creates public alarm. Almost every municipal incinerator tested since 1978 has shown traces of dioxin.
>
> Findings of an EPA study begun in 1984 suggest that trace amounts of dioxin are widely present in the environment, even in areas remote from industry and waste combustion. The dioxin source is by no means only waste incineration.
>
> We lack risk-free disposal solutions and know too little to evaluate risks. But no evidence so far suggests that we stop planning incinerators. We should, however, move cautiously, fully informing the public, installing pollution controls as needed, and searching for better solutions.

Notice that the essential message is still intact; related ideas are again combined and fewer supporting details are included. Clearly, length is adjustable according to your audience and purpose.

Forms of Summarized Information

In preparing a report, proposal, or other document, we might summarize works of others as part of our presentation. But we often summarize our own presentations as well. For instance, if our document extends to several pages, we usually include, near the end or the beginning, a summary. Depending on its location and its level of detail, this summarized information takes one of these three forms: **closing summary, informative abstract,** or **descriptive abstract.**[1] Figure 11.1 depicts these forms.

---

1. Adapted from Vaughan. Although I take liberties with his classification, Vaughan's insightful article helped clarify my thinking about the overlapping terminology that perennially seems to confound discussions of these distinctions.

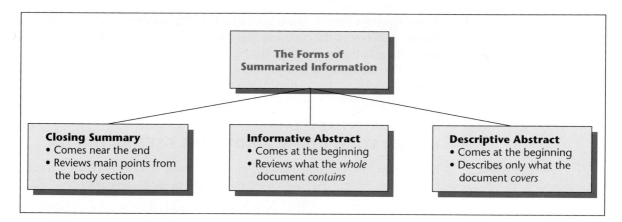

Figure 11.1   Summarized Information Assumes Various Forms

The Closing Summary

Summarized information at the end of a document's body section helps readers review and remember the main points or major findings from the presentation. This look back at "the big picture" helps readers appreciate and understand any conclusions and recommendations that follow.

The Informative Abstract

In addition to a closing summary, we might include an opening summary to show readers what the document is all about, to help them decide whether to read it, and to give them an orientation—a framework for understanding what follows. We call this opening summary an informative abstract, partly to differentiate it from its closing counterpart and partly because it is different: besides merely reviewing main points or major findings, the informative abstract identifies the issue or need that led to the report and includes condensed conclusions and recommendations.

Informative abstracts sometimes are referred to as **executive summaries,** but some writers argue that the executive summary has more of a persuasive emphasis, for readers who are executives rather than technical professionals. Executive summaries are very important to decision making—in cases when readers *expect* the writer to help guide their thinking. ("Tell me how to think about this" instead of merely "Help me understand this.")

The Descriptive Abstract

As we have seen above, the informative abstract *explains what the original document contains* (its origins, findings, and conclusions). But another, more compressed form of summarized information can precede a document: this version is called the descriptive abstract, and it merely *describes what the original is about* (its subject).

A descriptive abstract, then, conveys only the nature and extent of the document. It presents the broadest view and offers no major facts from the original. Whereas the informative abstract contains the meat of the original, the descriptive abstract has only its skeletal structure—a kind of summary of a summary. Compare, for instance, the abstract that follows with the article summary on page 229:

A DESCRIPTIVE ABSTRACT
INCINERATING TRASH: A HOT ISSUE GETTING HOTTER

As an alternative to landfill dumps, municipal trash incinerators offer environmental benefits, but they also create the danger of dioxin emissions.

Descriptive abstracts of articles often appear in magazine or journal tables of contents.

On the job, you might write informative abstracts for a boss who needs the information but who has no time to read the original. Or you might write descriptive abstracts to accompany a bibliography of works you are recommending to colleagues or clients (an *annotated bibliography*).

Placement of Summarized Information

Placement requirements for various report elements vary from company to company, but these general guidelines serve in many situations:

Revision Checklist for Summaries

Use this checklist to refine your summaries.
(Page numbers in parentheses refer to first page of discussion.)

CONTENT

☐ Does the summary contain only the essential message? (224)

☐ Does the summary make sense as an independent piece? (224)

☐ Is the summary accurate when checked against the original? (225)

☐ Is the summary free of any additions to the original? (224)

☐ Is the summary free of needless details? (224)

☐ Is the summary economical yet clear and comprehensive? (224)

☐ Is the source documented? (225)

☐ Does the descriptive abstract tell what the original is about? (231)

ORGANIZATION

☐ Is the summary coherent? (225)

☐ Are there enough transitions to reveal the line of thought? (225)

STYLE

☐ Is the summary's level of technicality appropriate for its audience? (224)

☐ Is the summary free of needless words? (225)

☐ Are all sentences clear, concise, and fluent? (261)

☐ Is the summary written in correct English? (Appendix)

- Place the closing summary in the concluding section of your report, usually preceding any conclusions and recommendations.
- Place a descriptive abstract single-spaced on the lower-third of your title page.
- Place an informative abstract on its own separate page, immediately following your table of contents.

Reports and proposals in later chapters illustrate the various placement options for summarized information.

☑ EXERCISES

1. Read each of these two paragraphs, and then list the significant ideas comprising each essential message. Write a summary of each paragraph.

In recent years, ski-binding manufacturers, in line with consumer demand, have redesigned their bindings several times in an effort to achieve a noncompromising synthesis between performance and safety. Such a synthesis depends on what appear to be divergent goals. Performance, in essence, is a function of the binding's ability to hold the boot firmly to the ski, thus enabling the skier to change rapidly the position of his or her skis without being hampered by a loose or wobbling connection. Safety, on the other hand, is a function of the binding's ability both to release the boot when the skier falls, and to retain the boot when subjected to the normal shocks of skiing. If achieved, this synthesis of performance and safety will greatly increase skiing pleasure while decreasing accidents.

Contrary to public belief, sewage-treatment plants do not fully purify sewage. The product that leaves the plant to be dumped into the leaching (sievelike drainage) fields is secondary sewage containing toxic contaminants such as phosphates, nitrates, chloride, and heavy metals. As the secondary sewage filters into the ground, this conglomeration is carried along. Under the leaching area develops a contaminated mound through which groundwater flows, spreading the waste products over great distances. If this leachate reaches the outer limits of a well's drawing radius, the water supply becomes polluted. And because all water flows essentially toward the sea, more pollution is added to the coastal regions by this secondary sewage.

2. Attend a campus lecture on a topic of interest and take notes on the significant points. Write a summary of the lecture's essential message.

3. Find an article about your major field or area of interest and write both an informative abstract and a descriptive abstract of the article.

4. Select a long paper you have written for one of your courses; write an informative abstract and a descriptive abstract of the paper.

5. After reading the article in Figure 11.2 prepare a descriptive abstract and an informative abstract, using the steps under "Critical Thinking in the Summary Process" (page 225) as a guide. Identify a specific audience and use for your material.

A possible scenario: You are assistant communications manager for a leading software development company. Part of your job involves publishing a monthly newsletter for employees. After coming across this article, you decide to summarize it for the upcoming issue. (Aspirin is a popular item in this company, given the headaches, stiff necks, and other medical problems that often result from prolonged computer work.) You have 350–375 words of newsletter space to fill. Consider carefully what this audience needs and doesn't need. In this situation, what information is most important?

ASPIRIN
A New Look at an Old Drug
by Ken Flieger

Americans consume an estimated 80 billion aspirin tablets a year. The *Physician's Desk Reference* lists more than 50 over-the-counter drugs in which aspirin is the principal active ingredient. Yet, despite aspirin's having been in routine use for nearly a century, both scientific journals and the popular media are full of reports and speculation about new uses for this old remedy.

Almost a century after its development aspirin is the focus of extensive laboratory research and some of the largest clinical trials ever carried out in conditions ranging from cardiovascular disease and cancer to migraine headache and high blood pressure in pregnancy.

How Does It Work?

The mushrooming interest in aspirin has come about largely because of fairly recent advances in understanding how it works. What is it about this drug that, at small doses, interferes with blood clotting, at somewhat higher doses reduces fever and eases minor aches and pains, and at comparatively large doses combats pain and inflammation in rheumatoid arthritis and several other related diseases?

The answer is not yet fully known, but most authorities agree that aspirin achieves some of its effects by inhibiting the production of prostaglandins. Prostaglandins are hormone-like substances that influence the elasticity of blood vessels, control uterine contractions, direct the functioning of blood platelets that help stop bleeding, and regulate numerous other activities in the body.

In the 1970s, a British pharmacologist, John Vane, Ph.D., noted that many forms of tissue injury were followed by the release of prostaglandins. In laboratory studies, he found that two groups of prostaglandins caused redness and fever, common signs of inflammation. Vane and his co-workers also showed that by blocking the synthesis of prostaglandins, aspirin prevented blood platelets from aggregating, one of the initial steps in the formation of blood clots.

This explanation of how aspirin and other nonsteroidal anti-inflammatory drugs (NSAIDs) produce their intriguing array of effects prompted laboratory and clinical scientists to form and test new ideas about aspirin's possible value in treating or preventing conditions in which prostaglandins play a role. Interest quickly focused on learning whether aspirin might prevent the blood clots responsible for heart attacks.

A heart attack or myocardial infarction (MI) results from the blockage of blood flow not *through* the heart, but *to* heart muscle. Without an adequate blood supply, the affected area of muscle dies and the heart's pumping action is either impaired or stopped altogether.

The most common sequence of events leading to an MI begins with the gradual build-up of plaque (atherosclerosis) in the coronary arteries. Circulation through these narrowed arteries is restricted, often causing the chest pain known as angina pectoris.

An acute heart attack is believed to happen when a tear in plaque inside a narrowed coronary artery causes platelets to aggregate, forming a clot that blocks the flow of blood. About 1,250,000 persons suffer heart attacks each year in the United States, and some 500,000 of them die. Those who survive a first heart attack are at greatly increased risk of having another.

Could Aspirin Help?

To learn whether aspirin could be helpful in preventing or treating cardiovascular disease, scientists have carried out numerous large randomized controlled clinical trials. In these studies, similar groups of hundreds or thousands of people are randomly assigned to receive either aspirin or a placebo, an inactive, look-alike tablet. The participants—and in double-blind trials the investigators, as well—do not know who is taking aspirin and who is swallowing a placebo.

Over the last two decades, aspirin studies have been conducted in three kinds of individuals: persons with a history of coronary artery or cerebral vascular disease, patients in the immediate, acute phases of a heart attack, and healthy men with no indication of current or previous cardiovascular illness.

The results of studies of people with a history of coronary artery disease and those in the immediate phases of a heart attack have proven to be of tremendous importance in the prevention and treatment of cardiovascular disease. The studies showed that aspirin substantially reduces the risk of death and/or non-fatal heart attacks in patients with a previous MI or unstable angina pectoris, which often occurs before a heart attack.

On the basis of such studies, these uses for aspirin (unstable angina, acute MI, and survivors of an MI) are described in the professional labeling of aspirin products, information provided to physicians and other

Figure 11.2   An Article To Be Summarized.   *Source: Excerpt from* FDA Consumer *Jan./Feb. 1994:19–21.*

health professionals. Aspirin labeling intended for the general public does not discuss its use in arthritis or cardiovascular disease because treatment of these serious conditions— even with a common over-the-counter drug—has to be medically supervised. The consumer labeling contains a general warning about excessive or inappropriate use of aspirin, and specifically warns against using aspirin to treat children and teenagers who have chickenpox or the flu because of the risk of Reye syndrome, a rare but sometimes fatal condition.

Aspirin for Healthy People?

Once aspirin's benefits for patients with cardiovascular disease were established, scientists sought to learn whether regular aspirin use would prevent a first heart attack in healthy individuals. The findings regarding that critical question have thus far been equivocal. The major American study designed to find out if aspirin can prevent cardiovascular deaths in healthy individuals was a randomized, placebo-controlled trial involving just over 22,000 male physicians between 40 and 84 with no prior history of heart disease. Half took one 325-milligram aspirin tablet every other day, and half took a placebo.

The trial was halted early, after about four-and-a-half years, and the findings quickly made public in 1988 when investigators found that the group taking aspirin had a substantial reduction in the rate of fatal and non-fatal heart attacks compared with the placebo group. There was, however, no significant difference between the aspirin and placebo groups in number of strokes (aspirin-treated patients did slightly worse) or in overall deaths from cardiovascular disease.

A similar study in British male physicians with no previous heart disease found no significant effect nor even a favorable trend for aspirin on cardiovascular disease rates. The British study of 5,100 physicians, while considerably smaller than the American study, reported three-quarters as many vascular "events." FDA scientists believe the results of the two studies are inconsistent.

The U.S. Preventive Services Task Force, a panel of medical-scientific authorities in health promotion and disease prevention, is one of many groups looking at new information on the role of aspirin in cardiovascular disease. In its *Guide to Clinical Preventive Services,* issued in 1989, the task force recommended that low-dose aspirin therapy "should be considered for men aged 40 and over who are at significantly increased risk for myocardial infarction and who lack contraindications" to aspirin use. A revised *Guide,* scheduled for publication in the fall of 1994, is expected to include a slightly revised recommendation concerning aspirin and cardiovascular disease but no major change in advice to physicians about aspirin's possible role in preventing heart attacks.

Better understanding of aspirin's myriad effects in the body has led to clinical trials and other studies to assess a variety of possible uses: preventing the severity of migraine headaches, improving circulation to the gums thereby arresting periodontal disease, preventing certain types of cataracts, lowering the risk of recurrence of colorectal cancer, and controlling the dangerously high blood pressure (called preeclampsia) that occurs in 5 to 15 percent of pregnancies.

None of these uses for aspirin has been shown conclusively to be safe and effective, and there is concern that people may be misusing aspirin on the basis of unproven notions about its effectiveness. Last October, FDA proposed a new labeling statement for aspirin products advising consumers to consult a doctor before taking aspirin for new and long-term uses. The proposed statement would read. "IMPORTANT: See your doctor before taking this product for your heart or for other new uses of aspirin because serious side effects could occur with self-treatment."

The Other Side of the Coin

While examining new possibilities for aspirin in disease treatment and prevention, scientists do not lose sight of the fact that even at low doses aspirin is not harmless. A small subset of the population is hypersensitive to aspirin and cannot tolerate even small amounts of the drug. Gastrointestinal distress—nausea, heartburn, pain—is a well-recognized adverse effect and is related to dosage. Persons being treated for rheumatoid arthritis who take large daily doses of aspirin are especially likely to experience gastrointestinal side effects.

Aspirin's antiplatelet activity apparently accounts for hemorrhagic strokes, caused by bleeding into the brain, in a small but significant percentage of persons who use the drug regularly. For the great majority of occasional aspirin users, internal bleeding is not a problem. But aspirin may be unsuitable for people with uncontrolled high blood pressure, liver or kidney disease, peptic ulcer, or other conditions that might increase the risk of cerebral hemorrhage or other internal bleeding.

New understanding of how aspirin works and what it can do leaves no doubt that the drug has a far broader range of uses than imagined [nearly a century ago]. The jury is still out, however, on a number of key questions about the best and safest ways to use aspirin. And until some critical verdicts are handed down, consumers are well-advised to regard aspirin with appropriate caution.

Figure 11.2 An Article To Be Summarized *Continued*

Bring your abstracts to class and exchange them with a classmate's for editing according to the revision checklist. Revise your edited copies before submitting them to your instructor.

☑ COLLABORATIVE PROJECT

Organize into small groups and choose a topic for discussion: an employment problem, a campus problem, plans for an event, suggestions for energy conservation, or the like. (A possible topic: Should employers have the right to require lie detector tests, drug tests, or AIDS tests for their employees?)

Discuss the topic for one class period, taking notes on significant points and conclusions. Afterward, organize and edit your notes in line with the directions for writing summaries. Next, write a summary of the group discussion in no more than 200 words. Finally, as a group, compare your individual summaries for accuracy, emphasis, conciseness, and clarity.

PART

III

Structural and Style Elements

Organizing for Readers

O NE of our biggest writing challenges is to transform our material into manageable form. First, we need to unscramble information to make sense of it for ourselves; then we need to shape it for the reader's understanding.

In order to follow our thinking, readers need a message organized in a way that makes sense to *them*. But data rarely materializes or thinking rarely occurs in a neat, predictable sequence. We cannot merely report ideas or data in the same random order they occur. Instead we *shape* this material into an organized unit of meaning. In trying to organize, we face questions like these:

Typical Questions in Organizing for Readers

- *What relationships do these data suggest?*
- *What do I want to emphasize?*
- *In which sequence will readers approach this material?*
- *What belongs where?*
- *What do I say first?*
- *What comes next?*
- *How do I end?*

Writers rely on the following strategies for organizing material: partitioning and classifying, outlining, paragraphing, and sequencing.

Partitioning and Classifying

Whenever we analyze something, we break it down to discover constituents, connections, similarities, trends, associations, correlations, relationships, and perspectives. Sometimes we analyze an item or an idea by dividing it into its parts; at other times, we analyze an assortment of items or ideas by dividing them into categories. Although these two activities often go hand in hand, each serves a distinct purpose: *Partition* deals with *one* thing only. Its purpose is to separate that thing into parts, pieces, sections, or categories—for closer examination (for example, a report divided into introduction, body, and conclusion). *Classification* deals with an *assortment* of things that share certain similarities. Its purpose is to group these things systematically—to show where everything belongs (for example, a CD collection sorted into categories—jazz, rock, country and western, classical, and so on).

How and when you use partition and classification will depend on your subject and purpose. To describe a microcomputer system to a novice, you might partition it into *terminal, keyboard, printer, power cord*, etc.; for a seasoned user who wants to install an expansion card, you might partition the system into *processor-direct slot, video-in slot, communication slot*, etc. On the other hand, if you have twenty-five computer programs to arrange so you easily can locate the one you want, you will have to group the pile into

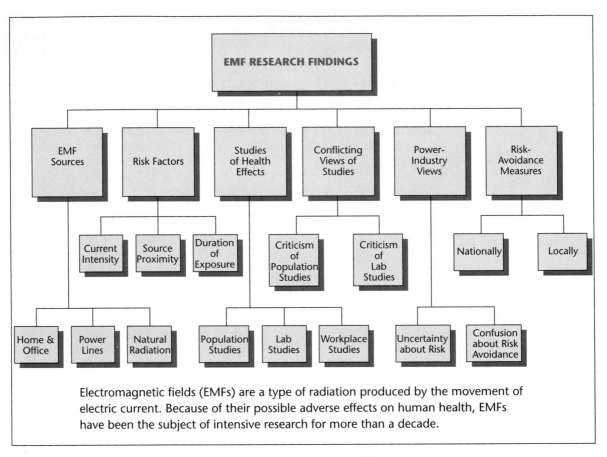

Electromagnetic fields (EMFs) are a type of radiation produced by the movement of electric current. Because of their possible adverse effects on human health, EMFs have been the subject of intensive research for more than a decade.

Figure 12.1   Assorted Items Classified into Categories

smaller categories. You might want to classify your programs according to function (*word processing, graphics, database management*) or according to expected frequency of use or relative ease of use.

Close examination of any complex problem usually requires both partition and classification. If you are designing a new supermarket, for example, you must first partition the whole market into parts: *display and shopping area, receiving and storage area, meat refrigeration and preparation area,* and so on. Next, you need to sort your inventory by grouping the thousands of items into smaller classes according to their similarities: *frozen foods, dairy products, meat, fish, and poultry,* and so on. In turn, you partition each of these sections, say dividing "meat" into three smaller groups: *beef, pork, lamb.* Under these headings, you will group the cuts of meat (*steaks, ribs,* and so on) in each category. You might carry the division further for some meat products, such as types of ground beef: *regular, lean, diet lean.*

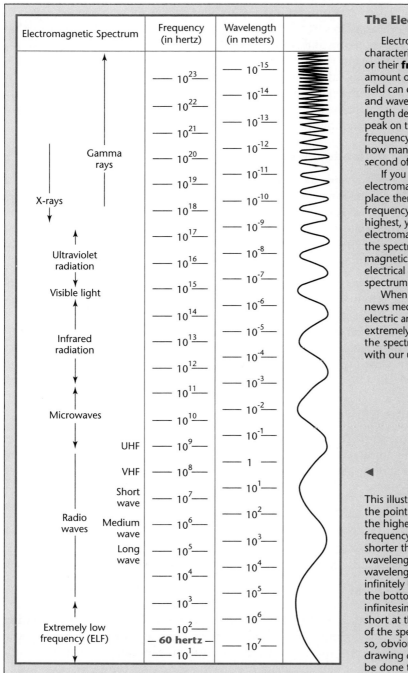

The Electromagnetic Spectrum

Electromagnetic fields can be characterized by either their **wavelength** or their **frequency,** which are related. The amount of energy an electric or magnetic field can carry depends on the frequency and wavelength of the field. The wave-length describes how far it is between one peak on the wave and the next peak. The frequency, measured in hertz, describes how many wave peaks pass by in one second of time.

If you take all the different kinds of electromagnetic fields we know about and place them on a chart, from the lowest frequency (i.e., lowest energy) to the highest, you have a chart of the electromagnetic spectrum. The low end of the spectrum includes electric and magnetic fields produced by everyday electrical appliances. At the top of the spectrum are X-rays and gamma-rays.

When you hear about "EMFs" in the news media, the term usually refers to electric and magnetic fields at the extremely low frequency (or ELF) end of the spectrum, such as those associated with our use of electric power.

◄

This illustrates the point that the higher the frequency, the shorter the wavelength. The wavelengths are infinitely long at the bottom and infinitesimally short at the top of the spectrum so, obviously, the drawing cannot be done to scale.

Figure 12.2 One Item Partitioned into Its Components

Source: U.S. Environmental Protection Agency, EMF in Your Environment. *Washington, DC: GPO, 1992: 4–5.*

This kind of dividing and grouping continues until you have enough categories or classes to sort the hundreds of crates and cartons of inventory that sit in your receiving and storage area. You have used division and classification to divide your store into parts and to sort your inventory into classes.

Although useful in describing concrete entities such as supermarkets or computer systems, partition and classification are especially valuable for analyzing and organizing more abstract information, ideas, or problems. Consider this example:

Data in Random Form

While researching the health effects of electromagnetic fields (EMFs), you encounter information about various radiation sources; ratio of risk to level of exposure; workplace studies, lab studies of cell physiology, biochemistry, and behavior; statistical studies of diseases in certain populations; conflicting expert views; views from local authorities, and so on. ■

Figure 12.1 depicts how classification might organize this random collection of EMF data into manageable categories. (Note that many of the Figure 12.1 categories might be divided further into subcategories, such as *kitchen sources, workshop sources, bedroom sources,* and so on.) Figure 12.2 depicts how partition might reveal the parts of a single concept (the electromagnetic spectrum).

In organizing their documents, writers use partition and classification routinely, in a process we know as *outlining*.

Outlining

With an outline you move from a random listing of items to a deliberate map that will guide readers from point to point. Readers more easily understand and remember material organized in a sequence they find logical. Organize in the sequence in which you expect readers to approach the material.

A Document's Basic Shape

How should you organize to make the document logical from your audience's point of view? Begin with the basics. Useful writing of any length—a book, chapter, news article, letter, or memo—typically follows this organizing pattern:

■ The *introduction* provides orientation by doing any of these things: explaining the topic's origin and significance and the document's purpose; identifying briefly your intended audience and your information sources; defining specialized terms or general terms that have special meanings in your document; accounting for limitations such as incomplete or questionable data; previewing the major topics to be discussed in the body section.

   Some introductions need to be long and involved; others, short and sweet. Reports too often waste readers' time with needless background.

If you don't know your readers well enough to give them only what they need, use subheadings so readers can choose what they need to know.

■ The *body* delivers on the promise implied in your introduction ("Show me!"). Here you present your data, discuss your evidence, lay out your case, or tell readers what to do and how to do it. Body sections come in all different sizes, depending on how much readers need and expect.

Body sections are titled to reflect their specific purpose: "Description and Function of Parts," for a mechanism description; "Required Steps," for a set of instructions; "Collected Data," for a feasibility analysis.

■ The *conclusion* of a document has assorted purposes: it might evaluate the significance of the report, reemphasize key points, take a position, predict an outcome, offer a solution, or suggest further study. If the issue is straightforward, the conclusion might be brief and definite. If the issue is complex or controversial, the conclusion might be lengthy and open ended. Whatever the conclusion's specific purpose, readers expect a clear perspective on the whole document.

Conclusions vary with the document. You might conclude a mechanism description by reviewing the mechanism's major parts and then briefly describing one operating cycle. You might conclude a comparison or feasibility report by offering judgments about the facts you've presented and then recommending a course of action.

Most workplace documents display this basic shape

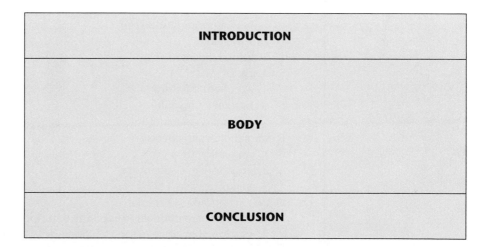

A suitable beginning, middle, and ending are essential, but alter your own outline as you see fit. No single form of outline should be followed slavishly by any writer. *The organization of any document ultimately is determined by its audience's needs and expectations.* In many cases, specific requirements about a document's organization and style are spelled out in a company's style guide.

The computer is especially useful for rearranging outlines until they reflect the sequence in which you expect readers to approach your message.

The Formal Outline

A simple list usually suffices for organizing short documents or as a tentative outline for longer documents. Seldom does an author or team begin by developing a formal outline when planning a manuscript. But at some stage (often a *latter* stage) in the writing process, a long, complex document usually calls for a more systematic, formal outline, to mark divisions and to show how categories relate. Here, for example, is a formal outline for the report examining the health effects of electromagnetic fields (pages 560–69):

A formal outline

Children Exposed to EMFs: A Risk Assessment

I. INTRODUCTION
    A. Definition of electromagnetic fields
    B. Background on the health issues
    C. Description of the local power-line configuration
    D. Purpose of this report[1]
    E. Brief description of data sources
    F. Scope of this inquiry
II. DATA SECTION [Body]
    A. Sources of EMF exposure
        1. home and office
            a. kitchen
            b. workshop[2] [and so on]
        2. power lines
        3. natural radiation
        4. risk factors
            a. current intensity
            b. source proximity
            c. duration of exposure
    B. Studies of health effects
        1. population surveys
        2. laboratory measurements
        3. workplace links
    C. Conflicting views of studies
        1. criticism of methodology in population studies
        2. criticism of overgeneralized lab findings

---

1. Long reports often begin directly with a statement of purpose. For the intended audience (i.e., generalists) of this report, however, the technical topic first must be defined for the readers' clear understanding of the context.

2. Note that each level of division yields at least two items. If you cannot divide a major item into at least two subordinate items, retain only your major heading.

D. Power industry views
  1. uncertainty about risk
  2. confusion about risk avoidance
E. Risk-avoidance measures
  1. nationally
  2. locally
III. CONCLUSION
  A. Summary and overall interpretation of findings
  B. Recommendations

A formal outline easily converts to a table of contents for the finished report, as shown in Chapter 16. (Because they serve mainly to guide the *writer*, minor outline headings, such as *a* and *b* under II.A.1 above, may be omitted from the table of contents or the report itself. Excessive headings make a document seem fragmented.)

In technical documents, the alphanumeric form of notation shown above often is replaced by decimal notation:

Decimal notation in a technical document

2.0 DATA SECTION
  2.1 Sources of EMF Exposure
    2.1.1 home and office
      2.1.1.1 kitchen
      2.1.1.2 workshop [and so on]
    2.1.2 power lines
    2.1.3 natural radiation
    2.1.4 Risk factors
      2.1.4.1 current intensity
      2.1.4.2 source proximity
      2.1.4.3 [and so on]

The decimal outline makes it easier to refer readers to specifically numbered sections of the document: ("See 2.1.2"). While both systems achieve the same organizing objective, decimal notation usually is preferred in business, government, and industry.

In some cases, you may wish to expand the above *topic outline* into a *sentence outline*, in which each sentence serves as a topic sentence for a paragraph in the report:

A sentence outline

2.0. DATA SECTION
  2.1. Although the 2 million miles of power lines crisscrossing the United States have been the focus of the EMF controversy, potentially harmful waves also are emitted by household wiring, appliances, electric blankets, and computer terminals.
    2.1.1. [and so on]

Sentence outlines are used mainly in collaborative projects in which various team members prepare different sections of a long document.

The Importance of Being Messy

The neat and ordered outline shown earlier represents the *product* of outlining—not the *process*. Beneath any finished outline (or any finished document) lie pages of scribbling and things crossed out, jumbled lists, arrows, and fragments of ideas. Writing begins in disorder. Messiness is a natural and often essential part of writing in its early stages.

A survey of the influence of computers on workplace writing found that traditional, formal outlining was giving way to outlining in the form of "notes on audience, purpose, direction, key content points, tone." These outlines were "flexible, sketchy, punctuated by arrows, numbers, or exclamation points; they looked more like lists" (Halpern 179). Outlines needn't be pretty, as long as they help you control your material.

Not until your final draft of a long document do you compose the finished outline, which serves as a model for your table of contents, as a check on your reasoning, and as a way of revealing to your readers a clear line of thinking.

Outlining and Reorganizing on a Computer

A word processing program such as Microsoft Word enables you to work on your document and your outline simultaneously. An "outline view" of the document enables you to see relationships among ideas at various levels, to create new headings, to add text beneath headings, and to move headings and their subtext (Figure 12.3). You also can *collapse* the outline view to display the headings only.

Switch between "normal view" (to compose your text) and "outline view" (to examine the arrangement of material). You can add or delete headings or text and reorganize whole sections of your document. (*Microsoft Word* 504–05).

As a visual alternative to traditional outlining, many computer graphics programs enable you to display prose outlines as tree charts.

Organizing for Cross-Cultural Audiences

Different cultures often have different expectations about how information should be organized. A document considered well organized by one culture may confuse or offend another. For instance, a paragraph in English typically begins with a main idea directly expressed as a topic or orienting sentence and followed by specific support; any digression from this main idea is considered harmful to the paragraph's *unity*. Some cultures, however, consider digression a sign of intelligence or politeness. To native readers of English, the long introductions and digressions in certain Spanish or Russian documents might seem tedious and confusing, but a Spanish or Russian reader might view the more direct organization of English as overly abrupt and simplistic (Leki 151).

Expectations can differ even among same-language cultures. British correspondence, for instance, typically expresses the bad news directly up front, instead of the indirect approach preferred in the United States. A bad-news letter or memo appropriate for U.S. readers could be considered evasive by British readers (Scott and Green 19).

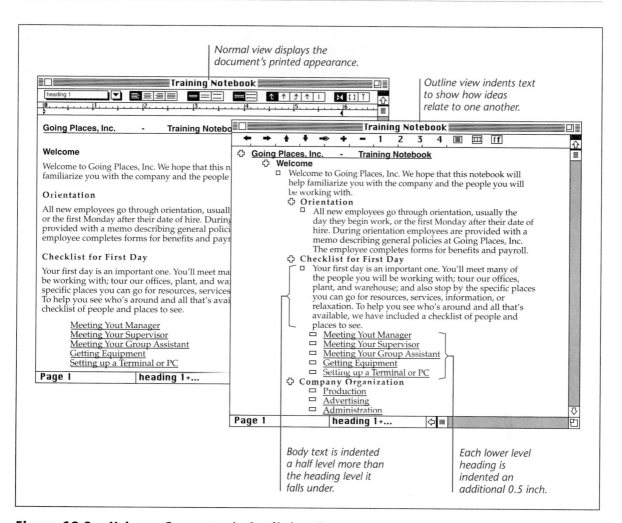

Figure 12.3 Using a Computer's Outlining Feature. *Source:* Microsoft Word User's Guide, *Macintosh version 5.0, 1992: 502. Reprinted by permission of Microsoft Corporation.*

Despite all our electronic communication tools, connecting with readers—especially in a global context—requires, above all, human sensitivity and awareness of audience.

The Report Design Worksheet

As an alternative to the audience and use profile sheet (page 65), the worksheet in Figure 12.4 can supplement your outline and help you focus on your audience and purpose.[3]

3. This version is based on a worksheet developed by Professor John S. Harris of Brigham Young University.

REPORT DESIGN WORKSHEET

Preliminary Information

What is to be done?   *A report on the health effects of electromagnetic radiation*

Whom is it to be presented to, and when?   *Town Meeting, April 1*

Audience Analysis	Primary Reader(s)	Secondary Reader(s)
Position and title:	*Town manager*	*Selectpersons, various town officials, school board, parents, colleagues, friends*
Relationship to author or organization:	*Employer*	
Technical expertise:	*Nontechnical (for this topic)*	*nontechnical*
Personal characteristics:	*highly efficient; expects results*	*most have strong views on this issue*
Attitude toward author or organization:	*Is preparing my annual performance review*	*friendly and respectful; selectboard will vote on my contract renewal and pay raise*
Attitude toward subject:	*extremely concerned*	*same*
Effect of report on readers or organization:	*will be read closely and acted upon*	*will be discussed at town meeting*

Reader's Purpose

Why has reader requested it?	*wants to address any potential hazards without delay*	
What does reader plan to do with it?	*use the data to make an informed decision about action*	*confer with the town manager about the decision*
What should reader know beforehand to understand it as written?	*nothing special; history of the issue is reviewed in report*	*same*
What does reader already know?	*has read and heard very general information about the issue*	*same*
What amount and kinds of detail will reader find significant?	*clear description of the issue and careful review of the evidence*	*same*
What should reader know and/or be able to do after reading it?	*make a decision based on the best evidence available*	*advise the town manager about her decision*

Figure 12.4   Report Design Worksheet

Writer's Purpose

Why am I writing it?   *to communicate my research findings*

What effect(s) do I wish to achieve?   *to have my readers conclude that, while we await further research, we should take immediate and inexpensive steps toward risk-avoidance and continue to assess the extent of EMF hazards throughout the school*

Design Specifications

Sources of data: *recently published research, including online and internet sources; interviews with local authorities*

Tone: *semiformal*

Point of view: *mostly third-person (except for recommendations)*

Needed visuals and supplements: *title page, letter of transmittal, table of contents, informative abstract charts, graphs, and tables*

Appropriate format (letter, memo, etc.):   *formal report format with full heading system*

Basic organization (problem-causes-solution, intro-instructions-summary, etc.):   *causes—possible effects—conclusions and recommendations*

Main items in introduction: *Definition of electromagnetic fields*
Background on the health issue
Description of the local power-line configuration
Purpose of report, and intended audience
Data sources
Scope of this inquiry

Main items in body: *Sources of EMF exposure*
Risk factors
Studies of health effects
Conflicting views of studies
Local power company views
Risk-avoidance measures

Main items in conclusion: *Summary of findings*
Overall interpretation of findings
Recommendations

Other Considerations: *no frills or complex technical data; these readers are all interested in the "bottom line" as far as what action they should take*

Figure 12.4   Report Design Worksheet   *Continued*

Paragraphing

Readers look for orientation, for shapes they can recognize. Beyond its larger shape (introduction, body, conclusion), a document depends on the smaller shapes of each paragraph.

Although paragraphs can have various structures and purposes (paragraphs of introduction, conclusion, or transition), our focus here is on *standard support paragraphs*. While part of the document's larger design, each of these middle blocks of thought usually can stand alone in meaning and emphasis.

The Standard Paragraph

All the sentences in a standard paragraph relate to the main point, which is expressed as the *topic sentence:*

Topic sentences

> Computer literacy has become a requirement for all "educated" people.
>
> A video display terminal can endanger the operator's health.
>
> Chemical pesticides and herbicides are both ineffective and hazardous.

Each topic sentence introduces an idea, judgment, or opinion. But in order to grasp the writer's exact meaning, readers need explanation. Consider the third statement:

> Chemical pesticides and herbicides are both ineffective and hazardous.

Imagine you are a researcher for the Epson Electric Light Company and have been asked to determine whether the company should (1) begin spraying pesticides and herbicides under its power lines, or (2) continue with its manual (and nonpolluting) ways of minimizing foliage and insect damage to lines and poles. If you simply responded with the preceding assertion, your employer would have questions:

- *Why, exactly, are these methods ineffective and hazardous?*
- *What are the problems?*
- *Can you explain?*

To answer these questions and to support your assertion, you need a fully developed paragraph:

Intro. (topic sent.)
Body (2–6)

> [1]**Chemical pesticides and herbicides are both ineffective and hazardous.** [2]Because none of these chemicals has permanent effects, pest populations invariably recover and need to be resprayed. [3]Repeated applications cause pests to develop immunity to the chemicals. [4]Furthermore, most of these products attack species other than the intended pest, killing off its natural predators, thus actually increasing the pest population. [5]Above all, chemical residues survive in the environment (and living tissue) for years, often carried hundreds of miles by wind and water. [6]This toxic legacy includes such biological effects as birth deformities, reproductive failures, brain damage, and cancer. [7]Although intended to control pest populations, these chemicals ironically

Conclusion (7–8)

threaten to make the human population their ultimate victims. [8]I therefore recommend we continue our present control methods.

Most standard paragraphs in technical writing have an introduction-body-conclusion structure. They begin with a clear topic (or orienting) sentence stating a generalization. Details in the body support the generalization.

The Topic Sentence

Readers look to a paragraph's opening sentences for a framework. When they don't know exactly what the paragraph is about, readers struggle to grasp your meaning. Read this next paragraph once only, and then try answering the questions that follow.

A paragraph with its topic sentence omitted

> Besides containing several toxic metals, it percolates through the soil, leaching out naturally present metals. Pollutants such as mercury invade surface water, accumulating in fish tissues. Any organism eating the fish—or drinking the water—in turn faces the risk of heavy metal poisoning. Moreover, acidified water can release heavy concentrations of lead, copper, and aluminum from metal plumbing, making ordinary tap water hazardous.

Can you identify the paragraph's main idea? Probably not. Without the topic sentence, you have no framework for understanding this information in its larger meaning. And you don't know where to place the emphasis: on polluted fish, on metal poisoning, on tap water?

Now, insert the following opening sentence and reread the paragraph:

The missing topic sentence

> Acid rain indirectly threatens human health.

With this orientation, the exact meaning becomes obvious.

The topic sentence should appear *first* in the paragraph, unless you have one good reason to place it elsewhere. Think of your topic sentence as the one sentence you would keep if you could keep only one (U.S. Air Force Academy 11). In some instances, a paragraph's main idea may require a "topic statement" consisting of two or more sentences, as in this example:

A topic statement can have two or more sentences

> The most common strip-mining methods are open-pit mining, contour mining, and auger mining. The specific method employed will depend on the type of terrain that covers the coal.

The topic sentence or topic statement should immediately tell readers what to expect. Don't write *Some pesticides are less hazardous and often more effective than others* when you mean *Organic pesticides are less hazardous and often more effective than their chemical counterparts.* The first topic sentence leads everywhere and nowhere; the second helps us focus, tells us what to expect from the paragraph. Don't write *Acid rain poses a danger,* leaving readers to decipher your meaning of *danger.* If you mean that *Acid rain is killing our lakes and polluting our water supplies,* say so. Uninformative topic sentences keep readers guessing.

Paragraph Unity

A paragraph is unified when all its material belongs there—when every word, phrase, and sentence directly supports the topic sentence.

A unified paragraph

> **Solar power offers an efficient, economical, and safe solution to the Northeast's energy problems.** To begin with, solar power is highly efficient. Solar collectors installed on fewer than 30 percent of roofs in the Northeast would provide more than 70 percent of the area's heating and air-conditioning needs. Moreover, solar heat collectors are economical, operating for up to twenty years with little or no maintenance. These savings recoup the initial cost of installation within only ten years. Most important, solar power is safe. It can be transformed into electricity through photovoltaic cells (a type of storage battery) in a noiseless process that produces no air pollution—unlike coal, oil, and wood combustion. In sharp contrast to its nuclear counterpart, solar power produces no toxic waste and poses no catastrophic danger of meltdown. Thus, massive conversion to solar power would ensure abundant energy and a safe, clean environment for future generations.

One way to damage unity in the paragraph above would be to discuss the differences between active and passive solar heating, or manufacturers of solar technology, or the advantages of solar power over wind power. Although these matters do *broadly* relate to the general issue of solar energy, none directly advances the meaning of *efficient, economical,* or *safe.*

Every topic sentence has a key word or phrase that carries the meaning. In the pesticide-herbicide paragraph (page 250), the key words are *ineffective* and *hazardous.* Anything that fails to advance their meaning throws the paragraph—and the readers—off track.

Paragraph Coherence

In a unified paragraph, everything belongs. In a coherent paragraph, everything sticks together: Topic sentence and support form a connected line of thought, like links in a chain. To convey precise meaning, a paragraph must be unified. To be readable, a paragraph must be coherent as well.

Paragraph coherence can be damaged by (1) short, choppy sentences, (2) sentences in the wrong sequence, or (3) insufficient transitions and connectors (Appendix) for linking related ideas. Here is how the solar energy paragraph might become incoherent:

An incoherent paragraph

> Solar power offers an efficient, economical, and safe solution to the Northeast's energy problems. Unlike nuclear power, solar power produces no toxic waste and poses no danger of meltdown. Solar power is efficient. Solar collectors could be installed on fewer than 30 percent of roofs in the Northeast. These collectors would provide more than 70 percent of the area's heating and air-conditioning needs. Solar power is safe. It can be transformed into electricity. This transformation is made possible by photovoltaic cells (a type of storage battery). Solar heat collectors are economical. The photovoltaic process produces no air pollution.

Here, in contrast, is the original, coherent paragraph with sentences numbered for later discussion and with transitions and connectors shown in boldface. Notice how this version reveals a clear line of thought:

A coherent paragraph

> [1]Solar power offers an efficient, economical, and safe solution to the Northeast's energy problems. [2]**To begin with,** solar power is highly efficient. [3]Solar collectors installed on fewer than 30 percent of roofs in the Northeast would provide more than 70 percent of the area's heating and air-conditioning needs. [4]**Moreover,** solar heat collectors are economical, operating for up to twenty years with little or no maintenance. [5]**These savings** recoup the initial cost of installation within only ten years. [6]**Most important,** solar power is safe. [7]**It** can be transformed into electricity through photovoltaic cells (a type of storage battery) in a noiseless process that produces no air pollution—unlike coal, oil, and wood combustion. [8]**In sharp contrast** to its nuclear counterpart, solar power produces no toxic waste and poses no danger of catastrophic meltdown. [9]**Thus,** massive conversion to solar power would ensure abundant energy and a safe, clean environment for future generations.

We easily can trace the sequence of thoughts in this paragraph.

    1. The topic sentence establishes a clear direction.
2–3. The first reason is given and then explained.
4–5. The second reason is given and explained.
6–8. The third and major reason is given and explained.
    9. The conclusion sums up and reemphasizes the main point.

Within this line of thinking, each sentence follows logically from the one before it. Readers know where they are at any place in the paragraph. To reinforce the logical sequence, related ideas are combined in individual sentences, and transitions and connectors signal clear relationships. The whole paragraph sticks together.

Paragraph Length

Paragraph length depends on the writer's purpose and the reader's capacity for understanding. Actual word count actually means very little. What matters is how thoroughly the paragraph makes your point. Consider these guidelines:

- In writing that carries highly technical information or complex instructions, short paragraphs (perhaps in a vertically displayed list) give readers plenty of breathing space. A clump of short paragraphs, however, can make a document seem choppy and poorly organized.

- In writing that explains concepts, attitudes, or viewpoints, support paragraphs generally run from 100 to 300 words. But long paragraphs can be tiring and hard to follow, especially if important ideas get buried in the middle.

- Long paragraphs can be broken into parts—using bullets, for example—to make the information more accessible to the reader.

- In letters, memos, or news articles, paragraphs of only one or two sentences focus the reader's attention. A short paragraph (even a single-sentence paragraph) can highlight an important idea in any document.

- Long paragraphs at the beginning or end of a document generally should be avoided because they can discourage the reader or obscure the emphasis.

Sequencing

Research demonstrates that readers more easily understand and remember material that is organized in a logical sequence (Felker et al. 11). Items in sequence follow some pattern that reveals a certain relationship: cause-and-effect, comparison-contrast, and so on. For instance, a progress report usually follows a *chronological* sequence (events presented in order of occurrence). An argument for a companywide exercise program likely would follow an *emphatic* sequence (benefits presented in order of importance—least to most, or vice versa).

A single paragraph usually follows one particular sequence. A longer document may use one particular sequence or a combination of sequences. Some common sequences are described below.

Spatial Sequence

A spatial sequence begins at one location and ends at another. It is most useful in describing a physical item or a mechanism. Describe the parts in the sequence in which readers would actually view them or in the order in which each part functions: left to right, inside to outside. This description of a hypodermic needle proceeds from the needle's base (hub) to its point:

"What does it look like?"

A hypodermic needle is a slender, hollow steel instrument used to introduce medication into the body (usually through a vein or muscle). It is a single piece composed of three parts, all considered sterile: the hub, the cannula, and the point. The hub is the lower, larger part of the needle that attaches to the necklike opening on the syringe barrel. Next is the cannula (stem), the smooth and slender central portion. Last is the point, which consists of a beveled (slanted) opening, ending in a sharp tip. The diameter of a needle's cannula is indicated by a gauge number; commonly, a 24–25 gauge needle is used for subcutaneous injections. Needle lengths are varied to suit individual needs. Common lengths used for subcutaneous injections are ⅜, ½, ⅝, and ¾ inch. Regardless of length and diameter, all needles have the same functional design.

Product and mechanism descriptions almost always have some type of visual to amplify the verbal description.

Chronological Sequence

Explanations of how to do something or how something happened generally are arranged according to a strict time sequence: first step, second step, and so on.

"How is it done?"

> Instead of breaking into a jog too quickly and risking injury, take a relaxed and deliberate approach. Before taking a step, spend at least ten minutes stretching and warming up, using any exercises you find comfortable. (After your first week, consult a jogging book for specialized exercises.) When you've completed your warmup, set a brisk pace walking. Exaggerate the distance between steps, taking long strides and swinging your arms briskly and loosely. After roughly 100 yards at this brisk pace, you should feel ready to jog. Immediately break into a very slow trot: lean your torso forward and let one foot fall in front of the other (one foot barely leaving the ground while the other is on the pavement). Maintain the slowest pace possible, just above a walk. *Do not bolt out like a sprinter!* The biggest mistake is to start fast and injure yourself. While jogging, relax your body. Keep your shoulders straight and your head up, and enjoy the scenery—after all, it is one of the joys of jogging. Keep your arms low and slightly bent at your sides. Move your legs freely from the hips in an action that is easy, not forced. Make your feet perform a heel-to-toe action: land on the heel; rock forward; take off from the toe.

The paragraph explaining how acid rain endangers human health (page 251) is another example of chronological sequence.

Effect-to-Cause Sequence

A sequence that first identifies a problem and then traces its causes is typically found in problem-solving analyses.

"How did this happen?"

> Modern whaling techniques have brought the whale population to the threshold of extinction. In the nineteenth century, invention of the steamboat increased hunters' speed and mobility. Shortly afterward, the grenade harpoon was invented so that whales could be killed quickly and easily from the ship's deck. In 1904, a whaling station opened on Georgia Island in South America. This station became the gateway to Antarctic whaling for the nations of the world. In 1924, factory ships were designed that enabled round-the-clock whale tracking and processing. These ships could reduce a ninety-foot whale to its by-products in roughly thirty minutes. After World War II, more powerful boats with remote sensing devices gave a final boost to the whaling industry. The number of kills had now increased far beyond the whales' capacity to reproduce.

Cause-to-Effect Sequence

A cause-to-effect sequence follows an action to its results. The topic sentence identifies the causes, and the remainder of the paragraph discusses its effects.

"What will happen if I do this?"

> Some of the most serious accidents involving gas water heaters occur when a flammable liquid is used in the vicinity. The heavier-than-air vapors of a flammable liquid such as gasoline can flow along the floor—even the length of a basement—and be explosively ignited by the flame of the water heater's pilot light or burner. Because the victim's clothing frequently ignites, the resulting burn injuries are commonly serious and extremely painful. They may require long hospitalization, and can result in disfigurement or death. *Never, under any circumstances, use a flammable liquid near a gas heater or any other open flame.* (Consumer Product Safety Commission)

Emphatic Sequence

Reasons offered in support of a specific viewpoint or recommendation often appear in workplace writing, as in the pesticide-herbicide paragraph on page 250 or the solar energy paragraph on page 252. For emphasis, the reasons or examples usually are arranged in decreasing or increasing order of importance.

"What should I remember about this?"

> Although strip mining is safer and cheaper than conventional mining, it is highly damaging to the surrounding landscape. Among its effects are scarred mountains, ruined land, and polluted waterways. Strip operations are altering our country's land at the rate of 5,000 acres per week. An estimated 10,500 miles of streams have been poisoned by silt drainage in Appalachia alone. If strip mining continues at its present rate, 16,000 square miles of U.S. land eventually will be stripped barren.

In this paragraph, the most dramatic example is saved for the end, for greatest emphasis.

Problem-Causes-Solution Sequence

The problem-solving sequence proceeds from description of the problem, through diagnosis, to solution. After outlining the cause of the problem, this next paragraph explains how the problem has been solved:

"How was the problem solved?"

> On all waterfront buildings, the unpainted wood exteriors had been severely damaged by the high winds and sandstorms of the previous winter. After repairing the damage, we took protective steps against further storms. First, all joints, edges, and sashes were treated with water-repellent preservative to protect against water damage. Next, three coats of nonporous primer were applied to all exterior surfaces to prevent paint from blistering and peeling. Finally, two coats of wood-quality latex paint were applied over the nonporous primer. To keep coats of paint from future separation, the first coat was applied within two weeks of the priming coats, and the second within two weeks of the first. Two weeks after completion, no blistering, peeling, or separation has occurred.

Comparison-Contrast Sequence

Evaluation of two or more items on the basis of their similarities or differences often appears in job-related writing.

*"How do these items
compare?"*

The ski industry's quest for a binding that ensures good performance as well as safety has led to development of two basic types. Although both bindings improve performance and increase the safety margin, they have different release and retention mechanisms. The first type consists of two units (one at the toe, another at the heel) that are spring-loaded. These units apply their retention forces directly to the boot sole. Thus the friction of boot against ski allows for the kind of ankle movement needed at high speeds over rough terrain, without causing the boot to release. In contrast, the second type has one spring-loaded unit at either the toe or the heel. From this unit extends a boot plate that travels the length of the boot to a fixed receptacle on its opposite end. With this plate binding, the boot has no part in release or retention. Instead, retention force is applied directly to the boot plate, providing more stability for the recreational skier, but allowing for less ankle and boot movement before releasing. Overall, the double-unit binding performs better in racing, but the plate binding is safer.

For comparing and contrasting more specific data on these bindings, two lists would be most effective.

The Salomon 555 offers the following features:

1. upward release at the heel and lateral release at the toe (thus eliminating 80 percent of leg injuries)
2. lateral antishock capacity of 15 millimeters, with the highest available return-to-center force
3. two methods of reentry to the binding: for hard and deep-powder conditions
4. five adjustments
5. (and so on)

The Americana offers these features:

1. upward release at the toe as well as upward and lateral release at the heel
2. lateral antishock capacity of 30 millimeters, with moderate return-to-center force
3. two methods of reentry to the binding
4. two adjustments, one for boot length and another for comprehensive adjustment for all angles of release and elasticity
5. (and so on)

Instead of this block structure (in which one binding is discussed and then the other), the writer might have chosen a point-by-point structure (in which points common to both items, such as "Reentry Methods" are listed together). The point-by-point comparison is favored in feasibility and recommendation reports because it offers readers a meaningful comparison between common points.

☑ EXERCISES

1. Locate, copy, and bring to class a paragraph that has the following features:

   - an orienting topic sentence
   - adequate development
   - unity
   - coherence
   - a recognizable sequence
   - appropriate length for its purpose and audience

   Be prepared to identify and explain each of these features in a class discussion.

2. For each of the following documents, indicate the most logical sequence. (For example, a description of a proposed computer lab would follow a spatial sequence.)

   - a set of instructions for operating a power tool.
   - a campaign report describing your progress in political fund raising
   - a report analyzing the weakest parts in a piece of industrial machinery
   - a report analyzing the desirability of a proposed oil refinery in your area
   - a detailed breakdown of your monthly budget to trim excess spending
   - a report investigating the reasons for student apathy on your campus
   - a report evaluating the effects of the ban on DDT in insect control
   - a report on any highly technical subject, written for a general reader
   - a report investigating the success of a no-grade policy at other colleges
   - a proposal for a no-grade policy at your college

☑ COLLABORATIVE PROJECTS

1. Organize into small groups. Choose *one* of these topics, or one your group settles on, and then brainstorm to develop a formal outline for the body section of a report. One representative from your group can write the final draft on the board, for class revision.

   - job opportunities in your career field
   - a physical description of the ideal classroom
   - how to organize an effective job search
   - how the quality of your higher educational experience can be improved
   - arguments for and against a formal grading system
   - an argument for an improvement you think this college needs most

2. Assume your group is preparing a report titled "The Negative Effects of Strip Mining on the Cumberland Plateau Region of Kentucky." After brainstorming and researching your subject, you all settle on these four major topics:
   - economic and social effects of strip mining
   - description of the strip-mining process
   - environmental effects of strip mining
   - description of the Cumberland Plateau

Arrange these topics in the most sensible sequence.

When your topics are arranged, assume that subsequent research and further brainstorming produce this list of subtopics:

   - method of strip mining used in the Cumberland Plateau region
   - location of the region
   - permanent land damage
   - water pollution
   - lack of educational progress
   - geological formation of the region
   - open-pit mining
   - unemployment
   - increased erosion
   - auger mining
   - natural resources of the region
   - types of strip mining
   - increased flood hazards
   - depopulation
   - contour mining

Arrange these subtopics (and perhaps some sub-subtopics) under appropriate topic headings. Use decimal notation to create the body section of a formal outline. Appoint one group member to present the outline in class.

Hint: Assume that this is your thesis: "Decades of strip mining (without reclamation) in the Cumberland Plateau have devastated this region's environment, economy, and social structure."

chapter 13

Revising for Readable Style

Y OU might write for a diverse or specific audience, or for experts or non-experts. But no matter how technically appropriate your document, audience needs are not served unless your style is readable.

A definition of style

What is *writing style,* and how does it influence reader response to a document? Your writing style is the product of

- the words you choose
- the way in which you put a sentence together
- the length of your sentences
- the way in which you connect sentences
- the tone you convey

Efficient writing style is neither fancy nor complicated; it is straightforward, easy to follow and understand—in a word, *readable.*

Efficient style requires much more than correct grammar, punctuation, and spelling. Granted, basic mechanical errors do distract readers; but correctness alone is no guarantee of a readable style. For example, this response to a job applicant is mechanically correct but inefficient:

Inefficient style

> We are in receipt of your recent correspondence indicating your interest in securing the advertised position. Your correspondence has been duly forwarded for consideration by the personnel office, which has employment candidate selection responsibility. You may expect to hear from us relative to your application as the selection process progresses. Your interest in the position is appreciated.

Notice how hard you have worked to extract information that could be expressed this simply:

More efficient

> Your application for the advertised position has been forwarded to our personnel office. As the selection process moves forward, we will be in touch. Thank you for your interest.

Inefficient style makes readers work harder than they should.

Style can be inefficient for many reasons, but it is especially inept when it

- makes the writing impossible to interpret
- takes too long to make the point
- reads like a Dick-and-Jane story from primary school
- uses imprecise or needlessly big words
- sounds stuffy and impersonal

Regardless of its cause, inefficient style results in writing that is less informative, less persuasive. Moreover, inefficient style can be unethical—by confusing or misleading readers.

To help your audience spend less time reading, you must spend more time revising for a style that is *clear, concise, fluent, exact,* and *likable.*

Revising for Clarity

A clear sentence conveys the writer's exact meaning on the first reading. It avoids ambiguous constructions, signals relationships among its parts, and emphasizes the main idea. The following guidelines will help you revise for clarity.

Avoid Ambiguous Phrasing. Workplace writing ideally has *one* meaning only, allows for *one* interpretation. Does one's "suspicious attitude" mean that one is "suspicious" or "suspect"?

Ambiguous phrasing	All managers are not required to submit reports. (*Are some or none required?*)
Revised	Managers are not all required to submit reports.

or

Managers are not required to submit reports.

Ambiguous phrasing	Most city workers strike on Friday.
Revised	Most city workers **are planning to strike** Friday.

or

Most city workers **typically strike** on Friday.

Make sure your writing conveys the meaning you intend.

Avoid Ambiguous Pronoun References. Whenever you use a pronoun (*he, she, it, their,* and so on), that pronoun must refer to one clearly identified noun. If the pronoun's referent (or antecedent) is vague, readers will be confused.

Ambiguous referent	Our patients enjoy the warm days while they last. (*Are the patients or the warm days on the way out?*)

Depending on whether the referent for *they* is *patients* or *warm days,* the sentence can be clarified.

Clear referent	While these warm days last, our patients enjoy them.

or

Our terminal patients enjoy the warm days.

Ambiguous referent	Jack resents his assistant because he is competitive. (*Who's the competitive one—Jack or his assistant?*)
Clear referent	Because his assistant is competitive, Jack resents him.

or

Because Jack is competitive, he resents his assistant.

Be sure readers can identify the noun your pronoun replaces. (See the Appendix for more on pronoun references, and page 300 for advice on avoiding sexist bias in pronoun use.)

Avoid Ambiguous Punctuation. A missing hyphen, comma, or other punctuation mark can obscure your meaning.

Missing hyphen	Replace the trailer's inner wheel bearings. (*The inner-wheel bearings or the inner wheel-bearings?*)
Missing comma	Does your company produce liquid hydrogen? If so, how[,] and where do you store it? (*Notice how the meaning changes with a comma after "how."*)
	Police surrounded the crowd[,] attacking the strikers. (*Without the comma, the crowd appears to be attacking the strikers.*)

Although missing hyphens and commas are prime culprits, other omissions can cause ambiguity as well. A missing colon after *kill* yields the headline "Moose Kill 200." A missing apostrophe after *Myers* creates this gem: "Myers Remains Buried in Portland." Punctuation *does* affect meaning.

Exercise 1.
Revise each sentence below to eliminate ambiguities in phrasing, pronoun reference, or punctuation.

a. Call me any evening except Tuesday after 7 o'clock.

b. The benefits of this plan are hard to imagine.

c. I cannot recommend this candidate too highly.

d. Visiting colleagues can be tiring.

e. Janice dislikes working with Claire because she's impatient.

f. Despite his efforts, Joe misinterpreted Sam's message.

g. Our division needs more effective writers.

h. Tell the reactor operator to evacuate and sound a general alarm.

i. If you don't pass any section of the test, your flying days are over.

j. Dial "10" to deactivate the system and sound the alarm.

Avoid Telegraphic Writing. Function words show relationships between the *content words* (nouns, adjectives, verbs, and adverbs) in a sentence. Some examples of function words:

- articles (*a, an, the*)
- prepositions (*in, of, to*)
- linking verbs (*is, seems, looks*)
- relative pronouns (*who, which that*)

Some writers mistakenly try to compress their writing by eliminating these function words.

Ambiguous	Proposal to employ retirees almost dead.
Revised	The proposal to employ retirees **is** almost dead.
Ambiguous	Uninsulated end pipe ruptured. (*What ruptured? The pipe or the end of the pipe?*)
Revised	**The** uninsulated end **of the** pipe ruptured.

or

The uninsulated pipe **on the** end ruptured.

Ambiguous	The reactor operator told management several times she expected an accident. (*Did she tell them once or several times?*)
Revised	The reactor operator told management several times **that** she expected an accident.

or

The reactor operator told management **that** several times she expected an accident.

Avoid Ambiguous Modifiers. Modifiers explain, define, or add detail to other words or ideas. If a modifier is too far from the words it modifies, the message can be ambiguous.

Misplaced modifier	**Only** press the red button in an emergency. (*Does **only** modify **press** or **emergency?***)
Revised	Press **only** the red button in an emergency.

or

Press the red button in an emergency **only.**

Another problem with ambiguity occurs when a modifying phrase has no word to modify.

Dangling modifier	**Being so well known in the computer industry,** I would appreciate your advice.

The writer intended to say that the *reader* is well known, but with no word to join itself to, the modifying phrase dangles. We can eliminate the confusion by adding a subject:

Revised	Because **you** are so well known in the computer industry, I would appreciate your advice.

See the Appendix for more on modifiers.

Exercise 2.
Revise each sentence below to repair telegraphic writing or to clarify ambiguous modifiers.

a. The manager claimed repeatedly she reported the danger.

b. I want the final Amex report written by your division.

    c. Replace main booster rocket seal.
    d. The president refused to believe any internal report was inaccurate.
    e. Our client failed to understand our recent proposal was preliminary.
    f. Only use this phone in a red alert.
    g. After offending our best client, I am deeply annoyed with the new manager.
    h. Send memo to programmer requesting explanation.
    i. Smith failed completely to explain the malfunction.
    j. Do not enter test area while contaminated.

Unstack Modifying Nouns. One noun can modify another noun (as in "software development"). But when two or more nouns modify a noun, the string of densely packed words becomes hard to read and ambiguous.

> Stacked   Be sure to leave enough time for a **training session participant** evaluation. (*Evaluation of the session or of the participants?*)

With no function words (articles, prepositions, verbs, relative pronouns) to break up the string of nouns, readers cannot see the relationships among the nouns. What modifies what?

Stacked nouns also deaden your style. Bring your style *and* your reader to life by using action verbs (*complete, prepare, reduce*) and prepositional phrases.

> Revised   Be sure to leave enough time **for** participants **to evaluate** the training session.
>
> *or*
>
> Be sure to leave enough time **to evaluate** participants **in the** training session.

No such problem with ambiguity occurs when *adjectives* are stacked in front of a noun.

> Clear   He was a **nervous, angry, confused,** but **dedicated** employee.

Readers can readily see that the adjectives modify *employee.*

Arrange Words for Coherence and Emphasis. In coherent writing, everything sticks together; each sentence builds on the preceding sentence and looks ahead to the following sentence. Sentences generally work best when the beginning looks back at familiar information and the end provides the new (or unfamiliar) information:

Familiar		*Unfamiliar*
My dog	has	fleas.
Our boss	just won	the lottery.
This company	is planning	a merger.

Besides helping a message stick together, the familiar-to-unfamiliar structure emphasizes the new information. Just as every paragraph has a key sentence, every sentence has a key word or phrase that sums up the new information. That key word or phrase usually is best emphasized at the end of the sentence.

Faulty emphasis	We expect a **refund** because of your error in our shipment.
Correct	Because of your error in our shipment, we expect a **refund.**
Faulty emphasis	In a business relationship, **trust** is a vital element.
Correct	In a business relationship, a vital element is **trust.**

One exception to placing key words last occurs with a statement in the imperative mood (a command, an order, an instruction), with the subject [you] understood. For instance, each step in a list of instructions contains an action verb (*insert, open, close, turn, remove, press*). To give readers a forecast, place the verb in that instruction at the beginning.

Correct	**Insert** the diskette before activating the system.
	Remove the protective seal.

With the key word at the beginning of the instruction, readers know immediately the action they need to take.

Exercise 3.
Revise each sentence below to unstack modifying nouns or to rearrange the word order for clarity and emphasis.

a. Develop online editing system documentation.
b. We need to develop a unified construction automation design.
c. Install a hazardous materials dispersion monitor system.
d. I recommend these management performance improvement incentives.
e. Our profits have doubled since we automated our assembly line.
f. Education enables us to recognize excellence and to achieve it.
g. In all writing, revision is required.
h. We have a critical need for technical support.
i. Sarah's job involves fault analysis systems troubleshooting handbook preparation.

Use Active Voice Often. The active voice ("I did it") is more direct, concise, and persuasive than the passive voice ("It was done by me"). In the active voice, the agent performing the action serves as subject:

Active	*Agent*	*Action*	*Recipient*
	Joe	lost	your report.
	Subject	*Verb*	*Object*

The passive voice reverses the pattern, making the recipient of an action serve as subject.

Passive	*Recipient*	*Action*	*Agent*
	Your report	was lost	by Joe.
	Subject	*Verb*	*Prepositional phrase*

Sometimes the passive eliminates the agent altogether:

Passive	Your report was lost. (*Who lost it?*)

Some writers mistakenly rely on the passive voice because they think it sounds more objective and important. But the passive voice often makes writing wordy, indecisive, evasive, and unethical. Consider the effect when an active statement is recast in the passive voice:

Concise and direct (active)	I underestimated labor costs for this project. (*7 words*)
Wordy and indirect (passive)	Labor costs for this project were underestimated by me. (*9 words*)
Evasive (passive)	Labor costs for this project were underestimated.

For economy, directness, and clarity, use the active voice in most of your writing.

Do not evade responsibility by hiding behind the passive voice:

Passive irresponsibles	A **mistake** was made in your shipment.
	It was decided not to hire you.
	A **layoff** is recommended.

Acknowledge responsibility for your actions:

Active	**I** made a mistake in your shipment.
	I decided not to hire you.
	Our committee recommends a layoff.

In reporting errors or bad news, use the active voice. Readers appreciate clarity and sincerity.

The passive voice creates a weak and impersonal tone.

Weak and impersonal	An offer will be made by us next week.
Strong and personal	We will make an offer next week.

Use the active voice when you want action. Otherwise, your statement will have no power.

Weak passive   If my claim is not settled by May 15, the Better Business Bureau will be contacted, and their advice on legal action will be taken.

This passive and tentative statement is unlikely to persuade readers to act: *Who* is making the claim? *Who* should settle the claim? *Who* will contact the Bureau? *Who* will take action? Nobody is here! Following is a more direct, concise, and forceful version, in the active voice:

Strong active   If you do not settle my claim by May 15, I will contact the Better Business Bureau for advice on legal action.

Notice how this active version emphasizes the new and significant information by placing it at the end.

Use the active voice for giving instructions.

Faulty passive   The bid should be sealed.
                  Care should be taken with the dynamite.

Correct active   **Seal** the bid.
                  Be careful with the dynamite.

Avoid shifts from active to passive voice in the same sentence.

Faulty shift   During the meeting, project members spoke and presentations were given.

Correct   During the meeting, project members spoke and gave presentations.

Unless you have a deliberate reason for choosing the passive voice, prefer the *active* voice for making forceful connections like the one described here:

By using the active voice, you direct the reader's attention to the subject of your sentence. For instance, if you write a job-application letter that is littered with passive verbs, you fail to achieve an important goal of that letter: to show the readers the important things you have done, and how prepared you are to do important things for them. That strategy requires active verbs, with clear emphasis on *you* and what you have done/are doing. (Pugliano 6)

Exercise 4.
The sentences below are wordy, weak, or evasive because of passive voice. Revise each sentence as a concise, forceful, and direct expression in the active voice, to identify the person or agent performing the action.

a. The evaluation was performed by us.

b. The report was written by our group.

    c. Unless you pay me within three days, my lawyer will be contacted.

    d. Hard hats should be worn at all times.

    e. It was decided to reject your offer.

    f. Gasoline was spilled on your Ferrari's leather seats.

    g. It is believed by us that this contract is faulty.

    h. Our test results will be sent to you as soon as verification is completed.

    i. The decision was made that your request for promotion should be denied.

Use Passive Voice Selectively. Passive voice can be quite appropriate in lab reports and other documents in which the agent's identity is immaterial to the message. Selected examples follow.

Use the passive when your audience does not need to know the agent.

| Correct passive | Mr. Jones was brought to the emergency room. |
| | The bank failure was publicized statewide. |

In the above contexts, readers would have little interest in knowing *who* brought Mr. Jones or *who* publicized the bank failure. Notice again how the passive voice focuses on the *recipient* rather than the *agent*.

Use the passive voice to focus on events or results when the agent is unknown, unapparent, or unimportant.

| Correct passive | All memos in the firm are filed in a database. |
| | Fred's article was published last week. |

The information that will interest readers here is *how* the memos are filed or *that* Fred's article was published.

Prefer the passive when you deliberately wish to be indirect or inoffensive (as in requesting the customer's payment or the employee's cooperation, or to avoid blaming someone—such as your boss) (Ornatowski 94).

Active but offensive	**You** have not paid your bill.
	You need to overhaul our filing system.
Inoffensive passive	**This bill** has not been paid.
	Our filing system needs to be overhauled.

By focusing on the recipient of the action rather than the agent, these passive versions help retain the audience's goodwill.

Use the passive voice if the person behind the action has reason for being protected.

| Correct passive | The criminal was identified. |
| | The embezzlement scheme was exposed. |

Here, the passive protects the innocent person who identified the criminal or who exposed the scheme.

Exercise 5.
The sentences below lack proper emphasis because of an improper use of the active voice. Revise each ineffective active as an appropriate passive, to emphasize the recipient rather than the actor.

a. Joe's company fired him.

b. A rockslide buried the mine entrance.

c. Someone on the maintenance crew has just discovered a crack in the nuclear-core containment unit.

d. A power surge destroyed more than 2,000 lines of our new applications program.

e. Your report confused me.

f. You are paying inadequate attention to worker safety.

g. You are checking temperatures too infrequently.

h. The tornado destroyed the barn.

i. The selection committee awarded Mary a Fulbright Scholarship.

j. You did a poor job editing this report.

Avoid Overstuffed Sentences. A sentence that crams too many ideas forces readers to struggle over its meaning:

> Overstuffed   Publicizing the records of a private meeting that took place three weeks ago to reveal the identity of a manager who criticized our company's promotion policy would be unethical.

Notice how the details are hard to remember, the relationships hard to identify. Clear things up by sorting out the relationships.

> Revised   In a private meeting three weeks ago, a manager criticized our company's policy on promotion. It would be unethical to reveal the manager's identity by publicizing the records of that meeting. (*Other versions are possible here, depending on the writer's intended meaning.*)

Give your readers only as much information as they can retain easily in one sentence.

Exercise 6.
Unscramble this overstuffed sentence by making shorter, clearer sentences:

A smoke-filled room causes not only teary eyes and runny noses but also can alter people's hearing and vision, as well as creating dangerous levels of carbon monoxide, especially for people with heart and lung ailments, whose health is particularly threatened by second-hand smoke.

Revising for Conciseness

Earlier sections showed that sometimes you have to add words to make meaning clear. But *needless* words can obscure meaning.

Writing can suffer from two kinds of wordiness: one kind occurs when readers are given information they don't need (think of an overly detailed weather report during local television news). The other kind of wordiness occurs when too many words are used conveying information readers *do* need (as in saying "a great deal of potential for the future" instead of "great potential").

Every word in the document should advance your meaning. We quote an expert on writing:

> Writing improves in direct ratio to the number of things we can keep out of it that shouldn't be there. (Zinsser 14)

A concise message conveys most information in fewest words. It gets right to the point without clutter.

Cluttered	These rafters weigh 300 pounds apiece, thereby exceeding by a total of 10 percent the load tolerance specified by the building inspector for this project.

However, conciseness does not mean omitting specific details necessary for clarity. A brief but vague message is useless.

Brief but vague	These rafters are too heavy.

Here is a concise version:

Brief but informative	These rafters weigh 300 pounds apiece, exceeding our specified load tolerance by 10 percent.

Be sure your information is adequate, but omit anything that adds no meaning.

Cluttered	At this point in time, I would like to say that we are ready to move ahead.
Concise	We are ready.

Use fewer words when fewer will do. But remember the difference between *clear writing* and *compressed writing* that is impossible to decipher.

Impenetrable	Give new vehicle air conditioner compression cut-off system specifications to engineering manager advising immediate action.

First drafts rarely are concise. The following strategies will help you revise for conciseness.

Avoid Needless Phrases. Don't use a whole phrase when one word will do. Instead of *in this day and age,* write *today.* Each needless phrase here can be reduced to one word—without loss in meaning:

at a rapid rate	=	rapidly
due to the fact that	=	because
the majority of	=	most
on a personal basis	=	personally
give instruction to	=	instruct
would be able to	=	could
readily apparent	=	obvious
a large number	=	many
prior to	=	before
aware of the fact that	=	know
conduct an inspection of	=	inspect

Eliminate Redundancy. A redundant expression says the same thing twice, in different words, as in *fellow colleagues.* Each boldfaced word below merely adds clutter, because its meaning is included in the other word.

a **dead** corpse	**end** result
completely eliminate	cancel **out**
basic essentials	consensus **of opinion**
enter **into**	**utter** devastation
mental awareness	**the month of** August
mutual cooperation	**utmost** perfection

Avoid Needless Repetition. Unnecessary repetition clutters writing and dilutes meaning.

Repetitious   In trauma victims, breathing is restored by **artificial respiration.** Techniques of **artificial respiration** include mouth-to-mouth **respiration** and mouth-to-nose **respiration.**

Repetition in that passage disappears when sentences are combined.

Concise   In trauma victims, breathing is restored by artificial respiration, either mouth-to-mouth or mouth-to-nose.

Repetition, of course, can be useful. Don't hesitate to repeat, or at least rephrase, material (even whole paragraphs in a longer document) if you feel that readers need reminders. Effective repetition helps avoid cross-references like this: "See page 23" or "Review page 10."

Exercise 7.
Revise each wordy sentence below to eliminate needless phrases, redundancy, and needless repetition.

a. I have admiration for Professor Jones.
b. Due to the fact that we made the lowest bid, we won the contract.
c. On previous occasions we have worked together.
d. She is a person who works hard.
e. We have completely eliminated the bugs from this program.
f. This report is the most informative report on the project.
g. Through mutual cooperation, we can achieve our goals.
h. I am aware of the fact that Sam is trustworthy.
i. This offer is the most attractive offer I've received.

Avoid There *Sentence Openers.* Save words, add force, and improve your emphasis by avoiding *There is* and *There are* to begin sentences—whenever your intended meaning allows.[1]

Weak	**There is** a coaxial cable connecting the antenna to the receiver.
Revised	A coaxial cable connects the antenna to the receiver.
Weak	**There is** a danger of explosion in Number 2 mineshaft.
Revised	Number 2 mineshaft is in danger of exploding.

Dropping these openers places the key words at sentence end, where they are best emphasized.

Avoid Some It *Sentence Openers.* Try not to begin a sentence with *It*—unless the *It* clearly points to a specific referent in the preceding sentence: "This document is excellent. It deserves special recognition."

Weak	**It** was his bad attitude that got him fired.
Revised	His bad attitude got him fired.
Weak	**It** is necessary to complete both sides of the form.
Revised	Please complete both sides of the form.

---

1. Of course, in some contexts, proper emphasis would call for a *there* opener.

Correct	People often have wondered about the rationale behind Boris's sudden decision. Actually, there are several good reasons for his dropping out of the program.

Most often, however, *there* openers are best dropped.

Delete Needless Prefaces. Don't keep readers waiting for the new information in your sentence. Get right to the point.

Wordy	**I am writing this letter because** I wish to apply for the position of copy editor.
Concise	Please consider me for the position of copy editor.
Wordy	**As far as artificial intelligence is concerned,** the technology is only in its infancy.
Concise	Artificial-intelligence technology is only in its infancy.

Exercise 8.

Revise each sentence below to eliminate *There* and *It* openers and needless prefaces.

a. There was severe fire damage to the reactor.

b. There are several reasons why Jane left the company.

c. It is essential that we act immediately.

d. It has been reported by Bill that several safety violations have occurred.

e. This letter is to inform you that I am pleased to accept your job offer.

f. The purpose of this report is to update our research findings.

Avoid Weak Verbs. Prefer strong verbs that express a definite action: *open, close, move, continue, begin.* These strong verbs advance your meaning. Avoid weak verbs that express no specific action: *is, was, are, has, give, make, come, take.* In some cases, such verbs are essential to your meaning: "Dr. Phillips is operating at 7 A.M." "Take me to the laboratory." But in other cases, weak verbs add words without advancing meaning. All forms of *to be* (*am, are, is, was, were, will, have been, might have been*) generally are weak. This next sentence achieves conciseness because of the strong verb *consider:*

Concise	Please **consider** my offer.

Here is what happens with a weak verb in the same sentence:

Weak and wordy	Please **take into consideration** my offer.

Don't disappear behind weak verbs and their baggage of needless nouns and prepositions.

Weak	My recommendation **is** for a larger budget.
Strong	I **recommend** a larger budget.

Strong verbs, or action verbs, suggest an assertive, positive, and confident writer. Here are some weak verbs converted to strong:

is in conflict with	=	conflicts
has the ability to	=	can
give a summary of	=	summarize
make an assumption	=	assume
come to the conclusion	=	conclude
take action	=	act
make a decision	=	decide

Exercise 9.
Revise each wordy and vague sentence below to eliminate weak verbs.

a. Our disposal procedure is in conformity with federal standards.

b. Please make a decision today.

c. We need to have a discussion about the problem.

d. I have just come to the realization that I was mistaken.

e. We certainly can make use of this information.

f. Your conclusion is in agreement with mine.

g. This manual gives instructions to end users.

Delete Needless To Be *Constructions.* The preceding section showed that forms of *to be* (*is, was, are,* and so on) are weak. Sometimes the *to be* form itself mistakenly appears behind such verbs as *appears, seems,* and *finds.*

> Wordy   Your product seems **to be** superior.
> I consider this employee **to be** highly competent.

Eliminating *to be* in these examples saves words while preserving meaning.

Avoid Excessive Prepositions. Earlier, we saw how prepositions can increase clarity by breaking up clusters of stacked modifying nouns. But prepositions combined with forms of *to be* can make wordy sentences.

> Wordy   The recommendation first appeared **in** the report written **by** the supervisor **in** January **about** that month's productivity.
>
> Concise   The recommendation first appeared in the supervisor's productivity report for January.

Here are some needlessly long prepositional phrases reduced to one or two words—without loss in meaning:

with the exception of	=	except for
in reference to	=	about (or regarding)
in order that	=	so
in the near future	=	soon

in the event that	=	if
at the present time	=	now
in the course of	=	during
in the process of	=	during (or in)

Fight Noun Addiction. Nouns manufactured from verbs (nominalizations) make sentences weak and wordy. Nominalizations often accompany weak verbs and needless prepositions.

Weak and wordy	We ask for the **cooperation** of all employees.
Strong and concise	We ask that all employees **cooperate.**
Weak and wordy	Give **consideration** to the possibility of a career change.
Strong and concise	**Consider** a career change.

Besides causing wordiness, a nominalization can be vague—by hiding the agent of an action.

| Wordy and vague | A **valid requirement** for immediate action exists. (*Who should take the action? We can't tell.*) |
| Precise | We **must act** immediately. |

Here are nominalizations restored to their verb forms:

conduct an investigation of	=	investigate
provide a description of	=	describe
conduct a test of	=	test
make a discovery of	=	discover

Along with weak verbs and needless prepositions, nominalizations drain the life from your style. In cheering for your favorite team, you wouldn't say "Blocking of that kick is a necessity!" instead of "Block that kick!"

Write as you would speak, but avoid slang or overuse of colloquialisms. Also avoid excessive economy. For example, "Employees must cooperate" would not be a desirable alternative to the first example in this section. But, for the final example, "Block that kick" would be.

Exercise 10.
Revise each sentence below to eliminate needless prepositions and *to be* constructions, and to cure noun addiction.

a. Igor seems to be ready for a vacation.

b. Our survey found 46 percent of users to be disappointed.

c. In the event of system failure, your sounding of the alarm is essential.

d. These are the recommendations of the chairperson of the committee.

    e. Our acceptance of the offer is a necessity.

    f. Please perform an analysis and make an evaluation of our new system.

    g. A need for your caution exists.

    h. I consider George to be an excellent technician.

    i. The appearance of this problem was just yesterday.

    j. Power surges are associated, in a causative way, with malfunctions of computers.

Make Negatives Positive. A positive expression is more easily understood than a negative one. As one expert on writing points out, "To understand the negative, we have to translate it into an affirmative, because the negative only implies what we should do by telling us what we shouldn't do. The affirmative states it directly" (Williams 54).

Indirect and wordy	Please do not be late in submitting your report.
Direct and concise	Please submit your report on time.

Readers have to work even harder to translate sentences with two or more negative expressions:

Confusing and wordy	Do **not** distribute this memo to employees who have **not** received a security clearance.
Clear and concise	Distribute this memo only to employees who have received a security clearance.

Besides the directly negative words (*no, not, never*), some words are indirectly negative (*except, forget, mistake, lose, uncooperative*). When these indirectly negative words combine with directly negative words, readers are forced to translate.

Confusing and wordy	**Do not neglect** to activate the alarm system.
	My diagnosis was **not inaccurate.**

The second example above shows how multiple negatives can make the writer seem evasive.

Clear and concise	**Be sure** to activate the alarm system.
	My diagnosis was **accurate.**

The positive versions are easier to understand *and* more persuasive.

Some negative expressions, of course, are perfectly correct, as in expressing disagreement.

Correct negatives	This is **not** the best plan.
	Your offer is **unacceptable.**
	This project **never** will succeed.

Prefer positives to negatives, however, whenever your meaning allows. Here are negative expressions translated into positive versions:

did not succeed	=	failed
does not have	=	lacks
did not prevent	=	allowed
not unless	=	only if
not until	=	only when
not absent	=	present

Clean Out Clutter Words. Clutter words stretch a message without advancing its meaning. Here are some of the commonest: *very, definitely, quite, extremely, rather, somewhat, really, actually, currently, situation, aspect, factor.*

Cluttered	**Actually,** one **aspect** of a business **situation** that could definitely make me **quite** happy would be to have a **somewhat** adventurous partner who **really** shared my **extreme** attraction to risks.
Concise	I seek an adventurous business partner who enjoys risks.

Use such words only when they *actually* advance your meaning.

Delete Needless Qualifiers. Qualifiers such as *I feel, It seems, I believe, In my opinion,* and *I think* soften the tone and impact of a statement. Use qualifiers to express uncertainty or to avoid seeming arrogant or overconfident.

Appropriate	Despite Frank's poor grades last year he will, **I think,** do well in college. Your product **seems** to be what we need.

When you are certain, eliminate the qualifier so as not to seem tentative or evasive.

Needless qualifiers	**It seems** that I've made an error. We **appear to** have exceeded our budget. **In my opinion,** this candidate is outstanding.

In communicating across cultures, keep in mind that a direct, forceful style might be seen as offensive (page 303).

Exercise 11.
Revise each sentence below to eliminate inappropriate negatives, clutter words, and needless qualifiers.

a. Our design must avoid nonconformity with building codes.

b. Never fail to wear protective clothing.

c. Do not accept any bids unless they arrive before May 1.

   d. I am not unappreciative of your help.
   e. We are currently in the situation of completing our investigation of all aspects of the accident.
   f. I appear to have misplaced the contract.
   g. Do not accept bids that are not signed.
   h. It seems as if I have just wrecked a company car.
   i. Our current situation is that we actually cannot offer you employment.

Revising for Fluency

Fluent sentences are easy to read because of clear connections, variety, and emphasis. Their varied length and word order eliminate choppiness and monotony. Fluent sentences enhance *clarity,* allowing readers to see what is most important, with no struggle to sort out relationships. Fluent sentences enhance *conciseness,* often replacing several short, repetitive sentences with one longer, economical sentence. The following strategies will help you write fluent sentences.

Combine Related Ideas. A series of short, disconnected sentences not only is choppy and wordy, but unclear as well.

Disconnected	Jogging can be healthful. You need the right equipment. Most necessary are well-fitting shoes. Without this equipment you take the chance of injuring your legs. Your knees are especially prone to injury. (*5 sentences*)
Clear, concise, and fluent	Jogging can be healthful if you have the right equipment. Shoes that fit well are most necessary because they prevent injury to your legs, especially your knees. (*2 sentences*)

Never force readers to figure out connections for themselves.

Most sets of information can be combined in different relationships, depending on what you want to emphasize. Imagine that this set of facts describes an applicant for a junior-management position with your company.

- Roy James graduated from an excellent management school.
- He has no experience.
- He is highly recommended.

Assume you are a personnel director, conveying to upper management your impression of this candidate. To convey a negative impression, you might combine the facts in this way:

Strongly negative emphasis	Although Roy James graduated from an excellent management school and is highly recommended, **he has no experience.**

The *independent* idea (in boldface) receives the emphasis. Earlier ideas are made dependent on (or subordinate to) the independent idea by the subordinating word *although*. When a sentence has two or more ideas of unequal importance, the less important idea is signaled by a subordinating word such as *often, as, because, if, unless, until,* or *while.*

To continue with our example: If you are undecided, but leaning in a negative direction, you might combine the information in this way:

> Strongly negative   Roy James graduated from an excellent management school
> emphasis   and is highly recommended, **but** he has no experience.

In the sentence above, the ideas before and after *but* are both independent. These independent ideas are joined by the coordinating word *but,* which suggests that both sides of the issue are equally important (or "coordinate"). Placing the negative idea last, however, gives it slight emphasis. When a sentence has two or more ideas equal in importance, their equality is signaled by coordinating words such as *and, but, for, nor, or, so,* or *yet.*

Consider once again our Roy James example. To emphasize strong support for the candidate, you could use this combination:

> Although Roy James has no experience, **he graduated from an excellent management school and is highly recommended.**

The above *although* subordinates the first idea, making the final ideas independent.

Caution: Combine sentences only to advance your meaning, to ease the reader's task. A sentence with too much information and too many connections can be impossible for readers to sort out.

> Overstuffed   Our night supervisor's verbal order from upper management
> to repair the overheated circuit was misunderstood by Leslie
> Kidd, who gave the wrong instructions to the emergency
> crew, thereby causing the fire within 30 minutes.
>
> Clearer   Upper management issued a verbal order to repair the over-
> heated circuit. When our night supervisor transmitted the
> order to Leslie Kidd, it was misunderstood. Kidd gave the
> wrong instruction to the emergency crew, and the fire began
> within 30 minutes.

Readability is affected less by the number of words in a sentence than by the amount of information. Even short sentences can be nearly impossible to interpret if they carry too many details:

> Overstuffed   Send three copies of Form 17-e to all six departments, unless
> Departments A or B or both request Form 16-w instead.

Although effective sentence combining enhances and streamlines meaning, overcombining makes meaning impenetrable.

Vary Sentence Construction and Length. Long and short sentences each have their purpose: to express ideas logically or forcefully.[2] We have just seen how related ideas often need to be linked in one sentence, so that readers can grasp the connections:

Disconnected	The nuclear core reached critical temperature. The loss-of-coolant alarm was triggered. The operator shut down the reactor.
Connected	As the nuclear core reached critical temperature, triggering the loss-of-coolant alarm, the operator shut down the reactor.

The ideas above have been combined to show that one action resulted from another. But an idea that should stand alone for emphasis needs a whole sentence of its own:

Correct	Core meltdown seemed inevitable.

Too much of anything loses effect. An unbroken string of long or short sentences can bore and confuse readers; so too can a series with identical openings:

Dreary	There are some drawbacks about diesel engines. **They** are difficult to start in cold weather. **They** cause vibration. **They** also give off an unpleasant odor. **They** cause sulfur dioxide pollution.
Varied	Diesel engines have some drawbacks. Most obvious are their noisiness, cold-weather starting difficulties, vibration, odor, and sulfur dioxide emission.

Notice how the use of parallel nouns in the above revision enables the listing of multiple items in a single sentence.

Opening sentences repeatedly with *The, This, He, She,* or *I* creates monotony. When you write in the first person, overusing *I* makes you appear self-centered. (Some organizations require use of the third person, avoiding the first person completely, for all manuals, lab reports, specifications, product descriptions, and so on.)

Do not avoid personal pronouns if they make the writing more readable (for example, by eliminating passive constructions). Instead, to avoid repetitious openings, combine ideas and shift word order.

Use Short Sentences for Special Emphasis. With all this talk about combining ideas, you might conclude that short sentences have no place in good writing. Wrong.

Whereas long sentences show connections and clarify relationships, short sentences (even one-word sentences) provide vivid emphasis. They stick in a reader's mind.

---

2. My thanks to Professor Edith K. Weinstein, University of Akron, for suggesting this distinction.

Exercise 12.

The sentence sets below[3] are disconnected, have no variety, or have no emphasis. Combine each set into *one* or *two* fluent sentences.

Choppy
: The world's forests are now disappearing. The rate of disappearance is 18 to 20 million hectares a year (an area half the size of California). Most of this loss occurs in humid tropical forests. These forests are in Asia, Africa, and South America.

Revised
: The world's forests are now disappearing at the rate of 18 to 20 million hectares a year (an area half the size of California). Most of the loss is occurring in the humid tropical forests of Africa, Asia, and South America.

a. The world's population will grow.
It will grow from 4 billion in 1975.
It will reach 6.5 billion in 2000.
This will be an increase of more than 50 percent.

b. In sheer numbers, population will be growing.
It will be growing faster after 2000 than it is today.
It will add 100 million people each year.
This figure compares with 75 million in 1975.

c. Energy prices are expected to increase.
Many less-developed countries will have increasing difficulty.
Their difficulty will be in meeting energy needs.

d. One-quarter of humanity depends primarily on wood.
They depend on wood for fuel.
For them, the outlook is bleak.

e. The world has finite fuel sources.
These include coal, oil, gas, oil shale, and uranium.
These resources theoretically are sufficient for centuries.
These resources are not evenly distributed.

f. Already the populations in parts of Africa and Asia have exceeded the carrying capacity of the immediate area.
This overpopulation has triggered erosion.
This erosion has reduced the land's capacity to support life.

---

3. Sample sentences are adapted from *Global Year 2000 Report to the President: Entering the 21st Century,* Washington, DC: GPO, 1980.

Exercise 13.

Combine each set of sentences below into one fluent sentence that provides the requested emphasis.

Sentence set
John is a loyal employee.
John is a motivated employee.
John is short-tempered with his colleagues.

Combined for positive emphasis
Even though John is short-tempered with his colleagues, he is a loyal and motivated employee.

Sentence set
This word processor has many features.
It includes a spelling checker.
It includes a thesaurus.
It includes a grammar checker.

Combined to emphasize thesaurus
Among its many features, such as spelling and grammar checkers, this word processor includes a thesaurus.

a. The job offers an attractive salary.
It demands long work hours.
Promotions are rapid.
(Combine for negative emphasis.)

b. The job offers an attractive salary.
It demands long work hours.
Promotions are rapid.
(Combine for positive emphasis.)

c. Our office software is integrated.
It has an excellent database management program.
Most impressive is its word processing capability.
It has an excellent spreadsheet program.
(Combine to emphasize the word processor.)

d. Company X gave us the lowest bid.
Company Y has an excellent reputation.
(Combine to emphasize Company Y.)

e. Superinsulated homes are energy efficient.
Superinsulated homes create a danger of indoor air pollution.
The toxic substances include radon gas and urea formaldehyde.
(Combine for a negative emphasis.)

f. Computers cannot *think* for the writer.
Computers eliminate many mechanical writing tasks.
They speed the flow of information.
(Combine to emphasize the first assertion.)

Finding the Exact Words

Too often, language can be a vehicle for *camouflage* rather than communication. People see many reasons to hide behind language, as when they

- speak for their company but not for themselves
- fear the consequences of giving bad news
- are afraid to disagree with company policy
- make a recommendation some readers will resent
- worry about making a bad impression
- worry about being wrong
- pretend to know more than they do
- avoid admitting a mistake or ignorance

Inflated and unfamiliar words, borrowed expressions, and needlessly technical terms camouflage meaning. Whether intentional or accidental, poor word choices have only one result: inefficient and often unethical writing that resists interpretation and frustrates the reader.

Following are strategies for finding words that are *convincing, precise,* and *informative.*

Use Simple and Familiar Words. If technical vocabulary is essential in your work, by all means use it. Yet be careful when writing for semitechnical or nontechnical readers. Don't replace technically precise words with nontechnical words that are vague or imprecise. Don't write *a part that makes the computer run* when you mean *central processing unit.* Use the precise term, and define it in a glossary for nontechnical readers:

Correct   Central processing unit: the part of the computer that controls information transfer and carries out arithmetic and logical instructions.

Certain technical words may be indispensable in certain contexts, but the nontechnical words usually can be simplified without any loss in meaning. Instead of *answering in the affirmative, say yes;* or instead of *endeavoring to promulgate* a new policy, *try to announce* it.

Unfamiliar words   Acoustically attenuate the food-consumption area.

Revised   Soundproof the cafeteria.

Don't use three syllables when one will do. Generally, trade for less:

aggregate	=	total
demonstrate	=	show
effectuate	=	cause
endeavor	=	effort, try
eventuate	=	result
frequently	=	often
initiate	=	begin
is contingent upon	=	depends on
multiplicity of	=	many
optimum	=	best
subsequent to	=	after
utilize	=	use

Count the syllables. Trim wherever you can. Most important, choose words you hear and use in everyday speaking—words that are universally familiar.

Don't write *I deem* when you mean *I think,* or *Keep me apprised* instead of *Keep me informed,* or *I concur* instead of *I agree,* or *securing employment* instead of *finding a job,* or *it is cost prohibitive* instead of *we can't afford it.* Experiments have shown that readers have to spend extra time on passages with unfamiliar or less familiar words (Bailey 70; Felker et al. 61).

Don't write like the author of a report from the Federal Aviation Administration, who recommended that manufacturers of the DC–10 reevaluate *the design of the entire pylon assembly to minimize design factors which are resulting in sensitive and/or critical maintenance and inspection procedures* (25 words, 50 syllables). A plain English translation: *Redesign the pylons so they are easier to maintain and inspect* (11 words, 18 syllables).

Besides the annoyance they cause, needlessly big or unfamiliar words can be *ambiguous.*

> Ambiguous   Make an improvement in the clerical situation.

Should we hire more clerical personnel or better personnel or should we train the personnel we have? Words chosen to impress readers too often confuse them instead. A plain style is more persuasive because "it leaves no one out" (Cross 6).

Of course, now and then the complex or more elaborate word is best—if it expresses your exact meaning. For instance, we would not substitute *end* for *terminate* in referring to something with an established time limit.

> Correct   Our trade agreement terminates this month.

If a complex word can replace a handful of simpler words—and can sharpen your meaning—use the complex word.

> Weak   Six rectangular grooves **around the outside edge** of the steel plate **are needed for** the pressure clamps **to fit into.**

Informative and precise	Six rectangular grooves on the steel plate **perimeter accommodate** the pressure clamps.
Weak	We need a **one-to-one exchange of ideas and opinions.**
Informative and precise	We need a **dialogue.**
Weak	Sexist language **contributes to the ongoing prevalence** of gender stereotypes.
Informative and precise	Sexist language **perpetuates** gender stereotypes.

Use the complex word *only* when your meaning and your audience demand it.

Exercise 14.
Revise each sentence below for straightforward and familiar language.

a. May you find luck and success in all endeavors.

b. I suggest you reduce the number of cigarettes you consume.

c. Within the copier, a magnetic-reed switch is utilized as a mode of replacement for the conventional microswitches that were in use on previous models.

d. A good writer is cognizant of how to utilize grammar in a correct fashion.

e. I will endeavor to ascertain the best candidate.

f. In view of the fact that the microscope is defective, we expect a refund of our full purchase expenditure.

g. I wish to upgrade my present employment situation.

Avoid Useless Jargon. Every profession has its own "shorthand." Among specialists, technical terms are a precise and economical way to communicate. For example, *stat* (from the Latin "statim" or "immediately") is medical jargon for *Drop everything and deal with this emergency.* For computer buffs, a *glitch* is a momentary power surge that can erase the contents of internal memory; a *bug* is an error that causes a program to run incorrectly. Such useful jargon conveys clear meaning to a knowledgeable audience.

Technical language, however, can be used appropriately or inappropriately. The latter is useless jargon, meaningless to insiders as well as outsiders. In the world of useless jargon people don't *cooperate* on a project; instead, they *interface* or *contiguously optimize their efforts.* Rather than *designing a model,* they *formulate a paradigm.* Instead of *observing limits* or *boundaries,* they *function within specific parameters.*

A popular form of useless jargon is adding *-wise* to nouns, as shorthand for *in reference to* or *in terms of.*

| Useless jargon | **Expensewise** and **schedulewise,** this plan is unacceptable. |
| Revised | In terms of expense and scheduling, this plan is unacceptable. |

Writers create another form of jargon when they invent verbs from nouns or adjectives by adding an *-ize* ending: Don't invent *prioritize* from *priority;* instead use *to rank priorities.*

Jargon's worst fault is that it makes the person using it seem stuffy and pretentious:

| Pretentious | Unless all parties interface synchronously within given parameters, the project will be rendered inoperative. |
| Possible translation | Unless we coordinate our efforts, the project will fail. |

Beyond reacting with frustration, readers often conclude that useless jargon is intended as camouflage by a writer who has something to hide.

Before using any jargon, think about your specific readers and ask yourself: "Can I find an easier way to say exactly what I mean?" Use jargon only if it *improves* your communication.

If your employer insists on needless jargon or elaborate phrasing, then you have little choice in the matter. What is best in matters of style is not always what some people consider appropriate. Determine what writing style your employer or organization expects. For short-term necessity, play by your organization's rules; but for long-term practice, remember that most documents that achieve superior results are written in plain English.

Use Acronyms Selectively. Acronyms are another form of specialized shorthand, or jargon. They are formed from the first letters of words in a phrase (as in *LOCA* from *l*oss *o*f *c*oolant *a*ccident) or from a combination of first letters and parts of words (as in *bit* from *b*inary dig*it* or *pixel* from *pic*ture *el*ement).

Computer technology has spawned countless acronyms, including:

ISDN	=	Integrated Services Digital Network
Telnet	=	Telephone Network
URL	=	Universal Resource Locator

Acronyms *can* communicate concisely—but only when the audience already knows their meaning, and only when you use the term often in your document. Always spell out the words from which the acronym is derived on first use.

An acronym defined

Modem ("modulator+demodulator"): a device that converts, or "modulates," computer data in electronic form into a sound signal that can be transmitted via phone line and then reconverted, or "demodulated," into electronic form for the receiving computer.

For lay audiences, try to avoid acronyms altogether or be sure to define the terms that make up the acronym.

Avoid Triteness. Writers who rely on tired old phrases (clichés) seem either too lazy or careless to find convincing or exact ways to say what they mean. Here are just a few of the countless expressions worn out by overuse:

make the grade	the chips are down
in the final analysis	not by a long shot
close the deal	last but not least
hard as a rock	welcome aboard
water under the bridge	over the hill
holding the bag	bite the bullet
up the creek	work like a dog

Exercise 15.
Revise each sentence below to eliminate useless jargon and triteness.

  a. For the obtaining of the X-33 word processor, our firm will have to accomplish the disbursement of funds to the amount of $6,000.

  b. To optimize your financial return, prioritize your investment goals.

  c. The use of this product engenders a 50-percent repeat consumer encounter.

  d. We'll have to swallow our pride and admit our mistake.

  e. We wish to welcome all new managers aboard.

  f. Not by a long shot will this plan succeed.

  g. Managers who make the grade are those who can take daily pressures in stride.

  h. Intercom utilization will be employed to initiate substitute employee operative involvement.

Avoid Misleading Euphemisms. Euphemisms are expressions aimed at politeness or at making unpleasant subjects seem less offensive. Thus, we *powder our nose* or *use the boys' room* instead of *using the bathroom;* we *pass away* or *meet our Maker* instead of *dying.* Euphemisms make the truth seem less painful.

When euphemisms avoid offending or embarrassing our audience, they are perfectly legitimate. Instead of telling a job applicant he or she is *unqualified,* we might say, *Your background doesn't meet our needs.* In addition, there are times when friendliness and interoffice harmony are more likely to be preserved with writing that is not too abrupt, bold, blunt, or emphatic (Mackenzie 2).

Euphemisms are unethical if they understate the truth when only the truth will serve. In the sugar-coated world of misleading euphemisms, bad news disappears:

  ■ Instead of being *laid off* or *fired,* workers are *surplused* or *deselected,* or the company is *downsized.*

- Instead of *lying* to the public, the government *engages in a policy of disinformation.*
- Instead of *wars* and *civilian casualties,* we have *conflicts* and *collateral damage.*

Language loses all meaning when *criminals* become *offenders,* when *rape* becomes *sexual assault,* and when people who are just plain *lazy* become *underachievers.* Plain talk is always better than deception. If someone offers you a job *with limited opportunity for promotion,* expect a *dead-end job.*

Avoid Overstatement. When they exaggerate to make a point, writers lose credibility. Be cautious when using words such as *best, biggest, brightest, most,* and *worst.*

Overstated	**Most** businesses have **no** loyalty toward their employees.
Revised	**Some** businesses have **little** loyalty toward their employees.
Overstated	You will find our product to be the **best.**
Revised	You will **appreciate the high quality** of our product.

Avoid Unsupported Generalizations. Base your conclusions on sufficient evidence.

Unsupported generalization	In 1983, twenty-one murderers were executed, and the murder rate dropped 8.1 percent for that year. These figures prove that the death penalty should be reinstated.

This single piece of evidence, from only one point in time and isolated from its larger social context, does not justify such a sweeping generalization. Establishing any kind of causal connection would require far more extensive data collected over many years and analyzed within a context that accounts for all sorts of variables: average age of the population, economic conditions, social conditions, and so on.

Unsupported generalizations harm your credibility because they have no way of being proved.

Be aware of the vast differences in meaning among these words:

few	never
some	rarely
many	sometimes
most	often
all	always

Unless you specify *few, some, many,* or *most,* readers can interpret your statement as meaning *all.*

Misleading	Assembly-line employees are doing shabby work.

Unless you mean *all* assembly-line employees, qualify your generalization with *some, most,* or another limiting word—even better, specify *20 percent.*

Exercise 16.
Revise each sentence below to eliminate euphemism, overstatement, or unsupported generalizations.

a. I finally must admit that I am an abuser of intoxicating beverages.
b. I was less than candid.
c. This employee is poorly motivated.
d. Your faulty machinery traumatically amputated my client's arm.
e. Most entry-level jobs are boring and dehumanizing.
f. Clerical jobs offer no opportunity for advancement.
g. Igor is the world's best employee.
h. Because of your absence of candor, we no longer can offer you employment.

Avoid Imprecise Words. Poorly chosen words can be offensive, embarrassing, misleading, or ambiguous. Be sure that what you say is what you mean.

Even words listed as synonyms in a thesaurus or dictionary carry different shades of meaning. Do you mean to say *I'm slender, You're thin, She's lean,* or *He's scrawny?* The wrong choice could be disastrous.

Just one wrong word can offend readers, as in this statement by a job applicant:

> Offensive   Another attractive feature of your company is its **adequate** training program.

While "adequate"might convey honestly the writer's intended meaning, the word seems inappropriate in this context (i.e., an applicant expressing a judgment about a program). Although the program may not have been highly ranked, our writer could have used any of several alternatives (*solid, respectable, growing*—or no modifier at all) without overstating the point or being offensive.

Poor word choice can be embarrassing, as in this example:

> Imprecise   **Chaos** is running this project.

Strictly speaking, *chaos* can't run anything!

Be especially aware of similar words with dissimilar meanings, as in these examples:

affect/effect	farther/further
all ready/already	fewer/less
almost dead/dying	healthy/healthful
among/between	imply/infer
continual/continuous	invariably/inevitably
eager/anxious	uninterested/disinterested
fearful/fearsome	worse/worst

Don't write *Skiing is healthy* when you mean that skiing promotes good health. Healthful things keep us healthy.

Be on the lookout for imprecisely phrased (and therefore illogical) comparisons.

Imprecise	Your bank's interest rate is higher than Citibank. (Can a rate be higher than a bank?)
Precise	Your bank's interest rate is higher than Citibank's.

Precision is above all essential to the informative value of your writing. Imprecise language can be misleading. Consider how meanings differ:

Differing meanings	Include **less** technical details in your report.
	Include **fewer** technical details in your report.

Imprecise language can result in ambiguous messages as well. For instance, is *send us more personal information* a request for more information that is personal or for information that is more personal?

Ambiguous	Loan payments are due **bimonthly.** (Every other month or twice monthly?)
Clear	Loan payments are due twice monthly.

or

Loan payments are due every other month.

Precision ultimately enhances conciseness, when one exact word replaces multiple inexact words.

Wordy and less exact	I have **put together** all the financial information.
	Keep doing this exercise for ten seconds.
Concise and more exact	I have **assembled** all the. . . .
	Continue this exercise. . . .

Be Specific and Concrete. General words name broad classes of things, such as *job, computer,* or *person.* Such terms usually need to be clarified by more specific ones.

job	=	senior accountant for Softbyte Press
computer	=	Macintosh Performa 636CD
person	=	Sarah Jones, production manager

The more specific your words, the sharper your meaning.

General	structure
	dwelling
	vacation home
	log cabin
Specific	log cabin in Vermont
	a three-room log cabin on the banks of the Battenkill River

Notice how the picture becomes more vivid as we move to lower levels of generality.

Abstract words name qualities, concepts, or feelings (*beauty, luxury, depression*) whose exact meaning has to be nailed down by *concrete* words—words that name things we can know through our five senses.

a **beautiful** view	=	snowcapped mountains, a wilderness lake, pink granite ledge, ninety-foot birch trees
a **luxurious** condominium	=	imported tiles, glass walls, oriental rugs
a **depressed** worker	=	suicidal urge, insomnia, feelings of worthlessness, no hope for improvement

Informative writing *tells* and *shows.*

> General   One of our **workers** was **injured** by a **piece of equipment recently.**

The boldface words only *tell* without showing.

> Specific   **Alan Hill** suffered a **broken thumb** while working on a **lathe yesterday.**

Choose informative words that express exactly what you mean. Don't write *thing* when you mean *lever, switch, micrometer,* or *disk.* Instead of evaluating an employee as *good, great, disappointing,* or *terrible,* use terms that are more concrete, such as *reliable, skillful, dishonest,* or *incompetent*—further clarified by examples, such as *never late for work.*

In some instances, of course, you may wish to generalize for the sake of diplomacy. Instead of writing *Bill, Mary, and Sam have been tying up the office phones with personal calls,* you might prefer to generalize: *Some employees have been.* The second version makes the point without accusing anyone in particular.

When you can, provide solid numbers and statistics that get your point across:

> General   In 1972, thousands of people were killed or injured on America's highways. Many families had at least one relative who was a casualty. After the speed limit was lowered to 55 miles per hour in late 1972, the death toll began to drop.
>
> Specific   In 1972, 56,000 people died on America's highways; 200,000 were injured; 15,000 children were orphaned. In that year, if you were a member of a family of five, chances are that someone related to you by blood or law was killed or injured in an auto accident. After the speed limit was lowered to 55 miles per hour in late 1972, the death toll dropped steadily to 41,000 in 1975.

Concrete and specific expressions not only are more informative; they are more persuasive as well.

Exercise 17.
Revise each sentence below to make it more precise or informative.

a. Our outlet does more business than Chicago.

b. Anaerobic fermentation is used in this report.

c. Confusion is in control of this office.

d. Your crew damaged a piece of office equipment.

e. His performance was admirable.

f. This thing bothers me.

Use Analogies to Sharpen the Image. Ordinary comparison shows similarities between two things *of the same class* (two computer keyboards, two technicians, two methods of cleaning dioxin-contaminated sites). Analogy, on the other hand, shows some essential similarity between two things of *different classes* (report writing and computer programming, computer memory and post office boxes).

Analogies are good for emphasizing a point (*Some rain is now as acidic as vinegar*). They are especially useful in translating something abstract, complex, or unfamiliar, as long as the easier subject is broadly familiar to readers. Analogy therefore calls for particularly careful analyses of audience.

Analogies can save words and convey vivid images. *Collier's Encyclopedia* describes the tail of an eagle in flight as "spread like a fan." The following sentence from a description of a trout feeder mechanism uses an analogy to clarify the positional relationship between two working parts:

> Analogy   The metal rod is inserted (and centered, **crosslike**) between the inner and outer sections of the clip.

Without the analogy *crosslike,* we would need something like this to visualize the relationship:

> Missing analogy   The metal rod is inserted, **perpendicular to the long plane and parallel to the flat plane,** between the inner and outer sections of the clip.

This second version is doubly inefficient: more words are needed to communicate, and more work is needed to understand the meaning.

Besides naming things vividly, analogies help *explain* things. The following analogy from the *Congressional Research Report* helps us understand something unfamiliar (dangerous levels of a toxic chemical) by comparing it to something more familiar (human hair).

> Analogy   A dioxin concentration of 500 parts per trillion is lethal to guinea pigs. One part per trillion is roughly equal to the thickness of a human hair compared to the distance across the United States.

Adjusting Your Tone

Your tone is your personal stamp—the personality that takes shape between the lines. The tone you create in any writing depends on (1) the distance you impose between yourself and the reader, and (2) the attitude you show toward the subject.

Assume that a friend is going to take over a job you've held. You've decided to write your friend instructions for parts of the job. Here is your first sentence:

Informal

> Now that you've arrived in the glamorous world of office work, put on your track shoes; this is no ordinary manager-trainee job.

What is the tone in that sentence? First, we notice that the sentence imposes little distance between the writer and the reader (it uses the direct address, "you," and the humorous suggestion to "put on your track shoes"). The ironic use of "glamorous" suggests that the writer means just the opposite: that the job holds little glamour.

For a different reader (the recipient of a company training manual, for example), the writer would have chosen some other opening:

Semiformal

> As a manager trainee at GlobalTech, you will work for many managers. In short, you will spend little of your day seated at your desk.

The tone now is serious, no longer intimate, and the writer expresses no distinct attitude toward the job. For yet another audience (those who will read a company pamphlet for clients or investors), the writer again might alter the tone:

Formal

> Manager trainees at GlobalTech are responsible for duties that extend far beyond desk work.

Here the businesslike, impersonal tone imposes greater distance between writer and audience, especially with the shift from second- to third-person address. The tone seems far too impersonal for any document addressed to the trainees themselves. Your tone changes in response to the situation and audience, even if the subject remains the same.

We already know how tone works in speaking. When you meet someone new, for example, you respond in a tone that defines your relationship:

Tone announces interpersonal distance

> Honored to make your acquaintance. [*formal tone—greatest distance*]
>
> How do you do? [*formal*]
>
> Nice to meet you. [*semiformal—medium distance*]
>
> Hello. [*semiformal*]
>
> Hi. [*informal—least distance*]
>
> What's happening? [*informal—slang*]

Your greeting (and tone) will depend on how much distance you decide is appropriate, and, in turn, the tone will determine how you come across. Each of these greetings is appropriate in some situations, inappropriate in others.

"What's happening" might be okay when you meet another student, but not the college or company president.

To decide on an appropriate distance from which to address a particular audience, follow these guidelines:

- Use a formal or semiformal tone in writing for superiors, professionals, or academics (depending on what you think the reader expects).
- Use a semiformal or informal tone in writing for colleagues and subordinates (depending on how close you feel to your readers).
- Use an informal tone when you want your writing to be conversational, or when you want it to sound like a person talking.
- Above all, find out what the preferences are in your organization.

Whichever tone you decide on, be consistent throughout your document.

> Inconsistent tone — My office isn't fit for a pig [*too informal*]; it is ungraciously unattractive [*too formal*].
>
> Revised — The shabbiness of my office makes it an unfit place to work.

In general, lean toward an informal tone without falling into slang.

In addition to setting the distance between writer and reader, your tone implies your *attitude* toward the subject. Consider the different attitudes expressed in these examples:

Tone announces attitude

> We dine at seven.
>
> Dinner is at seven.
>
> Let's eat at seven.
>
> Let's chow down at seven.
>
> Let's strap on the feedbag at seven.
>
> Let's pig out at seven.

The words we choose tell readers a great deal about where we stand.

If readers expect an impartial report, try to keep your own biases out of it. But for situations in which your opinion *is* expected, or in which you perceive some danger or ethics violation, let readers know where you stand.

> Plain English needed — Avoid prolix nebulosity.
>
> Revised — Don't be wordy and vague.

Say *I enjoyed the fiber optics seminar* instead of *My attitude toward the fiber optics seminar was one of high approval.* Say *Let's liven up our dull relationship* instead of *We should inject some rejuvenation into our lifeless liaison.* Say *Methane levels in No. 3 mineshaft pose the definite risk of explosion* instead of *Rising methane levels in No. 3 mineshaft should be evaluated.* Make sure your attitude is clear and appropriate for the situation.

Let your attitude reflect your relationship with the reader—and what you think the reader expects. In an upcoming meeting about the reader's job evaluation, does your reader expect to *discuss* the evaluation, *talk it over, have a chat,* or *chew the fat?* If the situation calls for a serious tone, don't use language that suggests a casual attitude—or vice versa. Use the following guidelines for making your tone conversational and appropriate.

Use an Occasional Contraction. Unless you have reason to be formal, use (but do not overuse) contractions to loosen the tone. Balance an *I am* with an *I'm*, a *you are* with a *you're*, an *it is* with an *it's* (as we've done throughout this book).

Missing contraction	Do not be wordy and vague.
Revised	Don't be wordy and vague.

Generally, use contractions only with pronouns, not with nouns or proper nouns (names). Otherwise, the constructions are awkward or ambiguous.

Awkward contractions	Barbara'll be here soon.
	Health's important
Ambiguous contractions	The dog's barking.
	Bill's skiing.

These ambiguous contractions easily can be confused with possessive constructions.

Address Readers Directly. Use the personal pronouns *you* and *your* to connect with readers. Otherwise, your writing sounds impersonal.

Impersonal tone	A writer should use **you** and **your** generously to connect with his or her readers.

Notice how distance *and* words increase. Direct address creates a more personal tone.

Impersonal tone	Students at our college will find the faculty always willing to help.
Personal tone	As a student at our college, **you** will find the faculty always willing to help.

Research shows that readers relate better to something they consider meaningful to them personally (Felker 33).

Caution: Use *you* and *your* only to correspond *directly* with the reader, as in a letter, memo, instructions, or some form of advice, encouragement, or persuasion. By using *you* and *your* when your subject and purpose call for first or third person, you might write something wordy and awkward like this:

Wordy and awkward	**When you** are in northern Ontario, **you** can see wilderness lakes everywhere around **you.**
Appropriate	Wilderness lakes are everywhere in northern Ontario.

Exercise 18.

The sentences below suffer from pretentious language, unclear expression of attitude, missing contractions, or indirect address. Adjust the tone.

a. Further interviews are a necessity to our ascertaining the most viable candidate.

b. This project is beginning to exhibit the characteristics of a loser.

c. We are pleased to tell you that you are a finalist.

d. Do not submit the proposal if it is not complete.

e. Employees must submit travel vouchers by May 1.

f. Persons taking this test should use the HELP option whenever they need it.

g. I am not unappreciative of your help.

h. My disapproval is far more than negligible.

Use I *and* We *When Appropriate.* Don't disappear behind your writing. Use *I* or *We* when referring to yourself or your organization.

> Distant   The writer would like a refund.
>
> Revised   I would like a refund.

A message can become doubly impersonal when both the writer and the reader disappear.

> Impersonal   The requested report will be sent next week.
>
> Personal   **We** will send the report **you** requested next week.

Avoid, of course, opening too many sentences with *I*. Combine ideas and shift word order instead.

Prefer the Active Voice. Because the active voice is more direct and economical than the passive voice, it generally creates a less formal tone. (Review pages 266–69.)

> Passive and impersonal   Travel expenses cannot be reimbursed unless receipts are submitted.
>
> Active and personal   We cannot reimburse your travel expenses unless you submit receipts.

Exercise 19.

These sentences have too few *I* or *We* constructions or too many passive constructions. Adjust the tone.

a. Payment will be made as soon as an itemized bill is received.

b. You will be notified.

c. Your help is appreciated.

d. Our reply to your bid will be sent next week.

e. Your request will be given our consideration.

f. My opinion of this proposal is affirmative.

g. This writer would like to be considered for your opening.

Emphasize the Positive. We all respond more favorably to encouragement than to criticism. Whenever you offer advice, suggestions, or recommendations, try to emphasize benefits rather than flaws.

Critical tone	Because of your division's lagging productivity, a management review may be needed.
Encouraging tone	A management review might help boost productivity in your division.

Avoid an Overly Informal Tone. We generally do not write in the same way we would speak to friends at the local burger joint or street corner. Achieving a conversational tone does not mean lapsing into substandard usage, slang, profanity, or excessive colloquialisms. *Substandard usage* ("He ain't got none," "I seen it today," "She brang the book") fails to meet standards of educated expression. *Slang* ("hurling," "belted," "bogus," "bummed") usually has specific meaning only for members of a particular in-group. *Profanity* ("This idea sucks," "What the hell") not only displays contempt for the audience but often earns contempt for the person using it. *Colloquialisms* ("O.K.," "a lot," "snooze," "in the bag") are understood more widely than slang, but tend to appear more in speaking than in writing.

Slang and profanity are almost always inappropriate in school or workplace writing. The occasional colloquial expression, however, helps soften the tone of any writing—as long as the situation calls for a measure of informality. Tone is considered offensive when it violates the reader's expectations: when it seems disrespectful or tasteless, or distant and aloof, or too chummy, casual, or otherwise inappropriate for the topic, the reader, and the situation.

A formal, or academic, tone, is perfectly appropriate in countless writing situations: a research paper, a job application, a report for the company president. In a history essay, for example, we would not refer to George Washington and Abraham Lincoln as "those dudes, George and Abe."

Whenever we begin with rough drafting or brainstorming, our tone might be overly informal and is likely to require some adjustment during subsequent drafts.

Avoid Bias. Biased people make judgments without examining the facts. Your responsibility as a writer is to report accurately and fairly without distorting the evidence or injecting loaded words that reflect personal biases. If you *are* asked to include your interpretations and conclusions, base them on the facts. Even controversial subjects deserve unbiased treatment.

Imagine you have been sent to investigate the causes of an employee-management confrontation at your company's Omaha branch. Your initial report, written for the New York central office, is intended simply to describe what happened. Here is how an unbiased description might read:

A factual account

> At 9:00 A.M. on Tuesday, January 21, eighty women employees set up picket lines around the executive offices of our Omaha branch, bringing business to a halt. The group issued a formal protest, claiming that their working conditions were repressive, their salary scale unfair, and their promotional opportunities limited. The women demanded affirmative action, insisting that the company's hiring and promotional policies and wage scales be revised. The demonstration ended when Garvin Tate, vice president in charge of personnel, promised to appoint a committee to investigate the group's claims and to correct any inequities.

Notice the absence of implied judgments; the facts are presented objectively. A less impartial version of the event, from a protestor's point of view, might read:

A biased version

> Last Tuesday, sisters struck another blow against male supremacy when eighty women employees paralyzed the company's repressive and sexist administration for more than six hours. The timely and articulate protest was aimed against degrading working conditions, unfair salary scales, and lack of promotional opportunities for women. Stunned executives watched helplessly as the group organized their picket lines, determined to continue their protest until their demands for equal rights were addressed. An embarrassed vice-president quickly agreed to study the group's demands and to revise the company's discriminatory policies. The success of this long-overdue confrontation serves as an inspiration to oppressed women employees everywhere.

Judgmental words (*male supremacy, degrading, paralyzed, articulate, stunned, discriminatory*) inject the writer's attitude, even though it isn't called for. In contrast to this bias, the following version patronizingly defends the status quo:

A biased version

> Our Omaha branch was the scene of an amusing battle of the sexes last Tuesday, when a group of irate feminists, eighty strong, set up picket lines for six hours at the company's executive offices. The protest was lodged against supposed inequities in hiring, wages, working conditions, and promotion for women in our company. The radicals threatened to surround the building until their demands for "equal rights" were met. A bemused vice-president responded to this carnival demonstration with patience and dignity, assuring the militants that their claims and demands—however inaccurate and immoderate—would receive just consideration.

Again, qualifying adjectives and superlatives slant the tone.

Being unbiased, of course, doesn't mean burying your head—and your values—in the sand. Remaining neutral about something you know to be wrong or dangerous is unethical (Kremers 59). You have an ethical responsibility to weigh the facts and make your views known. If, for instance, you conclude that the Omaha protest was clearly justified, don't hesitate to say so.

Avoid Sexist Usage. The way we as a culture use language reflects the way we think about ourselves. Usage that gives people in general a male identity allows no room for females. In fact, females become virtually invisible in a world of *policemen, congressmen, firemen, foremen, selectmen,* and *aldermen.*

Sexist usage refers to doctors, lawyers, and other professionals as *he* or *him,* while referring to nurses, secretaries, and homemakers as *she* or *her.* In this traditional stereotype, males do the jobs that really matter and that pay higher wages, whereas females serve only as support and decoration. When females do invade traditional "male" roles, we might express our surprise at their boldness by calling them *female executives, female sportscasters, female surgeons,* or *female hockey players.* Likewise, to demean males who have settled for "female" roles, we sometimes refer to *male secretaries, male nurses, male flight attendants,* or *male models.*

In the biased reality of sexist usage, the title "Mr." protects the privacy of a male who might be married or unmarried, while "Mrs." and "Miss" announce a female's marital status to the world. Moreover, an unmarried male fondly is referred to as *bachelor,* while his female counterpart is stigmatized as an *old maid* or a *spinster.*

Besides being misleading and demeaning, sexist usage is offensive. Instead of bringing writer and readers close together, sexist usage severs the human contact our writing otherwise might achieve.

To help eliminate sexism from your writing, follow these guidelines:

- Use neutral expressions:

chair, or chairperson	rather than	chairman
businessperson	rather than	businessman
supervisor	rather than	foreman
police officer	rather than	policeman
letter carrier	rather than	postman
homemaker	rather than	housewife
humanity, or humankind	rather than	mankind
actor	rather than	actor vs. actress

- Rephrase to eliminate the pronoun, if you can do so without changing your original meaning.

    Sexist   A writer will succeed if **he** revises.
    Revised   A writer who revises succeeds.

- Use plural forms. Instead of *Each doctor . . . he,* use *All doctors . . . they* (but not *Each doctor . . . they*).

    Sexist   A writer will succeed if **he** revises.
    Revised   Writers will succeed if **they** revise (but *not* A writer will succeed if **they** revise.)

When using a plural form, don't create an error in pronoun-referent agreement by having the *plural* pronoun *they* or *their* refer to a *singular* referent (as in **Each writer** *should do* **their** *best*).

- When possible (as in direct address) use *you:* **You** *will succeed if* **you** *revise.* But use this form *only* when addressing someone directly. (See page 296 for discussion.)
- Use occasional pairings (*him* or *her, she* or *he, his* or *hers*): *A writer will succeed if* **she or he** *revises.*

  Overuse of such pairings can be awkward: *A writer should do* **his or her** *best to make sure that* **he or she** *connects with* **his or her** *readers.* Most handbooks now encourage alternating use between the two pronouns, and discourage pairings and *he/she: An effective writer always focuses on* **her** *audience; the writer strives to connect with all* **his** *readers.*

- Drop diminutive endings such as *-ess* and *-ette* used to denote females (*poetess, drum majorette, actress,* etc.). Such endings seem to perpetuate an image of *the little woman.*
- Use *Ms.* instead of *Mrs.* or *Miss,* unless you know that person prefers one of the traditional titles. Or omit titles completely: *Roger Smith* and *Jane Kelly; Smith* and *Kelly.*
- In quoting sources that have ignored present standards for nonsexist usage, consider these options: Insert [*sic*] ("thus" or "so") following the first instance of sexist terminology in a particular passage. Use ellipsis to omit sexist phrasing. Paraphrase the material, instead of quoting it directly.

Not only do the words you choose reveal your way of seeing, but they also influence your reader's way of thinking. Sexist language carries built-in judgments, and, as renowned linguist S. I. Hayakawa reminds us: "Judgment stops thought." In a world unethically defined by sexist language, everyone remains frozen in her and his place.

Avoid Offensive Usages of All Types. Enlightened communication respects all people in reference to their specific cultural, racial, ethnic, and national background; sexual and religious orientation; age or physical condition. References to individuals and groups should be as neutral as possible; no matter how inadvertent, any expression that seems condescending or judgmental or that violates the reader's sense of appropriateness is offensive. Detailed guidelines for reducing biased usage appear in these two works, among others:

Schwartz, Marilyn, et al. *Guidelines for Bias-Free Writing.* Bloomington: Indiana UP, 1995.

Publication Manual of the American Psychological Association, 4th ed. Washington, DC: American Psychological Association, 1994.

Below is a sampling of suggestions adapted from the previous works:

- When referring to members of a particular culture, be as specific as possible about that culture's identity: Instead of *Latin American* or *Asian* or *Hispanic,* for instance, prefer *Cuban American* or *Korean* or *Nicaraguan.* Instead of *American workforce,* specify *U.S. workforce* when referring to the United States.

  Avoid judgmental expressions: Instead of *third-world* or *undeveloped nations* or the *Far East,* prefer *developing* or *newly industrialized nations* or *East Asia.* Instead of *nonwhites,* refer to *people of color.*

- When referring to someone who has a disability, avoid terms that could be considered pitying or overly euphemistic, such as *victims, unfortunates, challenged,* or *differently abled.* Focus on the individual instead of the disability: Instead of *blind person* or *amputee,* refer to a *person who is blind* or a *person who has lost an arm.*

  In general usage, avoid expressions that demean those who have medical conditions: *retard, mental midget, insane idea, lame excuse, the blind leading the blind, able-bodied workers,* and so on.

- When referring to members of a particular age group, prefer *girl* or *boy* for people of age fourteen or under; *young person, young adult, young man,* or *young woman* for those of high-school age; and *woman* or *man* for those of college age. (*Teenager* or *juvenile* carries certain negative connotations.) Instead of *the elderly,* prefer *older persons.*

Keep in mind that standards and conventions for bias-free communication continue to evolve. In addition to consulting relevant publications, keep up to date on new developments and always rely on careful and conscientious audience analysis for addressing a specific readers' expectations.

Exercise 20.
The sentences below suffer from negative emphasis, excessive informality, biased expressions, or offensive usage. Adjust the tone.

- *a.* If you want your workers to like you, show sensitivity to their needs.
- *b.* By not hesitating to act, you prevented my death.
- *c.* The union has won its struggle for a decent wage.
- *d.* The group's spokesman demanded salary increases.
- *e.* Each employee should submit his vacation preferences this week.
- *f.* While the girls played football, the men waved pom-poms.
- *g.* Aggressive management of this risky project will help you avoid failure.
- *h.* The explosion left me blind as a bat for nearly an hour.
- *i.* This dude would be an excellent employee if only he could learn to chill out.

Exercise 21.

Find examples of overly euphemistic language (such as "chronologically chal-lenged") or of insensitive language. Be prepared to discuss your examples in class.

Considering the Cultural Context

The style guidelines throughout this chapter apply specifically to standard English in a North American communications context. But practices and preferences can differ widely in different cultural contexts. Certain cultures might prefer long sentences and technical language, to convey an idea's full complexity. Other cultures value expressions of respect, politeness, praise, and gratitude more than mere efficiency (Hein 125–26; Mackin 349–50).

Some documents in other languages tend to be more formal than in English, and some rely heavily on passive voice (Weymouth 144). French audi-ences, for example, may prefer an elaborate style that reflects sophisticated and complex modes of thinking. In contrast, our "plain English," conversational style might connote simple-mindedness, disrespect, or incompetence (Thrush 277).

In translation or in a different cultural context, certain words carry offensive or unfavorable connotations. A few notable disasters (Gesteland 20; Victor 44):

- The Chevrolet *Nova*—meaning "don't go" in Spanish
- The Finnish beer *Koff*—for an English-speaking market
- Colgate's *Cue* toothpaste—an obscenity in French
- A brand of bicycle named *Flying Pigeon*—imported for a U.S. market

Idioms ("strike out," "over the top") hold no logical meaning for other cultures. Slang ("bogus," "fat city") and colloquialisms ("You bet," "Gotcha") can strike readers as too informal and crude.

Any form of offensive style (including profanity and inappropriate humor) can alienate audiences—toward your company *and* your culture (Sturges 32).

Avoiding Reliance on Automated Tools

Many of the strategies covered in this chapter could be executed rapidly with word-processing software. By using the *global search and replace function* in some programs, you can command the computer to search for ambiguous pronoun references, overuse of passive voice, *to be* verbs, *There* and *It* sen-tence openers, negative constructions, clutter words, needless prefaces and qualifiers, overly technical language, jargon, sexist language, and so on. With an online dictionary or thesaurus, you can check definitions or see a list of synonyms for a word you have used in your document.

Despite the increasing sophistication of style checkers, diction checkers, and other such editing aids, automation can be extremely imprecise. No amount of automation is likely to eliminate the writer's burden of *choice*. None of the rules or advice offered in this chapter applies universally. Our language confronts us with almost infinite choices that cannot be programmed into a computer. Ultimately, it is the informed writer's sensitivity to meaning, emphasis, and tone—the human contact—that determines the effectiveness of any document.

PART

IV

Graphic and Design Elements

Designing Visuals

Why Visuals Are Essential

When to Use a Visual

What Types of Visuals to Consider

How to Select Visuals for Your Purpose and Audience

Tables

Graphs

Charts

Graphic Illustrations

Computer Graphics

How to Avoid Visual Distortion

How to Incorporate Visuals with the Text

A VISUAL is any pictorial representation used to clarify a concept, emphasize a particular meaning, illustrate a point, or analyze ideas or data. Besides saving space and words, visuals help audiences process, understand, and remember information. Because they offer powerful new ways of looking at data, visuals reveal trends, problems, and possibilities that otherwise might remain buried in lists of facts and figures.

In printed or online documents, in oral presentations or multimedia programs, visuals are a staple of communication today. This chapter covers four main types of visuals: tables, graphs, charts, and illustrations.

Why Visuals Are Essential

Readers expect more than just raw information; they want the information processed for their understanding. Readers want to feel intelligent, to understand the message at a glance. Visuals help us answer many of the questions asked by readers as they process information:

Typical Reader Questions in Processing Information

- *Which information is most important?*
- *Where, exactly, should I focus?*
- *What do these numbers mean?*
- *What should I be thinking or doing?*
- *What should I remember about this?*
- *What does it look like?*
- *How is it organized?*
- *How is it done?*
- *How does it work?*

More receptive to images than to words, today's readers resist pages of mere printed text. Visuals help diminish a reader's resistance in several ways:

- *Visuals enhance comprehension by displaying abstract concepts in concrete, geometric shapes.* "How does the metric system work?" (Figure 14.1). "How do lasers work?" (Figure 14.33).

- *Visuals make meaningful comparisons possible.* "How do major industries compare in terms of environmental pollution?" (Figure 14.2). "How does one pound compare with one kilogram?" (Figure 14.1).

- *Visuals depict relationships.* "How does seasonal change affect the rate of construction in our county?" (see Figure 14.11). "What is the relationship between Fahrenheit and Celsius temperature?" (Figure 14.1).

- *Visuals serve as a universal language.* In the global workplace, carefully designed visuals can transcend cultural and language differences, and thus facilitate international communication (Figure 14.1).

- *Visuals provide emphasis.* To emphasize the change in death rates for heart disease and cancer since 1970, a table (Table 14.1) or bar graph (Figure 14.4) would be more vivid than a prose statement.

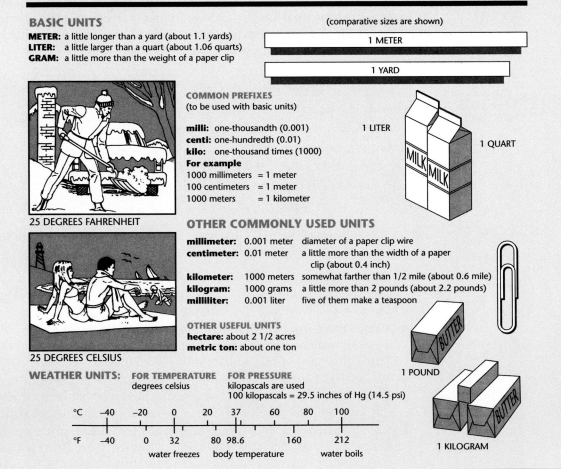

Figure 14.1   Visuals that Clarify and Simplify
Source: National Institute of Standards and Technology, 1992.

**Figure 14.2
A Graph
Displaying the
"Big Picture"**
*Source: U.S. Bureau of the
Census, Statistical Abstract
of the United States: 1994
(114th edition): 220.*

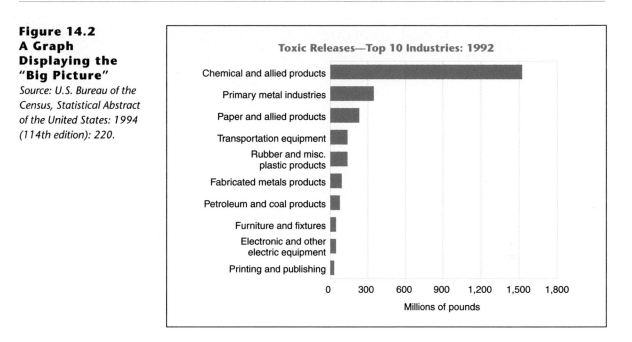

- *Visuals condense and organize information, making it easier to remember and interpret.* A simple table, for instance, can summarize a long and difficult printed passage, as in the example that follows.

Assume that you are researching recent death rates for heart disease and cancer. From various sources, you collect these data:

*Technical data in
printed form can be
hard to interpret*

1. In 1970, 419 males and 309 females per 100,000 people died of heart disease; 172 males and 135 females died of cancer.
2. In 1978, 371 males and 300 females per 100,000 people died of heart disease; 201 males and 160 females died of cancer.
3. In 1986. . . .

In the written form above, numerical information is repetitious, tedious, and hard to interpret. As the amount of numerical data increases, so does our difficulty in processing this material. Arranged in Table 14.1, these statistics become easier to compare and comprehend.

Along with your visual, analyze or interpret the important trends or the essential message you want your readers to see:

*A caption explaining the
numerical relationships*

As Table 14.1 indicates, both male and female death rates from heart disease decreased from 1970 to 1991, but males showed a sizably larger decrease. Cancer deaths during this period increased for both groups, with females showing the largest increase.

**TABLE 14.1
Data Displayed in
a Table**

Death Rates for Heart Disease and Cancer, 1970–1991				
Number of Deaths (per 100,000) Population[a]				
	Heart Disease		Cancer	
Year	Male	Female	Male	Female
1970	419	309	172	135
1978	371	300	201	160
1986	348	291	219	173
1991	293	279	222	187
Percent change, 1970–1991	−30.1	−9.7	+29.1	+39.2

[a] Figures are approximate.
Source: Adapted from *Statistical Abstract of the United States:* 1994 (114th edition). Washington: GPO: 98, 99.

Besides their value as presentation devices, visuals help us analyze information. Table 14.1 is one example of how visuals enhance critical thinking by helping us to identify and interpret crucial information and to discover meaningful connections.

When to Use a Visual

Translate your writing into visuals whenever they make your point more clearly than the prose.[1] Use visuals to *clarify* and to enhance your discussion, not to *decorate* it. Use a visual display to direct the audience's focus or to help them remember something, as in the following situations (Dragga and Gong 46–48):

Use visuals in situations like these

- when you want to instruct or persuade
- when you want to draw attention to something immediately important
- when you expect the document to be consulted randomly or selectively (e.g., a manual or other reference work) instead of being read in its original sequence (e.g., a memo or letter)
- when you expect the audience to be relatively less educated, less motivated, or less familiar with the topic
- when you expect the audience to be in a distracting environment

An effective visual advances the writer's purpose and the reader's understanding.

---

1. One alternative approach to the writing process is to begin with one or more key visuals and then compose the text to introduce and interpret the visual.

What Types of Visuals to Consider

Different types of visuals serve different functions. The following overview sorts visual displays into four categories: tables, graphs, charts, and graphic illustrations. Each type of visual offers readers a new way of seeing, a different perspective.

Tables Display Organized Lists of Data. Tables display data (as numbers or words) in rows and columns for comparison. Use tables to present exact numerical values and to organize data so that readers can sort out relationships for themselves. Complex tables usually are reserved for more specialized readers.

Numerical tables present data for analysis, interpretation, and exact comparison.

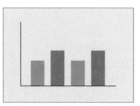

Prose tables organize verbal descriptions, explanations, or instructions for readers' access.

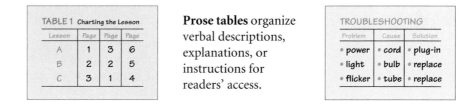

Graphs Display Numerical Relationships. Graphs translate numbers into shapes, shades, and patterns by plotting two or more data sets on a coordinate system. Use graphs to sort out or emphasize specific numerical relationships for readers. The visual representation helps readers grasp, at a glance, the approximate values, the point being made about those values, or the relationship being emphasized.

Bar graphs often show comparisons.

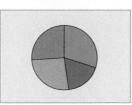

Line graphs often show changes over time.

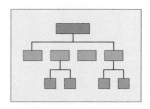

Charts Display the Parts of a Whole. Charts depict relationships without the use of a coordinate system by using circles, rectangles, arrows, connecting lines, and other designs.

Pie charts show the parts or percentages of a whole.

Organization charts show the links among departments, management structures, or other elements of a company.

Flowcharts trace the steps (or decisions) in a procedure or stages in a process.

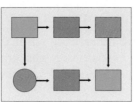

Tree charts show how the parts of an idea or a concept interrelate.

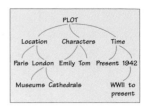

Gantt charts show when each phase of a project is to begin and end.

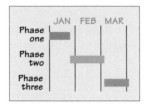

Pictorial charts (Pictograms) use icons (or isotypes) to symbolize the items being displayed or measured.

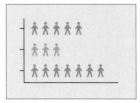

Graphic Illustrations Depict Actual or Virtual Views. Graphic illustrations are pictorial devices for helping readers visualize what something looks like, how it works, how it's done, how it happens, or where it's located. Certain diagrams present views that could not be captured by photographing or observing the object.

Representational diagrams present a realistic but simplified view, usually with essential parts labeled.

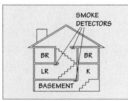

Source: Department of Energy.

Exploded diagrams show the item pulled apart, to reveal its assembly.

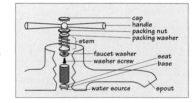

Cutaway diagrams eliminate outer layers to reveal inner parts.

Block or schematic diagrams present the conceptual elements of a principle, process, or system to depict *function* instead of appearance.

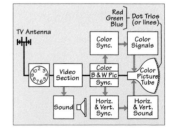

Maps enable readers to visualize a specific location or to comprehend data about a specific geographic region.

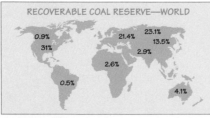

Source: Department of Energy.

Photographs present an actual picture of the item, process, or procedure.

Source: Superstock.

How to Select Visuals for Your Purpose and Audience

You usually will have more than one way to display information in a visual format. To select the most effective display among possible alternatives, consider carefully your specific purpose and the abilities and preferences of your audience.

Questions About a Visual's Purpose and Intended Audience

What is my purpose?

■ What do I want the audience to do or think (know facts and figures, follow directions, make a judgment, understand how something works, perceive a relationship, identify something, see what something looks like, pay attention, other)?

■ Do I want viewers to focus on one or more exact values, compare two or more values, or synthesize a range of approximate values?

Who is my audience?

■ What is their technical background on this topic?

■ What is their level of interest in this topic?

■ Would they prefer the raw data or interpretations of the data?

■ Are they accustomed to interpreting visuals?

Which type of visual might work best in this situation?

■ What forms of information should this visual depict (numbers, shapes, words, pictures, symbols)?

■ Which visual display would be most compatible with the type of judgment, action, or understanding I seek from this audience?

■ Which visual display would this audience find most accessible?

Choices to consider in selecting visuals

Here are a few examples of the choices you must consider in selecting visuals: if you just want the audience to know facts and figures, a table might be sufficient, but if you want them to make a particular judgment about these data, a bar graph, line graph, or pie chart might be preferable. To depict the operating parts of a mechanism, an exploded or cutaway diagram might be preferable to a photograph. Expert audiences tend to prefer numerical tables, flowcharts, schematics, and complex graphs or diagrams they can interpret for themselves. General audiences tend to prefer basic tables, graphs, diagrams, and other visuals that direct their focus, that interpret key points extracted from the data.

Although several alternatives might be possible, one particular type of visual (or a combination) usually is superior for a given purpose and audience. None of the above examples is, of course, immutable. Your particular audience or organization may express its own preferences. Or your choices may be limited by lack of equipment (software, scanners, digitizers), insufficient personnel (graphic designers, technical illustrators), or insufficient budget. In any case, your essential consideration is to ensure that the intended audience will be able to interpret the visual correctly.

Preferred Displays for Specific Visual Purposes

PURPOSE	PREFERRED VISUAL
■ Organize numerical data	Table
■ Show comparative data	Table, bar graph, line graph
■ Show a trend	Line graph
■ Interpret or emphasize data	Bar graph, line graph, pie chart, map
■ Introduce an unfamiliar object	Photo, representational diagram
■ Display a project schedule	Gannt chart
■ Show how parts are assembled	Photo, exploded diagram
■ Show how something is organized	Organization chart, map
■ Give instructions	Prose table, photo, diagrams, flowchart
■ Explain a process	Flowchart, block diagram
■ Clarify a concept or principle	Block or schematic diagram, tree chart
■ Describe a mechanism	Photo, representational diagram, or cutaway diagram

Tables

Tables can display exact quantities, compare sets of data, and present information systematically and economically. Numerical tables such as Table 14.1 present *quantitative information* (data that can be measured). In contrast, prose tables present *qualitative information* (brief descriptions, explanations or instructions). Table 14.2, for example, combines numerical data, probability estimates, comparisons, and instructions—all organized for the smoker's understanding of radon gas risk in the home.

No table should be overly complex for its intended audience. An otherwise impressive-looking table, such as Table 14.3, is hard for nonspecialists to interpret because it presents too much information at once. We can see how an unethical writer might use a complex table to bury numbers that are questionable or embarrassing (R. Williams 12). Can you discover any hidden facts in Table 14.3? (For instance, which industry has been slowest in cleaning up its act?)

Readers need to understand how the table is organized, where to find what they need, and how to interpret the information they find (Hartley 90).

Tables are constructed with various tools: (a) tab markers and tab keys on a typewriter or word processor, (b) row-and-column displays in a spreadsheet program, (c) the "Table" command in better word-processing programs. The "Table" command option offers a full range of table editing features: cut and paste, adjust spacing, insert text between rows, add rows or columns, adjust column width, and so on. Whichever options you employ, follow the general guidelines on page 315:

**TABLE 14.2
A Prose Table**
Source: Home Buyer's and
Seller's Guide to Radon.
Washington: GPO, 1993.

Radon Risk if You Smoke			
Radon level	If 1,000 people who smoked were exposed to this level over a lifetime . . .	The risk of cancer from radon exposure compares to . . .	WHAT TO DO: Stop Smoking and . . .
20 pCi/L[a]	About 135 people could get lung cancer	←100 times the risk of drowning	Fix your home
10 pCi/L	About 71 people could get lung cancer	←100 times the risk of dying in a home fire	Fix your home
8 pCi/L	About 57 people could get lung cancer		Fix your home
4 pCi/L	About 29 people could get lung cancer	←100 times the risk of dying in an airplane crash	Fix your home
2 pCi/L	About 15 people could get lung cancer	←2 times the risk of dying in a car crash	Consider fixing between 2 and 4 pCi/L
1.3 pCi/L	About 9 people could get lung cancer	(Average indoor radon level)	(Reducing radon levels below 2 pCi/L is difficult)
0.4 pCi/L	About 3 people could get lung cancer	(Average outdoor radon level)	

Note: If you are a former smoker, your risk may be lower.
[a]picocuries per liter

Guidelines for using tables

- Use a table only when you are reasonably sure it will enlighten—rather than frustrate—readers. For nonspecialized readers, use fewer tables and keep them simple.
- Try to limit the table to one page. Otherwise write "continued" at the bottom, and begin the second page with the full title, "continued," and the original column headings.
- If the table is too wide for the page, turn it 90 degrees and place its top toward the inside of the binding. Or divide the data into two tables. (Few readers may bother rotating the page to read the table broadside.)
- In your discussion, refer to the table by number, and explain what readers should be looking for. Or include a prose caption with the table. Specifically, introduce the table, show it, and then interpret it.

For more specific information about creating tables, see the Table Construction Guidelines accompanying Table 14.4.

Tables work well for displaying exact values, but for easier interpretation, readers prefer graphs or charts. Geometric shapes (bars, curves, circles) generally are easier to remember than lists of numbers (Cochran et al. 25).

Any visual other than a table usually is categorized as a *figure,* and so titled (*Figure 1 Aerial View of the Panhandle Site*). Figures covered in this chapter include graphs, charts, and illustrations.

TABLE 14.3
A Complex Table

Toxic Release Inventory, by Industry and Source: 1988 to 1992									
In millions of pounds. Based on reports from almost 23,000 manufacturing facilities which have 10 or more full-time employees and meet established thresholds for using the list of more than 300 chemicals covered.									
INDUSTRY	1987 SIC[a] code	1988	1989	1990	1991	1992			
						Total[b]	Air[c] point	Air[d] non-point	Water
Total	(X)	4,852.9	4,377.4	3,693.2	3,373.2	3,157.3	1,284.3	536.7	272.9
Food and kindred prod.	20	28.2	37.0	38.9	39.5	38.6	16.5	11.7	2.0
Tobacco products	21	1.8	1.8	2.5	2.3	2.0	1.9	0.1	(Z)
Textile mill products	22	38.2	32.2	27.1	25.0	21.5	15.3	5.8	0.3
Apparel and other textile prod.	23	1.1	1.4	1.3	1.4	1.6	1.1	0.5	(Z)
Lumber and wood products	24	32.9	37.8	35.6	32.6	32.4	25.2	7.0	0.1
Furniture and fixtures	25	66.8	65.3	61.7	55.9	55.1	47.0	7.7	(Z)
Paper and allied products	26	271.8	257.1	253.3	245.8	233.0	176.6	23.2	27.6
Printing and publishing	27	61.0	58.2	51.4	46.6	40.5	17.2	23.3	(Z)
Chemical and allied products	28	2,324.4	2,083.4	1,619.5	1,541.5	1,527.3	398.0	151.0	224.3
Petroleum and coal products	29	92.8	94.7	85.8	78.2	82.7	22.6	38.5	3.5
Rubber and misc. plastic prod.	30	170.4	184.0	177.5	149.8	134.4	93.3	40.1	0.5
Leather and leather products	31	15.8	13.5	12.7	9.9	10.5	6.7	3.5	0.3
Stone, clay, glass products	32	39.2	37.1	31.2	29.4	25.7	15.4	2.6	0.1
Primary metal industries	33	565.9	522.3	476.3	424.1	345.2	100.3	35.3	6.5
Fabricated metals products	34	136.9	137.3	128.2	111.0	101.2	61.6	38.3	0.2
Industrial machinery and equip.	35	60.6	57.8	48.8	38.6	33.7	19.6	13.5	0.1
Electronic, electric equip.	36	125.7	100.4	82.3	66.0	52.2	35.7	15.7	0.2
Transportation equipment	37	216.6	205.3	175.3	149.5	136.0	92.0	42.1	0.1

(X) Not applicable.   (Z) Less than 50,000 pounds.   [a]Standard Industrial Classification, see text, section 13.   [b]Includes other releases not shown separately.   [c]Stack.   [d]Fugitive.

Source: U.S. Environmental Protection Agency, 1992 Toxics Release Inventory Public Data Release.

Like all other components in the document, visuals are designed with audience and purpose in mind (Journet 3). An accountant doing an audit might need a table listing exact amounts, whereas the average public stockholder reading an annual report would prefer the "big picture" in an easily grasped bar graph or pie chart (Van Pelt 1). Similarly, an audience of scientists might find Table 14.3 perfectly appropriate, but a less specialized audience (say, environmental groups) might prefer the clarity and simplicity of Figure 14.2.

Graphs

Graphs translate numbers into pictures. Plotted as a set of points (a *series*) on a coordinate system, a graph shows the relationship between two variables.

Graphs have a horizontal and a vertical axis. The horizontal axis carries categories (the independent variables) to be compared, such as years within a period (1970, 1978, 1986). The vertical axis shows the range of values (the dependent variables) for comparing or measuring the categories, such as the number of people who died from heart failure in a specific year. A dependent variable changes according to activity in the independent variable (e.g., a decrease in quantity over a set time, as in Figure 14.3). In the equation $y = f(x)$, x is the independent variable and y is the dependent variable.

① TABLE 14.4 ■ Science and Engineering Graduates in 1988 and 1989: 1990 Career Status

COLUMN HEADS

| STUB HEAD ② DEGREE AND FIELD | Graduates 1988 and 1989 (1,000) | 1990—PERCENT DISTRIBUTION | | | | Median salary ($1,000) ③ |
| | | In school[a] | Employed | | Not employed | |
			In S&E[b]	In other		
All science fields	④ 494.5	⑤ 22	35	38	5	23.6
Computer science	69.3	6	77	13	4	30.1
⑥ Environmental science	7.3	30	49	15	5	23.7
Life sciences	111.2	32	34	28	6	⑦ 21.0
Math/statistics	35.2	18	50	26	6	23.6
Physical sciences	29.4	39	38	18	5	25.1
Psychology	85.7	21	20	54	5	18.6
Social sciences	156.4	⑧ X	19	55	5	21.9
All engineering fields[c]	148.7	11	73	12	3	33.0
Civil	15.2	10	78	9	3	30.1
Electrical/electronics	55.5	11	76	10	3	34.1
Industrial	12.3	5	72	18	5	31.1
Mechanical	30.0	11	76	10	4	34.0

ROW HEADS

SRC NOTE

⑨ [a]Full-time grad. students. [b]Science & engineering. [c]Other fields not shown. (X) Not available.

⑩ *Source: National Science Foundation/SRS,* Characteristics of Recent Science and Engineering Graduates: 1990. Statistical Abstract of the United States: 1994 (114th edition) Washington: GPO: 615.

1. Number the table in its order of appearance and provide a title that describes exactly what is being compared or measured.

2. Label stub, row, and column heads (*Degree and Field, Median salary, Computer Science*) so readers know what they are looking at.

3. Stipulate all units of measurement by using familiar symbols and abbreviations ($, hr., no.). Define specialized symbols or abbreviations (*A* for *angstrom, db* for *decibel*) in a footnote.

4. Compare data vertically (in columns) instead of horizontally (in rows). Columns are easier to scan than rows. Try to include row or column averages or totals, as reference points for comparing individual values.

5. Use horizontal rules to separate headings from data. In a complex table, use vertical rules to separate columns. In a simple table, use as few rules as clarity allows.

6. List the items in a logical order (alphabetical, chronological, decreasing cost). Space listed items so they are not cramped or too far apart for easy comparison. Keep prose entries as brief as clarity allows.

7. Convert fractions to decimals, and align decimals vertically. Keep decimal places for all numbers equal. Round insignificant decimals to the nearest whole number.

8. Use *x, NA,* or a dash to signify any omitted entry, and explain the omission in a footnote ("Not available," "Not applicable").

9. Use footnotes to explain entries, abbreviations, or omissions. Label footnotes with lowercase letters so readers do not confuse the notation with the numerical data.

10. Cite data sources beneath any footnotes. When adapting or reproducing a copyrighted table for a work to be published, obtain written permission from the copyright holder.

Graphs are especially useful for displaying comparisons, changes over time, patterns, or trends. When you decide to use a graph, choose the best type for your purpose: bar graph or line graph.

Bar Graphs

Easily understood by most readers, bar graphs show discrete comparisons, as on a year-by-year or month-by-month basis. Each bar represents a specific quantity. Use bar graphs to help readers focus on one value or compare values that change over equal time intervals (expenses calculated at the end of each month, sales figures totaled at yearly intervals). Use a bar graph only to compare values that are noticeably different. Otherwise, all the bars will appear almost identical.

Simple Bar Graphs. The simple bar graph in Figure 14.3 displays one relationship taken from the data in Table 14.1, the rate of male deaths from heart disease. To aid interpretation, you can record exact values above each bar—but only if readers need exact numbers.

**Figure 14.3
A Simple Bar
Graph**

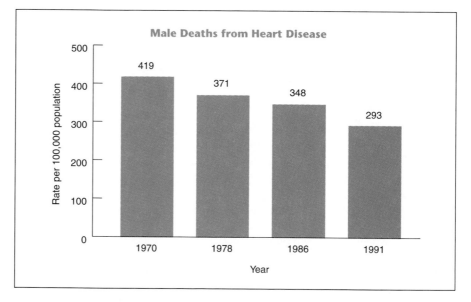

Multiple-Bar Graphs. A bar graph can display two or three relationships simultaneously, each relationship plotted as a separate series. Figure 14.4 displays two comparisons from Table 14.1, the rate of male deaths from both heart disease and cancer.

Whenever a graph shows more than one relationship (or series), each series of numbers is represented by a different pattern, color, shade, or symbol, and the patterns are identified by a *legend*.

The more relationships a graph displays, the harder it is to interpret. As a rule, plot no more than three series of numbers on one graph.

**Figure 14.4
A Multiple-Bar
Graph**

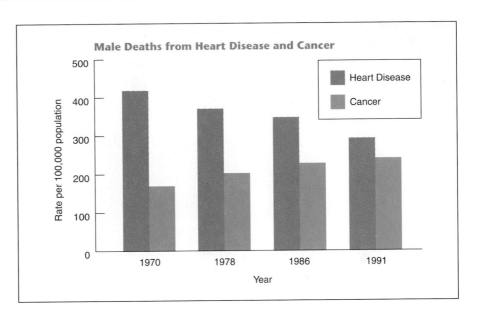

Horizontal-Bar Graphs. To make a horizontal-bar graph, turn a vertical-bar graph (and scales) on its side. Horizontal-bar graphs are good for displaying a large series of bars arranged in order of increasing or decreasing value, as in Figure 14.5. The horizontal format leaves room for labeling the categories horizontally (*Service, and so on*). A vertical-bar graph would leave no room for horizontal labeling.

**Figure 14.5
A Horizontal-Bar
Graph**
*Source: Bureau of Labor
Statistics.*

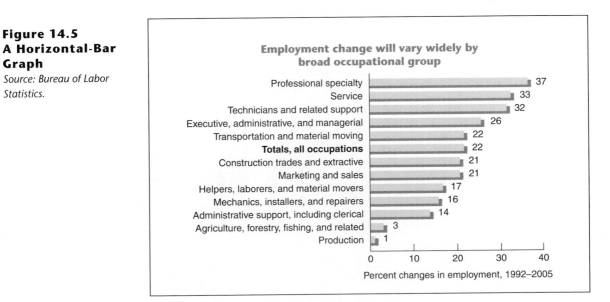

**Figure 14.6
A Stacked-Bar
Graph**

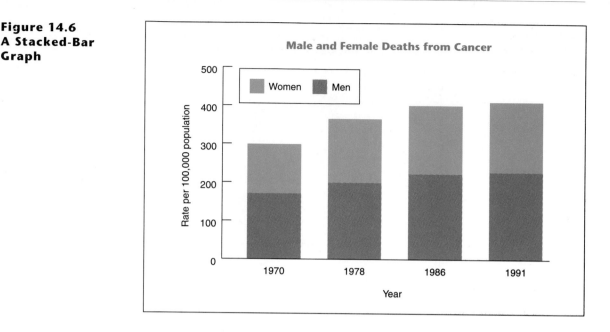

Stacked-Bar Graphs. Instead of side-by-side clusters of bars, you can display multiple relationships by stacking bars. Stacked-bar graphs are especially useful for showing how much each item contributes to the whole. Figure 14.6 displays other comparisons from Table 14.1.

Display no more than four or five relationships in a stacked-bar graph. Excessive subdivisions and patterns create visual confusion.

**Figure 14.7
A 100-percent Bar
Graph**

Source: U.S. Bureau of the Census.

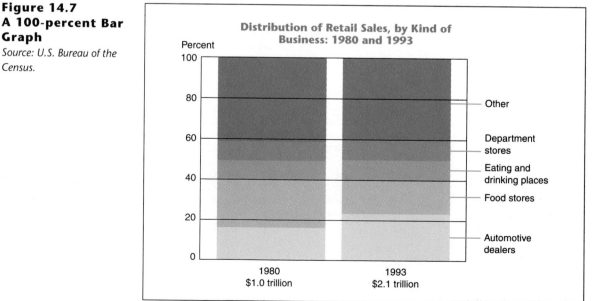

100-Percent Bar Graph. A type of stacked bar graph, the 100-percent bar graph is useful for showing the relative values of the parts that make up the 100-percent value, as in Figure 14.7. Like any bar graph, the 100-percent graph can have either horizontal or vertical bars.

Notice how bar graphs become harder to interpret as bars and patterns increase. For less sophisticated readers, the two, separate data depictions in Figure 14.7 might be easier to interpret in pie charts (pages 326–27).

Deviation Bar Graphs. The deviation bar graph can display both positive and negative values, as in Figure 14.8. Notice how the vertical axis extends to the negative side of the zero baseline, following the same incremental division as above the baseline.

Figure 14.8
A Deviation Bar Graph

Source: Bureau of Labor Statistics.

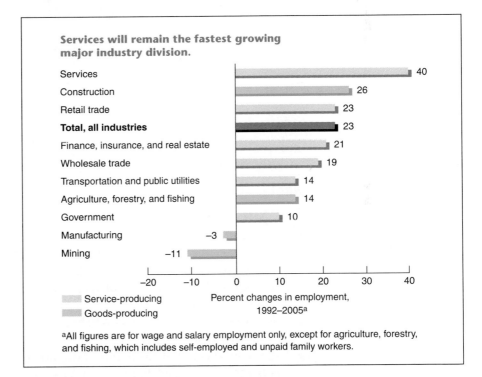

3-D Bar Graphs. Graphics software enables us to shade and rotate images and produce 3-dimensional views. The 3-D perspectives in Figure 14.9 engage our attention and add visual emphasis to the data.

Although 3-D graphs can enhance and dramatize a presentation, an overly complex graph can be almost impossible to interpret. Never sacrifice clarity and simplicity for the sake of visual effect.

Bar Graph Guidelines. Once you decide on a type of bar graph, follow the suggestions on page 322 for presenting the graph to your audience.

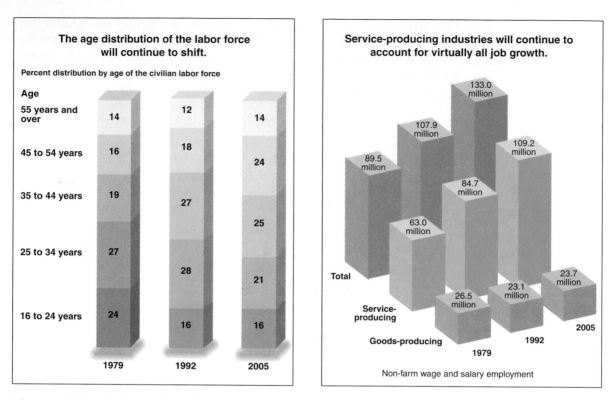

The age distribution of the labor force will continue to shift.

Percent distribution by age of the civilian labor force

Age	1979	1992	2005
55 years and over	14	12	14
45 to 54 years	16	18	24
35 to 44 years	19	27	25
25 to 34 years	27	28	21
16 to 24 years	24	16	16

Service-producing industries will continue to account for virtually all job growth.

Total: 89.5 million (1979), 107.9 million (1992), 133.0 million (2005)
Service-producing: 63.0 million (1979), 84.7 million (1992), 109.2 million (2005)
Goods-producing: 26.5 million (1979), 23.1 million (1992), 23.7 million (2005)

Non-farm wage and salary employment

Figure 14.9   3-D Bar Graphs.   *Source: Bureau of Labor Statistics.*

Bar graph guidelines

- Keep the graph simple and easy to read. Avoid plotting more than three types of bars in each cluster. Avoid needless visual details.
- Number your scales in units the audience will find familiar and easy to follow. Units of 1 or multiples of 2, 5, or 10 are best (Lambert 45). Space the numbers equally.
- Label both scales to show what is being measured or compared. If space allows, keep all labels horizontal for easier reading.
- Use *tick marks* to show the points of division on your scale. If the graph has many bars, extend the tick marks into *grid lines* to help readers relate bars to values.

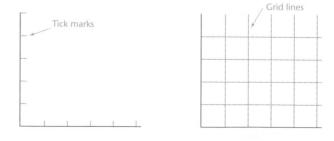

Tick marks

Grid lines

- To avoid confusion, make all bars the same width (unless you are overlapping them). If you must produce your graphs by hand, use graph paper to keep bars and increments evenly spaced.
- In a multiple-bar graph, use a different pattern, color, or shade for each bar in a cluster. Provide a legend identifying each pattern, color, or shade.
- If you are trying for emphasis, be aware that darker bars are seen as larger, closer, and more important than lighter bars of the same size (Lambert 93).
- Cite data sources beneath the graph. When adapting or reproducing a copyrighted graph for a work to be published, you must obtain written permission from the copyright holder.
- In your discussion, refer to the graph by number ("Figure 1"), and explain what readers should be looking for. Or include a prose caption along with the graph.

Many computer graphics programs automatically employ most of the design features discussed above. Anyone producing visuals, however, should know all the conventions.

Line Graphs

A line graph can accommodate many more data points than a bar graph (e.g., a twelve-month trend, measured monthly). Line graphs help readers synthesize large bodies of information in which exact quantities need not be emphasized. Whereas bar graphs display quantitative differences among items (cities, regions, yearly or monthly intervals), line graphs display data whose value changes over time, as in a trend, forecast, or other change during a specified time (profits, losses, growth). Some line graphs depict cause-and-effect relationships (e.g., how seasonal patterns affect sales or profits).

**Figure 14.10
A Simple Line
Graph**

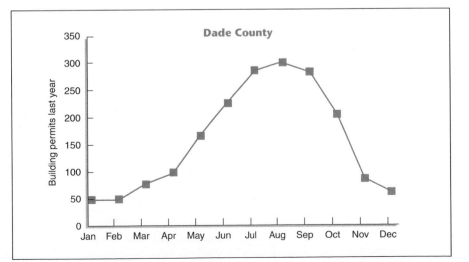

**Figure 14.11
A Multiple-Line
Graph**

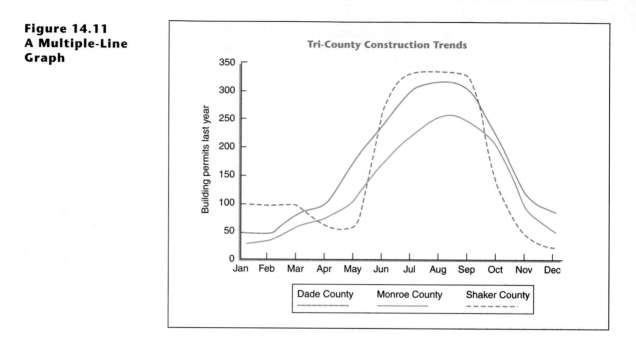

Simple Line Graphs. A simple line graph, as in Figure 14.10, uses one line to plot time intervals on the horizontal scale and values on the vertical scale. The relationship depicted here would be much harder to express in words alone.

Multiple-Line Graphs. A multiple-line graph displays several relationships simultaneously, as in Figure 14.11.

**Figure 14.12
A Deviation Line
Graph**

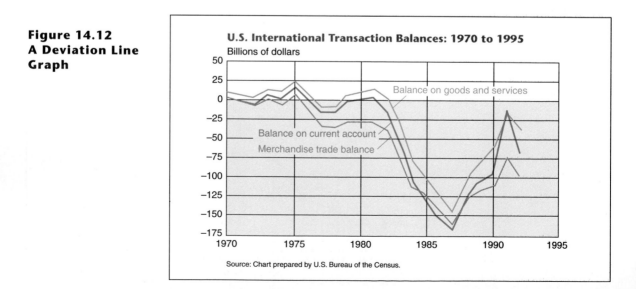

For legibility, use no more than three or four curves in a single graph. Explain the relationships readers are supposed to see.

A caption explaining the visual relationships

> Building permits in all three counties increased steadily as the weather warmed, but Shaker's increase was more erratic. Its permits declined for April–May, but then surpassed Dade's and Monroe's for June–September.

Deviation Line Graphs. Extend your vertical scale below the zero baseline to display positive and negative values in one graph, as in Figure 14.12. Mark values below the baseline in intervals parallel to those above it.

Band or Area Graphs. For emphasis and appeal, fill the area beneath each plotted line with a pattern. Figure 14.13 is a version of the Figure 14.10 line graph.

The multiple-bands in Figure 14.14 depict relationships among sums instead of the direct comparisons depicted in the equivalent Figure 14.11 line graph.

Despite their visual appeal, multiple-band graphs are easy to misinterpret: In a multiple-line graph, each line depicts its own distance from the zero baseline. But in a multiple-band graph, the very top line depicts the *total,* each band below it being a part of that total (like stacked-bar graph segments). Always clarify these relationships for viewers.

Line Graph Guidelines. Follow bar graph guidelines (page 322), with these additions:

Line graph guidelines

- Display no more than three or four lines on one graph.
- Mark each individual data point used in plotting each line.
- Make each line visually distinct (using color, symbols, and so on).
- Label each line so readers will know what it represents.
- Avoid grid lines that readers could mistake for plotted lines.

**Figure 14.13
A Simple Band
Graph**

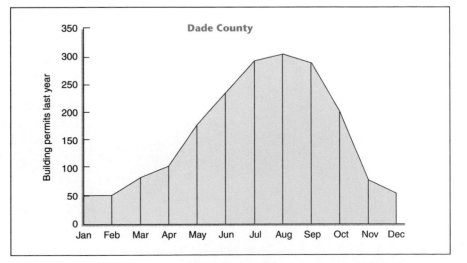

Figure 14.14
A Multiple-Band
Graph

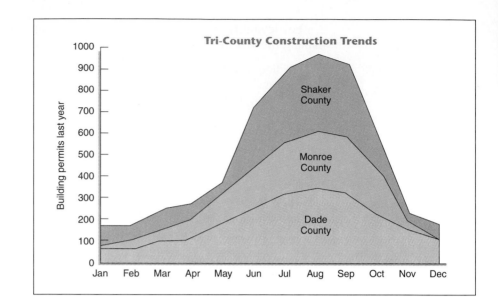

Charts

The terms *chart* and *graph* often appear interchangeably. But a chart is more precisely a figure that displays relationships (quantitative or cause-and-effect) that are not plotted on a coordinate system. Commonly used charts include pie charts, organizational charts, flowcharts, tree charts, Gantt charts, and pictorial charts (pictograms).

Figure 14.15
A Simple Pie Chart

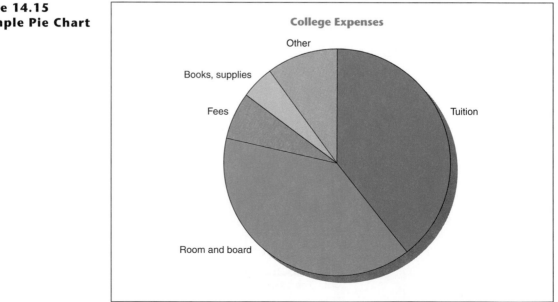

**Figure 14.16
Two Other
Versions of Figure
14.15**

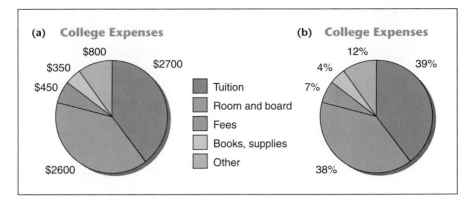

Pie Charts

Considered easy for readers to understand, a pie chart depicts the percentages or proportions of the parts that make up a whole. In a pie chart, readers can compare the parts to each other as well as to the whole (to show how much was spent on what, how much income comes from which sources, and so on). Figure 14.15 shows a pie chart. Figure 14.16 shows two other versions of the pie chart in Figure 14.15. Version (a) displays dollar amounts, and version (b) the percentage relationships among these dollar amounts.

For constructing pie charts, follow these suggestions:

Pie chart guidelines

- Be sure the parts add up to 100 percent.
- If you must produce your charts by hand, use a compass and protractor for precise segments. Each 3.6-degree segment equals 1 percent. Include any number from two to eight segments. A pie chart containing more

**Figure 14.17
An Organization
Chart for One
Corporation**

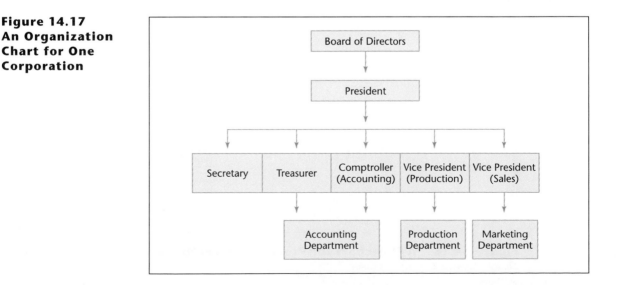

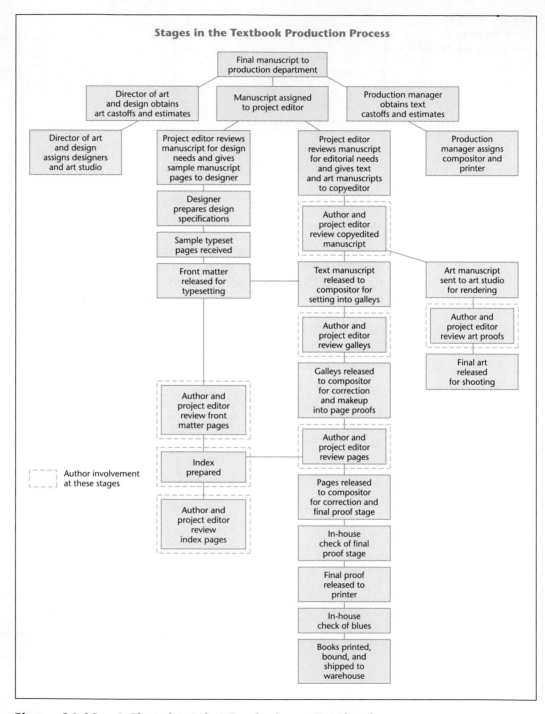

Figure 14.18   A Flowchart for Producing a Textbook

Source: Harper & Row Author's Guide.

than eight segments can be hard to interpret, especially if the segments are small (Hartley 96).

■ Combine small segments under the heading "Other."
■ Locate your first radial line at 12 o'clock and then move clockwise from large to small (except for "Other," usually the final segment).
■ For easy reading, keep all labels horizontal.

Organization Charts

An organization chart divides an organization into its administrative or managerial parts. Each part is ranked according to its authority and responsibility in relation to other parts and to the whole, as in Figure 14.17.

Flowcharts

A flowchart traces a procedure or process from beginning to end. In displaying the steps in a manufacturing process, the flowchart would begin at the raw materials and proceed to the finished product. Figure 14.18 traces the procedure for producing a textbook. (Other flowchart examples appear on pages 106, 130, and elsewhere throughout the text.)

Tree Charts

Whereas flowcharts display the steps in a process, tree charts show how the parts of an idea or concept relate to each other. Figure 14.19 displays the parts of an outline for this chapter so that readers can better visualize relationships. The tree version seems clearer and more interesting than the prose listing.

**Figure 14.19
An Outline
Converted to a
Tree Chart**

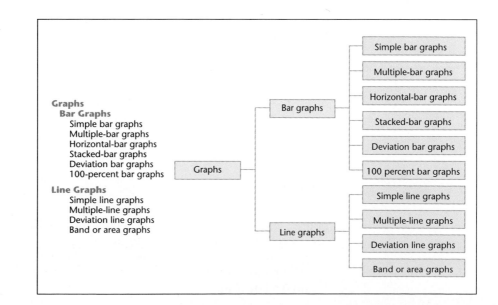

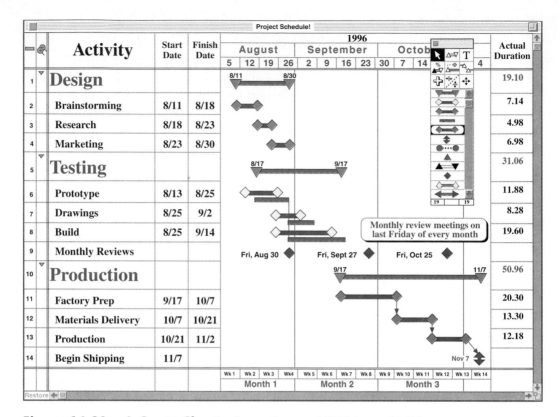

Figure 14.20   A Gantt Chart.   *Source: Courtesy of AEC Software, ©1996.*

Gantt Charts

Named for engineer H. L. Gantt (1861–1919), a Gantt chart depicts progress as a function of time. A series of bars or lines (time lines) indicates start-up and completion dates for each phase or task in a project, relative to the other phases or tasks. Gantt charts especially are useful for planning a project (as in a proposal) and for tracking it (as in a progress report). Notice that the Gantt chart in Figure 14.20 depicts tasks whose time lines can be simultaneous, overlapping, or consecutive.

Pictograms

Pictograms depict numerical relationships with icons or symbols (cars, houses, smokestacks) of the items being measured, instead of using bars or lines. Each symbol represents a stipulated quantity, as in Figure 14.21. Many graphics programs provide an assortment of predrawn symbols.

Use pictograms when you want to make information more interesting for nontechnical audiences.

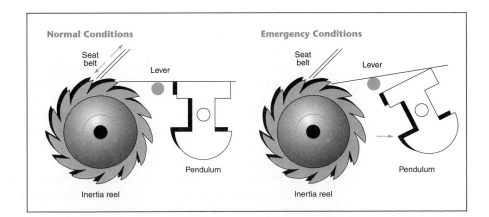

**After increasing between 1973 and 1984, the college-age population
declined over the 1984–1995 period.**

Total U.S. population (Each figure = 1 million)

Age	
18–24	
1973	26,635,000
1984	28,939,000
1995	23,702,000
25–34	
1973	29,375,000
1984	41,107,000
1995	40,520,000
35–44	
1973	22,810,000
1984	30,718,000
1995	41,997,000

Graphic Illustrations

Illustrations consist of diagrams, maps, drawings, and photographs depicting relationships that are physical rather than numerical. Good illustrations help readers understand and remember the material (Hartley 82). Consider this information from a government pamphlet, explaining the operating principle of the seat belt:

> The safety-belt apparatus includes a tiny pendulum attached to a lever, or locking mechanism. Upon sudden deceleration, the pendulum swings forward, activating the locking device to keep passengers from pitching into the dashboard.

**Figure 14.22
A Diagram of a
Safety-Belt
Locking
Mechanism**

Source: Safety Belts. *U.S. Department of Transportation.*

Normal Conditions

Emergency Conditions

**Figure 14.23
An Exploded
Diagram of a
Brace for a
Basketball Hoop**
*Source: Courtesy of
Spalding.*

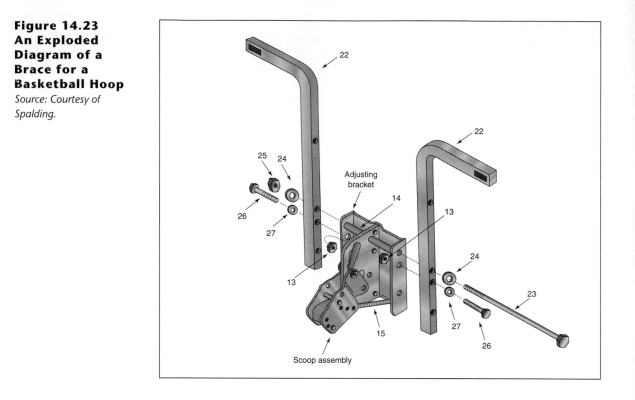

Without the illustration in Figure 14.22 we have difficulty visualizing the mechanism. Clear and uncluttered, a good diagram eliminates unnecessary details and focuses only on material useful to the reader. The following pages sample some commonly used diagrams.

Diagrams

Exploded diagrams, like that of a brace for an adjustable basketball hoop in Figure 14.23, show how the parts of an item are assembled; they often appear in repair or maintenance manuals. Notice how all parts are numbered for the reader's easy reference in the written instructions.

Cutaway diagrams show the item with its exterior layers removed in order to reveal the interior structure, as in Figure 14.24. Unless the specific viewing perspective is immediately recognizable (as in Figure 14.24), define for readers the angle of vision: "top view," "side view," and so on.

Block diagrams are simplified sketches that represent the relationship between the parts of an item, principle, system, or process. Because block diagrams are designed to illustrate *concepts* (such as current flow in a circuit), the parts are represented as symbols or shapes. The block diagram in Figure 14.25 illustrates how any process can be controlled automatically through a feedback mechanism. Figure 14.26 shows the feedback concept applied as the cruise-control mechanism on a motor vehicle.

**Figure 14.24
Cutaway Diagram
of a Surgical
Procedure**

Source: Transsphenoidal
Approach for Pituitary
Tumor, ©1986 by The
Ludann Co., Grand
Rapids, MI.

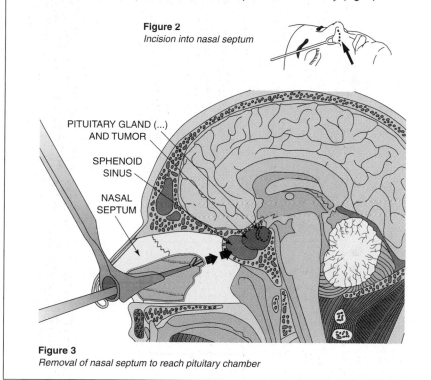

THE OPERATION

Incision

Transsphenoidal surgery is performed with the patient under general anesthesia and positioned on his back. The head is fixed in a special headrest, and the operation is monitored on a special x-ray machine (fluoroscope).

In the approach illustrated here (not used by all surgeons), a small incision is made in one side of the nasal septum **(Fig. 2)**. Part of the septum is then removed to provide access to the sphenoid sinus cavity **(Fig. 3)**.

Figure 2
Incision into nasal septum

PITUITARY GLAND (...)
AND TUMOR

SPHENOID
SINUS

NASAL
SEPTUM

Figure 3
Removal of nasal septum to reach pituitary chamber

Increasingly available are electronic drawing programs, clip-art programs, image banks, and other resources for creating visuals or downloading pre-drawn images. Specialized diagrams, however, often require the services of graphic artists or technical illustrators. The client requesting or commissioning the visual provides the art professional with an *art brief* (often prepared by writers and editors) that spells out the visual's purpose and specifications for the visual. For example, part of the brief addressed to the medical illustrator for Figure 14.24 might read as follows:

*An art brief for Figure
14.24*

- **Purpose:** to illustrate transsphenoidal adenomectomy for laypersons
- **View:** full cutaway, axial

- **Range:** descending from cranial apex to a horizontal plane immediately below the upper jaw and second cervical vertebra
- **Depth:** medial cross section
- **Structures omitted:** cranial nerves, vascular and lymphatic systems
- **Structures included:** gross anatomy of bone, cartilage, and soft tissue—delineated by color, shading, and texture
- **Structures highlighted:** nasal septum, sphenoid sinus, and sella turcica, showing the pituitary embedded in a 1.5 cm tumor invading the sphenoid sinus via an area of erosion at the base of the sella

**Figure 14.25
A Block Diagram
Illustrating the
Concept of
Feedback**

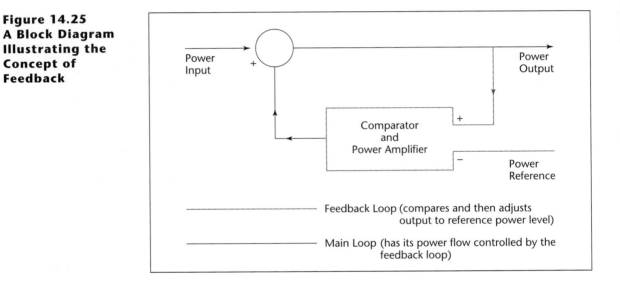

**Figure 14.26
A Block Diagram
Illustrating a
Cruise-Control
Mechanism**

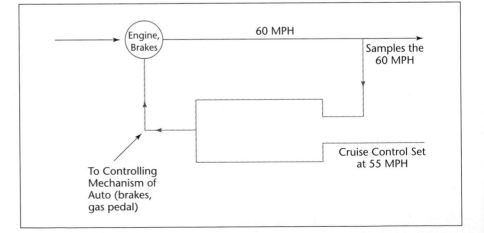

Maps

Besides being visually engaging and easily remembered, maps are especially useful for showing comparisons and for enabling readers to *visualize* position, location, and relationships among complex data. Consider how Figure 14.27 conveys important statistical information in a format that is both accessible and understandable. Color enhances the percentage comparisons.

**Figure 14.27
A Map Rich in
Statistical
Significance**
*Source: Bureau of the
Census.*

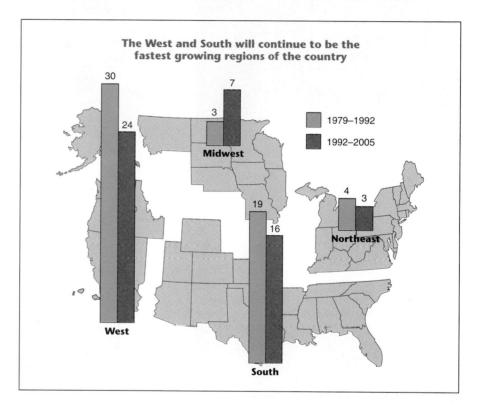

Photographs

Photographs are especially useful for showing what something looks like (Figure 14.28) or how something is done (Figure 14.29). But no matter how visually engaging, a photograph is hard to interpret if it includes needless details or fails to identify or emphasize the important material. One graphic design expert offers this practical advice for technical documents:

> To use pictures as tools for communication, pick them for their capacity to carry meaning, not just for their prettiness as photographs, . . . [but] for their inherent significance to the [document]. (White, *Great Pages* 110, 122)

Specialized photographs often require the services of a professional who knows how to use angles, lighting, and special film to obtain the desired focus and emphasis.

FIGURE 14.28   A Photograph that Shows the Appearance of Something

A Fixed-Platform Oil Rig   *Source: SuperStock.*

FIGURE 14.29   A Photograph that Shows How Something Is Done

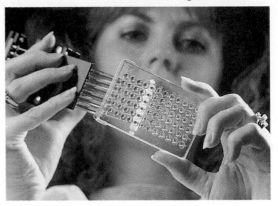

Antibody Screening Procedure   *Source: SuperStock.*

FIGURE 14.30   A Photograph that Shows a Realistic Angle of Vision

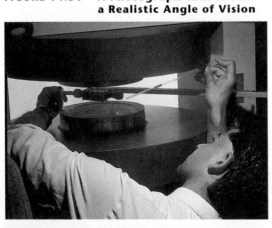

Titration in Measuring Electron-Spin Resonance
Source: SuperStock.

FIGURE 14.31   A Photograph that Needs To Be Cropped

Replacing the Microfilter Activation Unit
Source: SuperStock.

FIGURE 14.32   The Cropped Version of Figure 14.31

Source: SuperStock.

FIGURE 14.34   A Simplified Diagram of Figure 14.33

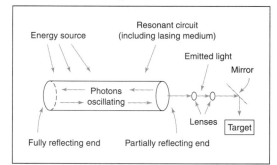

Major Parts of the Laser

FIGURE 14.33   A Photograph of a Complex Mechanism

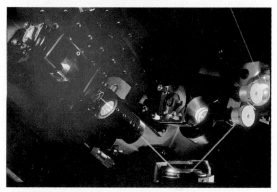

Sapphire Tunable Laser   Source: SuperStock.

FIGURE 14.35   A Photograph with Essential Features Labeled

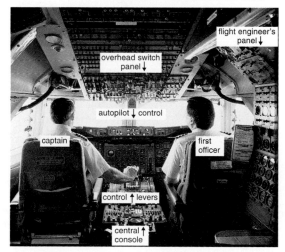

Standard Flight Deck for a Long-Range Jet
Source: SuperStock.

Whenever you plan to include photographs in a document or a presentation, observe these guidelines:

Guidelines for using photographs

■ Try to simulate the approximate angle of vision readers would have in identifying or viewing the item or, for instructions, in doing the procedure (Figure 14.30).

■ Trim (or crop) the photograph to eliminate needless details (Figures 14.31 and 14.32).

- For emphasizing selected features of a complex mechanism or procedure, consider using diagrams in place of photographs or as a supplement (Figures 14.33 and 14.34).
- Label all the parts readers need to identify (Figure 14.35).
- For an image unfamiliar to readers, provide a sense of scale by including a person, a ruler, or a familiar object (such as a hand) in the photo.
- If your document will be published, obtain a signed release from any person depicted in the photograph and written permission from the copyright holder. Beneath the photograph, cite the photographer and the copyright holder.
- In your discussion, refer to the photograph by figure number and explain what readers should be looking for. Or include a prose caption.

Digital-imaging technology allows photographs to be scanned and stored electronically. These stored images then can be retrieved, edited, and altered. Such capacity for altering photographic content creates unlimited potential for distortion and raises questions about the ethics of digital manipulation (Callahan 64–65).

Computer Graphics

Computer technology transfers many of the tasks formerly performed by graphic designers and technical illustrators to individuals with little formal training in graphic design. In whatever career you anticipate, you probably will be expected to produce high-quality graphics for conferences, presentations, and in-house publications.

Today's computer systems create sophisticated, multicolor, graphic displays and multimedia presentations. Among the virtually countless types of computer-generated visuals are these examples:

- With an electronic stylus (a pen with an electronic signal), you can draw pictures on a graphics tablet to be displayed on the monitor, stored, or sent to other computers.
- You can create three-dimensional effects, showing an object from different angles through the use of shading, shadows, on-screen rotation, background lighting, or other techniques.
- You can recreate the visual effect of a mathematical model, as in writing equations to explain what happens when high winds strike a tall building. (As the wind deforms the structure, the equations change. Then you can take those new equations and represent them visually.)
- You can create a design, build a model, simulate the physical environment, and let the computer forecast what will happen with different variables.
- You can integrate computer-assisted design (CAD) with computer-assisted manufacturing (CAM), so that the design will direct the machinery that makes the parts themselves (CAD/CAM).

- You can create animations, to see how bodies move (as in a car crash or in athletics).
- You can practice dealing with toxic chemicals, operating sophisticated machines, or making other rapid decisions in medical or technical environments, without the cost or danger in actual situations.
- Through various types of scientific visualization, you can do "what-if" projections and explore countless ways of conceptualizing and understanding your data. Because the computer can generate and evaluate many possibilities rapidly, it enables you to test hypotheses without doing the calculations.

Selecting Design Options

Given all applications for personal computers, graphics production is among the most rapidly expanding. Computer graphics systems allow you to experiment with scales, formats, colors, perspectives, and patterns. Most systems offer WYSIWYG (What You See Is What You Get) packages, in which the on-screen display is almost identical to the hard-copy image that will be printed out. By testing design options on the screen, you can revise and enhance your visual repeatedly until it achieves your exact purpose.

Here is a sampling of design options:

- Update charts and graphs whenever the data change. The software will calculate the new data and plot the relationships.
- Edit your graphics on the screen, adding, deleting, or moving material as needed.
- Create your image at one scale, and later specify a different scale for the same image.
- Annotate and label, and create multiple typefaces within one visual.
- Overlay images in one visual.
- Adjust bar width and line thickness.
- Fill a shape with a color or pattern.
- Scan, edit, and alter pages, photos, or other images.

Most of these options are available via simple keystrokes or pull-down menus.

Using Clip Art

Clip art is a generic term for collections of ready-to-use images (of computer equipment, maps, machinery, medical equipment, and so on), all stored on computer disks. Various clip-art packages enable you to import into your document countless images like the one in Figure 14.36. By running the image through a drawing program, you can enlarge, enhance, or customize it, as in Figure 14.37.

Clip-art software offers images of all kinds from science, architecture, business, and technology.

Figure 14.36 A Clip-art Image
Source: Desktop Art®; Business 1
© Dynamic Graphics, Inc.

Figure 14.37 A Customized Image
Source: Professor R. Armand Dumont.

One form of clip art especially useful in technical writing is the icon (an image with all nonessential background removed). Icons convey a specific idea visually as in Figure 14.38. Icons appear routinely in computer documentation and in other types of instructions because the image provides readers with an immediate signal of the desired action.

Whenever you use an icon, be sure it is "intuitively recognizable" to your readers ("Using Icons" 3). Otherwise, readers are likely to misinterpret its meaning—in some cases with disastrous results.

Figure 14.38 Icons
Source: Desktop Art®; Business 1 and Health Care 1, © Dynamic Graphics, Inc.

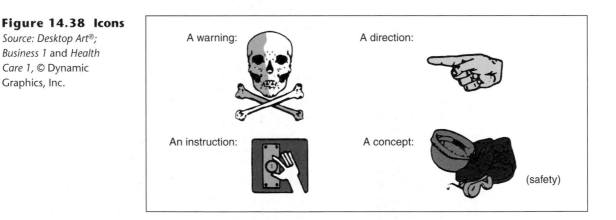

Using Color

Color often makes a presentation more esthetically pleasing, more interesting to look at. But color serves purposes beyond visual appeal. Used effectively in a visual, color draws and directs readers' attention and helps them identify the various elements. In Figure 14.20, for example, color helps readers sort out the key schedule elements of a Gantt chart for a major project: activities, time lines, durations, and meetings.

**Figure 14.39
Color Used as a
Visualizing Tool**

*Source: Courtesy of
National Audubon Society.*

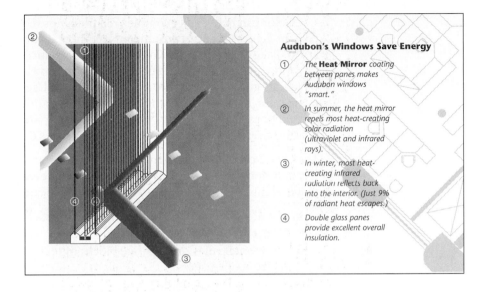

Audubon's Windows Save Energy

① The **Heat Mirror** coating between panes makes Audubon windows "smart."

② In summer, the heat mirror repels most heat-creating solar radiation (ultraviolet and infrared rays).

③ In winter, most heat-creating infrared radiation reflects back into the interior. (Just 9% of radiant heat escapes.)

④ Double glass panes provide excellent overall insulation.

Color can help clarify a concept or dramatize how something works. In Figure 14.39, for example, the use of bright colors against a darker, duller background enables readers to *visualize* the "heat-mirror" concept.

**Figure 14.40
Colors Used to
Show Relationships**

*Source: Courtesy of
Scientific American,
October 1995; 32D.
Copyright Rodger Doyle
©1995.*

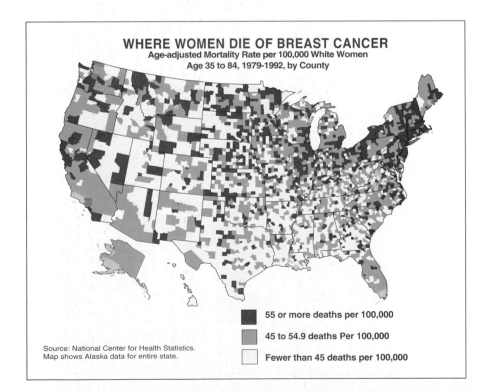

WHERE WOMEN DIE OF BREAST CANCER
Age-adjusted Mortality Rate per 100,000 White Women
Age 35 to 84, 1979-1992, by County

55 or more deaths per 100,000

45 to 54.9 deaths Per 100,000

Fewer than 45 deaths per 100,000

Source: National Center for Health Statistics.
Map shows Alaska data for entire state.

Color can help depict complex relationships in ways that are understandable. In Figure 14.40, an area map using three distinctive colors enables readers to make any number of comparisons at a glance.

Along with shape, type style, and position of elements on a printed page, color can guide readers through the material. Used effectively on a printed page, color helps organize the reader's understanding, provides orientation, and emphasizes important material. Following are just a few among many possible uses of color (White, *Color* 39–44; Keyes 647–49).

Use Color to Organize. Readers look for ways of organizing their understanding of a document (Figure 14.41). Color can reveal structure and break up material into discrete blocks that are easier to locate, process, and digest:

- A color background screen can set off like elements such as checklists, instructions, or examples.
- Horizontal rules can separate blocks of text, such as sections of a report or areas of a page.
- Vertical rules can set off examples, quotations, captions, and so on.

Use Color to Orient. Readers look for clear bearings and signposts that help them find their place and find what they need (Figure 14.42):

- Color can help headings stand out from the text and differentiate major from minor headings.
- Color tabs and boxes can serve as location markers.
- Color sidebars (for marginal comments), callouts (for labels), and leader lines (for connecting a label to its referent) can guide the eyes.

Use Color to Emphasize. Readers look for places to focus their attention in a document (Figure 14.43).

- Color typeface can highlight key words or ideas.

**Figure 14.41
Color Used to
Organize Page
Elements**

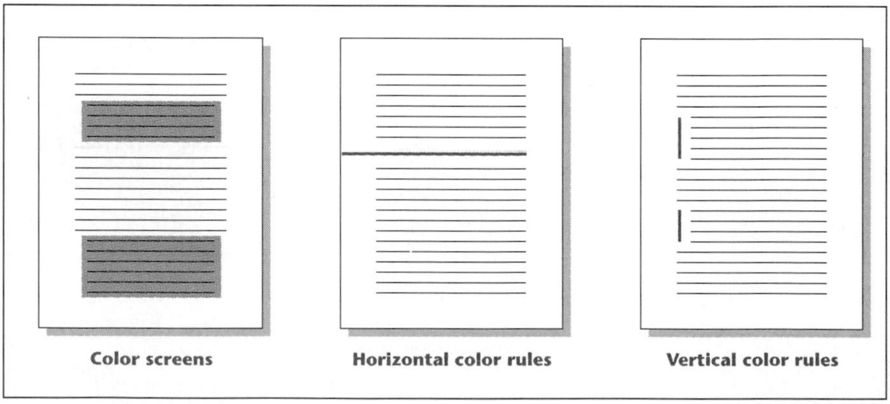

Color screens   Horizontal color rules   Vertical color rules

**Figure 14.42
Color Used as an
Orientation Device**

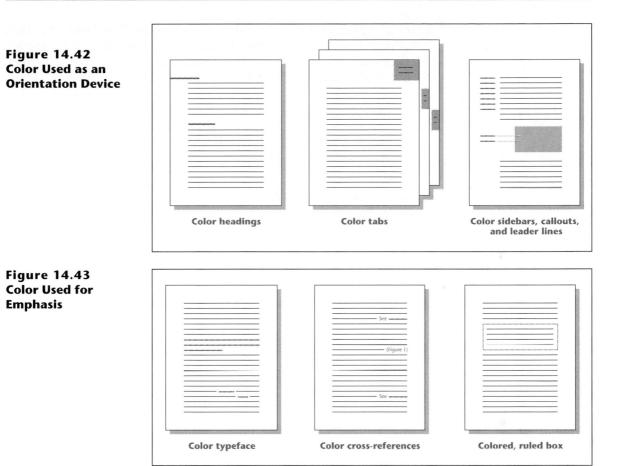

**Figure 14.43
Color Used for
Emphasis**

- Color can call attention to cross-references.
- A colored, ruled box can frame a warning, caution, note, or hint.

Guidelines for Using Color:

- "Color gains impact when it is used selectively. It loses impact when it is overused." (*Aldus Guide* 39). Use color sparingly, and use no more than three or four distinct colors—including black and white (White, *Great Pages* 76).
- Apply color consistently to elements throughout your document. Inconsistent use of color can distort readers' perception of the relationships (Wickens 117).
- Make color redundant. Be sure all elements are first differentiated in black and white: by shape, location, texture, type style, or typesize. Different readers perceive colors differently or, in some cases, not at all. A sizable percentage of readers have impaired color vision (White, *Great Pages* 76).

- Use a darker color to make a stronger statement. The darker the color the more important the material. Readers perceive differently the sizes of variously colored objects. Darker items can seem larger and closer than lighter objects of identical size.

- Make colored type larger than body type. Try to avoid color for body type, or use a high-contrast color (dark against a light background). Color is less visible on the page than black ink on a white background. The smaller the image or the thinner the rule, the stronger or brighter the color (White, *Editing* 229, 237).

- For contrast in a color screen, use a very dark type against a very light background, say a 10–20 percent screen (Gribbons 70). The bigger the area of the screen, the paler the background color (Figure 14.44).

- A color's connotations can vary from culture to culture. In the United States for example, red signifies danger and green traditionally signifies safety. But in Ireland, green or orange carry political connotations in certain contexts. In Muslim cultures, green is a holy color (Cotton 169).

**Figure 14.44
A Color-Density
Chart**

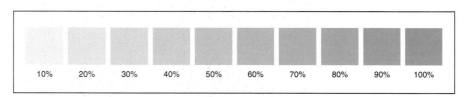

Using Web Sites for Graphics Support

The World Wide Web offers a growing array of visual resources. Following is a sample of useful Web sites (Martin 135–36).

- Image banks such as Graphics Web, Photo Web, and the Internet Font directory offer clip-art, photo, and font databases that can be browsed and from which material can be purchased online and downloaded to a personal computer.

- Some royalty-free images are available from the Digital Picture Archive.

- Samples of computer-generated and traditional art work can be browsed and purchased via Artists on Line.

- Online versions of graphics magazines offer helpful design suggestions.

- Graphic design firms offer samples of work that often embody innovative design ideas.

Whether you seek design ideas, tools for creating your own visuals, or electronic catalogs of completed artwork, the Web is a good bet.

Be extremely cautious about downloading visuals (or any material, for that matter) from the Web and then using it. Review the copyright law (pages 175, 176). Originators of any work on the Web own the work and the copyright.

How to Avoid Visual Distortion

Although you are perfectly justified in presenting data in its best light, you are ethically responsible for avoiding misrepresentation. Any one set of data can support contradictory conclusions. Even though your numbers may be accurate, their visual display could be misleading.

Present the Real Picture

Visual relationships in a graph should portray accurately the numerical relationships they represent. Begin the vertical scale at zero. Never compress the scales to reinforce your point.

Notice how visual relationships in Figure 14.45 become distorted when the value scale is compressed or fails to begin at zero. In version A, the bars accurately depict the numerical relationships measured from the value scale. In version B, item Z (400) is depicted as three times X (200). In version C, the scale is overly compressed, causing the shortened bars to understate the quantitative differences.

Deliberate distortions are unethical because they imply conclusions contradicted by the actual data.

Present the Complete Picture

Without bogging down in needless detail, an accurate visual includes all essential data. Figure 14.46 shows how distortion occurs when data that would provide a complete picture are selectively omitted. Version A accurately depicts the numerical relationships measured from the value scale. In version B, too few points are plotted.

Decide carefully what to include and what to leave out of your visual display.

**Figure 14.45
An Accurate Bar
Graph and Two
Distorted Versions**

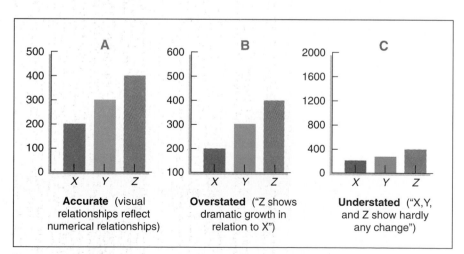

Accurate (visual relationships reflect numerical relationships)

Overstated ("Z shows dramatic growth in relation to X")

Understated ("X,Y, and Z show hardly any change")

**Figure 14.46
An Accurate Line
Graph and a
Distorted Version**

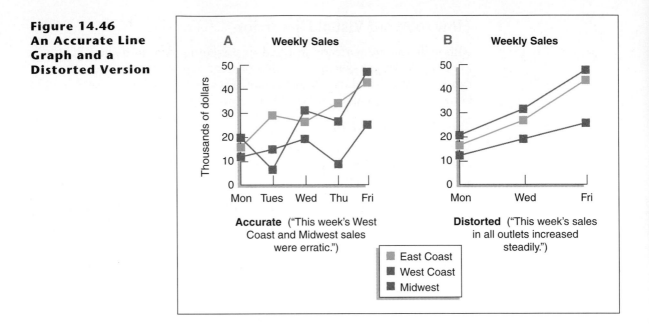

Never Mistake Distortion for Emphasis

When you want to emphasize a point (a sales increase, a safety record, etc.), be sure your data support the conclusion implied by your visual. For instance, don't use inordinately large visuals to emphasize good news or small ones to downplay bad news (Williams 11). When using clip art, pictograms, or drawn images to dramatize a comparison, be sure the relative size of the images or icons reflects the quantities being compared.

A visual accurately depicting a 100-percent increase in phone sales at your company might look like version A in Figure 14.47. Version B overstates

**Figure 14.47
An Accurate
Pictogram and a
Distorted Version**

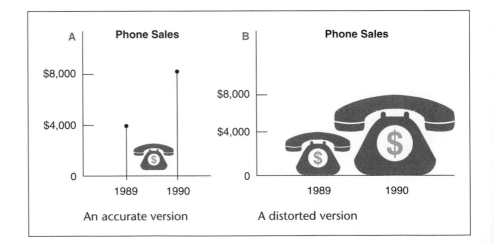

the good news by depicting the larger image four times the size, instead of twice the size, of the smaller. Although the larger image is twice the height, it is also twice the *width,* so the total area conveys the visual impression that sales have *quadrupled.*

Visuals have their own rhetoric and persuasive force, which we can use to advantage—for positive or negative purposes, for the reader's benefit or detriment (Van Pelt 2). Avoiding visual distortion is ultimately a matter of ethics.

How to Incorporate Visuals with the Text

An effective visual enables readers to locate and extract the information they need. To simplify the reader's task, visual and verbal elements in a document should complement each other. For example, a visual should be able to stand alone in meaning—without being isolated from the verbal text.

Following are specific guidelines for incorporating visual and verbal elements effectively.

Guidelines for fitting visuals with text

- *Place the visual where it will best serve your readers.* If it is central to your discussion, place the visual as close as possible to the material it clarifies. (Achieving proximity often requires that you ignore the traditional "top or bottom" design rule for placement of visuals on a page.) If the visual is peripheral to your discussion or of interest to only a few readers, place it in an appendix so that interested readers can refer to it as they wish. Tell readers when to consult the visual and where to find it.

- *Never refer to a visual that readers cannot easily locate.* In a long document, don't be afraid to repeat a visual if you discuss it again later.

- *Never crowd a visual into a cramped space.* Set your visual off by framing it with plenty of white space, and position it on the page for balance. To save space and to achieve proportion with the surrounding text, consider carefully the size of each visual and the amount of space it will occupy.

- *Number the visual and give it a clear title and labels.* Your title should tell readers what they are seeing. Label all the important material.

- *Match the visual to your audience.* Don't make it too elementary for specialists or too complex for nonspecialists. Be sure your intended audience will be able to interpret the visual correctly.

- *Introduce and interpret the visual.* In your introduction, tell readers what to expect:

Informative   As shown in Table 2, operating costs have increased 7 percent annually since 1980.

Uninformative   See Table 2.

Visuals alone make ambiguous statements (Girill 35); pictures need to be interpreted. Instead of leaving readers to struggle with a page of raw data, explain the relationships displayed. Follow the visual with a discussion of its important features:

Focusing on Your Purpose

- What is this visual's purpose (to instruct, persuade, create interest)? _____

- What forms of information (numbers, shapes, words, pictures, symbols) will this visual depict?
  _____

- What kind of relationship(s) will the visual depict (comparison, cause-effect, connected parts, sequence of steps)? _____

- What judgment, conclusion, or interpretation is being emphasized (that profits have increased, that toxic levels are rising, that X is better than Y, that time is being wasted)?
  _____

- Is a visual needed at all? _____

Focusing on Your Audience

- Is this audience accustomed to interpreting visuals? _____

- Is the audience interested in specific numbers or an overall view? _____

- Should the audience focus on one exact value, compare two or more values, or synthesize a range of approximate values (Wickens 121)? _____

- Which type of visual will be most accurate, representative, accessible, and compatible with the type of judgment, action, or understanding expected from the audience? _____
  _____

- In place of one complicated visual, would two or more straightforward ones be preferable?
  _____

Focusing on Your Presentation

- What enhancements, if any, will increase audience interest (colors, patterns, legends, labels, varied typefaces, shadowing, enlargement or reduction of some features)? _____
  _____

- Which medium—or combination of media—will be most effective for presenting this visual (slides, transparencies, handouts, large-screen monitor, flip chart, report text)? _____
  _____

- To achieve the greatest utility and effect, where in the presentation does this visual belong?
  _____

Figure 14.48   A Planning Sheet for Preparing Visuals

Informative   This cost increase means that . . . .

Always tell readers what to look for and what it means.

- *Use prose captions to explain important points made by the visual.* Captions help readers interpret a visual (as in Table 14.1). When possible, use a smaller typesize so that captions don't compete with text type (*Aldus Guide* 35).
- *Never include excessive information in one visual.* Any visual that contains too many lines, bars, numbers, colors, or patterns will overwhelm readers, causing them to ignore the display. In place of one complicated visual, use two or more straightforward ones.
- *Be sure the visual's meaning can stand alone.* Even though it repeats or augments information already in the text, the visual should contain everything readers will need to interpret it correctly.

The Visual Plan Sheet, Figure 14.48, and the Revision Checklist for Visuals on page 350 will help ensure that your visuals enhance your meaning.

☑ EXERCISES

1. The following statistics are based on data from three colleges in a large western city. They give the number of applicants to each college over six years.
   - In 1991, X College received 2,341 applications for admission, Y College received 3,116, and Z College 1,807.
   - In 1992, X College received 2,410 applications for admission, Y College received 3,224, and Z College 1,784.
   - In 1993, X College received 2,689 applications for admission, Y College received 2,976, and Z College 1,929.
   - In 1994, X College received 2,714 applications for admission, Y College received 2,840, and Z College 1,992.
   - In 1995, X College received 2,872 applications for admission, Y College received 2,615, and Z College 2,112.
   - In 1996, X College received 2,868 applications for admission, Y College received 2,421, and Z College 2,267.

Illustrate these data in a line graph, a bar graph, and a table. Which version seems most effective for a reader who (a) wants exact figures, (b) wonders how overall enrollments are changing, or (c) wants to compare enrollments at each college in a certain year? Include a caption interpreting each of these versions.

2. Devise a flowchart for a process in your field or area of interest. Include a title and a brief discussion.

3. Devise an organization chart showing the lines of responsibility and authority in an organization where you hold a part-time or summer job.

4. Devise a pie chart to depict your yearly expenses. Title and discuss the chart.

5. Obtain enrollment figures at your college for the past five years by sex, age, race, or any other pertinent category. Construct a stacked-bar graph to illustrate one of these relationships over the five years.

6. Keep track of your pulse and respiration at thirty-minute intervals over a four-hour period of changing activities. Record your findings in a line graph, noting the times and specific activities below your horizontal coordinate. Write a prose interpretation of your graph and give it a title.

7. In textbooks or professional journal articles, locate each of these visuals: a table, a multiple-bar graph, a

Revision Checklist for Visuals

(Numbers in parentheses refer to the first page of discussion.)

CONTENT

❑ Does the visual serve a legitimate purpose (clarification, not mere ornamentation) in the document? (310)

❑ Is the visual titled and numbered? (347)

❑ Is the level of complexity appropriate for the audience? (347)

❑ Are all patterns in the visual identified by label or legend? (323)

❑ Are all values or units of measurement specified (grams per ounce, millions of dollars)? (317)

❑ Are the numbers accurate and exact? (317)

❑ Do the visual relationships represent the numerical relationships accurately? (345)

❑ Are explanatory notes added as needed? (317)

❑ Are all data sources cited? (317)

❑ Has written permission been obtained for reproducing or adapting a visual from a copyrighted source in any type of work to be published? (317)

❑ Is the visual introduced, discussed, interpreted, integrated with the text and referred to by number? (347)

❑ Can the visual itself stand alone in meaning? (349)

ARRANGEMENT

❑ Is the visual easy to locate? (349)

❑ Are all design elements (title, line thickness, legends, notes, borders, white space) positioned for balance? (347)

❑ Is the visual positioned on the page to achieve balance? (347)

❑ Is the visual set off by adequate white space or borders? (347)

❑ Does the top of a wide visual face the inside binding? (315)

❑ Is the visual in the best report location? (347)

STYLE

❑ Is this the best type of visual for your purpose and audience? (347)

❑ Are all decimal points in each column vertically aligned? (317)

❑ Is the visual uncrowded and uncluttered? (347)

❑ Is the visual engaging (patterns, colors, shapes), without being too busy? (349)

❑ Is the visual in good taste? (344)

❑ Is the visual ethically acceptable? (345)

---

multiple-line graph, a diagram, and a photograph. Evaluate each according to the revision checklist, and discuss the most effective visual in class.

8. We have discussed the importance of choosing an appropriate scale for your graph, and the most effective form for presenting your data. Study this presentation carefully:

> Strong evidence now indicates that not only nicotine and tar in cigarette smoke can be lethal, but also carbon monoxide. Much of cigarette smoke is carbon monoxide. The bar graph in Figure 14.49 lists the ten leading U.S. cigarette brands according to the carbon monoxide given off per pack of inhaled cigarettes.

Is the scale effective? If not, why not? Can these data be presented effectively in a bar graph? Present the same data in some other form that seems most effective.

9. Choose the most appropriate visual for illustrating each of these relationships. Justify each choice in a short paragraph.

   a. A comparison of three top brands of fiberglass skis, according to cost, weight, durability, and edge control.

**Figure 14.49
Ten Leading U.S.
Cigarette Brands in
Order of CO Content**

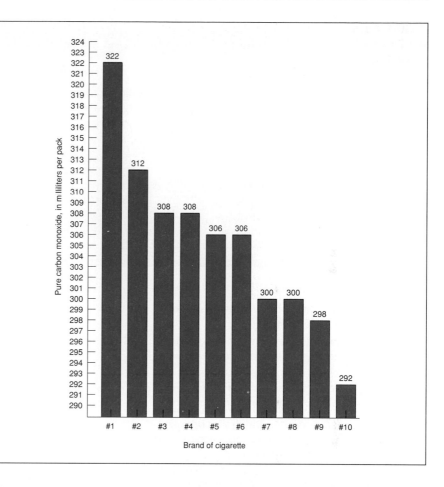

b. A breakdown of your monthly budget.

c. The changing cost of an average cup of coffee, as opposed to that of an average cup of tea, over the past three years.

d. The percentage of college graduates finding desirable jobs within three months after graduation, over the last ten years.

e. The percentage of college graduates finding desirable jobs within three months after graduation, over the last ten years—by gender.

f. An illustration of automobile damage for an insurance claim.

g. A breakdown of the process of radio wave transmission.

h. A comparison of five cereals on the basis of cost and nutritive value.

i. A comparison of the average age of students enrolled at your college in summer, day, and evening programs, over the last five years.

j. Comparative sales figures for three items made by your company.

10. *Computer graphics:* Compose and enhance one or more visuals electronically. You might begin by looking through a recent edition of the *Statistical Abstract of the United States* (in the government documents, reserve, or reference section of your library). From the *Abstract,* or from a source you prefer, select a body of numerical data that will interest your classmates. After completing the Visual Plan Sheet, compose one or more visuals to convey a message about your data, to make a point, as in these examples:

TABLE 14.5
An Example of a
Poor Layout

Educational Attainment of Persons 25 Years Old and Over		
	Highest level completed	
Year	High school (4 years or more)	College (4 years or more)
1970	52.3164	10.7431
1980	66.5432	16.2982
1982	71.0178	17.7341
1984	73.3124	19.1628
1993	80.2431	21.9316

Source: Adapted from *Statistical Abstract of the United States: 1994 (114 edition)*. Washington, DC: GPO: 157.

- Consumer buying power has increased or decreased since 1980.
- Defense spending, as a percentage of the federal budget, has increased or decreased since 1980.
- Average yearly temperatures across the United States are rising or falling.

Experiment with formats and design options, and enhance your visual(s) as appropriate. Add any necessary prose explanations.

Be prepared to present your visual message in class, using either an overhead or opaque projector or a large-screen monitor.

11. Revise the layout of Table 14.5 according to the guidelines on page 317, and explain to readers the significant comparisons in the table. (*Hint:* The unit of measurement is percentage.)

12. Display each of these sets of information in the visual format most appropriate for the stipulated audience. Complete the Visual Plan Sheet for each visual. Explain why you selected the type of visual as most effective for that audience. Include with each visual a brief prose passage interpreting and explaining the data.

   a. (For general readers.) The 1994 *Statistical Abstract of the United States* breaks down energy consumption (by source) into these percentages: In 1970, coal, 18.5; natural gas, 32.8; hydro and geothermal, 3.1; nuclear, 1.2; oil, 44.4. In 1980, coal, 20.3; natural gas, 26.9; hydro and geothermal, 3.8; nuclear, 4.0; oil, 45.0. In 1990, coal, 23.5; natural gas, 23.8; hydro and geothermal, 7.3; nuclear, 4.1; oil, 41.3.

   b. (For experienced investors in rental property.) As an aid in estimating annual heating and air-conditioning costs, here are annual maximum and minimum temperature averages from 1961 to 1990 for five Sunbelt cities (in Fahrenheit degrees): In Jacksonville, the average maximum was 78.7; the minimum was 57.2. In Miami, the maximum was 82.6; the minimum was 67.8. In Atlanta, the maximum was 71.3; the minimum was 51.1. In Dallas, the maximum was 76.9; the minimum was 55. In Houston, the maximum was 77.5; the minimum was 57.4. (From U.S. National Oceanic and Atmospheric Administration.)

   c. (For the student senate.) Among the students who entered our university four years ago, here are the percentages of those who graduated, withdrew, or are still enrolled: In Nursing, 71 percent graduated; 27.9 percent withdrew; 1.1 percent are still enrolled. In Engineering, 62 percent graduated; 29.2 percent withdrew; 8.8 percent are still enrolled. In Business 53.6 percent graduated; 43 percent withdrew; 3.4 percent are still enrolled. In Arts and Sciences, 27.5 percent graduated; 68 percent withdrew; 4.5 percent are still enrolled.

d. (For the student senate.) Here are the enrollment trends from 1987 to 1996 for two colleges in our university. In Engineering: 1987, 455 students enrolled; 1988, 610; 1989, 654; 1990, 758; 1991, 803; 1992, 827; 1993, 1,046; 1994, 1,200; 1995, 1,115; 1996, 1,075. In Business: 1987, 922; 1988, 1,006; 1989, 1,041; 1990, 1,198; 1991, 1,188; 1992, 1,227; 1993, 1,115; 1994, 1,220; 1995, 1,241; 1996, 1,366.

13. Anywhere on campus or at work, locate at least one visual that needs revision for accuracy, clarity, appearance, or appropriateness. Look in computer manuals, lab manuals, newsletters, financial aid or admissions or placement brochures, student or faculty handbooks, newspapers, or textbooks. Use the Visual Plan Sheet and the Revision Checklist as guides to revise and enhance the visual. Submit to your instructor a copy of the original, along with a memo explaining your improvements. Be prepared to discuss your revision in class.

✅ COLLABORATIVE PROJECTS

1. Assume that your technical writing instructor is planning to purchase five copies of a graphics software package for students to use in designing their documents. The instructor has not yet decided which general-purpose package would be most useful. Your group's task is to test one package and to make a recommendation.

   In small groups, visit your school's microcomputer lab and ask for a listing of the graphics packages that are available to students and faculty. Select one package and learn how to use it. Design at least four representative visuals. In a memo or presentation to your instructor and classmates, describe the package briefly and tell what it can do. Would you recommend purchasing five copies of this package for general-purpose use by writing students? Explain. Submit your report, along with the sample graphics you have composed. Appoint one member to present your group's recommendation in class.

Do the same assignment, comparing various clip-art packages. Which package offers the best image selection for writers in your specialty?

2. Compile a list of twelve World Wide Web sites that offer graphics support by way of advice, image banks, design ideas, artwork catalogs, and the like. Provide the address for each site, along with a description of the resources offered and their appropriate cost. Report your findings in a format stipulated by your instructor.

3. Creating Icons: The U.S. Department of Energy has created a facility in New Mexico for storing nuclear defense byproducts and waste materials that will remain radioactive for more than 10,000 years. How can future generations be warned against drilling in this area or disturbing the material buried more than 2,000 feet underground—especially when these distant generations might differ radically in use of language and symbols or in technological and scientific awareness? What system of icons would most likely endure through the centuries and convey a clear warning to any culture at anytime? Among the solutions proposed:

   • Covering the entire site with giant concrete blocks (30–40 feet square), placed close enough together to prevent any useful human activity in the spaces. Furthermore, black-dyed concrete would absorb the desert sun, creating unbearable heat.

   • Covering the site with huge stone "thorns."

   • Erecting 50-foot-high stone pillars over the site, with an entrance chamber at the surface that would contain an array of warning symbols, icons, and drawings.

   Your group's assignment is to improve on these proposals by devising a system of symbolic obstacles and visual warnings. Prepare a detailed proposal that includes a full set of visuals depicting your plan.

   Note: You may wish to consult the article from which this assignment was adapted: Kliewer, Gary. "The 10,000-Year Warning." *The Futurist* Sept. 1992: 17–19.

Designing Pages and Documents

·············

P AGE design determines the *look* of a page, the layout of words and graphics. Well-designed pages invite readers in, guide them through the material, and help them understand and remember it.

Readers' *first* impression of a document tends to be a purely visual, esthetic judgment. They are attracted by documents that appear inviting and accessible.

Page Design in Workplace Writing

Page design becomes especially significant when we consider these realities about writing and reading in the workplace:

1. *Technical information generally is designed differently from material in novels, news stories, and other forms of writing.* To be accessible, a technical document requires more than just an unbroken sequence of paragraphs. To find their way through complex material, readers may need the help of charts, diagrams, lists, various typesizes and fonts, different headings, and other page-design elements.

2. *Technical documents rarely get readers' undivided attention.* Readers may be skimming the document while they jot down ideas, talk on the phone, or drink coffee. Or they may refer to sections of the document during a meeting or presentation. Amid frequent distractions, readers must be able to leave the document and return easily.

3. *People read work-related documents only because they have to.* Novels, general newspapers, and magazines are read for relaxation, but work-related documents (trade magazines, reports, newsletters) often require the reader's *labor*. The more complex the document, the harder readers have to labor. If these readers had other ways of getting the information, they might not read at all.

4. *As computers generate more and more paper, any document is forced to compete for readers' attention.* Even brilliant writing is useless unless it gets read by its intended audience. Suffering from information overload, today's readers resist any document that appears overwhelming. They want formats that will help them find the information they really need. A user-friendly document has an accessible format: At a glance, readers can see how the document is organized, where they are in the document, which items matter most, and how the items relate.

Having decided at a glance whether your document is inviting and accessible, readers will draw conclusions about the value of your information, the quality of your work, and your credibility. For an otherwise useful document, carefully crafted pages become your stamp of professionalism.

Figure 15.1 Ineffective Page Design

Source: U.S. Department of Energy. Sunspaces and Solar Greenhouses. Washington: GPO, 1984.

Sunspaces

Either as an addition to a home or as an integral part of a new home, sunspaces have gained considerable popularity.

A sunspace should face within 30 degrees of true south. In the winter, sunlight passes through the windows and warms the darkened surface of a concrete floor, brick wall, water-filled drums, or other storage mass. The concrete, brick, or water absorbs and stores some of the heat until after sunset, when the indoor temperature begins to cool. The heat *not* absorbed by the storage elements can raise the daytime air temperature inside the sunspace to as high as 100 degrees Fahrenheit. As long as the sun shines, this heat can be circulated into the house by natural air currents or drawn in by a low-horsepower fan.

In order to be considered a passive solar heating system, any sunspace must consist of these parts: a collector, such as a double layer of glass or plastic; an absorber, usually the darkened surface of the wall, floor, or water-filled containers inside the sunspace; a storage mass, normally concrete, brick, or water, which retains heat after it has been absorbed; a distribution system, the means of getting the heat into and around the house by fans or natural air currents; and a control system, or heat-regulating device, such as movable insulation, to prevent heat loss from the sunspace at night. Other controls include roof overhangs that block the summer sun, and thermostats that activate fans.

Even though the information in Figure 15.1 is worthwhile and accurate, its format resists our attention. Without visual cues, we have no way of grouping this information into organized units of meaning. Figure 15.2 shows the same information after a design overhaul. As readers, we take good page design for granted—that is, we hardly notice it *unless* it is ineffective.

Sunspaces

Either as an addition to a home or as an integral part of a new home, sunspaces have gained considerable popularity.

How Sunspaces Work

A sunspace should face within 30 degrees of true south. In the winter, sunlight passes through the windows and warms the darkened surface of a concrete floor, brick wall, water-filled drums, or other storage mass. The concrete, brick, or water absorbs and stores some of the heat until after sunset, when the indoor temperature begins to cool.

The heat *not* absorbed by the storage elements can raise the daytime air temperature inside the sunspace to as high as 100 degrees Fahrenheit. As long as the sun shines, this heat can be circulated into the house by natural air currents or drawn in by a low-horsepower fan.

The Parts of a Sunspace

In order to be considered a passive solar heating system, any sunspace must consist of these parts:

1. A *collector,* such as a double layer of glass or plastic.

2. An *absorber,* usually the darkened surface of the wall, floor, or water-filled containers inside the sunspace.

3. A *storage mass,* normally concrete, brick, or water, which retains heat after it has been absorbed.

4. A *distribution system,* the means of getting the heat into and around the house (by fans or natural air currents).

5. A *control system* (or heat-regulating device), such as movable insulation, to prevent heat loss from the sunspace at night. Other controls include roof overhangs that block the summer sun, and thermostats that activate fans.

Figure 15.2   Effective Page Design

Desktop Publishing

Planning, drafting, and writing at their workstations, writers themselves are increasingly responsible for all stages in document preparation. Because they combine word processing, typesetting, and graphics, *desktop publishing* (DTP) systems (Figure 15.3) mean less reliance on secretaries, print shops, and graphic artists.

Using page-design software, optical scanners, and laser printers, writers at their desks can control the entire production cycle: designing, illustrating, laying out, and printing the final document.

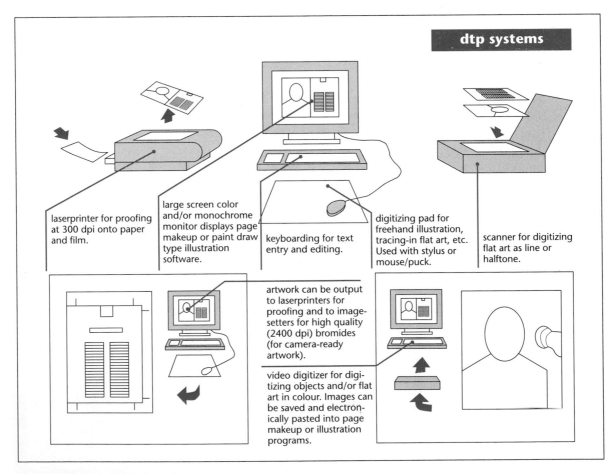

Figure 15.3   Equipment in a Desktop Publishing System
Source: Cotton, Bob, ed. The New Guide to Graphic Design. *Secaucus, NJ: Chartwell, 1990: 35. Copyright © 1990. Reprinted by permission of Quarto Publishing.*

Some specific features of DTP systems (Cotton 36–47):

- Text can be typed or scanned into the program and then edited, checked for spelling and grammar, displayed in columns or other spatial arrangements, set in a variety of sizes and fonts—or sent to a typesetter electronically.
- Page highlights and orienting devices can be added: headings, ruled boxes, vertical or horizontal rules, colored background screens, marginal sidebars or labels, page-locator tabs, shadowing, shading, and so on.
- Images can be drawn directly or imported into the program via scanners; charts, graphs, and diagrams can be created via graphics programs. These visuals then can be enlarged, reduced, cropped, and pasted electronically on the text pages.
- At any point in the process, entire pages can be viewed and evaluated for visual appeal, accessibility, and emphasis, then revised as needed.
- All work at all stages can be stored in the computer for later use, adaptation, or upgrading.

You might work alone or in collaboration with colleagues and graphic design professionals. In either case, you will need to understand basic principles of effective page design.

As automated design continues to improve the look of workplace writing, audiences raise their expectations for inviting and accessible documents.

Page-Design Guidelines

Approach your design decisions from the top down. First, consider the overall look of your pages; next, the shape of each paragraph; and finally, the size and style of individual letters and words (Kirsh 112). Figure 15.4 depicts how design considerations follow a top-down sequence, moving from large matters to small. (All design considerations are influenced by the size of the budget for a publication. For instance, adding a single color, say, to major heads, can double the printing cost.)

If your organization has no prescribed style guidelines, the following design principles should satisfy most readers' expectations.

Shaping the Page

In shaping a page, we consider its look, feel, and overall layout. The following suggestions will help you shape your pages for appeal and accessibility.

Use the Right Paper and Ink. For routine documents (memos, letters, in-house reports) type or print in black ink, on $8\frac{1}{2}$-by-11-inch low-gloss, white paper. Use rag-bond paper (20 pound or heavier) with a high fiber content (25 percent minimum). Shiny paper produces glare and tires the eyes. Flimsy or waxy paper feels inferior.

**Figure 15.4
A Flowchart for
Decisions in Page
Design**

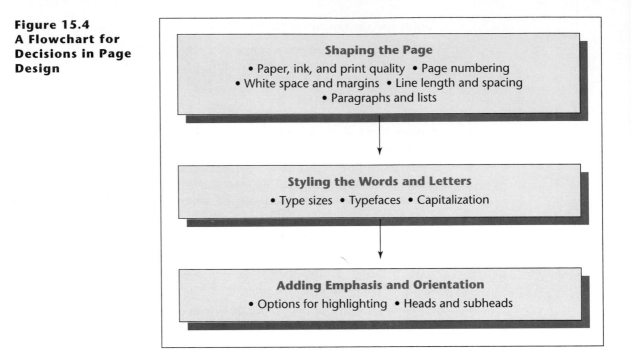

For documents that will be published (manuals, marketing literature, and so on), the grade and quality of paper are important considerations. Paper varies in weight, grain, and finish—from low-cost newsprint, with noticeable wood fiber, to wood-free, specially coated paper with custom finishes. Specific choice of paper finally depends on the artwork to be included, the type of printing, and the intended esthetic effect: for example, specially coated, heavy-weight, glossy—for an elegant effect in an annual report (Cotton 73).

Use High-Quality Type or Print. Typewriter erasures with opaquing fluid or opaquing film (correction tape) are neatest. Use a fresh ribbon and keep typewriter keys clean. On a computer, print your hard copy on a letter-quality printer, a laser printer, or a dot-matrix printer in the near-letter-quality mode. Many older dot-matrix printers produce copy that is hard to read or looks unprofessional.

Number Pages Consistently. For a long document, count your title page as page i, without numbering it, and number all front-matter pages, including your table of contents and abstract, with lowercase roman numerals (ii, iii, iv). Number the first page of your report and subsequent pages with arabic numerals (1, 2, 3). Place all page numbers in the upper-right corner, or centered in the top or bottom margin. Word processing software will number pages automatically, but you must specify placement (e.g., top right, bottom center).

Use Adequate White Space. White space is all the space not filled by text. White space divides printed areas into small, digestible chunks. For instance, it separates sections in a document, headings and visuals from text, paragraphs on a page. White space can be designed to enhance a document's appearance, clarity, and emphasis.

Well-designed white space imparts a shape to the whole document, a shape that orients readers and lends a distinctive visual form to the printed matter by:

1. keeping related elements together
2. isolating and emphasizing important elements
3. providing breathing room between blocks of information

Use white space to orient the readers

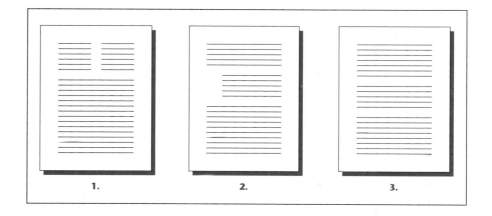

1.   2.   3.

Pages that look uncluttered, inviting, and easy to follow convey an immediate sense of user-friendliness.

Provide Ample and Appropriate Margins. Small margins make a page look crowded and difficult. On your 8½-by-11-inch page, leave margins of at least 1 inch or 1½ inches. For a document that will be bound, widen the left margin to two inches.

On the computer, you can choose between *unjustified* text (uneven or "ragged" right margins) and *justified* text (even right margins). Each arrangement creates its own "feel."

Justified lines

To make the right margin even in justified text, the spaces vary between words and letters on a line, sometimes creating channels or rivers of white space. The reader's eyes are then forced to adjust continually to these space variations within a line or paragraph. Because each line ends at an identical vertical space, the eyes must work harder to differentiate one line from another (Felker 85). Moreover, to preserve the even margin, words at line's end are often hyphenated, and frequently hyphenated line endings can be distracting.

Unjustified lines　　Unjustified text, on the other hand, uses equal spacing between letters and words on a line, and an uneven right margin (as traditionally produced by a typewriter). For some readers, a ragged right margin makes reading easier. These differing line lengths can prompt the eye to move from one line to another (Pinelli 77). In contrast to justified text, an unjustified page seems to look less formal, less distant, and less official.

Justified text seems preferable for books, annual reports, and other formal materials. Unjustified text seems preferable for more personal forms of communication such as letters, memos, and in-house reports.

Keep Line Length Reasonable. Long lines tire the eyes. The longer the line, the harder it is for readers to return to the left margin and locate the next line (White, *Visual Design* 25).

Notice how your eye labors to follow the apparently endless message that here seems to stretch in lines that continue long after your eye was prepared to move down to the next line. After reading more than a few of these lines, you begin to feel tired and bored and annoyed, without hope of ever reaching the end.

Short lines force the eyes back and forth (Felker 79). "Too-short lines disrupt the normal horizontal rhythm of reading" (White 25).

> Lines that are too
> short cause your eye
> to stumble from one
> fragment to another
> at a pace that too
> soon becomes
> annoying, if not
> nauseating.

A reasonable line length is 50–60 characters (or 9 to 12 words) per line for an 8½-by–11-inch single-column page. The number of characters will depend on print size on your typewriter or word processor. Longer lines call for larger type and wider spacing between them (White, *Great Pages* 70).

Line length, of course, is affected by the number of columns (vertical blocks of print) on your page. Two-column pages often appear in newsletters and brochures, but research indicates that single-column pages work best for complex, specialized information (Hartley 148).

Keep Line Spacing Consistent. For any document likely to be read completely (letters, memos, instructions), single-space within paragraphs and double-space between. Instead of indenting the first line of single-spaced paragraphs, separate them with a line of space.

For longer documents likely to be read selectively (proposals, formal reports), and prepared on a computer, increase line spacing within paragraphs. Indent these paragraphs or separate them with an extra line of space.

This open spacing enables readers to scan a long document, and to locate quickly what they need.

Tailor Each Paragraph to Its Purpose. Readers often skim a long document to find what they want. Most paragraphs therefore begin with a topic sentence forecasting the content.

Shape each paragraph

Use a long paragraph (no more than fifteen lines) for clustering material that is closely related (such as history and background, or any body of information best understood in one block).

Use short paragraphs for making complex material more digestible, for giving step-by-step instructions, or for emphasizing vital information.

Instead of indenting a series of short paragraphs, separate them by inserting an extra line of space (as here).

Avoid "orphan" lines: the paragraph's opening line on the bottom of a page, or the paragraph's closing line on the top line of a page.

Make Lists for Easy Reading. Readers prefer information in list form rather than in continuous prose paragraphs (Hartley 51). Whenever you have a paragraph or long sentence containing a series of related items, consider displaying these items one by one.

Types of items you might list: advice or examples, conclusions and recommendations, criteria for evaluation, errors to avoid, materials and equipment for a procedure, parts of a mechanism, or steps or events in a sequence. Notice how the preceding information becomes easier to grasp and remember when displayed in the list below.

Types of items you might list:

- advice or examples
- conclusions and recommendations
- criteria for evaluation
- errors to avoid
- materials and equipment for a procedure
- parts of a mechanism
- steps or events in a sequence

A list of brief items usually needs no punctuation at the end of each line. A list of full sentences or questions requires appropriate punctuation after each item.

Depending on the list's contents, set off each item with some kind of visual or verbal signal. If the items require a strict sequence (as in a series of steps, or parts of a mechanism), use arabic numbers (1, 2, 3) or the words *First, Second, Third,* and so on. If the items require no strict sequence (as in some examples or advice), use dashes, asterisks, or bullets. For a checklist use open boxes.

Introduce your list with an explanation. Phrase all listed items in parallel grammatical form. If the items suggest no strict sequence, try to impose some logical ranking (most important to least important, alphabetical, or some such). Set off the list with extra white space above and below.

Keep in mind that a document with too many lists appears busy, disconnected, and splintered (Felker 55). And long lists could be used by unethical writers to camouflage bad or embarrassing news (Williams 12). As with all format options, use restraint and good judgment.

Use lists to help readers organize their understanding

| | Bulleted List | Numbered List | Checklist |

Styling the Words and Letters

In styling words and letters, we consider typographic choices that will make the text easy to read.

Use Standard Type Sizes. Word-processing programs offer a wide variety of type sizes:

Select the appropriate type size

9 point

10 point

12 point

14 point

18 point

24 point

The standard typesize for most documents is 10 to 12 point. Use larger or smaller sizes only for certain headings, titles, captions (brief explanation of a visual), sidebars (marginal comments), or special emphasis. Use the same typesize for similar elements throughout the document. Certain fonts (see the following) are easier to read than others, making smaller sizes possible.

Select Appropriate Fonts. A font, or typeface, is the style of individual letters and characters. Each font has its own *personality:* "The typefaces you select for . . . [heads], subheads, body copy, and captions affect the way readers experience your ideas" (*Aldus Guide* 24)

An effective font makes for easier reading. In fact, depending on the font, reading speed can vary by as much as 30 percent (Chauncey 26).

Word-processing programs offer a variety of fonts like the examples below, listed by name.

Select a font for its personality

11-point New York

`11-point Courier`

11-point Palatino

11-point Geneva

11-point Monaco

11-point Chicago

11-point Helvetica

11-point Times

Can you assign a personality to each of the above fonts?

For greater visual unity, try to use different sizes and versions (**bold,** *italic,* e x p a n d e d, condensed) of the same font throughout your document—with the possible exception of headings, captions, sidebars, or visuals.

All fonts divide into two broad categories: *serif* and *sans serif.* Serifs are the fine lines that extend horizontally from the main strokes of a letter.

Serif type makes body copy more readable because the horizontal lines "bind the individual letters" and thereby guide the reader's eyes from letter to letter—as in the type you are now reading (White, *Visual Design* 14).

Decide between serif and sans serif type

In contrast, sans serif type is purely vertical (like this). Clean looking and "businesslike," sans serif is considered ideal for technical material (numbers, equations, etc.), marginal comments, headings, examples, tables, and captions to pictures and visuals, and any other material set off from the body copy (White 16).

European readers generally prefer sans serif type throughout their documents, and other cultures have their own preferences as well. Learn all you can about the design conventions of the culture you are addressing.

Under no circumstance should typography in a routine document call attention to itself. Adopt consistent style guidelines for the whole document. Except for special emphasis, stick to the more conservative styles and avoid ornate ones altogether.

Avoid Sentences in Full Caps. Sentences or long passages in full capitals (uppercase letters) are hard to read because all uppercase letters and words have the same visual outline (Felker 87). The longer the passage, the harder readers must work to sort things out and grasp your intended emphasis.

MY DOG HAS MANY FLEAS.

My dog has many fleas.

FULL CAPS *are good for emphasis but hard to read*

Hard   ACCORDING TO THE NATIONAL COUNCIL ON RADIA-TION PROTECTION, YOUR MAXIMUM ALLOWABLE DOSE OF LOW-LEVEL RADIATION IS 500 MILLIREMS PER YEAR.

Easier   According to the National Council on Radiation Protection, your MAXIMUM allowable dose of low-level radiation is 500 millirems per year.

Lowercase letters take up less space, and the distinctive shapes make each word easier to recognize and remember (Benson 37).

Use full caps as section headings (INTRODUCTION) or to highlight a word or phrase (WARNING: NEVER TEASE THE ALLIGATOR). As with other highlighting options discussed in the next section, use full caps sparingly in your document.

Highlighting for Emphasis

Effective highlighting helps readers distinguish the important from the less important elements. Highlighting options include fonts, type sizes, white space, and other graphic devices that:

- emphasize key points
- make headings prominent
- separate sections of a long document
- set off examples, warnings, and notes

On a typewriter, you can highlight with underlining, FULL CAPS, dashes, parentheses, and asterisks.

> You can indent to set off examples, explanations, or any material that should be distinguished from other elements in your document.

Using ruled (or typed) horizontal lines, you can separate sections in a long document:

Using ruled lines, broken lines, or ruled boxes, you can set off crucial information such as a warning or a caution:

Caution: A document with too many highlights can appear confusing, disorienting, and tasteless.

In adding emphasis and orientation, we consider design elements that will direct readers' focus and help them navigate the text. See pages 342–43 for more on background screens, ruled lines, and ruled boxes.

Word processing software offers highlighting options that might include **boldface,** *italics,* SMALL CAPS, varying type sizes and fonts, and color. For specific highlighted items, some options are better than others:

Not all highlighting is equal

Boldface works well for emphasizing a single sentence or brief statement, and is perceived by readers as being "authoritative" (*Aldus Guide* 42).

Italics suggest a more subtle or "refined" emphasis than boldface (Aldus Guide 42). Italics can highlight words, phrases, book titles, or anything else you would have underlined on a typewriter. But multiple lines (like these) of italic type are hard to read.

SMALL CAPS WORK FOR HEADINGS AND SHORT PHRASES. BUT ANY LONG STATEMENT ALL IN CAPS IS HARD TO READ.

Small type sizes (usually sans serif) work well for captions and as labels for visuals or to set off other material from the body copy.

Large type sizes and dramatic typefaces are both hard to miss and hard to digest. Be conservative— unless you really need to convey forcefulness.

Color is appropriate only in some documents, and only when used sparingly. (Pages 340–44 discuss how color can influence readers' perception and interpretation of a message.)

Whichever highlights you select for a document, be consistent. Make sure that all headings at one level are highlighted identically, that all warnings and cautions are set off identically, and so on. And *never* combine too many highlights.

Using Headings for Access and Orientation

Most long documents are read not like novels but like textbooks or reference books. Readers look back or jump ahead to sections that interest them most. Without headings, readers don't know how the document is organized, and they can't find what they need. Headings make your writing task easier as well; like guideposts or points on a road map, they keep you on course and provide transitions.

Headings signal readers that something is ending and something else is beginning. They make a document more attractive and less intimidating by dividing it into accessible blocks or "chunks." An informative heading can help a reader decide whether a section is worth reading (Felker 17). Besides cutting down on reading and retrieval time, headings help readers remember information (Hartley 15).

Make Headings Informative. A heading should be long enough to be informative without being wordy. Informative headings orient readers, showing them what to expect. Vague or general headings can be more misleading than no headings at all (Redish et al. 144). Whether your heading takes the form of a phrase, a statement, or a question, be sure it advances thought.

> Uninformative heading   Document Formatting

What should we expect here: specific instructions, an illustration, a discussion of formatting policy in general? We can't tell.

> Informative versions   How to Format Your Document
> Format Your Document in This Way
> How Do I Format My Document?

When you use questions as headings, phrase the questions in the same way as readers might ask them.

Make Headings Specific as Well as Comprehensive. Focus the heading on a specific topic. Do not preface a discussion of the effects of acid rain on lake trout with a broad heading such as "Acid Rain." Use instead "The Effects of Acid Rain on Lake Trout."

Also, provide enough headings to contain each discussion section. If chemical, bacterial, and nuclear wastes are three *separate* discussion items, provide a heading for each. Do not simply lump them under the sweeping heading "Hazardous Wastes." If you have prepared an outline for your document, adapt major and minor headings from it.

Make Headings Grammatically Consistent. All major topics or all minor topics in a document share equal rank; to emphasize this equality, express topics at the same level in identical—or parallel—grammatical form.

> Nonparallel headings   How to Avoid Damaging Your Disks:
> 1. Clean Disk-Drive Heads
> 2. Keep Disks Away from Magnets
> 3. Writing on Disk Labels with a Felt-Tip Pen
> 4. It is Crucial that Disks Be Kept Away from Heat
> 5. Disks Should Be Kept Out of Direct Sunlight
> 6. Keep Disks in Their Protective Jackets

```
                          SECTION HEADING

     Write section headings in full caps, centered horizontally,
one extra line space below any preceding text. Do not underline
or further highlight these heads.


Major Topic Heading
     Begin each word (except articles and prepositions) in major
topic heads with an uppercase letter. Abut the left margin (flush
left), one extra line space below preceding text. Underline.


     Minor Topic Heading
     Begin each important word in minor topic heads with an upper-
case letter. Indent the heads five spaces from the left margin,
one extra line space below preceding text. Underline these heads.


Subtopic Heading. Begin each important word in subtopic heads with
an uppercase letter. Instead of indenting, place the heads flush
left, one extra line space below preceding text, and on the same
line as following text (separated from the text by a period).
```

Figure 15.5   Recommended Format for Typewritten Headings

In items 3, 4, and 5, the lack of parallelism (no verbs in the imperative mood) obscures the relationship between individual steps and confuses readers. This next version emphasizes the equal rank of these items.

Parallel headings

3. Write on Disk Labels with a Felt-Tip Pen
4. Keep Disks Away from Heat
5. Keep Disks Out of Direct Sunlight

Parallelism helps make a document readable and accessible.

Make Headings Visually Consistent. To understand how your document is organized, readers consider not only what the headings *say* but also how they *look:* "Wherever heads are of equal importance, they should be given similar visual expression, because the regularity itself becomes an understandable symbol" (White, *Visual Design* 104). Use identical type size and typeface for all headings at a given rank.

Lay Out Headings by Rank. Like a road map, your headings should identify clearly the large and small segments in your document. A long document ordinarily has section headings to mark its introduction, body, and conclusion. The body section in particular can have several lower ranks of headings: major and minor topic headings, and perhaps subtopic headings as well. (Use the logical divisions from your outline as a model for heading layout.) Think of each heading at a particular rank as an "event in a sequence" (White *Visual Design* 95).

Figure 15.5 shows how headings (typewritten) vary in positioning and highlighting, depending on their rank. The layout in the figure embodies these guidelines:

- *Ordinarily, use no more than four levels of heading (section, major topic, minor topic, subtopic).* Excessive heads and subheads can make a document seem cluttered or fragmented.

- *To divide logically, be sure each higher-level heading yields at least two lower-level headings.*

- *Insert one additional line of space above your heading.* For double-spaced text, triple-space before the heading, and double-space after; for single-spaced text, double-space before the heading, and single-space after. The spacing indicates whether the heading belongs with the text preceding it or following it.

- *Never begin the sentence right after the heading with "this," "it," or some other pronoun referring to the heading.* Make the sentence's meaning independent of the heading.

Four Levels of Headings   **Poor Heading Format (Floating Head)**

Section Heading

On a word processor, section headings in boldface and enlarged type are more appealing and readable than heads in full caps. Use a type size roughly 4 points larger than body copy (say, 16-point section heads for 12-point body copy). Avoid *overly* large heads, and use no other highlights. Set these and all lower heads one extra line space below any preceding text.

Major Topic Heading

Major topic heads abut the left margin (flush left), and each important word begins with an uppercase letter. Use boldface and a type size roughly 2 points larger than body copy, with no other highlights.

Minor Topic Heading

Minor topic heads also are set flush left. Use boldface, italics (optional), and the same type size as in the body copy, with no other highlights.

Subtopic Heading. Instead of indenting the first line of body copy, place subtopic heads flush left on the same line as following text, and set off by a period. Use boldface and the same type size as in the body copy, with no other highlights.

Figure 15.6   Recommended Format for Word-Processed Headings

- *Never leave a heading floating on the final line of page.* Unless two lines of text can fit below the heading, carry it over to the top of the next page.
- *If possible, use different type sizes to reflect the ranks of heads.* Readers tend to equate large type size with importance (White, *Visual Design* 95; Keyes 641). On a typewriter, set major heads in full caps. On a word processor set major heads in a larger type size, and perhaps in boldface (as in Figure 15.6).

When headings show the relationships among all the parts, readers can grasp at a glance how a document is organized.

Audience Considerations in Page Design

Like any writing decisions, page design choices are by no means random. An effective writer designs a document for specific use by a specific audience.

How do you decide on a format? Your best bet is to work from a detailed audience and use profile (Wight 11). Know who your readers are and how they will use your information. Design a document to meet their particular needs and expectations, as in these examples:

- If readers will use your document for reference only (as in a repair manual), make sure you have plenty of headings.
- If readers will follow a sequence of steps, show that sequence in a numbered list.
- If readers will need to evaluate something, give them a checklist of criteria (as in this book at the end of most chapters).
- If readers need a warning, highlight the warning so that it cannot possibly be overlooked.
- If readers have asked for your one-page résumé, save space by using the 10-point type size.
- If readers will be facing complex information or difficult steps, widen the margins, increase all white space, and shorten the paragraphs.

How effectively you combine your options will depend mostly on how carefully you have analyzed your audience. But regardless of the audience, never make the document look "too intellectually intimidating" (White, *Visual Design* 4).

Consider also the specific cultural expectations of your audience. For instance, Arabic and Persian text is written from right to left instead of left to right (Leki 149). In other cultures, readers move up and down the page, instead of across. Be aware that a particular culture might be offended by certain icons or by a typeface that seems too plain or too fancy (Weymouth 144). Ignoring a culture's design conventions can be interpreted as a sign of disrespect.

Finally, keep in mind that even the most brilliant page design cannot redeem a document whose content is worthless, organization chaotic, or style unreadable. The value of a document ultimately depends on elements that are more than skin deep.

REVISION CHECKLIST FOR PAGE DESIGN

(*Numbers in parentheses refer to the first page of discussion.*)

❑ Is the paper white, low-gloss, rag bond, with black ink? (359)

❑ Is all type or print neat and legible? (360)

❑ Does the white space adequately orient the readers? (361)

❑ Are the margins ample? (361)

❑ Is line length reasonable? (362)

❑ Is the right margin unjustified? (361)

❑ Is line spacing appropriate and consistent? (362)

❑ Does each paragraph begin with a topic sentence? (363)

❑ Does the length of each paragraph suit its subject and purpose? (363)

❑ Are all paragraphs free of "orphan" lines? (363)

❑ Do parallel items in strict sequence appear in a numbered list? (363)

❑ Do parallel items of any kind appear in a list whenever a list is appropriate? (363)

❑ Are pages numbered consistently? (360)

❑ Is the body type size 10 to 12 points? (364)

❑ Do full caps highlight only words or short phrases? (365)

❑ Is the highlighting consistent, tasteful, and subdued? (366)

❑ Are all format patterns distinct enough so that readers will find what they need? (355)

❑ Are there enough headings for readers to know where they are in the document? (367)

❑ Are headings informative, comprehensive, specific, parallel, and visually consistent? (368)

❑ Are headings clearly differentiated according to rank? (370)

❑ Is the overall design inviting without being overwhelming? (372)

❑ Does this design respect the cultural conventions of my audience? (372)

✓ EXERCISES

1. Find an example of effective page design, in a textbook or elsewhere. Photocopy a selection (two or three pages), and attach a memo explaining to your instructor and classmates why this design is effective. Be specific in your evaluation. Now do the same for an example of poor formatting. Bring your examples and explanations to class, and be prepared to discuss why you chose them.

2. These are headings from a set of instructions for listening. Rewrite the headings to make them parallel.

   - You Must Focus on the Message
   - Paying Attention to Nonverbal Communication
   - Your Biases Should Be Suppressed
   - Listen Critically

   - Listening for Main Ideas
   - Distractions Should Be Avoided
   - Provide Verbal and Nonverbal Feedback
   - Making Use of Silent Periods
   - Are You Allowing the Speaker Time to Make His or Her Point?
   - Keeping an Open Mind Is Important

3. Using the above revision checklist, redesign an earlier assignment or a document you've prepared on the job. Submit to your instructor the revision and the original, along with a memo explaining your improvements. Be prepared you discuss your format design in class.

4. Anywhere on campus or at work, locate a document with a design that needs revision. Candidates include career counseling handbooks,

financial aid handbooks, student or faculty handbooks, software or computer manuals, medical information, newsletters, or registration procedures. Redesign the document or a two- to-five page selection from it. Submit to your instructor a copy of the original, along with a memo explaining your improvements. Be prepared to discuss your revision in class.

☑ COLLABORATIVE PROJECT

Working in small groups, redesign a document you select or your instructor provides. Prepare a detailed explanation of your group's revision. Appoint a group member to present your revision to the class, using an opaque or overhead projector, a large-screen computer monitor, or photocopies.

Adding Document Supplements

Purpose of Supplements

Cover

Title Page

Letter of Transmittal

Table of Contents

List of Tables and Figures

Informative Abstract

Glossary

Appendixes

Documentation

S UPPLEMENTS are reference items generally added to a long report or proposal to make the document more accessible. A document's supplements accommodate readers with various interests: The title page, letter of transmittal, table of contents, and abstract give summary information about the content of the document. The glossary, appendixes, and list of works cited can either provide supporting data or help readers follow technical sections. According to their needs, readers can refer to one or more of these supplements or skip them altogether. All supplements, of course, are written only after the document itself has been completed.

Some companies and organizations require that a full range of supplements routinely accompany any long document. Others do not. For situations in which your audience has not stipulated its requirements, select only those supplements that enhance the informative value of your particular document. Avoid using any supplements merely as decoration.

Purpose of Supplements

Documents must be accessible to varied readers for many purposes. Supplements address these workplace realities:

- *Confronted by a long document, many readers will try to avoid reading the whole thing.* Instead they look for the least information they need to complete the task, make the decision, or take some other action.
- *Different readers often use the same document for different purposes.* Some look for an overview; others want details; others want only conclusions and recommendations, or the "bottom line." Technical personnel might focus on the body of a highly specialized report and on the appendixes for supporting data (maps, formulas, calculations). Executives and managers might read only the transmittal letter and the abstract. If the latter audience reads any of the report proper, they are likely to focus on conclusions and recommendations.

A document with carefully planned supplements accommodates the needs of diverse audiences.

Report supplements can be classified in two groups:

1. *supplements that precede your report* (front matter): cover, title page, letter of transmittal, table of contents (and figures), and abstract
2. *supplements that follow your report* (end matter): glossary, appendix(es), footnotes, endnote pages

Cover

Use a sturdy, plain cover with page fasteners. With the cover on, the open pages should lie flat. Use covers only for long documents.

Center the report title and your name four to five inches below the upper edge of your page:

THE FEASIBILITY OF A TECHNICAL MARKETING CAREER:
AN ANALYSIS
by
Richard B. Larkin Jr.

Title Page

The title page lists the report title, author's name, name of person(s) or organization to whom the report is addressed, and date of submission.

How to prepare a title page

Title. Your title announces the report's purpose and subject. The previous title (given as an example for the cover) is clear, accurate, comprehensive, and specific. But even slight changes can distort this title's signal.

> An unclear title   A TECHNICAL MARKETING CAREER

The version above is unclear about the report's purpose. Is the report *describing* the career, *proposing* the career, *giving instructions* for career preparation, or *telling one person's career story?* Insert descriptive words ("analysis," "instructions," "proposal," "feasibility," "description," "progress") that accurately state your purpose.

To be sure that your title forecasts what the report delivers, write its final version *after* completing the report.

Placement of Title Page Items. Do not number your title page but count it as page i of your prefatory pages. Center the title horizontally, three to four inches below the upper edge. Place other items in the spacing and order shown in Figure 16.1, the title page to a report. Or devise your own system, as long as your page is balanced.

Letter of Transmittal

For college reports, the letter of transmittal usually follows the title page and is bound as part of the report. For workplace reports, the letter usually precedes the title page. Include a letter of transmittal with any formal report or proposal addressed to a specific reader. Your letter adds a note of courtesy and gives you a place for personal remarks. For instance, your letter might

What to include in a letter of transmittal

- acknowledge those who helped with the report
- refer to sections of special interest: unexpected findings, key visuals, major conclusions, special recommendations, and the like
- discuss the limitations of your study, or any problems gathering data
- discuss the need and approaches for follow-up investigations
- describe any personal (or off-the-record) observations
- suggest some special uses for the information
- urge the reader to immediate action

Feasibility Analysis
of a Career
in Technical Marketing

for

Professor J. M. Lannon

Technical Writing Instructor

University of Massachusetts

North Dartmouth, Massachusetts

by

Richard B. Larkin, Jr.

English 266 Student

May 1, 19XX

Figure 16.1   Title Page for a Formal Report

165 Hammond Way
Hyannis, MA 02457
April 29, 19XX

John Fitton
Placement Director
University of Massachusetts
North Dartmouth, MA 02747

Dear Mr. Fitton:

Here is my analysis to determine the feasibility of a career in technical marketing and sales. In preparing my report, I've learned a great deal about the requirements and modes of access to this career, and I believe my information will help other students as well.

Although committed to their specialities, some technical and science graduates seem interested in careers in which they can apply their technical knowledge to customer and business problems. Technical marketing may be an attractive choice of career for those who know their field, who can relate to different personalities, and who are good communicators.

Technical marketing is competitive and demanding, but highly rewarding. In fact, it is an excellent route to upper-management and executive positions. Specifically, marketing work enables one to develop a sound technical knowledge of a company's products, to understand how these products fit into the marketplace, and to perfect sales techniques and interpersonal skills. This is precisely the kind of background that paves the way to top-level jobs.

I've enjoyed my work on this project, and would be happy to answer any questions.

Sincerely,

Richard B. Larkin, Jr.

Figure 16.2   Letter of Transmittal for a Formal Report

The letter of transmittal can be tailored to a particular reader, as is Richard Larkin's in Figure 16.2. If a report is being sent to a number of people who are variously qualified and bear various relationships to the writer, individual letters of transmittal may vary, within the following basic structure:

Introduction. Open with reference to the reader's original request. Briefly review the reasons for your report or include a brief descriptive abstract. Maintain a confident and positive tone throughout. Indicate pride and satisfaction in your work. Avoid implied apologies, such as "I hope this report meets your expectations."

Body. In the letter body, include items from our prior list of possibilities (acknowledgments, special problems). Although your informative abstract will summarize major findings, conclusions, and recommendations, your letter gives a brief and personal overview of the *entire project.*

Conclusion. State your willingness to answer questions or discuss findings. End positively with something like "I believe that the data in this report are accurate, that they have been analyzed rigorously and impartially, and that the recommendations are sound."

Table of Contents

Your table of contents serves as a road map for readers and a checklist for you. Compose this supplement by assigning page numbers to headings from your outline. Keep in mind, however, that not all levels of outline headings appear in your table of contents or your report. Excessive headings can fragment the discussion.

Follow these guidelines:

How to prepare a table of contents

- List front matter (transmittal letter, abstract), numbering the pages with small roman numerals. (The title page, though not listed, is counted page i.) List glossary, appendix, and endnotes; number these pages with arabic numerals, continuing the page sequence of your report proper, in which page 1 is the first page of report text.
- Include no headings in the table of contents not listed as headings or subheadings in the report; the report may, however, contain subheadings not listed in the table of contents.
- Phrase headings in the table of contents exactly as in the report.
- To reflect their rank, list headings at various levels in varying typeface and indention.
- Use *leader lines* (. . . . .) to connect heading to page number. Align rows of dots vertically, each above the other.

Figure 16.3 shows the table of contents for Richard Larkin's feasibility analysis.

iii

CONTENTS

page

Figure 16.3   Table of Contents for a Formal Report

TABLES AND FIGURES

Figure 16.4   A List of Tables and Figures for a Formal Report

List of Tables and Figures

Following the table of contents is a list of tables and figures, if needed. When a report has more than four or five visuals, place this table on a separate page. Figure 16.4 shows the table of tables and figures for Larkin's report.

Informative Abstract

Some readers have neither time nor willingness to read your entire report. For these readers, the informative abstract (page 383) is the most important part of the document. The abstract is always written *after* your report proper.

Writing the abstract gives you the chance to measure your own control over the material. Because you are condensing your message, you must establish explicit connections. If you cannot effectively summarize your report, it probably needs revision.

Follow these guidelines[1] for your abstract:

How to prepare an informative abstract

- Make your abstract able to stand alone in meaning.
- Write for general readers. Readers of the abstract are likely to vary in expertise, perhaps more than those who read the report itself; translate all technical data.
- Add no new information. Simply summarize the report.
- Present your information in the following sequence:

---

1. My thanks to Professor Edith K. Weinstein for these suggestions.

ABSTRACT

The feasibility of technical marketing as a career is based on a college graduate's interests, abilities, and expectations, as well as on possible entry options.

Technical marketing is a feasible career for anyone who is motivated, who can communicate well, and who knows how to get along. Although this career offers job diversity and excellent potential for income, it entails almost constant travel, competition, and stress.

College graduates enter technical marketing through one of four options: entry-level positions that offer hands-on experience, formal training programs in large companies, prior experience in one's specialty, or graduate programs. The relative advantages and disadvantages of each option can be measured in resulting immediacy of income, rapidity of advancement, and long-term potential.

Anyone considering a technical marketing career should follow these recommendations:

• Speak with people who work in the field.
• Weigh carefully the implications of each entry option.
• Consider combining two or more options.
• Choose options for personal as well as professional benefits.

Figure 16.5   Informative Abstract

a. Identify the issue or need that led to the report.
b. Offer the major findings from the report body.
c. Include a condensed conclusion and recommendations, if any.

The informative abstract in Figure 16.5 accompanies Richard Larkin's report.

Glossary

A glossary is an alphabetical listing of specialized terms and their definitions, following your report. Specialized reports often contain glossaries, especially when written for both technical and nontechnical readers. A glossary makes key definitions available to nontechnical readers without interrupting technical readers. If fewer than five terms need defining, place them instead in the report introduction as working definitions, or use footnote definitions. If you use a separate glossary, inform readers of its location: "(see the glossary at the end of this report)."

GLOSSARY

Analgesic: a medication given to relieve pain during the first stage of labor.

Cervix: the neck-shaped anatomical structure that forms the mouth of the uterus.

Dilation: cervical expansion occurring during the first stage of labor.

Episiotomy: an incision of the outer vaginal tissue, made by the obstetrician just before the delivery, to enlarge the vaginal opening.

First stage of labor: the stage in which the cervix dilates and the baby remains in the uterus.

Induction: the stimulating of labor by puncturing the membranes around the baby or by giving an oxytoxic drug (uterine contractant), or both.

Figure 16.6   A Glossary (Partial)

Follow these guidelines for a glossary:

How to prepare a glossary

- Define all terms unfamiliar to a general reader (an intelligent layperson).
- Define all terms that have a special meaning in your report (e.g., "In this report, a small business is defined as").
- Define all terms by giving their class and distinguishing features, unless some terms need expanded definitions.
- List your glossary and its first page number in your table of contents.
- List all terms in alphabetical order. Underline each term and use a colon to separate it from its single-spaced definition.
- Define only terms that need explanation. In doubtful cases, overdefining is safer than underdefining.
- On first use, place an asterisk in the text by each item defined in the glossary.

Figure 16.6 shows part of a glossary for a comparative analysis of two techniques of natural childbirth, written by a nurse practitioner for expectant mothers and student nurses.

Appendixes

An appendix follows the text of your report. The appendix expands items discussed in the report without cluttering the report text. Typical items in an appendix include:

APPENDIX A

Table 1 Allocations and Performance of Five Massachusetts College Newspapers

	Stonehorse College	Alden College	Simms University	Fallow State	UMD
Enrollment	1,600	1,400	3,000	3,000	5,000
Fee paid (per year)	$65.00	$85.00	$35.00	$50.00	$65.00
Total fee budget	$88,000	$119,000	$105,000	$150,000	$334 429.28
Newspaper budget	$10,000	$6,000	$25.300	$37,000	$21,500 $25.337.14[a]
Yearly cost per student	$6.25	$4.29	$8.43	$12.33	$4.06 $5.20[a]
Format of paper	Weekly	Every third week	Weekly	Weekly	Weekly
Average no. of pages	8	12	18	12	20
Average total pages	224	120	504	336	560 672[a]
Yearly cost per page	$44.50	$50.00	$50.50	$110.11	$38.25 $38.69[a]

[a]These figures are next year's costs for the UMD *Torch.*

Source: Figures were quoted by newspaper business managers in April 19XX.

Figure 16.7   An Appendix

What an appendix might include

- complex formulas
- details of an experiment
- interview questions and responses
- long quotations (one or more pages)
- maps
- material more essential to secondary readers than to primary readers
- photographs
- related correspondence (letters of inquiry, and so on)
- sample questionnaires and tabulated responses
- sample tests and tabulated results
- some visuals occupying more than one full page

- statistical or other measurements
- texts of laws and regulations

The appendix is a catch-all for items that are important but difficult to integrate within your text. Figure 16.7 shows an appendix to a budget proposal.

Do not stuff appendixes with needless information. Do not use them unethically for burying bad or embarrassing news that belongs in the report proper. Follow these guidelines:

How to prepare an appendix

- Include only material that is relevant.
- Use a separate appendix for each major item.
- Title each appendix clearly: "Appendix A: Projected Costs."
- Do not use appendixes excessively. Four or five appendixes in a ten-page report would indicate a poorly organized document.
- Limit an appendix to a few pages, unless more length is essential.
- Mention your appendix early in your introduction, and refer readers to it at appropriate points in the report: "(see Appendix A)."

Use an appendix for any material that is essential but might harm the unity and coherence of your report. Remember that readers should be able to understand your report without having to turn to the appendix. Distill the essential facts from your appendix and place them in your report text.

Improper reference	The whale population declined drastically between 1976 and 1977 (see Appendix B for details).
Proper reference	The whale population declined by 16 percent from 1976 to 1977 (see Appendix B for statistical breakdown).

Documentation

The endnote or works cited pages list each of your outside references in the same numerical order as they are cited in the report proper. See Chapter 10 for a discussion of documentation.

☑ EXERCISES

1. These titles are intended for investigative, research, or analytical reports. Revise each inadequate title to make it clear and accurate.

   a. The Effectiveness of the Prison Furlough Program in Our State

   b. Drug Testing on the Job

   c. The Effects of Nuclear Power Plants

   d. Woodburning Stoves

   e. Interviewing

   f. An Analysis of Vegetables (for a report assessing the physiological effects of a vegetarian diet)

   g. Wood as a Fuel Source

   h. Oral Contraceptives

   i. Lie Detectors and Employees

2. Prepare a title page, letter of transmittal (for a definite reader who can use your information in a definite way), table of contents, and informative abstract for a report you have written earlier.

3. Find a short but effective appendix in one of your textbooks. In a journal article in your field, or in a report from your workplace. In a memo to your instructor and classmates, explain how the appendix is used, how it relates to the main text, and why it is effective. Attach a copy of the appendix to your memo. Be prepared to discuss your evaluation in class.

PART

V

Specific Documents and Applications

Definitions

CHAPTER

17

Purpose of Definitions

Elements of Definition

    Plain English

    Basic Properties

    Objectivity

Types of Definition

    Parenthetical Definition

    Sentence Definition

    Expanded Definition

Expansion Methods

    Etymology

    History and Background

    Negation

    Operating Principle

    Analysis of Parts

    Visuals

    Comparison and Contrast

    Required Materials or Conditions

    Example

Sample Situations

Placement of Definitions

T O DEFINE a term is to explain the precise meaning you intend by using that term. Clear writing depends on definitions that both reader and writer understand. Unless you are sure readers know the exact meaning you intend, always define the term upon first use.

Purpose of Definitions

Every specialty has its own technical language. Engineers, architects, or programmers talk about "torque," "tolerances," or "microprocessors"; lawyers, real estate brokers, and investment counselors discuss "easements," "liens," "amortization," or "escrow accounts." Whenever such terms are unfamiliar to an audience, they need defining.

For colleagues, you rarely have to define specialized terms (unless the term is new), but reports often are written for the layperson—the client or some other general reader. When you write for nonspecialists, clarify your meaning with definitions.

Most of the specialized terms previously mentioned are concrete and specific. Once "microprocessor" has been defined for the reader, its meaning will not differ appreciably in another context. When a term is highly technical, a writer can figure out that it should be defined for some readers. However, familiar terms like "disability," "guarantee," "tenant," "lease," or "mortgage" acquire very specialized meanings in specialized contexts. Here definition becomes crucial. What "guarantee" means in one situation is not necessarily what it means in another. Contracts are detailed (and legal) definitions of the specific terms of an agreement.

Assume you're shopping for disability insurance to protect your income in case of injury or illness. Besides comparing prices, you want each company to define "physical disability." Although Company A offers the cheapest policy, it might define physical disability as inability to work at *any* job. Therefore, if a neurological disorder prevents you from continuing work as designer of electronic devices, without disabling you for some menial job, you might not qualify as "disabled." In contrast, Company B's policy, although more expensive, might define physical disability as inability to work at your *specific* job. Both companies use the term "physical disability," but each defines it differently. Because they are legally responsible for the documents they prepare, all communicators rely on the technique of clear definition.

Definition carries *ethical* requirements, too. Chapter 5, for instance, explains how the shuttle *Challenger* explosion resulted from faulty definition: namely, the definition of "acceptable risk" finally was determined not by the technical facts, but by social pressures to launch on schedule. This social definition of acceptable risk was passed forward and influenced the top decision makers to approve the disastrous launch. Agreeing on meaning in such cases may not be easy, but we are ethically bound to communicate definitions based on a legitimate interpretation of the facts as we understand them.

Growth in technology, specialization, and global communication makes definition increasingly vital. Know for whom you're writing, and why. If unable to pinpoint your audience, assume a general readership and define generously.

Elements of Definition

For all definitions, use these guidelines:

Plain English

Clarify meaning by using language readers understand.

Unclear	A tumor is a neoplasm.
Better	A tumor is a growth of cells that occurs independently of surrounding tissue and serves no useful function.
Unclear	A solenoid is an inductance coil that serves as a tractive electromagnet. *(A definition appropriate for an engineering manual, but too specialized for general readers.)*
Better	A solenoid is an electrically energized coil that converts electrical energy to magnetic energy capable of performing mechanical functions.

Basic Properties

Convey the properties of an item that differentiate it from all others. A thermometer has a singular function; it measures temperature. Without this essential information, a definition would have no real meaning for uninformed readers. Any other data about thermometers (types, special uses, materials used in construction) are secondary. A book, on the other hand, cannot be defined according to functional properties because books have multiple functions. A book can be used to write in or to display pictures, to record financial transactions, to read, and so on. Also, other items (individual sheets of paper, posters, newspapers, picture frames) serve the same functions. The basic property of a book is physical: it is a bound volume of pages. Readers would have to know this *first,* to understand what a book is.

Objectivity

Unless readers understand that your purpose is to persuade, omit your opinions from a definition. "Bomb" is defined as "an explosive weapon detonated by impact, proximity to an object, a timing mechanism, or other predetermined means." If you define a bomb as "an explosive weapon devised and perfected by hawkish idiots to blow up the world," you are editorializing, *and* ignoring a bomb's basic property.

Likewise, in defining "diesel engine," simply tell what it is and how it works. You might think that diesels are too noisy and sluggish for automobiles, but omit these judgments from your definition.

Types of Definition

Definitions vary greatly in length and detail: from a few words in parentheses, to one or more complete sentences, to multiple paragraphs or pages.

Your choice of definition type depends on what information readers need, and that, in turn, depends on why they need it. "Carburetor," for instance, could be defined in one sentence, briefly telling readers what it is and how it works. But this definition would be expanded for the student mechanic who needs to know the origin of the term, how the device was developed, what it looks like, how it is used, and how its parts interact. Audience needs should guide your choice.

Parenthetical Definition

A parenthetical definition explains the term in a word or phrase, often as a synonym in parentheses following the term:

Parenthetical definitions

> The effervescent (bubbling) mixture is highly toxic.
>
> The leaching field (sievelike drainage area) requires crushed stone.

Another option is to express your definition as a clarifying phrase:

> The trees on the site are mostly deciduous; that is, they shed their foliage at season's end.

Use parenthetical definitions to convey the general meaning of specialized terms so that readers can follow the discussion where these terms are used. A parenthetical definition of "leaching field" might be adequate in a progress report to a client whose house you are building. But a public-health report titled "Groundwater Contamination from Leaching Fields" would call for expanded definition.

Sentence Definition

A definition may require one or more sentences with this structure: (1) the item or term being defined, (2) the class (specific group) to which the term belongs, and (3) the features that differentiate the term from all others in its class.

Elements of sentence definitions

Term	Class	Distinguishing features
carburetor	a mixing device	in gasoline engines that blends air and fuel into a vapor for combustion within the cylinders
transit	a surveying instrument	that measures horizontal and vertical angles
diabetes	a metabolic disease	caused by a disorder of the pituitary gland or pancreas and characterized by excessive urination, persistent thirst, and decreased ability to metabolize sugar

Term	Class	Distinguishing features
liberalism	a political concept	based on belief in progress, the essential goodness of people, and the autonomy of the individual, and advocating protection of political and civil liberties
brief	a legal document	containing all the facts and points of law pertinent to a specific case, and filed by an attorney before the case is argued in court
stress	an applied force	that strains or deforms a body
laser	an electronic device	that converts electrical energy to light energy, producing a bright, intensely hot, and narrow beam of light
fiber optics	a technology	that uses light energy to transmit voices, video images, and data through hair-thin glass fibers

These elements can be combined into one or more sentences.

A complete sentence definition

> Diabetes is a metabolic disease caused by a disorder of the pituitary gland or pancreas. This disease is characterized by excessive urination, persistent thirst, and decreased ability to metabolize sugar.

Sentence definition is especially useful for stipulating the precise working meaning of a term that has several possible meanings. State your working definitions at the beginning of your report:

A working definition

> Throughout this report, the term "disadvantaged student" means . . .

Classifying the Term. Be specific and precise in your classification. The narrower your class, the more specific your meaning. "Transit" is correctly classified as a "surveying instrument," not as a "thing" or an "instrument." "Stress" is classified as "an applied force"; to say that stress "takes place when . . ." or "is something that . . ." fails to reflect a specific classification. Be sure to select precise terms of classification: "Diabetes" is precisely classified as "a metabolic disease," not as "a medical term."

Differentiating the Term. Differentiate the term by separating the item it names from every other item in its class. Make these distinguishing features narrow enough to pinpoint the item's unique identity and meaning, yet broad enough to be inclusive. A definition of "brief" as "a legal document introduced in a courtroom" is too broad because the definition doesn't differentiate "brief" from all other legal documents (wills, written confessions, etc.). Conversely, differentiating "carburetor" as "a mixing device used in automobile engines" is too narrow because it ignores the carburetor's use in all other gasoline engines.

Also, avoid circular definitions (repeating, as part of the distinguishing features, the word you are defining). Thus, "stress" should not be defined as "an applied force that places stress on a body." The class and distinguishing features must express the item's basic property ("an applied force that strains or deforms a body").

Expanded Definition

An expanded definition can include parenthetical and sentence definitions, but it provides greater detail for readers who need it. The sentence definition of "solenoid" on page 392 is good for a general reader who simply needs to know what a solenoid is. But a manual for mechanics or mechanical engineers would define solenoid in detail (as on pages 401–03); these readers need to know also how a solenoid works and how to use it.

The problem with defining an abstract and general word, such as "condominium" or "loan," is different. "Condominium" is a vaguer term than "solenoid" (solenoid A is pretty much like solenoid B) because the former refers to many types of ownership agreements.

Concrete, specific terms such as "diabetes," "transit," and "solenoid" often can be defined in a sentence, and they require expanded definition only for certain audiences. But terms such as "disability" and "condominium" require expanded definition for almost any audience. The more general or abstract the term, the more need for expanded definition.

An expanded definition may be a single paragraph (as for a simple tool) or may extend to scores of pages (as for a digital dosimeter—a device for measuring radiation exposure); sometimes the definition itself *is* the whole report.

This excerpt from an automobile insurance policy defines coverage for "bodily injury to others." Its style and and detail make this definition clear to general readers. Instead of the "legalese" in some policies, this policy is written in plain English.

An expanded definition

PART I. BODILY INJURY TO OTHERS

Under this Part, we will pay damages to people injured or killed by your auto in Massachusetts accidents. Damages are the amounts an injured person is legally entitled to collect for bodily injury through a court judgment or settlement. We will pay only if you or someone else using your auto with your consent is legally responsible for the accident. The most we will pay for injuries to any one person as a result of any one accident is $5,000. The most we will pay for injuries to two or more people as a result of any one accident is a total of $10,000. This is the most we will pay as the result of a single accident no matter how many autos or premiums are shown on the Coverage Selections page.

We will *not* pay:

1. For injuries to guest occupants of your auto.
2. For accidents outside of Massachusetts or in places in Massachusetts where the public has no right of access.
3. For injuries to any employees of the legally responsible person if they are entitled to Massachusetts workers' compensation benefits.

Expansion Methods

How you expand a definition depends on the questions you think readers need answered, as shown in Figure 17.1. Begin with a sentence definition, and then use only those expansion strategies which serve your reader's needs.

Etymology

A word's origin (its development and changing meanings) can clarify its definition. *Biological control* of insects is derived from the Greek "bio," meaning *life* or *living organism,* and the Latin "contra," meaning *against* or *opposite.* Biological control, then, is the use of living organisms against insects. College dictionaries contain etymological information, but your best bet is *The Oxford English Dictionary* and encyclopedic dictionaries of science, technology, and business.

**Figure 17.1
Directions in
Which a Definition
Can Be Expanded**

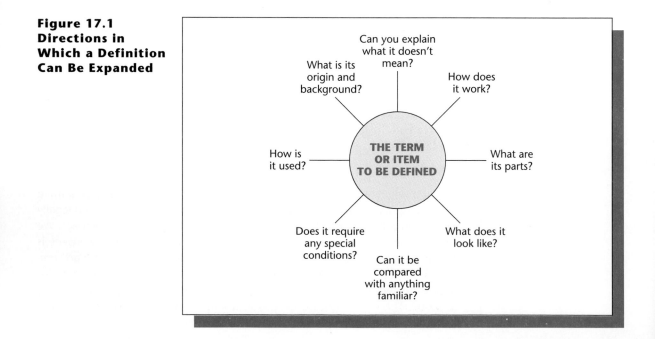

Some technical terms are acronyms, derived from the first letters or parts of several words. *Laser* is an acronym for *light amplification by stimulated emission of radiation.*

Sometimes a term's origin can be colorful as well as informative. *Bug* (jargon for *programming error*) is said to derive from an early computer at Harvard that malfunctioned because of a dead bug blocking the contacts of an electrical relay. Because programmers, like many of us, were reluctant to acknowledge error, the term became a euphemism for *error.* Correspondingly, *debugging* is the correcting of errors in a program.

History and Background

The meaning of specialized terms such as "radar," "bacteriophage," "silicon chips," or "x-ray" often can be clarified through a background discussion: discovery or history of the concept, development, method of production, applications, and so on. Specialized encyclopedias are a good background source.

"Where did it come from?"

> The idea of lasers . . . dates back as far as 212 B.C., when Archimedes used a [magnifying] glass to set fire to Roman ships during the siege of Syracuse. (Gartaganis 22)

"How was it perfected?"

> The early researchers in fiber optic communications were hampered by two principal difficulties—the lack of a sufficiently intense source of light and the absence of a medium which could transmit this light free from interference and with a minimum signal loss. Lasers emit a narrow beam of intense light, so their invention in 1960 solved the first problem. The development of a means to convey this signal was longer in coming, but scientists succeeded in developing the first communications-grade optical fiber of almost pure silica glass in 1970. (Stanton 28)

Negation

Readers can grasp some meanings by understanding clearly what the term *does not* mean. The insurance policy excerpt on page 395 defines coverage for "bodily injury to others" partly by using negation: "We will *not* pay: 1. For injuries to guest occupants of your auto." In this next example, negation clarifies the definition of "lasers": "Lasers are not merely weapons from science fiction."

Operating Principle

Most items work according to an operating principle, whose explanation should be part of your definition:

"How does it work?"

> A clinical thermometer works on the principle of heat expansion: As the temperature of the bulb increases, the mercury inside expands, forcing a mercury thread up into the hollow stem.

> Air-to-air solar heating involves circulating cool air, from inside the home, across a collector plate (heated by sunlight) on the roof. This warmed air is then circulated back into the home.

> Basically, a laser [uses electrical energy to produce] coherent light, light in which all the waves are in phase with each other, making the light hotter and more intense. (Gartaganis 23)

> [A fiber optics] system works as follows: An electrical charge activates the laser . . . , and the resulting light . . . energy passes through the optical fiber. At the other end of the fiber, a . . . receiver . . . converts this light signal back into electrical impulses. (Stanton 28)

Even abstract concepts or processes can be explained on the basis of their operating principle:

> Economic inflation is governed by the principle of supply and demand: If an item or service is in short supply, its price increases in proportion to its demand.

Analysis of Parts

When your subject can be divided into parts, identify and explain them:

"What are its parts?"

> The standard frame of a pitched-roof wooden dwelling consists of floor joists, wall studs, roof rafters, and collar ties.

> Psychoanalysis is an analytic and therapeutic technique consisting of four parts: (1) free association, (2) dream interpretation, (3) analysis of repression and resistance, and (4) analysis of transference.

In discussing each part, of course, you would further define specialized terms such as "floor joists" and "repression."

Analysis of parts is particularly useful for helping nontechnical readers understand a technical subject. This next analysis helps explain the physics of lasing by dividing the process into three discrete parts:

> 1. [Lasers require] a source of energy, [such as] electric currents or even other lasers.
> 2. A resonant circuit . . . contains the lasing medium and has one fully reflecting end and one partially reflecting end. The medium—which can be a solid, liquid, or gas—absorbs the energy and releases it as a stream of photons [electromagnetic particles that emit light]. The photons . . . vibrate between the fully and partially reflecting ends of the resonant circuit, constantly accumulating energy—that is, they are amplified. After attaining a prescribed level of energy, the photons can pass through the partially reflecting surface as a beam of coherent light and encounter the optical elements.
> 3. Optical elements—lenses, prisms, and mirrors—modify size, shape, and other characteristics of the laser beam and direct it to its target. (Gartaganis 23)

Figure 1 (page 399) shows the three parts of a laser.

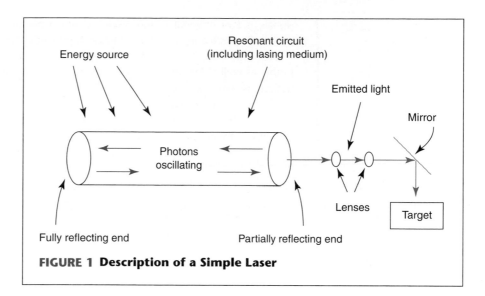

FIGURE 1 Description of a Simple Laser

Visuals

Well-labeled visuals (such as the laser description) are excellent for clarifying definitions. Always introduce your visual and explain it. If your visual is borrowed, credit the source. Unless the visual takes up one whole page or more, do not place it on a separate page. Include the visual near its discussion.

Comparison and Contrast

Comparisons and contrasts help readers understand. Analogies (a type of comparison) to something familiar can help explain the unfamiliar:

"Does it resemble anything familiar?"

> To visualize how a simplified earthquake starts, imagine an enormous block of gelatin with a vertical knife slit through the middle of its lower half. Gigantic hands are slowly pushing the right side forward and pulling the left side back along the slit, creating a strain on the upper half of the block that eventually splits it. When the split reaches the upper surface, the two halves of the block spring apart and jiggle back and forth before settling into a new alignment. Inhabitants on the upper surface would interpret the shaking as an earthquake. ("Earthquake Hazard Analysis" 8)

> The average diameter of an optical cable is around two-thousandths of an inch, making it about as fine as a hair on a baby's head (Stanton 29–30).

Here is a contrast between optical fiber and conventional copper cable:

"How does it differ from comparable things?"

> Beams of laser light coursing through optical fibers of the purest glass can transmit many times more information than the present communications systems. . . . A pair of optical fibers has the capacity to carry more than 10,000 times as many

signals as conventional copper cable. A ½-inch optical cable can carry as much information as a copper cable as thick as a person's arm. . . .

Not only does fiber optics produce a better signal, [but] the signal travels farther as well. All communications signals experience a loss of power, or attenuation, as they move along a cable. This power loss necessitates placement of repeaters at one- or two-mile intervals of copper cable in order to regenerate the signal. With fiber, repeaters are necessary about every thirty or forty miles, and this distance is increasing with every generation of fiber. (Stanton 27–28)

Here is a combined comparison and contrast:

"How is it both similar and different?"

Fiber optics technology results from the superior capacity of lightwaves to carry a communications signal. Sound waves, radio waves, and light waves can all carry signals; their capacity increases with their frequency. Voice frequencies carried by telephone operate at 1000 cycles per second, or hertz. Television signals transmit at about 50 million hertz. Light waves, however, operate at frequencies in the hundreds of trillions of hertz. (Stanton 28)

Required Materials or Conditions

Some items or processes need special materials and handling, or they may have other requirements or restrictions. An expanded definition should include this important information.

"What is needed to make it work (or occur)?"

Besides training in engineering, physics, or chemistry, careers in laser technology require a strong background in optics (study of the generation, transmission, and manipulation of light).

Abstract concepts might also be defined in terms of special conditions:

To be held guilty of libel, a person must have defamed someone's character through written or pictorial statements.

Example

Familiar examples showing types or uses of an item can help clarify your definition. This example shows how laser light is used as a heat-generating device:

"How is it used or applied?"

Lasers are increasingly used to treat health problems. Thousands of eye operations involving cataracts and detached retinas are performed every year by ophthalmologists. . . . Dermatologists treat skin problems. . . . Gynecologists treat problems of the reproductive system, and neurosurgeons even perform brain surgery—all using lasers transmitted through optical fibers. (Gartaganis 24–25)

The next example shows how laser light is used to carry information:

The use of lasers in the calculating and memory units of computers, for example, permits storage and rapid manipulation of large amounts of data. And audiodisc players use lasers to improve the quality of the sound they reproduce. The use of optical cable to transmit data also relies on lasers. (Gartaganis 25)

And this final example shows how optical fiber can relay a video signal:

> Acting, in essence, as tiny cameras, optical fibers can be inserted into the body and relay an image to an outside screen. (Stanton 28)

Examples are a most powerful communication tool—as long as you tailor the examples to the readers' level of understanding.

Whichever expansion strategies you use, be sure to document your information sources.

Sample Situations

The following definitions employ expansion strategies appropriate to their audiences' needs. Specific strategies are labeled in the margin. Each definition, like a good essay, is unified and coherent: Each paragraph is developed around one main idea and logically connected to other paragraphs. Visuals are incorporated. Transitions emphasize the connection between ideas. Each definition is at a level of technicality that connects with the intended audience.

To illustrate the importance of audience analysis in a writer's decision about "How much is enough?" this example, like many throughout the text, is preceded by an audience and use profile based on the worksheet on page 65.

An Expanded Definition for Semitechnical Readers

AUDIENCE AND USE PROFILE. The intended readers of this material are beginning student mechanics. Before they can repair a solenoid, they will need to know where the term comes from, what a solenoid looks like, how it works, how its parts operate, and how it is used. This definition is designed as merely an *introduction,* so it offers only a general (but comprehensive) view of the mechanism.

Because the intended readers are not engineering students, they do *not* need details about electromagnetic or mechanical theory (e.g., equations or graphs illustrating voltage magnitudes, joules, lines of force). ■

EXPANDED DEFINITION: SOLENOID

Formal sentence definition

A solenoid is an electrically energized coil that forms an electromagnet capable of performing mechanical functions. The term "solenoid" is derived from the word "sole," which in reference to electrical equipment means "a part of," or

Etymology

"contained inside, or with, other electrical equipment." The Greek word *solenoides* means "channel," or "shaped like a pipe."

Description and analysis of parts

A simple plunger-type solenoid consists of a coil of wire attached to an electrical source, and an iron rod, or plunger, that passes in and out of the coil along the axis of the spiral. A return spring holds the rod outside the coil when the current is de-energized, as shown in Figure 1.

Special conditions and operating principle

When the coil receives electric current, it becomes a magnet and thus draws the iron rod inside, along the length of its cylindrical center. With a lever attached to its end, the rod can transform electrical energy into mechanical force. The

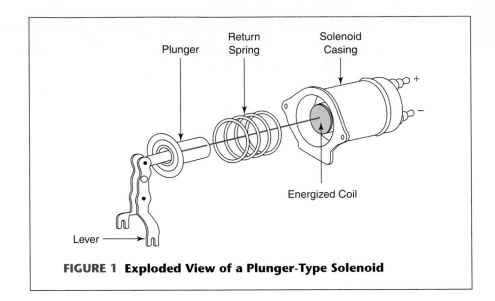

FIGURE 1 Exploded View of a Plunger-Type Solenoid

Example and analysis of parts

Explanation of visual

amount of mechanical force produced is the product of the number of turns in the coil, the strength of the current, and the magnetic conductivity of the rod.

The plunger-type solenoid in Figure 1 is commonly used in the starter motor of an automobile engine. This type is 4 1/2 inches long and 2 inches in diameter, with a steel casing attached to the casing of the starter motor. A linkage (pivoting lever) is attached at one end to the iron rod of the solenoid, and at the other end to the drive gear of the starter, as shown in Figure 2. When the ignition key is turned, current from the battery is supplied to the solenoid coil, and the iron rod is drawn inside the coil, thereby shifting the attached linkage. The linkage, in turn, engages the drive gear, activated by the starter motor, with the flywheel (the main rotating gear of the engine).

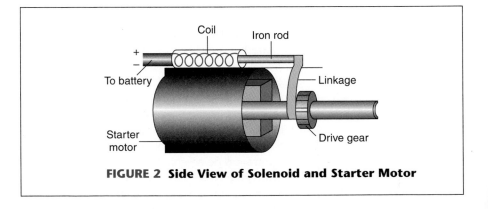

FIGURE 2 Side View of Solenoid and Starter Motor

Comparison of sizes and applications

Because of the solenoid's many uses, its size varies according to the work it must do. A small solenoid will have a small wire coil, hence a weak magnetic field. The larger the coil, the stronger the magnetic field; in this case, the rod in the solenoid can do harder work. An electronic lock for a standard door would, for instance, require a much smaller solenoid than one for a bank vault.

The audience for the following definition (an entire community) is too diverse to define precisely, so the writer wisely addresses the lowest level of technicality—to ensure that all readers will understand.

An Expanded Definition for Nontechnical Readers

AUDIENCE AND USE PROFILE. The following definition is written for members of a community whose water supply (all obtained from wells, because the town has no reservoir) is doubly threatened: (1) by chemical seepage from a recently discovered toxic dump site, and (2) by a two-year drought that has severely depleted the water table. This definition forms part of a report that analyzes the severity of the problems and explores possible solutions.

To understand the problems, these readers first need to know what a water table is, how it is formed, what conditions affect its level and quality, and how it figures into town planning decisions. The concepts of *recharge* and *permeability* are vital to readers' understanding of the problem here, so these terms are defined parenthetically. These readers have no interest in geological or hydrological (study of water resources) theory. They simply need a broad picture. ▪

EXPANDED DEFINITION: WATER TABLE

Formal sentence definition

Example

The water table is the level below the earth's surface at which the ground is saturated with water. Figure 1 shows a typical water table that might be found in the East. Wells driven into such a formation will have a water level identical to that of the water table.

Operating principle

The world's fresh-water supply comes almost entirely as precipitation that originates with the evaporation of sea and lake water. This precipitation falls to earth and follows one of three courses: It may fall directly onto bodies of water, such as rivers or lakes, where it is directly used by humans; it may fall onto land, and either evaporate or run over the ground to the rivers or other bodies of water; or it may fall onto land, be contained, and seep into the earth. The latter precipitation makes up the water table.

Comparison

Similar in contour to the earth's surface above, the water table generally has a level that reflects such features as hills and valleys. Where the water table intersects the ground surface, a stream or pond results.

Operating principle

A water table's level, however, will vary, depending on the rate of recharge (replacement of water). The recharge rate is affected by rainfall or soil permeability (the ease with which water flows through the soil). A water table then is never static; rather, it is the surface of a body of water striving to maintain a

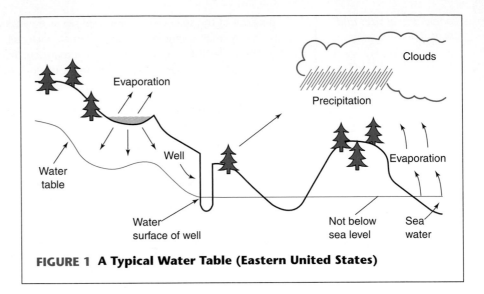

FIGURE 1 A Typical Water Table (Eastern United States)

Example

balance between the forces which deplete it and those which replenish it. In areas of Florida and some western states where the water table is depleted, the earth caves in, leaving sinkholes.

Special conditions and examples

The water table's depth below ground is vital in water resources engineering and planning. It determines an area's suitability for wastewater disposal, or a building lot's ability to handle sewage. A high water table could become contaminated by a septic system. Also, bacteria and chemicals seeping into a water table can pollute an entire town's water supply. Another consideration in water-table depth is the cost of drilling wells. These conditions obviously affect an industry's or homeowner's decision on where to locate.

Special conditions

The rising and falling of the water table give an indication of the pumping rate's effect on a water supply (drawn from wells) and of the sufficiency of the recharge rate in meeting demand. This kind of information helps water resources planners decide when new sources of water must be made available.

Placement of Definitions

Poorly placed definitions interrupt the information flow. If you have only a few parenthetical definitions, place them in parentheses after the terms. Any more than a few definitions per page will be disruptive. Rewrite them as sentence definitions and place them in a "Definitions" section of your introduction, or in a glossary.

If your sentence definitions are few, place them in a "Definitions" section of your introduction; otherwise, in a glossary. Definitions of terms in the report's title belong in your introduction.

Place expanded definitions in one of three locations:

*Where to place
definitions*

1. If the definition is essential to the reader's understanding of the *entire* document, place it in the introduction. A report titled "The Effects of Aerosol on the Earth's Ozone Shield" would require expanded definitions of "aerosol" and "ozone shield" early.

2. When the definition clarifies a major part of your discussion, place it in that section of your report. In a report titled "How Advertising Influences Consumer Habits," "operant conditioning" might be defined early in the appropriate section. Too many expanded definitions *within* a report, however, can be disruptive.

3. If the definition serves only as a reference, place it in an appendix. A report on fire safety in a public building might have an expanded definition of "carbon monoxide detectors" in an appendix.

Electronic documents pose special problems for placement of definitions. In a hypertext document, for instance, each reader explores the material differently (Chapter 6). One option for making definitions available when they are needed is the "pop-up note": The term to be defined is highlighted in the text, to indicate that its definition can be called up and displayed in a small window on the actual text screen (Horton 25).

REVISION CHECKLIST FOR DEFINITIONS

Use this list to revise your definition. (Numbers in parentheses refer to the first page of discussion.)

CONTENT

❑ Is the type of definition (parenthetical, sentence, expanded) suited to its purpose and readers' needs? (393)

❑ Does the definition convey the basic property of the item? (392)

❑ Is the definition objective? (392)

❑ Is the expanded definition adequately developed? (396)

❑ Are all data sources documented? (401)

❑ Are visuals employed adequately and appropriately? (399)

ARRANGEMENT

❑ Does the sentence definition describe features that distinguish the item from other items in the same class? (393)

❑ Is the expanded definition unified and coherent (like an essay)? (401)

❑ Are transitions between ideas adequate? (401)

❑ Does the definition appear in the appropriate location? (404)

STYLE AND PAGE DESIGN

❑ Is the definition in plain English? (392)

❑ Will the level of technicality connect with the audience? (401)

❑ Are sentences clear, concise, and fluent? (261)

❑ Is word choice precise? (284)

❑ Is the definition grammatical? (Appendix)

❑ Is the definition ethically acceptable? (391)

❑ Is the page design inviting and accessible? (355)

☑ EXERCISES

1. Sentence definitions require precise classification and differentiation. Is each of these definitions adequate for a general reader? Rewrite those which seem inadequate. Consult dictionaries and encyclopedias as needed.

   a. A bicycle is a vehicle with two wheels.

   b. A transistor is a device used in transistorized electronic equipment.

   c. Surfing is when one rides a wave to shore while standing on a board specifically designed for buoyancy and balance.

   d. Bubonic plague is caused by an organism known as *pasteurella pestis.*

   e. Mace is a chemical aerosol spray used by the police.

   f. A Geiger counter measures radioactivity.

   g. A cactus is a succulent.

   h. In law, an indictment is a criminal charge against a defendant.

   i. A prune is a kind of plum.

   j. Friction is a force between two bodies.

   k. Luffing is what happens when one sails into the wind.

   l. A frame is an important part of a bicycle.

   m. Hypoglycemia is a medical term.

   n. An hourglass is a device used for measuring intervals of time.

   o. A computer is a machine that handles information with amazing speed.

   p. A Ferrari is the best car in the world.

   q. To meditate is to exercise mental faculties in thought.

2. Standard dictionaries define for the general reader, whereas specialized reference books define for the specialist. Choose an item in your field and copy the definition (1) from a standard dictionary and (2) from a technical reference book. For the technical definition, label each expansion strategy. Rewrite the specialized definition for a general reader.

3. Using reference books as necessary, write sentence definitions for these terms or for terms from your field.

biological insect control	gyroscope
generator	coronary bypass
dewpoint	oil shale
microprocessor	chemotherapy
capitalism	estuary
local area network	Boolean logic
marsh	classical conditioning
artificial intelligence	hypothermia
economic inflation	thermistor
anorexia nervosa	aquaculture
low-impact camping	nuclear fission
hemodialysis	modem

4. Select an item from the list in Exercise 3 or from an area of interest. Identify an audience and purpose. Complete an audience and use profile sheet (page 65). Begin with a sentence definition of the term. Then write an expanded definition for a first-year student in that field. Next, write the same definition for a layperson (client, patient, or other interested party). Leave a margin at the left side of your page to list expansion strategies. Use at least four expansion strategies in each version, including at least one visual or an art brief (page 333) and a rough diagram. In preparing each version, consult no fewer than four outside references. Cite and document each source, using one of the documentation styles discussed in Chapter 10. Submit, with your two versions, an explanation of your changes from the first version to the second.

5. Figure 17.2 shows a page from a brochure titled *Cogeneration.* The brochure provides an expanded definition for potential users of fuel conservation systems engineered and packaged by Ewing Power Systems. The intended readers are plant engineers and other technical experts unfamiliar with cogeneration.

   Another page of the brochure is designed in a question and answer format. Figure 17.3 shows parts of that page.

   Identify each expansion strategy in Figures 17.2 and 17.3. Is the definition appropriate for a technical audience? Why, or why not? Be prepared to discuss your analysis and evaluation in class.

☑ COLLABORATIVE PROJECTS

1. Divide into small groups on the basis of academic majors or interests. Appoint one person as group manager. Decide on an item, concept, or process that would require expanded definition for a layperson.

   Examples

   From computer science: an algorithm, an applications program, artificial intelligence, binary coding, top-down procedural thinking, or systems analysis

   From nursing: a pacemaker, coronary bypass surgery, or natural childbirth

   Complete an audience and use profile (page 65).
   Once your group has decided on the appropriate expansion strategies (etymology, negation, etc.), the group manager will assign each member to work on one or two specific strategies as part of the definition. As a group, edit and incorporate the collected material into an expanded definition, revising as often as needed.
   The group manager will assign one member to present the definition in class, using either opaque or overhead projection, a large-screen monitor, or photocopies.

2. Beyond informative definition, your group's goal in this next project is to develop a *persuasive* definition.
   Assume you work for an organization that has recently formulated a policy to eliminate sexual harassment. Your charge as the communications group is to develop printed material that will publicize this new policy. Your specific task is to develop an expanded definition of *sexual harassment* to be published in the company newsletter.

   At this stage, the company is plagued by confusion, misunderstanding, resentment, and paranoia about the harassment issue. Therefore, beyond compliance with legal requirements, your organization seeks to improve gender relations between coworkers, in the hope of boosting productivity.

   Thus, you face an informative, persuasive, and ethical challenge: to move beyond the usual matters of clarity so that your definition promotes real understanding and reconciliation. In short, you need to amplify the legal definition so that employees are able to recognize harassment and to understand clearly what does and what does not constitute harassment. But unless you also do something to change their us-against-them *attitude,* you will only create greater tension and overreaction.

   In other words, you want your definition not merely to *inhibit* behavior, but to *enlighten* the readers. Insensitive readers, of course, are likely to change their behavior only if they feel coerced. But appeals to fear almost always have limited success, and people generally tend to be reasonable in the long run. You want your definition to have rational appeal, to cause readers to *internalize* the values that underlie the issue. No sermons, please.

TECHNICAL CONSIDERATIONS

Turbine generator sets make electricity by converting a steam pressure drop into mechanical power to spin the generator. Conceptually, steam turbines work much the same way as water turbines. Just as water turbines take the energy from water as it flows from a high elevation to a lower elevation, steam turbines take the energy from steam as it flows from high pressure to low pressure. The amount of energy that can be converted to electricity is determined by the difference between the inlet pressure and the exhaust pressure (pressure drop) and the volume of steam flowing through the turbine.

Steam turbines have been used in industry in a variety of applications for decades and are the most common way utilities generate electricity. Exactly how a steam turbine generator can be used in your plant depends upon your circumstances.

IF YOU USE WASTE AS A BOILER FUEL

If you use wood waste or incinerator waste as a boiler fuel you can afford to condense turbine exhaust steam in a condenser. This allows you to convert waste fuel into electricity.

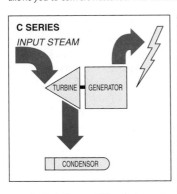

The simplest form of a condensing turbine generator set is the Ewing Power Systems C Series. All surplus steam enters the turbine at high pressure and exhausts to a condenser at a very low pressure, usually a vacuum. Because of the very low exhaust pressure, the pressure drop through the turbine is greater and more energy is extracted from each pound of steam. This is the same basic design as utilities use to produce power. The condenser can be either air or water cooled. In water cooled systems the "cooling" water can be hot enough for use as process hot water or for space heat.

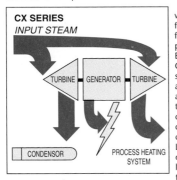

In situations where there is surplus fuel and also a need for low pressure process steam, the Ewing Power Systems CX Series is the system of choice. This arrangement includes a back pressure turbine and a condensing turbine connected to a common generator. Low pressure process or space heating loads are met with the back pressure turbine while surplus steam is directed to the condensing turbine to maximize power production.

IF YOU PURCHASE BOILER FUEL SUCH AS OIL OR GAS

If oil or gas is used as boiler fuel the best use of a turbine is as a replacement for a steam pressure reducing valve. Many plants produce steam at high pressure and then use some or all of the steam at low pressure after passing it through a pressure reducing valve. Other plants have high pressure boilers but run them at low pressure because they do not need high pressure steam for their process. In either case, a turbine generator can turn the pressure drop energy potential into electricity.

The Ewing Power Systems BP Series turbine generator sets are designed for pressure reducing (back pressure) applications. Very little energy is consumed by the turbine, so most of the inlet steam is available for process. The turbine generator uses about 3631 BTU's per hour for each kilowatt-hour produced. At 40 cents per gallon for No. 6 fuel oil and 85% boiler efficiency, it will cost about 1.1 cents per kilowatt-hour to generate your own power with a BP Series turbine. Generating costs for gas-fired boilers are similar.

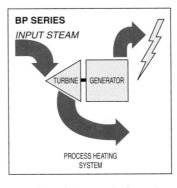

To generate power at this very low cost, all the exhaust steam must be used productively. Generator output is therefore completely governed by process steam demand. For example, if steam is used for space heating you will make more electricity on cold days than on warmer days because more steam will flow through the turbine.

ELECTRICAL CONSIDERATIONS

In most cases the generator will be connected to your plant electrical system and to your utility. This means that you will not give up the security of utility power. It also means that you do not have to generate *all* your own power; most cogeneration systems provide only part of the plant load. The more power the generator is making, the less you buy from the utility. If you make more power than you use, you will be able to sell the excess to the utility. If your generator is off-line for any reason, you will be able to buy power from the utility, just as you do now.

There are two primary generator designs: induction and synchronous. Induction generators are similar to induction motors and are much simpler than synchronous generators. Synchronous generators require more elaborate controls and are usually more expensive but offer the advantage of stand-alone capability. Whereas induction generators cannot operate unless they are connected to a utility grid, synchronous sets can be operated in isolation as emergency units or when it is economically advantageous to avoid interconnection with the utility.

All turbine generator sets from Ewing Power Systems include a complete electrical control panel. Our standard panels meet most utility interconnection requirements and we will customize the panel to meet unusual requirements. Synchronous panels can be built for full utility paralleling, stand-alone capability or both.

Figure 17.2   Expanded Definition in a Technical Brochure
Source: Courtesy of Ewing Power Systems, So. Deerfield, MA 01373.

Q. What is cogeneration?

A. It is the simultaneous production of electricity and useful thermal energy. This means that you can generate electricity with the same steam you are now using for heating or process. *You can use the same steam twice.* In modern usage cogeneration has also come to mean using waste fuel for in-plant electricity generation.

Q. How does it save money?

A. Cogeneration saves money by allowing you to produce your own electricity for a fraction of the cost of utility power. Cogenerated power is cheaper because cogeneration systems are much more efficient than central utility plants. By using the same steam twice, cogeneration systems can achieve efficiencies of up to 80%, whereas the best utilities can do is about 40%.

Q. Are there other benefits to cogeneration?

A. Yes. For companies using waste fuel, elimination of waste disposal can be a very important benefit. Depending upon design, a cogeneration system can provide emergency standby power and can smooth out boiler load swings.

Q. Is cogeneration new?

A. No. It's been done ever since the beginning of the electrification of industrial America. Originally, most electric power was cogenerated by individual manufacturers, not the utilities. In the 1920s and '30s, as cheaper utility electricity became available, cogeneration waned. With cheap oil available, power rates continued to decline through the 1950s and '60s. Then came the 1973–74 Arab Oil Embargo. Everything changed abruptly. Since then electricity prices have risen. Further upward pressure was produced by some utilities' nuclear power plant building programs. Now many companies are getting back to their original source of power: cogeneration.

Figure 17.3   Expanded Definition in a Technical Brochure

Descriptions and Specifications

D ESCRIPTION (creating a picture with words) is part of all writing. But technical descriptions convey information about a product or mechanism to someone who will use it, buy it, operate it, assemble it, manufacture it, or to someone who has to know more about it. Any item can be visualized from countless different perspectives. Therefore, *how* you describe—your perspective—depends on your purpose and the needs of your audience.

Purpose of Description

Manufacturers use descriptions to sell products; banks require detailed descriptions of any business or construction venture before approving a loan; medical personnel maintain daily or hourly descriptions of a patient's condition and treatment.

No matter what the subject of description, readers expect answers to as many of these questions as are applicable: *What is it? What does it do? What does it look like? What is it made of? How does it work? How was it put together?* The description in Figure 18.1, part of an installation and operation manual, answers applicable questions for do-it-yourself homeowners.

Specifications

Airplanes, bridges, smoke detectors, and countless other items are produced according to certain *specifications*. A particularly exacting type of description, specifications (or "specs") prescribe standards for performance, safety, and quality. For almost any product, specifications spell out:

- the methods for manufacturing, building, or installing the product
- the materials and equipment to be used
- the size, shape, and weight of the product

Because these requirements define an acceptable level of quality, specifications have ethical and legal implications. Any product "below" specifications provides grounds for a lawsuit. When injury or death results (as in a bridge collapse caused by inferior reinforcement), the contractor, subcontractor, or supplier who cut corners is criminally liable.

Federal and state regulatory agencies routinely issue specifications to ensure safety. The Consumer Product Safety Commission specifies that power lawn mowers be equipped with a "kill switch" on the handle, a blade guard to prevent foot injuries, and a grass thrower that aims downward to prevent eye and facial injury. This same agency issues specifications for baby products, as in governing the fire retardancy of pajama material. Passenger airline specifications for aisle width, seat-belt configurations, and emergency equipment are issued by the Federal Aviation Administration. State and local agencies issue specifications in the form of building codes, fire codes, and other standards for safety and reliability.

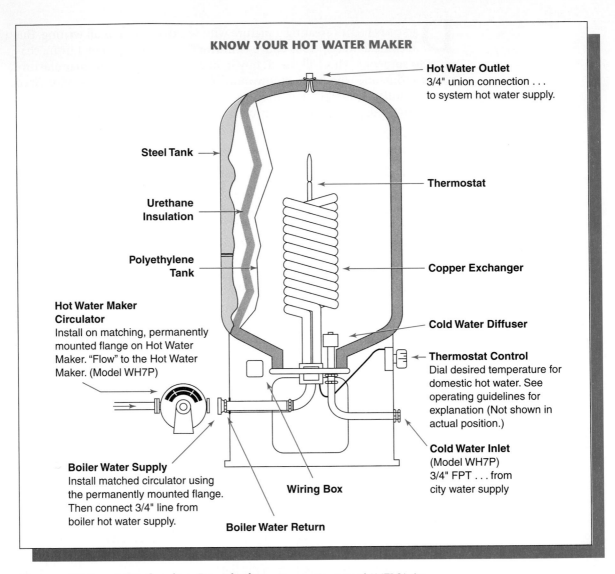

Figure 18.1   A Mechanism Description.   *Source: Courtesy of AMTROL. Inc.*

Government departments (Defense, Interior, etc.) issue specifications for all types of military hardware and other equipment. A set of NASA specifications for spacecraft parts can be hundreds of pages long, prescribing the standards for even the smallest nuts and bolts, down to screw-thread depth and width in millimeters.

The private sector issues specifications for countless products or projects, to help ensure that customers get *exactly* what they want. Figure 18.2 shows

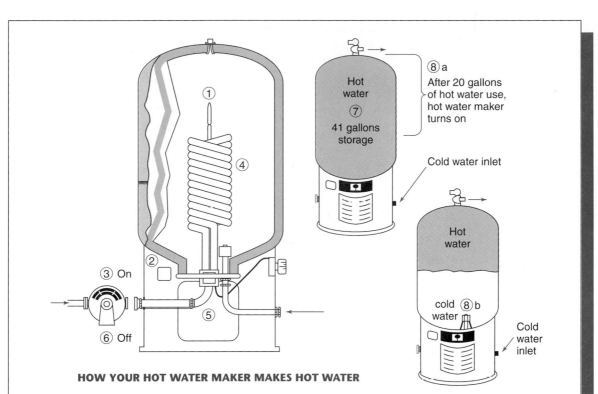

HOW YOUR HOT WATER MAKER MAKES HOT WATER

1. The thermostat calls for energy to make hot water in your Hot Water Maker.

2. The built-in relay signals your boiler/burner to generate energy by heating boiler water.

3. The Hot Water Maker circulator comes on and circulates hot boiler water through the inside of the Hot Water Maker heat exchanger.

4. Heat energy is transferred, or "exchanged" from the boiler water inside the exchanger to the water surrounding it in the Hot Water Maker.

5. The boiler water after the maximum of heat energy is taken out of it, is returned to the boiler so it can be reheated.

6. When enough heat has been exchanged to raise the temperature in your Hot Water Maker to the desired temperature, the thermostat will de-energize the relay and turn off the Hot Water Maker circulator and your boiler/burner. This will take approximately 23 minutes—when you first

start up the Hot Water Maker. During use, reheating will be approximately 9–12 minutes.

7. You now have 41 gallons of hot water in storage . . . ready for use in washing machines, showers, sinks, etc. This 41 gallons of hot water will stay hot up to 10 hours, if you don't use it, without causing your boiler/burner to come on. (Unless, of course, you need it for heating your home in the winter.)

8. When you do use hot water, you will be able to use approximately 20 gallons, before the Hot Water Maker turns on. Then you will still have 21 gallons of hot water left for use, as your Hot Water Maker "recoups" 20 gallons of cold water. This means, during normal use (3 1/2 GPM Flow), you will never run out of hot water.

9. You can expect substantial energy savings with your Hot Water Maker, as its ability to store hot water and efficiently transfer energy to make more hot water will keep your boiler off for longer periods of time.

Figure 18.1   A Mechanism Description *Continued*

partial specifications drawn up by an architect for a building that will house a small medical clinic. This section of the specs covers only the structure's "shell." Other sections detail the requirements for plumbing, wiring, and interior finish work.

Specifications like those in Figure 18.2 must be clear enough for *identical* interpretation by the widest possible range of readers (Glidden 258–59):

- *the customer*, who has the big picture of what is needed and who wants the best product at the best price
- *the designer* (architect, engineer, computer scientist, etc.), who must translate the customer's wishes into the actual specifications
- *the contractor or manufacturer*, who won the job by making the lowest bid, and so must preserve profit by doing *only* what is prescribed
- *the supplier*, who must provide the exact materials and equipment
- *the workforce*, who will do the actual assembly, construction, or installation (managers, supervisors, subcontractors, and workers—some working on only one part of the product, such as plumbing or electrical)
- *the inspectors* (such as building, plumbing, or electrical inspectors), who evaluate how well the product conforms to the specifications

Each of these parties has to understand and agree on exactly *what* is to be done and *how* it is to be done. In the case of a lawsuit over failure to meet specifications, the readership broadens to include judges, lawyers, and jury. Figure 18.3 depicts how a clear set of specifications unifies all readers (their various viewpoints, motives, and levels of expertise) in a shared understanding.

In addition to guiding a product's design and construction, specifications can facilitate the product's use and maintenance. For instance, specifications in a computer manual include the product's performance limits, or *ratings:* its power requirements, work or processing or storage capacity, environment requirements, the makeup of key parts, and so on. Product support literature for appliances, power tools, and other items routinely contains ratings to help readers select a good operating environment or replace worn or defective parts (Riney 186). The ratings in Figure 18.4 are taken from the owner's manual for a dot-matrix printer.

Specifications play a persuasive role in technical marketing by spelling out the product and service quality that potential customers can expect in return for their investment. Figure 18.5 (pages 429–30), displays marketing literature that accompanies the expanded definition in Figures 17.2 and 17.3 (pages 408, 409). After a brief introduction to the C Series Cogeneration System, the description focuses on that system's major components and specifications as well as the support services offered by the vendor. Representational diagrams and system schematics depict the system's general appearance and operating principle.

Ruger, Filstone, and Grant Architects

SPECIFICATIONS FOR THE POWNAL CLINIC BUILDING

Foundation
    footings: 8" x 16" concrete (load-bearing capacity: 3000 lbs. per sq. in.)
    frost walls: 8" x 4' @ 3000 psi
    slab: 4" @ 3000 psi, reinforced with wire mesh over vapor barrier

Exterior Walls
    frame: eastern pine #2 timber frame with exterior partitions set inside posts
    exterior partitions: 2" x 4" kiln-dried spruce set at 16" on center
    sheathing: 1/4" exterior-grade plywood
    siding: #1 red cedar with a 1/2" x 6' bevel
    trim: finished-pine boards ranging from 1" x 4" to 1" x 10"
    painting: 2 coats of Clear Wood Finish on siding; trim primed and finished with one
            coat of bone-white, oil base paint

Roof system
    framing: 2" x 12" kiln-dried spruce set at 24" on center
    sheathing: 5/8" exterior-grade plywood
    finish: 240 Celotex 20-year fiberglass shingles over #15 impregnated felt roofing paper
    flashing: copper

Windows
    Anderson casement and fixed-over-awning models, with white exterior cladding,
      insulating glass and screens, and wood interior frames

Landscape
    driveway: gravel base, with 3" traprock surface
    walks: timber defined, with traprock surface
    cleared areas: to be rough graded and covered with wood chips
    plantings: 10 assorted lawn plants along the road side of the building

Figure 18.2   Specifications for a Building Project (Partial)

**Figure 18.3
Readers and
Potential Readers
of Specifications**

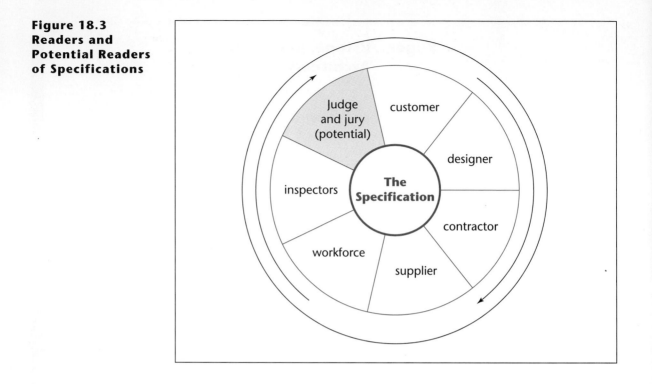

Objectivity in Description

Descriptions are mainly *subjective* or *objective:* based on feelings or fact. Subjective description emphasizes the writer's attitude toward the thing, whereas objective description emphasizes the thing itself.

Essays describing "An Unforgettable Person" or "A Beautiful Moment" express opinions, a personal point of view. Subjective description aims at expressing feelings, attitudes, and moods. You create an *impression* of your subject ("The weather was miserable"), more than communicating factual information about it ("All day, we had freezing rain and gale-force winds").

Objective description shows an impartial view, filtering out personal impressions and focusing on observable details.

Except for promotional writing, descriptions on the job should be impartial, if they are to be ethical. Pure objectivity is, of course, humanly impossible. Each writer filters the facts and their meaning through her or his own perspective. Nonetheless, we are expected to communicate the facts as we know them and understand them. One writer offers this useful distinction: "All communication requires us to leave something out, but we must be sure that what is left out is not essential to our [reader's] understanding of what is put in" (Coletta 65).

**Figure 18.4
Specifications for
the Imagewriter™ II
Printer.** *Source: Reprinted
by permission of Apple
Computer, Inc.*

Printer Specifications

Paper Types:	Single sheets Pin-feed paper (hole centers 4.0–9.5 inches)
Ribbon:	Cassette containing black inked fabric ribbon 13 mm wide by 13000 mm long, continuous Four color ribbon optional 21 mm wide by 18000 mm long, continuous

Power Options:	American 120 volts AC ± 10%, 60 hertz
	Universal 100 volts AC ± 10%, 50/60 hertz
	             120 volts AC ± 10%, 50/60 hertz
	             140 volts AC ± 10%, 50/60 hertz
	             200 volts AC ± 10%, 50/60 hertz
	             220 volts AC ± 10%, 50/60 hertz
	             240 volts AC ± 10%, 50/60 hertz

Power Consumption:	Operating: 180 watts maximum Standby: 20 watts maximum
Data Interface:	8-bit serial
Weight:	11.36 kilograms (25 pounds)
Dimensions:	Width Depth Height 431.8 304.8 127.0 millimeters 17.0 12.0 5.0 inches

Ambient Temperature:
  Operating
  Storage
	
	10 to 40 degrees Celsius (50 to 104 degrees F.)
	–40 to +47 degrees Celsius (–40 to +116 degrees F.)

Humidity:
  Operating
  Storage
	
	20% to 95% relative humidity, noncondensing
	10% to 95% relalive humidity, noncondensing

An ethical writer "is obligated to express her or his opinions of products, as long as these opinions are based on objective and responsible research and observation" (Mackenzie 3). Being "objective" does not mean forsaking personal evaluation in cases in which a product may be unsafe or unsound. Even positive claims made in promotional writing (for example, "reliable," "rugged," and so on in Figure 18.5) should be based on objective and verifiable evidence.

Here are guidelines for remaining impartial.

Record the Details That Enable Readers to Visualize the Item. Ask these questions: What could any observer recognize? What would a camera record?

Subjective His office has an *awful* view, *terrible* furniture, and a *depressing* atmosphere.

The italicized words only *tell;* they do not *show.*

Objective   His office has broken windows looking out on a brick wall, a rug with a six-inch hole in the center, chairs with bottoms falling out, missing floorboards, and a ceiling with plaster missing in three or four places.

Use Precise and Informative Language. Use high-information words that enable readers to *visualize.* Name specific parts without calling them "things," "gadgets," or "doohickeys." Avoid judgmental words ("impressive," "poor"), unless your judgment is requested and can be supported by facts. Instead of "large," "long," and "near," give exact measurements, weights, dimensions, and ingredients.

Use words that specify location and spatial relationships: "above," "oblique," "behind," "tangential," "adjacent," "interlocking," "abutting," and "overlapping." Use position words: "horizontal," "vertical," "lateral," "longitudinal," "in cross-section," "parallel."

Indefinite	*Precise*
a late-model car	a 1997 Ford Taurus sedan
an inside view	a cross-sectional, cutaway, or exploded view
next to the foundation	adjacent to the right side
a small red thing	a red activator button with a 1-inch diameter and a concave surface

Do not confuse precise language, however, with overly complicated technical terms or needless jargon. Don't say "phlebotomy specimen" instead of "blood," or "thermal attenuation" instead of "insulation," or "proactive neutralization" instead of "damage control." The clearest writing uses precise but plain language. General readers prefer nontechnical language, as long as the simpler words do the job. Always think about your specific readers' needs.

Elements of Description

Clear and Limiting Title

Promise exactly what you will deliver—no more and no less. "A Description of a Velo Ten-Speed Racing Bicycle" promises a complete description, down to the smallest part. If you intend to describe the braking mechanism only, be sure your title so indicates: "A Description of the Velo's Center-Pull Caliper Braking Mechanism."

Overall Appearance and Component Parts

Let readers see the big picture before you describe each part.

The standard stethoscope is roughly 24 inches long and weighs about 5 ounces. The instrument consists of a sensitive sound-detecting and amplifying device whose flat surface is pressed against a bodily area. This amplifying device is attached to rubber and metal tubing that transmits the body sound to a listening device inserted in the ear.

Seven interlocking pieces contribute to the stethoscope's Y-shaped appearance: (1) diaphragm contact piece, (2) lower tubing, (3) Y-shaped metal piece, (4) upper tubing, (5) U-shaped metal strip, (6) curved metal tubing, and (7) hollow ear plugs. These parts form a continuous unit.

Visuals

Use drawings, diagrams, or photographs generously. Our overall description of the stethoscope is greatly clarified by Figure 1, below.

Function of Each Part

Explain what each part does and how it relates to the whole.

The diaphragm contact piece is caused to vibrate by body sounds. This part is the heart of the stethoscope that receives, amplifies, and transmits the sound impulse.

Appropriate Details

Give enough detail for a clear picture, but do not burden readers needlessly. Identify your readers and their reasons for reading your description.

Assume you set out to describe a specific bicycle model. The picture you create will depend on the details you select. How will your reader use this description? What is the reader's level of technical understanding? Is this a customer interested in the bike's appearance—its flashy looks and racy style? Is it a repair technician who needs to know how parts operate? Or is it a helper

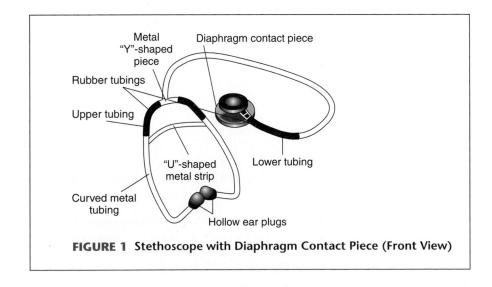

FIGURE 1 Stethoscope with Diaphragm Contact Piece (Front View)

in your bicycle shop who needs to know how to assemble this bike? If you had designed the bike, you would give the manufacturer detailed specifications.

The description of the hot-water maker in Figure 18.1 focuses on what this model looks like, what it's made of, and how it works. Its intended audience of do-it-yourselfers will know already what a hot-water maker is and what it does. That audience needs no background. A description of how it was put together appears with the installation and maintenance instructions later in the manual.

Specifications for readers who will manufacture the hot-water maker would describe each part in exacting detail (e.g., the steel tank's required thickness and pressure rating as well as required percentages of iron, carbon, and other constituents in the steel alloy).

Clearest Descriptive Sequence

Any item usually has its own logic of organization, based on (1) the way it appears as a static object, (2) the way its parts operate in order, or (3) the way its parts are assembled. We describe these relationships, respectively, in a spatial, functional, or chronological sequence.

Spatial Sequence. Part of all physical descriptions, a spatial sequence answers these questions: *What is it? What does it do? What does it look like? What parts and material is it made of?* Use this sequence when you want readers to visualize the item as a static object or mechanism at rest (a house interior, a document, the Statue of Liberty, a plot of land, a chainsaw, or a computer keyboard). Can readers best visualize this item from front to rear, left to right, top to bottom? (What logical path do the parts create?) A retractable pen would logically be viewed from outside to inside. The specifications in Figure 18.2 proceed from the ground upward.

Functional Sequence. The functional sequence answers: *How does it work?* It is best used in describing a mechanism in action, such as a 35-millimeter camera, a nuclear warhead, a smoke detector, or a car's cruise-control system. The logic of the item is reflected by the order in which its parts function. Like the hot-water maker in Figure 18.1, a mechanism usually has only one functional sequence. The stethoscope description on page 419 follows the sequence of parts through which sound travels.

In describing a solar home-heating system, you would begin with the heat collectors on the roof, moving through the pipes, pumping system, and tanks for the heated water, to the heating vents in the floors and walls—from source to outlet. After this functional sequence of operating parts, you could describe each part in a spatial sequence.

Chronological Sequence. A chronological sequence answers: *How has it been put together?* The chronology follows the sequence in which the parts are assembled.

Use the chronological sequence for an item that is best visualized by its assembly (such as a piece of furniture, an umbrella tent, or a prehung window or door unit). Architects might find a spatial sequence best for describing a proposed beach house to clients; however, they would use a chronological sequence (of blueprints) for specifying to the builder the prescribed dimensions, materials, and construction methods at each stage.

Combined Sequences. The description of a bumper jack on pages 424–26 alternates among all three sequences; a spatial sequence (bottom to top) for describing the overall mechanism at rest, a chronological sequence for explaining the order in which the parts are assembled, and a functional sequence for describing the order in which the parts operate.

A General Model for Description

Description of a complex mechanism almost invariably calls for an outline. This model is adaptable to any description.

    I. Introduction: General Description[1]
        A. Definition, Function, and Background of the Item
        B. Purpose (and Audience—for classroom only)
        C. Overall Description (with general visuals, if applicable)
        D. Principle of Operation (if applicable)
        E. List of Major Parts
    II. Description and Function of Parts
        A. Part One in Your Descriptive Sequence
            1. Definition
            2. Shape, dimensions, material (with specific visuals)
            3. Subparts (if applicable)
            4. Function
            5. Relation to adjoining parts
            6. Mode of attachment (if applicable)
        B. Part Two in Your Descriptive Sequence (and so on)
    III. Summary and Operating Description
        A. Summary (used only in a long, complex description)
        B. Interrelation of Parts
        C. One Complete Operating Cycle

This outline is tentative, because you might modify, delete, or combine certain parts to suit your subject, purpose, and reader.

Introduction: General Description

Give readers only as much background as they need to get the picture.

---

1. In most descriptions, the subdivisions in the introduction can be combined and need not appear as individual headings in the document.

A Description of the Standard Stethoscope

Introduction

Definition and function

The stethoscope is a listening device that amplifies and transmits body sounds to aid in detecting physical abnormalities.

History and background

This instrument has evolved from the original wooden, funnel-shaped instrument invented by a French physician, R. T. Lennaec, in 1819. Because of his female patients' modesty, he found it necessary to develop a device, other than his ear, for auscultation (listening to body sounds).

Purpose and audience

This report explains to the beginning paramedical or nursing student the structure, assembly, and operating principle of the stethoscope. [*Omit this section if you submit to your instructor an audience and use profile or if you write for a workplace audience.*]

Finally, give a brief, overall description of the item, discuss its principle of operation, and list its major parts, as in the overall stethoscope description on pages 418–19.

Description and Function of Parts

The body of your text describes each major part. After arranging the parts in sequence, follow the logic of each part. Provide only as much detail as your readers need.

Readers of this description will use a stethoscope daily, so they need to know how it works, how to take it apart for cleaning, and how to replace worn or broken parts. (Specifications for the manufacturer would require many more technical details—dimensions, alloys, curvatures, tolerances, and so on.)

Diaphragm Contact Piece

Definition, size, shape, and material

The diaphragm contact piece is a shallow metal bowl, about the size of a silver dollar (and twice its thickness), which is caused to vibrate by various body sounds.

Subparts

Three, separate parts make up the piece: hollow steel bowl, plastic diaphragm, and metal frame, as shown in Figure 2.

The stainless steel metal bowl has a concave inner surface, with concentric ridges that funnel sound toward an opening in the tapered base, then out through the hollow appendage. Lateral threads ring the outer circumference of the bowl to accommodate the interlocking metal frame. A fitted diaphragm covers the bowl's upper opening.

The diaphragm is a plastic disk, 2 millimeters thick, 4 inches in circumference, with a molded lip around the edge. It fits flush over the metal bowl and vibrates sound toward the ridges. A metal frame that screws onto the bowl holds the diaphragm in place.

The stainless steel frame fits over the disk and metal bowl. A $\frac{1}{4}$-inch ridge between the inner and outer edge accommodates threads for screwing the

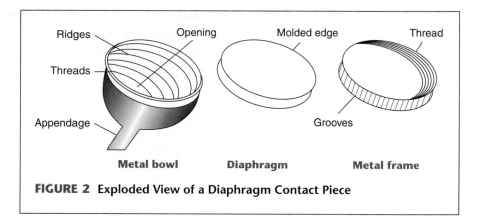

Metal bowl   **Diaphragm**   **Metal frame**

FIGURE 2 **Exploded View of a Diaphragm Contact Piece**

frame to the bowl. The frame's outside circumference is notched with equally spaced, perpendicular grooves—like those on the edge of a dime—to provide a gripping surface.

Function and relation to adjoining parts

Mode of attachment

   The diaphragm contact piece is the heart of the stethoscope that receives, amplifies, and transmits sound through the system of attached tubing. The piece attaches to the lower tubing by an appendage on its apex (narrow end), which fits inside the tubing.

Each part of the stethoscope, in turn, is described according to its own logic of organization.

Summary and Operating Description

Conclude by explaining how the parts work together to make the whole item function.

Summary and Operating Description

How parts interrelate

The seven major parts of the stethoscope provide support for the instrument, flexibility of movement for the operator, and ease in auscultation.

One complete operating cycle

   In an operating cycle, the diaphragm contact piece, placed against the skin, picks up sound impulses from the body surface. These impulses cause the plastic diaphragm to vibrate. The amplified vibrations, in turn, are carried through a tube to a dividing point. From here, the amplified sound is carried through two separate but identical series of tubes to hollow ear plugs.

A Sample Situation

The following description of an automobile jack, aimed toward a general audience, follows our outline model.

A Mechanism Description for Nontechnical Readers

AUDIENCE AND USE PROFILE. Some readers of this description (written for an owner's manual) will have no mechanical background. Before they can follow instructions for *using* the jack safely, they will have to learn what it is, what it looks like, what its parts are, and how, generally, it works. They will *not* need precise dimensions (e.g., "The rectangular base is 8 inches long and $6\frac{1}{2}$ inches wide, sloping upward $1\frac{1}{2}$ inches from the front outer edge to form a secondary platform 1 inch high and 3 inches square"). The engineer who designed the jack might include such data in specifications for the manufacturer. Laypersons, however, need only the dimensions that will help them recognize specific parts and understand their function, for safe use and assembly.

Also, this audience will need only the broadest explanation of how the leverage mechanism operates. Although the physical principles (*torque, fulcrum*) would interest engineers, they would be of little use to readers who simply need to operate the jack safely. ■

DESCRIPTION OF A STANDARD BUMPER JACK

Introduction—General Description

Definition, purpose, and function

The standard bumper jack is a portable mechanism for raising the front or rear of a car through force applied with a lever. This jack enables even a frail person to lift one corner of a 2-ton automobile.

Overall description (spatial sequence)

The jack consists of a molded steel base supporting a free-standing, perpendicular, notched shaft (Figure 1). Attached to the shaft are a leverage mechanism, a bumper catch, and cylinder for insertion of the jack handle. Except for the main shaft and leverage mechanism, the jack is made to be dismantled and to fit neatly in the car's trunk.

Operating principle

The jack operates on a leverage principle, with a human hand traveling 18 inches and the car only $\frac{3}{8}$ of an inch during a normal jacking stroke. Such a device requires many strokes to raise the car off the ground but may prove a lifesaver to a motorist on some deserted road.

List of major parts

Five main parts make up the jack: base, notched shaft, leverage mechanism, bumper catch, and handle.

Description of Parts and Their Function

(chronological sequence)

Base

First major part

Definition, shape, and material

Function and mode of attachment

The rectangular base is a molded steel plate that provides support and a point of insertion for the shaft (Figure 2). The base slopes upward to form a platform containing a 1-inch depression that provides a stabilizing well for the shaft. Stability is increased by a 1-inch cuff around the well. As the base rests on its flat surface, the bottom end of the shaft is inserted into its stabilizing well.

Shaft

Second major part, etc.

The notched shaft is a steel bar (32 inches long) that provides a vertical track for the leverage mechanism. The notches, which hold the mechanism in its position on the shaft, face the operator.

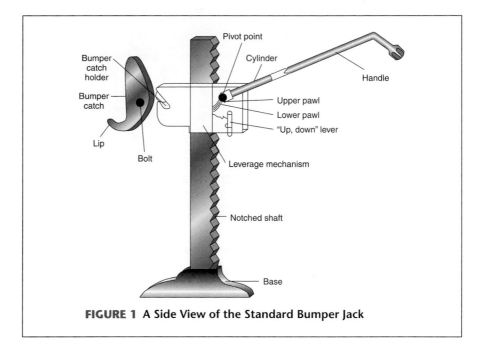

FIGURE 1 A Side View of the Standard Bumper Jack

The shaft vertically supports the raised automobile, and attached to it is the leverage mechanism, which rests on individual notches.

Leverage Mechanism

The leverage mechanism provides the mechanical advantage needed for its operator to raise the car. It is made to slide up and down the notched shaft. The main body of this pressed-steel mechanism contains two units: one for transferring the leverage and one for holding the bumper catch.

The leverage unit has four major parts: the cylinder, connecting the handle and a pivot point; a lower pawl (a device that fits into the notches to allow forward and prevent backward motion), connected directly to the cylinder; an upper pawl, connected at the pivot point; and an "up-down" lever, which applies or releases pressure on the upper pawl by means of a spring (Figure 1). Moving the cylinder up and down with the handle causes the alternate release of the pawls, and thus movement up or down the shaft—depending on the setting of the "up-down" lever. The movement is transferred by the metal body of the unit to the bumper-catch holder.

The holder consists of a downsloping groove, partially blocked by a wire spring (Figure 1). The spring is mounted in such a way as to keep the bumper catch in place during operation.

Bumper Catch

The bumper catch is a steel device that attaches the leverage mechanism to the bumper. This 9-inch molded plate is bent to fit the shape of the bumper.

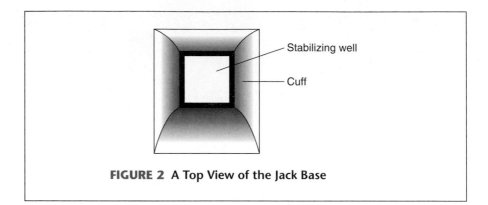

FIGURE 2 **A Top View of the Jack Base**

Its outer $\frac{1}{2}$-inch is bent up to form a lip (Figure 1), which hooks behind the bumper to hold the catch in place. The two sides of the plate are bent back 90 degrees to leave a 2-inch bumper-contact surface, and a bolt is riveted between them. This bolt slips into the groove in the leverage mechanism and provides the attachment between the leverage unit and the car.

Jack Handle

The jack handle is a steel bar that serves both as lever and lug-bolt remover. This round bar is 22 inches long, $\frac{5}{8}$ inch in diameter, and is bent 135 degrees roughly 5 inches from its outer end. Its outer end is a wrench made to fit the wheel's lug bolts. Its inner end is beveled to form a blade-like point for prying the wheel covers and for insertion into the cylinder on the leverage mechanism.

Conclusion and Operating Description

Assembly

One quickly assembles the jack by inserting the bottom of the notched shaft into the stabilizing well in the base, the bumper catch into the groove on the leverage mechanism, and the beveled end of the jack handle into the cylinder. The bumper catch is then attached to the bumper, with the lever set in the "up" position.

One complete operating cycle (functional sequence)

As the operator exerts an up-down pumping motion on the jack handle, the leverage mechanism gradually climbs the vertical notched shaft until the car's wheel is raised above the ground. When the lever is in the "down" position, the same pumping motion causes the leverage mechanism to descend the shaft.

REVISION CHECKLIST FOR DESCRIPTIONS

Use this list to refine the content, arrangement, and style of your description. (Numbers in parentheses refer to first page of discussion.)

CONTENT

❏ Does the title promise exactly what the description delivers? (418)

❏ Are the item's overall features described, as well as each part? (418)

❏ Is each part defined before it is discussed? (421)

❏ Is the function of each part explained? (419)

❏ Do visuals appear whenever they can provide clarification? (419)

❏ Will readers be able to visualize the item? (417)

❏ Are any details missing, needless, or confusing for this audience? (419)

❏ Is the description ethically acceptable? (416)

ARRANGEMENT

❏ Does the description follow the clearest possible sequence? (420)

❏ Are relationships among the parts clearly explained? (421)

STYLE AND PAGE DESIGN

❏ Is the description sufficiently impartial? (416)

❏ Is the language informative and precise? (418)

❏ Will the level of technicality connect with the audience? (419)

❏ Is the description written in plain English? (418)

❏ Is each sentence clear, concise, and fluent? (261)

❏ Is the description written in grammatical English? (Appendix)

❏ Is the page design inviting and accessible? (355)

☑ EXERCISES

1. Select an item from this list or a device used in your major field. Using the general outline as a model, develop an objective description. Include (a) all necessary visuals or (b) a brief (page 333) and a rough diagram for each visual or (c) a "reference visual" (a copy of a visual published elsewhere) with instructions for adapting your visual from that one. (If you borrow visuals from other sources, provide full documentation.) Write for a specific use by a specified audience. Attach your written audience profile (based on the worksheet, page 65) to your document.

soda-acid fire extinguisher	algorithm
breathalyzer	Skinner box
sphygmomanometer	radio
transit	distilling apparatus
saber saw	bodily organ
hazardous waste site	brand of woodstove
photovoltaic panel	catalytic converter

Remember, you are simply describing the item, its parts, and its function: *do not* provide directions for its assembly or operation.

As an optional assignment, describe a place you know well. You are trying to convey a visual image, not a mood; therefore, your description should be impartial, discussing only the observable details.

2. The bumper-jack description in this chapter is aimed toward a general reading audience. Evaluate it by using the revision checklist. In one or two paragraphs, discuss your evaluation, and suggest revisions.

3. Figure 18.5 is designed to promote as well as describe a technical product. Answer the following questions about the document:

• When read in conjunction with Figures 17.2 and 17.3, is 18.5 an effective introduction to this particular product for its intended audience of engineers and other technical experts? Why or why not?

- Is the overall page design effective? Why or why not? Be specific. (See Chapter 15 and "Using Color," pages 340–44).
- Are the visuals adequate and appropriate? Why or why not? (See Chapter 14). Why aren't the diagrams on page one labeled more extensively?
- Is this a sufficiently impartial description? Why or why not? Given its purpose as a marketing document, is the description ethically appropriate? (See Chapter 5.)

Be prepared to discuss your analysis and evaluation in class.

4. Locate a description and specifications for a particular brand of automobile or some other consumer product. Evaluate this material for promotional and descriptive value and ethical appropriateness.

5. Figure 18.6 shows a description from an owner's manual for a telephone answering machine. Study the description and prepare *detailed* answers to these questions:

- Is this description impartial? Explain.
- Compare Figure 18.6 with Figure 18.1. One description focuses on the workings of the *internal* mechanism itself, while the other describes the function of *external* parts. Which is which, and why?
- Are the details in Figures 18.1 and 18.6 adequate and appropriate for the respective audiences? Why?
- What is the descriptive sequence in Figure 18.6? How does it differ from the sequence in 18.1? Why is each sequence appropriate?

☑ COLLABORATIVE PROJECTS

1. Divide into small groups. Assume that your group works in the product-development division of a large and diversified manufacturing company.

   Your division has just invented an idea for an inexpensive consumer item with a potentially vast market. (Choose one from the following list.)

   flashlight

   nail clippers

   retractable ballpoint pen

   scissors

   stapler

   any simple mechanism—as long as it has moving parts

   Your group's assignment is to prepare three descriptions of this invention:

   a. for company executives who will decide whether to produce and market the item

   b. for the engineers, machinists, and so on, who will design and manufacture the item

   c. for the customers who might purchase and use the item

   Before writing for each audience, be sure to collectively complete audience and use profiles (page 424).

   Appoint a group manager, who will assign tasks (visuals, typing, etc.) to members. When the descriptions are fully prepared, the group manager will appoint one member to present them in class. The presentation should include explanations of how the descriptions differ for the different audiences.

2. Assume your group is an architectural firm designing buildings at your college. Develop a set of specifications for duplicating the interior of the classroom in which this course is held. Focus only on materials, dimensions, and equipment (chalkboard, desk, etc.) and use visuals as appropriate. Your audience includes the firm that will construct the classroom, teachers, and school administrators. Use the same format as in Figure 18.2, or design a better one. Appoint one member to present the completed specifications in class. Compare versions from each group for accuracy and clarity.

3. As a group, select a particular product (sound system, laptop computer, video game, or the like) for which descriptions and specifications are available (in product manuals, brochures, and so on). Using Figure 18.5 as a model, design a marketing document that describes and promotes this product in a one-page, double-sided format. Include (a) all necessary visuals or (b) a brief (page 333) and a rough diagram for each visual or (c) a "reference visual" with instructions for adapting your visual from that one. (If you borrow visuals from other sources, provide full documentation.)

C Series System Description

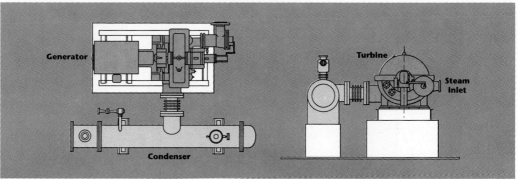

The Ewing Power Systems C Series is a complete single stage condensing turbine generator package designed for use where maximum electricity production is desired and surplus steam is available. High pressure steam passes through the turbine and exits at a vacuum pressure to a close-coupled condenser. The condenser may be either water cooled or air cooled. In water cooled systems the cooling water can be used for heating or the heat can be dissipated in a cooling tower. This series is ideal for converting waste fuel into valuable electricity. It is generally not suited for applications where oil or gas is the primary boiler fuel unless the condenser cooling water will be used for heating.

Features and Specifications

Turbine Features
- Coppus RLHA turbine, fully proven in world-wide applications
- Integrated steam control system including all transmitters, actuators and controllers
- Dual electronic and mechanical overspeed trip mechanisms
- Handvalves for maximum operating efficiency
- Very low maintenance
- Rugged, reliable design, 20 year minimum service life
- Meets all applicable NEMA and API specifications

Steam Specifications
- Recommended Inlet Pressure
  - Maximum: 700 psig
  - Minimum: 14 psig
- Recommended Exhaust Pressure
  - Maximum: 0 psig (14.7 psia)
  - Minimum: –10 psig (4.7 psia)
- Steam Flow
  2,500 pounds of steam per hour (75 boiler horsepower) or greater

Generator Features
- Louis Allis induction generator, renowned for high efficiency and dependability*
- Models from 55 kW to 800 kW continuous duty at 480 volts
- Models to 2,000 kW at higher voltages
- Extra high efficiency design is standard

*Synchronous generators also available

Standard Prewired Electrical Controls
- Shunt trip, 3-pole, motor operated circuit breaker with stored energy trip mechanism
- Utility Grade Protective Relays
  - Over/under Voltage
  - Over/under Frequency
  - Ground overcurrent
- Stator thermostats
- Time delay relay to disconnect on motoring
- Pilot lights for operating and trip status
- Ammeter and voltmeter
- Digital tachometer
- Kilowatt meter
- Synchronous panels available

NOTE: *We will customize our control panels to meet the interconnection requirements of any utility.*

Condenser Features
- Air cooled models
  - Specially designed to minimize power consumption
  - Freeze protection system
  - Standard design is for 97°F ambient, higher temperatures available
- Water cooled models
  - Includes cooling tower
  - Steel shell and tubesheet and admiralty brass tubes
  - Integral hot-well
- Both models include:
  - Steam ejector to remove non-condensables
  - Condensate pump

EWING POWER SYSTEMS

Figure 18.5   Specifications in Marketing Literature

429

Typical System Schematic

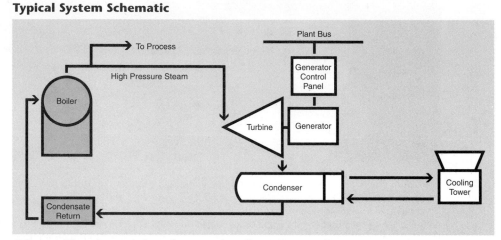

C Series with water cooled condenser and cooling tower.

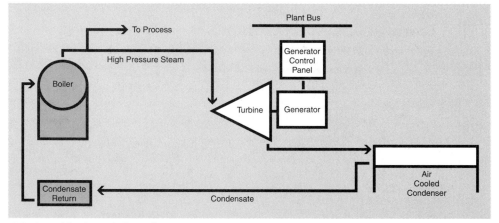

C Series with air cooled condenser.

Services

In addition to supplying the finest available equipment, Ewing Power Systems also provides complete engineering services. We recognize that you may not be familiar with many of the engineering details of cogeneration so we go to extra lengths to assure our systems will meet your specific requirements for steam and electricity and will provide detailed prints and installation supervision. We will also work with your local utility to assure that your system is designed to take maximum advantage of their rate structure and will meet all their technical requirements. These engineering services, together with our complete economic analysis, assure that your cogeneration system will provide the shortest possible payback.

EWING POWER SYSTEMS

5 North Street • South Deerfield, MA 01373-0566

Figure 18.5   Specifications in Marketing Literature *Continued*

Guide To Model 2200 Controls

In the installation and operating instructions that follow, you'll be directed to switches, keys, indicator lights and other controls on your Model 2200. For orientation and easy reference, they're all mapped out here.

1. Tape Counter
Three-digit counter keeps track of tape used for incoming messages. Button resets tape counter to zero.

2. Microphone
Use it to record your outgoing messages or dictate memos.

3. POWER key
Turns the unit on or off. Red light above key indicates "power on."

4. REW (rewind) key
Rewinds your message cassette. Green light above key indicates use.

5. PLAY key
Plays your incoming or outgoing message cassette. Green light above key indicates use.

6. FFW (fast forward) key
Scans the message cassette quickly during playback. Green light above key indicates use.

7. STOP key
Stops record, play, rewind or fast forward functions on incoming message cassette.

8. RECORD key
Records your announcements on the outgoing cassette. Records memos on the incoming cassette. Red light above key indicates use.

9. Function Switch
Set to AUTO to have your calls answered automatically. Set to MESSAGE to play back your incoming messages. Set to ANNOUNCE to record and test your outgoing announcements.

10. AUTO indicator
Red indicator glows steadily when your Model 2200 is set to answer calls automatically.

It flashes to signal that messages have been received.

11. VOLUME control
Controls speaker volume during message playback.

(At the back of unit:)

12. AC cord

13. MESSAGE SELECT switch
Incoming message length can be set to a 30-second or 270-second limit. The ANS ONLY (answer only) position can be used when you want your callers to hear your outgoing announcement, but you do not want to record any incoming messages.

14. RING SELECT switch
Program the unit to answer calls on the second or fourth ring.

15. Modular telephone line cord
Connects your Model 2200 to a modular wall jack.

16. Modular telephone jack
Accepts modular plug of a telephone or related equipment.

(Not visible in diagrams)

Cassette compartment
Under the unit's cover. The incoming message cassette goes on the left. The outgoing message cassette goes on the right.

Speaker
On the bottom of the unit behind circular grillwork. Plays back messages.

Identification plate
On the bottom of the unit. Includes the Model 2200 FCC registration number, the ringer equivalence number, and the remote control Touch-Tone codes.

Figure 18.6   Description of an Answering Machine
Source: Reprinted by permission of AT&T. Copyright © 1985 AT&T.

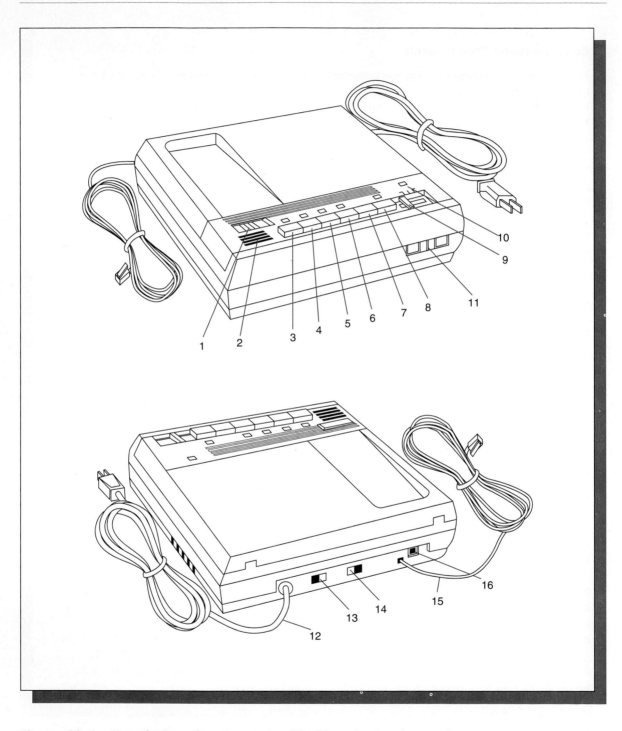

Figure 18.6   Description of an Answering Machine *Continued*

Procedures and Processes

A PROCESS is a series of actions or changes leading to a product or result. A procedure is a way of carrying out a process.

Purpose of Process-Related Explanation

Writers explain procedures and processes as a way of answering these two questions for readers: (1) *How do I do it?* (2) *How does it happen?* In the workplace, you might instruct a colleague or customer in a specific procedure: how to test a soil sample for chlordane contamination, how to access a database, how to ship radioactive waste. Besides procedural instructions, you might have to explain how something happens: how the budget for various departments is determined, or how the electronic mail system in your company works. This chapter discusses both types of process-related explanation: instructions and process analysis.

Elements of Instruction

Almost anyone with a responsible job writes and reads instructions. The new employee uses instructions for operating office equipment or industrial machinery; the employee going on vacation writes instructions for the person filling in. The person who buys a computer reads the manuals (or *documentation*) for instructions on connecting a printer or running a program.

Instructions carry profound ethical and legal implications. As many as 10 percent of workers are injured yearly on the job (Clement 14). Countless injuries also result from misuse of consumer products such as power tools and car jacks—misuse often caused by defective instructions.

A reader injured because of inaccurate, incomplete, or unclear instructions can sue the writer. The court has ruled that a writing defect in product support literature carries liability, as would a design or manufacturing defect in the product itself (Girill 37). Some legal experts argue that defects in the instructions carry even greater liability than product defects because they are more easily demonstrated to a nontechnical jury (Bedford and Stearns 28).

To ensure that your own instructions meet professional and legal requirements for accuracy, completeness, and clarity, observe the following guidelines.

Clear and Limiting Title

Make your title promise exactly what your instructions deliver—no more and no less. The title "Instructions for Cleaning the Drive Head of a Laptop Computer" tells readers what to expect: instructions for a specific procedure on a selected part. But the title "The Laptop Computer" gives no forecast; a document so titled might contain a history of the laptop, a description of each part, or a wide range of related information.

Informed Content

Make sure you know *exactly* what you're talking about. Ignorance on your part makes you no less liable for instructions that are faulty or inaccurate:

*Never count on
ignorance as an excuse*

> If the author of [a car repair] manual had no experience with cars, yet provided faulty instructions on the repair of the car's brakes, the home mechanic who was injured when the brakes failed may recover from the author. (Walter and Marsteller 165)

Do not write instructions unless you know the procedure in detail and unless you actually have performed it.

Visuals

In addition to showing what to do, instructional visuals attract the attention of today's graphic-oriented readers and help keep words to a minimum. Instructions also often include a persuasive dimension: to motivate interest, commitment, or action.

Types of visuals especially suited to instructions include icons, representational and schematic diagrams, flowcharts, photographs, and prose tables.

Illustrate any step that might be hard for readers to visualize. Show the same angle of vision the reader will have when doing the activity or using the equipment—and name the angle (*side view, top view*) if you think readers will have trouble figuring it out for themselves.

The less specialized your audience, the more visuals they are likely to need. But do not illustrate any action simple enough for readers to visualize on their own, such as "Press RETURN" for any user familiar with a keyboard.

Figure 19.1 depicts an array of visuals and their specific instructional functions. Virtually each of these visuals is easily constructed and some could be further enhanced, depending on your production budget and graphics capability. Writers and editors often provide a *brief* (page 333) and a rough sketch describing the visual and its purpose for the graphic designer or art department.

Appropriate Level of Technicality

Unless you know your readers have the relevant background and skills, write for general readers, and do three things:

1. Give them enough background to understand *why* they need your instructions.
2. Give them enough detail to understand *what* to do.
3. Give them enough examples to visualize the procedure clearly.

Page 438 shows how you might adapt instructions titled "How to Copy a Disk Using a Two-Drive Macintosh System" for a general audience.

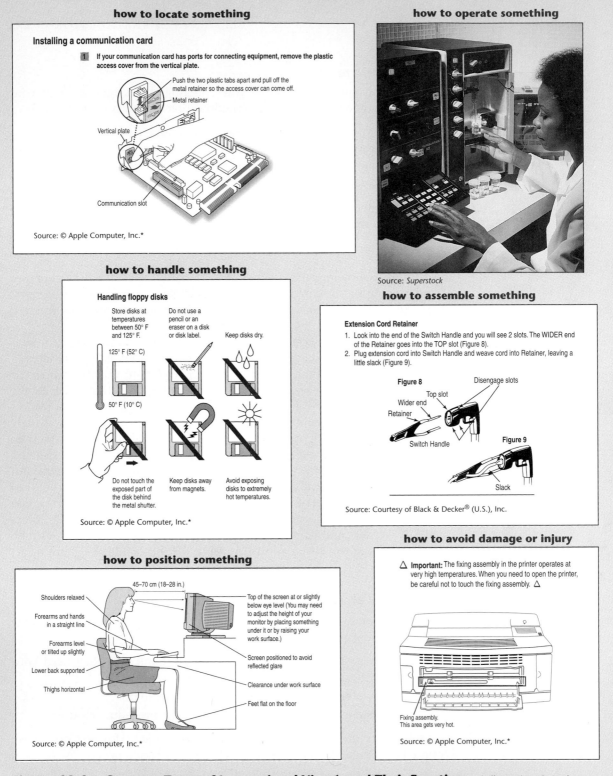

how to locate something

Installing a communication card

1 If your communication card has ports for connecting equipment, remove the plastic access cover from the vertical plate.

Push the two plastic tabs apart and pull off the metal retainer so the access cover can come off.

Metal retainer

Vertical plate

Communication slot

Source: © Apple Computer, Inc.*

how to operate something

Source: *Superstock*

how to handle something

Handling floppy disks

Store disks at temperatures between 50° F and 125° F.

Do not use a pencil or an eraser on a disk or disk label.

Keep disks dry.

125° F (52° C)

50° F (10° C)

Do not touch the exposed part of the disk behind the metal shutter.

Keep disks away from magnets.

Avoid exposing disks to extremely hot temperatures.

Source: © Apple Computer, Inc.*

how to assemble something

Extension Cord Retainer

1. Look into the end of the Switch Handle and you will see 2 slots. The WIDER end of the Retainer goes into the TOP slot (Figure 8).
2. Plug extension cord into Switch Handle and weave cord into Retainer, leaving a little slack (Figure 9).

Figure 8

Disengage slots

Top slot

Wider end

Retainer

Switch Handle

Figure 9

Slack

Source: Courtesy of Black & Decker® (U.S.), Inc.

how to avoid damage or injury

△ **Important:** The fixing assembly in the printer operates at very high temperatures. When you need to open the printer, be careful not to touch the fixing assembly. △

Fixing assembly. This area gets very hot.

Source: © Apple Computer, Inc.*

how to position something

45–70 cm (18–28 in.)

Shoulders relaxed

Forearms and hands in a straight line

Forearms level or tilted up slightly

Lower back supported

Thighs horizontal

Top of the screen at or slightly below eye level (You may need to adjust the height of your monitor by placing something under it or by raising your work surface.)

Screen positioned to avoid reflected glare

Clearance under work surface

Feet flat on the floor

Source: © Apple Computer, Inc.*

Figure 19.1   Common Types of Instructional Visuals and Their Functions. * *Illustrations © Apple Computer, Inc. 1993. Used with permission. Apple, the Apple logo and Power Macintosh are registered trademarks ™ of Apple Computer, Inc. All rights reserved.*

how to diagnose and solve problems

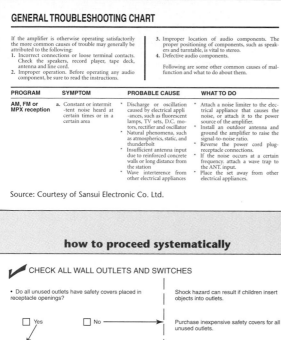

GENERAL TROUBLESHOOTING CHART

If the amplifier is otherwise operating satisfactorily the more common causes of trouble may generally be attributed to the following:
1. Incorrect connections or loose terminal contacts. Check the speakers, record player, tape deck, antenna and line cord.
2. Improper operation. Before operating any audio component, be sure to read the instructions.

3. Improper location of audio components. The proper positioning of components, such as speakers and turntable, is vital to stereo.
4. Defective audio components.

Following are some other common causes of malfunction and what to do about them.

PROGRAM	SYMPTOM	PROBABLE CAUSE	WHAT TO DO
AM, FM or MPX reception	a. Constant or intermit -tent noise heard at certain times or in a certain area	* Discharge or oscillation caused by electrical appli -ances, such as fluorescent lamps, TV sets, D.C. mo-tors, rectifier and oscillator * Natural phenomena, such as atmospherics, static, and thunderbolt * Insufficient antenna input due to reinforced concrete walls or long distance from the station * Wave interference from other electrical appliances	* Attach a noise limiter to the elec-trical appliance that causes the noise, or attach it to the power source of the amplifier. * Install an outdoor antenna and ground the amplifier to raise the signal-to-noise ratio. * Reverse the power cord plug-receptacle connections. * If the noise occurs at a certain frequency. attach a wave trap to the ANT. input. * Place the set away from other electrical appliances.

Source: Courtesy of Sansui Electronic Co. Ltd.

how to proceed systematically

☑ CHECK ALL WALL OUTLETS AND SWITCHES

• Do all unused outlets have safety covers placed in receptacle openings?

Shock hazard can result if children insert objects into outlets.

☐ Yes ☐ No → Purchase inexpensive safety covers for all unused outlets.

• Are all outlets and switches working properly?

Improperly operating outlets or switches indicate an unsafe wiring condition may exist.

☐ Yes ☐ No → Have an electrician check them.

• Are all outlets and switches cool to the touch?

Unusually warm outlet or switch may indicate an unsafe wiring condition exists.

☐ Yes ☐ No → Unplug any cord or stop using the switch and have an electrician check.

• Do electrical plugs fit snugly into all outlets?

Loose-fitting plugs can cause overheating.

☐ Yes ☐ No → Have the outlet replaced.

Source: U.S. Consumer Product Safety Commission

how to make the right decisions

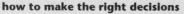
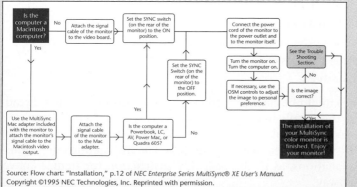

Source: Flow chart: "Installation," p.12 of *NEC Enterprise Series MultiSync® XE User's Manual*. Copyright ©1995 NEC Technologies, Inc. Reprinted with permission.

how to identify safe or acceptable limits

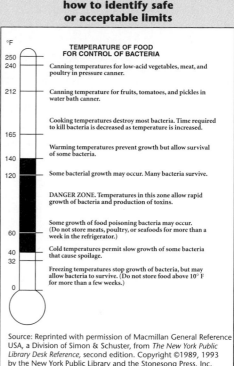

°F

TEMPERATURE OF FOOD FOR CONTROL OF BACTERIA

250
240 Canning temperatures for low-acid vegetables, meat, and poultry in pressure canner.

212 Canning temperature for fruits, tomatoes, and pickles in water bath canner.

Cooking temperatures destroy most bacteria. Time required to kill bacteria is decreased as temperature is increased.

165

Warming temperatures prevent growth but allow survival of some bacteria.

140

120 Some bacterial growth may occur. Many bacteria survive.

DANGER ZONE. Temperatures in this zone allow rapid growth of bacteria and production of toxins.

Some growth of food poisoning bacteria may occur. (Do not store meats, poultry, or seafoods for more than a week in the refrigerator.)

60

40 Cold temperatures permit slow growth of some bacteria that cause spoilage.

32

Freezing temperatures stop growth of bacteria, but may allow bacteria to survive. (Do not store food above 10° F for more than a few weeks.)

0

Source: Reprinted with permission of Macmillan General Reference USA, a Division of Simon & Schuster, from *The New York Public Library Desk Reference*, second edition. Copyright ©1989, 1993 by the New York Public Library and the Stonesong Press, Inc.

why action is important

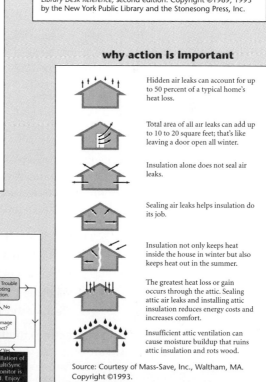

Hidden air leaks can account for up to 50 percent of a typical home's heat loss.

Total area of all air leaks can add up to 10 to 20 square feet; that's like leaving a door open all winter.

Insulation alone does not seal air leaks.

Sealing air leaks helps insulation do its job.

Insulation not only keeps heat inside the house in winter but also keeps heat out in the summer.

The greatest heat loss or gain occurs through the attic. Sealing attic air leaks and installing attic insulation reduces energy costs and increases comfort.

Insufficient attic ventilation can cause moisture buildup that ruins attic insulation and rots wood.

Source: Courtesy of Mass-Save, Inc., Waltham, MA. Copyright ©1993.

Figure 19.1   Common Types of Instructional Visuals and Their Functions *Continued*

Background Information. Begin by explaining the purpose of the procedure.

Tell readers why they
are doing this

You might easily lose information stored on a floppy disk if:

1. the disk is damaged by direct sunlight, extreme temperature, or moisture.
2. the disk is erased by a faulty disk drive, a power surge, or a user error.
3. the stored information is scrambled by a nearby magnet (telephone, computer terminal, or the like).

Always make a backup copy of any disk that contains important material.

Also, state your assumptions about your reader's level of technical understanding.

Tell them what they
should know already

To follow these instructions, you should be able to identify these parts of a Macintosh system: computer, monitor, keyboard, mouse, disk drive, and a 3.5-inch floppy disk.

Define any specialized terms that appear in your instructions.

Working Definition

Tell them what each key
term means

Initialize: Before you can store or retrieve information on a disk, you must initialize the blank disk. Initializing creates a format the computer can understand—a directory of specific memory spaces (like post office boxes) on the disk, where you can store and retrieve information as needed.

When your reader understands *what* and *why,* you are ready to explain *how* the reader can carry out the procedure.

Adequate Detail. Explain the procedure in enough detail for readers to know exactly what to do. Vague instructions result from the writer's failure to consider the readers' needs, as in these unclear instructions for giving first aid to an electrical shock victim:

Inadequate detail for
general readers

1. Check vital signs.
2. Establish an airway.
3. Administer external cardiac massage as needed.
4. Ventilate, if cyanosed.
5. Treat for shock.

These instructions might be clear to medical experts, but not to general readers. Not only are the details inadequate, but terms such as "vital signs" and "cyanosed" are too technical for laypersons. Such instructions posted for workers in a high-voltage area would be useless. The instructions need illustrations and explanations, as in the page 439 instruction for item 2 above, establishing an airway:

Adequate detail for general readers

MOUTH-TO-MOUTH BREATHING

If there are no signs of breathing, place one hand under the victim's neck and gently lift. At the same time, push with the other hand on the victim's forehead. This will move the tongue away from the back of the throat to open the airway.

Source: Reprinted with permission of Macmillan General Reference USA, a Division of Simon & Schuster, from *The New York Public Library Desk Reference,* second edition. Copyright ©1989, 1993 by the New York Public Library and the Stonesong Press, Inc.

It's easy to overestimate what people already know, especially when the procedure is almost automatic for you. (Think about when a relative or friend was teaching you to drive a car, or perhaps you tried to teach someone else.) Always assume the reader knows less than you. A colleague will know at least a little less; a layperson will know a good deal less—maybe nothing—about this procedure.

Exactly how much information is enough? These suggestions can help you find an answer:

How to provide adequate detail

- Give everything readers need, so the instructions can stand alone.
- Give only what readers need. Don't tell them how to build a computer when they need to know only how to copy a disk.
- Instead of focusing on the *product* ("How does it work?"), focus on the *task* ("How do I use it?" or "How do I do it?") (Grice, "Focus" 32).
- Omit steps (*Seat yourself at the computer*) obvious to readers.
- Adjust the information rate ("the amount of information presented in a given page," Meyer 17) to the reader's background and the difficulty of the task. For complex or sensitive steps, slow the information rate. Don't make readers do too much too fast.
- Reinforce the prose with visuals. Don't be afraid to repeat information if it saves readers from flipping pages.
- When writing instructions for consumer products, assume "a barely literate reader" (Clement 151). Simplify.
- Recognize the persuasive dimension of the instructions. You may need to persuade readers that this procedure is necessary or beneficial, or that they can complete this procedure with relative ease and competence.

Examples. Procedures require specific examples (how to load a program, how to order a part), to help readers follow the steps correctly.

Use plenty of examples

To load your program, type this command:

```
Load "Style Editor"
```

Then press RETURN.

Like visuals, examples *show* readers what to do. Examples in fact often appear as visuals.

In the sample procedure that follows, careful use of background, detailed explanation, and visual examples create a user-friendly level of technicality for readers just learning to use a computer.

First Step: How to Initialize Your Blank Disk

Before you can copy or store information on a blank disk, you must initialize the disk. Follow this procedure:

Begin each instruction with an action verb

1. Switch the computer on.
2. Insert your application disk in the main disk drive.
3. Insert your blank disk in the external disk drive. Unable to recognize this new disk, the computer will respond with a message asking whether you wish to initialize the disk (Figure 1).

Let the visual repeat, restate, or reinforce the prose

FIGURE 1 The "Initialize" Message

4. Using your mouse, place the tip of the on-screen pointer (a small arrow) inside the "Initialize" box.
5. Press and quickly release the mouse button. Within 15–20 seconds the initializing will be completed, and a message will appear, asking you to name your disk (Figure 2).

Place the visual close to the step

FIGURE 2 The "Disk-naming" Message

Second Step: How to Name Your Initialized Disk (and so on)

In the sample, notice that instructions and illustrations repeat the same information—they are redundant (Weiss 100). You may recall from Chapter 13 that writers avoid *style redundancy* (extra words that give no extra information). For instructions, however, a writer deliberately seeks *content redundancy*, giving the same information in prose and then in visuals. Granted, this prose-visual redundancy makes a longer document, but such repetition helps prevent misinterpretation. When you can't be sure how much is enough, risk overexplaining rather than underexplaining.

Logically Ordered Steps

Instructions not only divide the procedure into steps; they also guide users through the steps *in chronological order.* They organize the facts and explanations in ways that make sense to the reader.

Show how steps are connected

> You can't splice two wires to make an electrical connection until you have removed the insulation. To remove the insulation, you will need . . . .

Try to keep all information for one step close together.

Warnings, Cautions, and Notes

Here are the only items that should interrupt the steps in a set of instructions (Van Pelt 3):

- A *note* clarifies a point, emphasizes vital information, or describes options or alternatives.

  > NOTE: The computer will not initialize a disk that is scratched or imperfect. If your blank disk is rejected, try a new disk.

- A *caution* prevents possible mistakes that could result in injury or equipment damage:

  > CAUTION: A momentary electrical surge or power failure will erase the contents of internal memory. To avoid losing your work, every few minutes save on disk what you have just typed into the computer.

- A *warning* alerts users against potential hazards to life or limb:

  > WARNING: To prevent electrical shock, always disconnect your printer from its power source before cleaning internal parts.

- A *danger* notice identifies an immediate hazard to life or limb:

  > DANGER: The red canister contains **DEADLY** radioactive material. **Do not break the safety seal** under any circumstances.

In addition to prose warnings, attract readers' attention and help them identify hazards by using symbols, or icons (Bedford and Stearns 128):

Use symbols to alert readers

Do not enter   **Radioactivity**   **Fire danger**

Preview the warnings, cautions, and notes in your introduction, and place them, *clearly highlighted*, immediately before the respective steps.

NOTE: A recent study found that product users were six times more likely to comply with warnings included with the directions for using the product instead of on a separate warning label ("Notes" 2).

Use notes, warnings, and cautions only when needed; overuse will dull their effect, and readers may overlook their importance.

Appropriate Words, Sentences, and Paragraphs

Of all communications, instructions have the strictest requirements for clarity, because they lead to *immediate action*. Readers are impatient, and often will not read the entire instructions before plunging into the first step. Poorly phrased and misleading instructions can be disastrous.

Like descriptions, instructions name parts, use location and position words, and state exact measurements, weights, and dimensions. Instructions additionally require your strict attention to phrasing, sentence structure, and paragraph structure.

Transitions to Mark Time and Sequence. Transitional words provide a bridge between related ideas. Some transitions ("in addition," "next," "meanwhile," "finally," "ten minutes later," "the next day," "before") mark time and sequence. They help readers understand the step-by-step process:

Use transitions to mark time and sequence

> **Preparing the Ground for a Tent**
>
> Begin by clearing and smoothing the area that will be under the tent. This step will prevent damage to the tent floor and eliminate the discomfort of sleeping on uneven ground. *First,* remove all large stones, branches, or other debris within a level 10 x 13-foot area. Use your camping shovel to remove half-buried rocks that cannot easily be moved by hand. *Next,* fill in any large holes with soil or leaves. *Finally,* make several light surface passes with the shovel or a large, leafy branch to smooth the area.

Carefully Shaped Paragraphs and Sentences. Much of the introductory and explanatory material in instructions take the form of standard prose paragraphs, with enough sentence variety to keep readers interested. But the steps themselves have unique paragraph and sentence requirements. Unless the procedure consists of simple steps (as in "Preparing the Ground for a Tent," shown above), separate each step by using a numbered list—one step for one activity. If the activity is especially complicated, use a new line (not indented) to begin each sentence in the step.

Instructions ordinarily employ short sentences. But brief is not always best, especially when readers have to fill in the information gaps. Never telegraph your message by omitting articles (*a, an, the*), as shown on page 263.

Unlike many types of documents, instructions call for very little sentence variety. Use sentences with similar structure ("Do this. Then do that.") to avoid confusing readers trying to follow the procedure.

If a single step covers two related actions, follow the sequence of actions that are required:

> Confusing   Insert the disk in the drive before switching on the computer.
> sequence

The second action required is mistakenly given before the first.

> Logical   Before switching on the computer, insert the disk in the drive.
> sequence

Make your explanations easier to follow by using a familiar-to-unfamiliar sequence:

> Hard   You must initialize a blank disk before you can store information on it.

This sentence is clearer if the familiar material comes first:

> Easier   Before you can store information on a blank disk, you must initialize the disk.

Shape every sentence and every paragraph for the reader's access.

Active Voice and Imperative Mood. Use the active voice and imperative mood ("Insert the disk") to address the reader directly. Otherwise, your instructions can lose authority ("You should insert the disk"), or become ambiguous ("The disk is inserted"). In the ambiguous example, we can't tell if the disk is to be inserted or if it already has been inserted.

> Indirect or   The user types in his or her access code.
> confusing   You should type in your access code.
>   It is important to type in the access code.
>   The access code is typed in.
>
> Direct and   Type in your access code.
> clear

The imperative makes instructions more definite and easier to understand because the action verb—the crucial word that tells what the next action will be—comes first. Instead of burying your verb in midsentence, *begin* with an action verb ("raise," "connect," "wash," "insert," "open"), to give readers an immediate signal.

Affirmative Phrasing. Research shows that readers respond more quickly and efficiently to instructions phrased affirmatively rather than negatively (Spyridakis and Wenger 205).

Weaker	Verify that your disk is not contaminated with dust.
Stronger	Examine your disk for dust contamination.

Parallel Phrasing. Like any items in a series, steps should be in identical grammatical form. Parallelism is important in all writing but in instructions especially, because repeating grammatical forms emphasizes the step-by-step organization of the instructions.

Not parallel   To log on to the VAX 20, follow these steps:

1. Switch the terminal to "on."
2. The CONTROL key and C key are pressed simultaneously.
3. Typing LOGON, and pressing the ESCAPE key.
4. Type your user number, and press the ESCAPE key.

Parallel   To log on to the DEC 20, follow these steps:

1. Switch the terminal to "on."
2. Press the CONTROL key and C key simultaneously.
3. Type LOGON, and press the ESCAPE key.
4. Type your user number, and press the ESCAPE key.

Parallelism increases readability and lends continuity to the instructions.

Effective Page Design

Instructions rarely get undivided attention. The reader, in fact, is doing two things more or less at once: interpreting the instructions and performing the task. Effective instructional design is *accessible* and *inviting*. An effective design conveys the sense that the task is within a qualified user's range of abilities.

Here are suggestions for designing instructions that help users find, recognize, and remember what they need:

How to design your instructions

- Use informative headings that tell readers what to expect, that emphasize what is most important, and that serve as an aid to navigation. A heading such as "How to Initialize Your Blank Disk" is more informative than "Disk Initializing." Also, the second version sounds less like a person speaking and more like a robot!
- Arrange all steps in a numbered list.
- Single-space within steps and double-space between, to separate steps visually.
- Double-space to signal a new paragraph, instead of indenting.
- Use white space and highlighting to separate discussion from step.

   Set off your discussion on a separate line (like this), indented or highlighted or both. On a typewriter, you can highlight with

> underlining, capitals, dashes, parentheses, and asterisks. On a computer, you can use **boldface**, *italics,* varying type sizes, and varying **typefaces.**

■ Set off warnings, cautions, and notes in ruled boxes or use highlighting and plenty of white space.

■ Keep the visual and the step close together. If room allows, place the visual right beside the step; if not, right after the step. Set off the visual with plenty of white space.

■ Strive for format variety that is appealing but not overwhelming or inconsistent. Readers can be overwhelmed by a page with excessive or inconsistent graphic patterns.

The more accessible and inviting the design, the more likely your readers will follow the instructions. Don't be afraid to experiment until you find a design that works.

Measurements of Usability

When we evaluate how well a document achieves its objectives, we measure its **usability.** In measuring the usability of instructions, we ask this question: *How effectively do these instructions enable users to complete the intended task safely, efficiently, and appropriately?*

Consumers look for products that are easy to use. As a result, companies routinely measure the usability of their products—including the product's **documentation** (assembly, operating, or troubleshooting instructions, warnings, manuals). To win customer goodwill—and to avoid lawsuits—a company will go to great lengths to anticipate all the ways a product might be used or misused.

Usability testing is a complex process in which testers have to know (1) how to design tests that are valid and reliable, and (2) how to interpret their findings accurately. Primary research methods from Chapter 8 are employed in usability testing (interviews, questionnaires, observation, experiment).

Ideally, usability tests are conducted in a situation that simulates the actual situation, with people who actually will use the document (Redish and Schell 67; Ruhs 8). But even in a classroom setting we can make reasonable assessments of a document's effectiveness on the basis of usability criteria.

Usability criteria for a specific document are defined by its intended use and by characteristics of the user and the environment that affect its use, as depicted in Figure 19.2.

Intended Use

A first step in evaluating usability is to define what we want users to be able to do as a result of reading the document:

■ Will users merely carry out a desired action ("What do I do next?"), as in copying a disk correctly? Or will they navigate a complex task that requires decisions ("What option should I choose?") or judgments ("Is this good enough?"), as in diagnosing or treating an ailment? What kinds of knowledge and skills does the task require?

■ Besides procedural knowledge (how to do things), what kinds of declarative knowledge (knowing facts) or conceptual knowledge (understanding principles or theories) underlie the procedure? How much learning versus performing is required (Mirel et al. 79; Wickens 243)?

■ Are we trying to impart perceptual motor skills (swinging a golf club, making a surgical incision), which are less easily forgotten—as opposed to procedural skills (using a computer program, cleaning a fuel-injection system), which are more easily forgotten (Wickens 250)?

■ Will users need to remember this information for a short or long period (Wickens 232)? Will they always have the document in front of them while performing the task?

■ Do we need to *persuade* the user to pay attention, be careful, or the like?

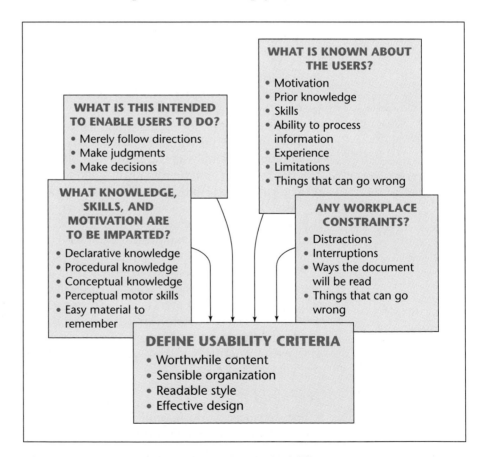

Figure 19.2   Assessing a Document's Usability

Human Factors

Those characteristics of the user and the environment that enhance or limit job performance are known as **human factors.** These include the user's attitudes, abilities, and limitations, and the constraints of the work environment (Wickens 3).

- Who are the users? What do they know already? How informed are they in general? How accustomed to such procedures? How educated? How motivated?
- Will readers be scanning the document, studying it or memorizing it, or merely referring to it periodically and randomly?
- Can we anticipate ways in which the document could be misinterpreted or misunderstood (Boiarsky 100)?
- Can we anticipate possible distractions or interruptions from the work environment?

For a closer look at human factors, review the Audience and Use Profile Sheet (page 65).

Usability Criteria for the Document

Only after identifying its intended use and the possible human factors can we decide on criteria for evaluating the document's effectiveness.

- Is everything easy to find, appropriate to this reader for this task, easy to understand, and complete and accurate (Haynes 239)?
- Is the document organized in a way the readers are expected to proceed? Is everything easy to read and follow (style, format, visuals)?

The Revision Checklist (pages 462–63) offers one approach for a detailed assessment of usability.

A General Model for Instructions

You can adapt the introduction-body-conclusion structure to any instructions. Here are the possible sections to include:

    I. INTRODUCTION
        A. Definition, Benefits, and Purpose of the Procedure
        B. Intended Audience (usually omitted for workplace audiences)
        C. Prior Knowledge and Skills Needed by the Audience
        D. Brief Overall Description of the Procedure
        E. Principle of Operation
        F. Materials, Equipment (in order of use), and Special Conditions
        G. Working Definitions (always in the introduction)
        H. Warnings, Cautions, Notes (previewed here and spelled out at steps)

I. List of Major Steps
II. REQUIRED STEPS
    A. First Major Step
        1. Definition and purpose
        2. Materials, equipment, and special conditions for this step
        3. Substeps (if applicable)
            a.
            b.
    B. Second Major Step (and so on)
III. CONCLUSION
    A. Review of Major Steps (for a complex procedure only)
    B. Interrelation of Steps
    C. Troubleshooting or Follow-up Advice (as needed)

This outline is only tentative; you might modify, delete, or combine some elements, depending on your subject, purpose, and audience.

IN BRIEF

Usability Issues with Online or Multimedia Documents

In contrast to paper documents, online or multimedia documents create unique usability issues (Holler 25; Humphreys 754–55):

- Online documents tend to focus more on *doing* than on detailed explanations. Workplace readers typically use online documents for reference or training rather than for study or memorizing.

- Users of online instructions rarely need persuading (e.g., to pay attention or follow instructions) because they are guided interactively through each step of the procedure.

- Visuals play an increasingly essential role in online instruction.

- Online documents typically are organized to be read interactively and selectively rather than in linear sequence. Users move from place to place, depending on their immediate needs. Organization therefore is flexible and modular, with small bits of easily accessible information that can be combined to suit a particular user's needs and interests.

- Online readers easily can lose their bearings. Unable to shuffle or flip through a stack of printed pages in linear order, readers need constant orientation (to retrieve some earlier bit of information or to compare something on one page with another). In the absence of page or chapter numbers, index, or table of contents, "Find," "Search," and "Help" options need to be plentiful and complete.

- Studies often suggest that reading an online document is slower than a paper document.

As increasing applications materialize for online and multimedia instruction, usability criteria will continue to be defined.

☆ ☆ ☆

Introduction

The introduction should enable readers to begin "doing" as soon as they are able to proceed safely, effectively, and confidently (Van der Meij and Carroll 245–46). Most readers are interested primarily in "how to use it or fix it," and will require only a general understanding of "how it works." You don't want to bury readers in a long introduction, nor to set them loose on the procedure without adequate preparation. Know your audience—what they need and don't need.

Following is an introduction from instructions for users of the college library. Some users will be computer experts; some will be novices—but all will require a detailed introduction to computerized literature searches before they conduct a search on their own.

Clear and limiting title	**HOW TO USE THE OCLC TERMINAL TO SEARCH FOR A BOOK**
	Introduction
Definition of the procedure	Our library's OCLC (On-line Computer Library Center) terminal offers an efficient way to search for books, journals, government publications, and other printed materials. This terminal is connected to the OCLC database in Columbus, Ohio.
Definition (continued)	The Ohio database contains more than 8 million records of books and other published materials in libraries nationwide. Each record lists the information found on a catalog card, and identifies the libraries holding the work.
Purpose	These instructions will enable you to determine whether a book you seek can be found in our library or in other libraries throughout the country.
Audience and required skills	Any library patron can use the OCLC terminal to search for a book. To operate the terminal, you need only be familiar with the keyboard (Figure 1) and to have an operator's manual handy for general reference.
General description of the procedure	After logging on the system, you can search for a book by TITLE, AUTHOR, or AUTHOR and TITLE. Once you have viewed the catalog entry for the book you seek, you can use the terminal to determine the libraries from which you might borrow the book. These instructions show a search by TITLE only.
Materials and equipment	The only additional equipment you need is a copy of the *Manual of OCLC Participating Institutions,* to find a listing of library names according to their OCLC symbol displayed on your terminal screen.
Working definitions	Specialized terms that appear in these instructions are here defined:

    Cursor: a small, blinking rectangle that indicates the screen position of the next character you will type.

    HOME Position: the cursor location at the extreme top left of your screen.

        HOME position is where you will type most of your messages to the computer, and where the computer will display its messages to you.

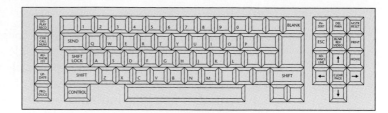

FIGURE 1 The OCLC Terminal Keyboard

Cautions and notes

Pay close attention to the NOTES that accompany the logging-on step and to the CAUTION preceding the logging-off step.

List of major steps

The major steps in using OCLC to search for a book by its TITLE are (1) logging on the system, (2) initiating the search, (3) viewing the information on your title, (4) locating the book, and (5) logging off.

Body

In your body section (labeled *Required Steps*), give each step and substep in order. Insert warnings, cautions, and notes as needed. Begin each step with its definition or purpose or both. Readers who understand the reasons for a step will do a better job. A numbered list (like the following) is an excellent way to segment the steps.

Section heading

Heading previews the step

Capitals and spacing set off note

Headings 1–5 offer an overview of the process

Explanation begins new line

Required Steps

1. *Logging on the OCLC System*

   To activate the system, you must first log on.

   a. Flick on the red power switch at the keyboard's right edge.

      NOTE: Always press HOME before typing, to make sure your cursor is at HOME position.

   b. Type the authorization number (07–34–6991).

      NOTE: If you make a typing error, place the cursor over the incorrect character, and type the correct character. To move the cursor one space, press and release one of the four keys marked with an arrow (lower right of keyboard).

   c. Press SEND, and wait for the ⎡ HELLO ⎤ response.

2. *Initiating the Search*

   To initiate a computer search, you must enter your title's exact code name.

   a. Type in the code for your desired title, excluding articles (*a, an, the*) when they appear as the first word of the title. Always type your entry exactly as in the following example:

      Suppose your title is *Presentation Graphics on the Apple Macintosh;* you would type

Example shows what to do

Single-spacing within steps; double-spacing between

Visual reinforces the prose

Visual set off by white space and ruled box

All steps have parallel phrasing

Result of step begins new line

                                  pre,gr,on,t

    b. Press DISPLAY/RECD and then SEND.

The computer will display a numbered list of titles that match the code you have sent (Figure 2).

```
1 Presentation Graphics on the Apple Macintosh: how to
  use Microsoft Chart to create dazzling graphics for
  professional and corporate applications/Lambert,
  Steve. Microsoft Press: Distributed in the U.S. and
  Canada by Simon and Schuster, c 1984.
2 Presentation Graphics on the IBM PC and compatibles:
  how to use Microsoft Chart to create dazzling graph-
  ics for professional and corporate applications/
  Lambert, Steve. Microsoft Press: Distributed in the
  U.S. and Canada by Simon and Schuster, c 1986.
```

FIGURE 2 **The Computer's List of Titles that Match the Code You Typed**

3. *Viewing the Information on Your Title*

Now that you have found your title, check to see if our library has the book. To view the full information on your title, follow these steps:

    a. Type in the number of the title you seek.

    b. Press DISPLAY/RECD and then SEND.

The computer will display a catalog entry for your title (Figure 3).

```
NO HOLDINGS IN SMU-FOR HOLDINGS ENTER dh DEPRESS DIS-
PLAY RECD. SEND OCLC 10753933
 3  020  09148511X (pbk. : cover) : $18.95
 5  050  T385b.L34 1984
 9  100  Lambert, Steve, 1945-
10  245  Presentation graphics on the Apple Macintosh:
          how to use Microsoft Chart to create dazzling
          graphics for professional and corporate ap-
          plications / c Steve Lambert.
11  260  Bellevue, Wash.: Microsoft Press; distributed
          in the U.S. and Canada by Simon and Schuster,
          New York; c 1984.
```

FIGURE 3 **The Partial Catalog Entry for Your Title**

Because SMU does not have this title, you will have to determine where it can be found.

4. *Locating the Book*

"Then" signals
transition between
substeps
Steps begin with action
verbs
Example shows results of
the step

To determine which OCLC libraries hold the title you seek, follow these steps:

 a. Type dh and then press DISPLAY/RECD and then SEND. The computer will list symbols of holding libraries by state (Figure 4).

 b. Identify the closest library by looking up symbols in the *Manual of OCLC Participating Institutions.*

Your title, *Presentation Graphics,* can be found in these Massachusetts libraries: (LDV) Lotus Development Corporation, (LIN) Massachusetts Institution of Technology, and so on.

 c. For reference, obtain a printout of the library listing by pressing PRINT.

```
REGIONAL LOCATIONS-FOR OTHER HOLDINGS ENTER dhs, dha,
OR dh. DISPLAY RECD. SEND; FOR BIBLIOGRAPHIC RECORD EN-
TER bib. DISPLAY RECD. SEND
STATE      LOCATIONS
  CT       ETN
  MA       LDV LIN MRK
  NY       NYP VYC VYM YQR ZBM ZNC ZQP ZSA
  VT       VTU
```

FIGURE 4 Display of Symbols for Regional Libraries Holding Your Title

5. *Logging off*

Once you have your printout, you can log off the system.

CAUTION: Before logging off, press CLEAR PAGE simultaneously with CONTROL. This step will clear any images that might damage the screen's phosphorus coating.

To log off, type *end* and then press SEND. The computer will respond with GOODBYE.

Conclusion

The conclusion of a set of instructions has several possible functions:

- You might summarize the major steps in a long and complex procedure, to help readers review their performance.
- You might describe the results of the procedure.
- You might offer follow-up advice, about what could be done next or refer the reader to further sources of documentation.
- You might give troubleshooting advice about what to do if anything goes wrong.

You might do all these things—or none of them. If your procedural section has given readers all they need, omit the conclusion altogether.

In the case of our OCLC instructions, these concluding remarks are appropriate:

Conclusion

Follow-up advice

Once you locate the title you seek, ask a reference librarian for help requesting the book via Interlibrary Loan. (Books usually arrive within two days.)

Troubleshooting advice

In case the first library is unable to supply the book requested, you initially should choose several libraries from your printout. The Interlibrary Loan system will automatically forward your request to each lender you have specified, until one lender indicates it can supply the book.

A Sample Situation

The instructions for *doing something* (felling a tree) in Figure 19.3 are patterned after our general outline, shown earlier. They will appear in a series of brochures for forestry students about to begin summer jobs with the Idaho Forestry Service.

Instructions for Semitechnical Readers

AUDIENCE AND USE PROFILE. I'm writing these instructions for partially informed readers who know how to use chainsaws, axes, and wedges but who are approaching this dangerous procedure for the first time. Therefore, I'll include no visuals of cutting equipment (chainsaws and so on) because the audience already knows what these items look like. I can omit basic information (such as what happens when a tree binds a chainsaw) because the audience already has this knowledge also. Likewise, these readers need no definition of general forestry terms such as *culling* and *thinning,* but they *do* need definitions of terms that relate specifically to tree felling (*undercut, holding wood,* and so on).

To ensure clarity, I will illustrate the final three steps with visuals. The conclusion, for these readers, will be short and to the point—a simple summary of major steps with emphasis on safety. I'll experiment with formats until I create a design that is most accessible. ∎

A General Model for Process Analysis

An explanation of how things work or happen divides the process into its parts or principles. Colleagues and clients need to know how stock and bond prices are governed, how your bank reviews a mortgage application, how an optical fiber conducts an impulse, and so on. Process analysis emphasizes *the process itself* so that readers can follow it as "observers."

INSTRUCTIONS FOR FELLING A TREE   **1**

INTRODUCTION

Forestry Service personnel fell (cut down) trees to cull or thin a forested plot, to eliminate the hazard of dead trees standing near power lines, to clear an area for construction, and the like.

These instructions explain how to remove sizable trees for personnel who know how to use a chainsaw, axe, and wedge safely.

When you set out to fell a tree, expect to spend most of your time planning the operation and preparing the area around the tree. To fell the tree, you will make two chainsaw cuts, severing the stem from the stump. Depending on the direction of the cuts, the weather, and the terrain, the severed tree will fall into a predetermined clearing.

> **WARNING:** Although these instructions cover the basic procedure, felling is very dangerous. Trees, felling equipment, and terrain vary greatly. Even professionals sometimes are killed or injured because of judgment errors or misuse of tools.
>   Your main concern is safety. Be sure to have an expert demonstrate this procedure before you try it. Also, pay attention to warnings in steps 1 and 4.

To fell sizable trees, you will need this equipment:
   —a 3- to 5-horsepower chainsaw with a 20-inch blade
   —a single-blade splitting axe
   —two or more 12-inch steel wedges

The major steps in felling a tree are (1) choosing the lay, (2) providing an escape path, (3) making the undercut, and (4) making the backcut.

REQUIRED STEPS

1. *How to Choose the Lay*

   The "lay" is where you want the tree to fall. On level ground in an open field, which way you direct the fall makes little difference. But such ideal conditions are rare.

   Consider ground obstacles and topography, surrounding trees, and the condition of the tree to be felled. Plan your escape path and the location of your cuts depending on surrounding houses, electrical wires, and trees. Then follow these steps:

   a. Make sure the tree still is alive.

Figure 19.3   A Set of Instructions

2

b. Determine the direction and amount of lean.

> **WARNING:** If the tree is dead and leaning substantially, do not try to fell it without professional help. Many dead, leaning trees have a tendency to split along their length, causing a massive slab to fall spontaneously.

c. Find an opening into which the tree can fall in the direction of lean or as close to the direction of lean as possible.

Because most trees lean downhill, try to direct the fall downhill. If the tree leans slightly away from your desired direction, use wedges to direct the fall.

2. *How to Provide an Escape Path*
Falling trees are unpredictable. Avoid injury by planning a definite escape path. Follow these steps:

a. Locate the path in the direction opposite of the fall (Figure 1).

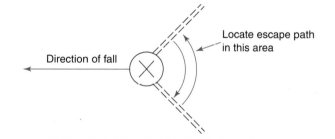

FIGURE 1 Escape-path Location

b. Clear a path 2 feet wide extending beyond where the top of the tree could land.

3. *How to Make the Undercut*
The undercut is a triangular slab of wood cut from the trunk on the side toward which you want the tree to fall. Follow these steps:

a. Start the chainsaw.

b. Holding the saw with blade parallel to the ground, make a first cut 2 to 3 feet above ground. Cut horizontally, to no more than 1/3 the tree's diameter (Figure 2).

Figure 19.3   A Set of Instructions *Continued*

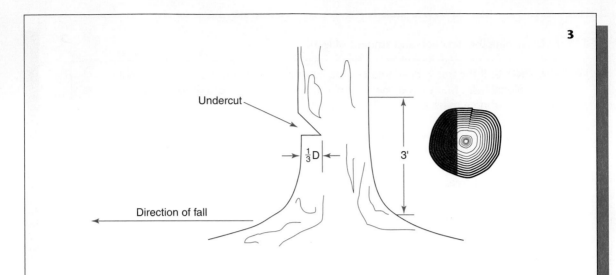

FIGURE 2 Making the Undercut

   c. Make a downward-sloping cut, starting 4 to 6 inches above the first so that the cuts intersect at 1/3 the diameter (Figure 2).

4. *How to Make the Backcut*
   After completing the undercut, you make the backcut to sever the stem from the stump. This step requires good reflexes and absolute concentration.

> **WARNING:** Observe tree movement closely during the backcut. If the tree shows any sign of falling in your direction, drop everything and move out of its way.
>   Also, do not cut completely through to the undercut; instead, leave a narrow strip of "holding wood" as a hinge, to help prevent the butt end of the falling tree from jumping back at you.

   To make your backcut, follow these steps:

   a. Holding the saw with the blade parallel to the ground, start your cut about 3 inches above the undercut, on the opposite side of the trunk. Leave a narrow strip of "holding wood" (Figure 3).

Figure 19.3   A Set of Instructions *Continued*

4

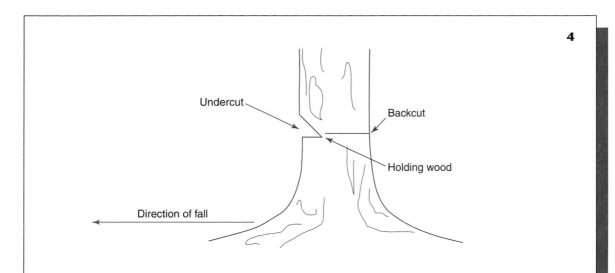

FIGURE 3 Making the Backcut

If the tree begins to bind the chainsaw, hammer a wedge into the backcut with the blunt side of the axe head. Then continue cutting.

b. As soon as the tree begins to fall, turn off the chainsaw, withdraw it, and step back immediately—the butt end of the tree could jump back toward you.

c. Move rapidly down the escape path.

CONCLUSION

Felling is a complex and dangerous procedure. Choosing the lay, providing an escape path, making the undercut, and making the backcut are the basic steps—but trees, terrain, and other circumstances vary greatly.

For the safest operation, seek professional advice and help whenever you foresee *any* complications whatsoever.

Figure 19.3   A Set of Instructions *Continued*

Like instructions, a process analysis must be detailed enough to enable readers to follow the process step by step. Because it emphasizes the process itself, rather than the reader's role, process analysis is written in the third person. Instead of leading to immediate action, process analysis helps readers gain understanding. Format and paragraph structure, therefore, resemble those in a traditional essay.

Much of your college writing explains how things happen. Your audience is the professor, who will evaluate what you have learned. Because this informed reader knows *more* than you do about the subject, you often discuss only the main points, omitting details.

But your real challenge comes in analyzing a process for readers who know *less* than you, and who are neither willing nor able to fill in the blank spots; you then become the teacher, and your readers become the students.

For an uninformed audience, introduce your analysis by telling what the process is, and why, when, and where it happens. In the body, tell how it happens, analyzing each stage in sequence. In the conclusion, summarize the stages, and describe one full cycle of the process.

Sections from the following general outline can be adapted to any process analysis:

I. INTRODUCTION
   A. Definition, Background, and Purpose of the Process
   B. Intended Audience (usually omitted for workplace readers)
   C. Prior Knowledge Needed to Understand the Process
   D. Brief Description of the Process
   E. Principle of Operation
   F. Special Conditions Needed for the Process to Occur
   G. Definitions of Special Terms
   H. Preview of Stages

II. STAGES IN THE PROCESS
   A. First Major Stage
      1. Definition and purpose
      2. Special conditions needed for the specific stage
      3. Substages (if applicable)
         a.
         b.
   B. Second Stage (and so on)

III. CONCLUSION
   A. Summary of Major Stages
   B. One Complete Process Cycle

Adapt this outline to the structure of the process you are explaining.

Following is a document patterned after the sample outline. Our writer, Bill Kelly, belongs to an environmental group studying the problem of acid rain in their Massachusetts community. (Massachusetts is among the states most affected by acid rain.) To gain community support, the environmentalists must educate citizens about the problem. Bill's group is publishing and mailing a series of brochures. The first brochure explains how acid rain is formed.

A Process Analysis for Nontechnical Readers

AUDIENCE AND USE PROFILE. My audience will consist of general readers. Some already will be interested in the problem; others will have no awareness (or interest). Therefore, I'll keep my explanation at the lowest level of technicality (no chemical formulas, equations). But my explanation needs to be vivid enough to appeal to less aware or less interested readers. I'll use visuals to create interest and to illustrate simply. To give an explanation thorough enough for broad understanding, I'll divide the process into three chronological steps: how acid rain develops, spreads, and destroys. ∎

HOW ACID RAIN DEVELOPS, SPREADS, AND DESTROYS

Introduction

Definition

Acid rain is environmentally damaging rainfall that occurs after fossil fuels burn, releasing nitrogen and sulfur oxides into the atmosphere. Acid rain, simply stated, increases the acidity level of waterways because these nitrogen and sulfur oxides combine with the air's normal moisture. The resulting rainfall is far more acidic than normal rainfall. Acid rain is a silent threat because its effects, although slow, are cumulative. This analysis explains the cause, the distribution cycle, and the effects of acid rain.

Purpose

Brief description of the process

Most research shows that power plants burning oil or coal are the primary cause of acid rain. The burnt fuel is not completely expended, and some residue enters the atmosphere. Although this residue contains several potentially toxic elements, sulfur oxide and, to a lesser extent, nitrogen oxide are the major problem, because they are transformed when they combine with moisture. This chemical reaction forms sulfur dioxide and nitric acid, which then rain down to earth.

Preview of stages

The major steps explained here are (1) how acid rain develops, (2) how acid rain spreads, and (3) how acid rain destroys.

The Process

How Acid Rain Develops

Once fossil fuels have been burned, their usefulness is over. Unfortunately, it is here that the acid rain problem begins.

First stage

Fossil fuels contain a number of elements that are released during combustion. Two of these, sulfur oxide and nitrogen oxide, combine with normal moisture to produce sulfuric acid and nitric acid. (Figure 1 illustrates how acid rain develops.) The released gases undergo a chemical change as they combine with atmospheric ozone and water vapor. The resulting rain or snowfall is more acid than normal precipitation.

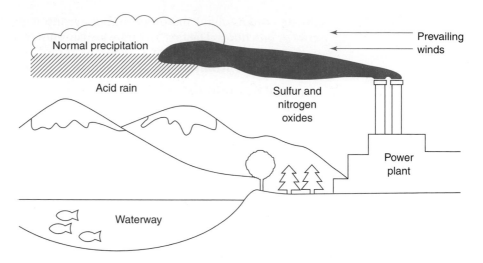

FIGURE 1 **How Acid Rain Develops**

Acid level is measured by pH readings. The pH scale runs from 0 through 14—a pH of 7 is considered neutral. (Distilled water has a pH of 7.) Numbers above 7 indicate increasing degrees of alkalinity. (Household ammonia has a pH of 11.) Numbers below 7 indicate increasing acidity. Movement in either direction on the pH scale, however, means multiplying by 10. Lemon juice, which has a pH value of 2, is 10 times more acidic than apples, which have a pH of 3, and is 1,000 times more acidic than carrots, which have a pH of 5.

Because of carbon dioxide (an acid substance) normally present in air, unaffected rainfall has a pH of 5.6. At this time, the pH of precipitation in the northeastern United States and Canada is between 4.5 and 4. In Massachusetts, rain and snowfall have an average pH reading of 4.1. A pH reading below 5 is considered to be abnormally acidic, and therefore a threat to aquatic populations.

How Acid Rain Spreads

Although it might seem that areas containing power plants would be most severely affected, acid rain can in fact travel thousands of miles from its source. Stack gases escape and drift with the wind currents. The sulfur and nitrogen oxides are thus able to travel great distances before they return to earth as acid rain.

For an average of two to five days after emission, the gases follow the prevailing winds far from the point of origin. Estimates show that about 50 percent of the acid rain that affects Canada originates in the United States; at the same time, 15 to 25 percent of the U.S. acid rain problem originates in Canada.

The tendency of stack gases to drift makes acid rain a widespread menace. More than 200 lakes in the Adirondacks, hundreds of miles from any industrial center, are unable to support life because their water has become so acidic.

Third stage

Substage

Substage

How Acid Rain Destroys

Acid rain causes damage wherever it falls. It erodes various types of building rock such as limestone, marble, and mortar, which are gradually eaten away by the constant bathing in acid. Damage to buildings, houses, monuments, statues, and cars is widespread. Some priceless monuments and carvings already have been destroyed, and even trees of some varieties are dying in large numbers.

More important, however, is acid rain damage to waterways in the affected areas. (Figure 2 illustrates how a typical waterway is infiltrated.) Because of its high acidity, acid rain dramatically lowers the pH in lakes and streams. Although its effect is not immediate, acid rain eventually can make a waterway so acidic it dies. In areas with natural acid-buffering elements such as limestone, the dilute acid has less effect. The northeastern United States and Canada, however, lack this natural protection, and so are continually vulnerable.

The pH level in an affected waterway drops so low that some species cease to reproduce. In fact, a pH level of 5.1 to 5.4 means that fisheries are threatened; once a waterway reaches a pH level of 4.5, no fish reproduction occurs. Because each creature is part of the overall food chain, loss of one element in the chain disrupts the whole cycle.

In the northeastern United States and Canada, the acidity problem is compounded by the runoff from acid snow. During the cold winter months, acid

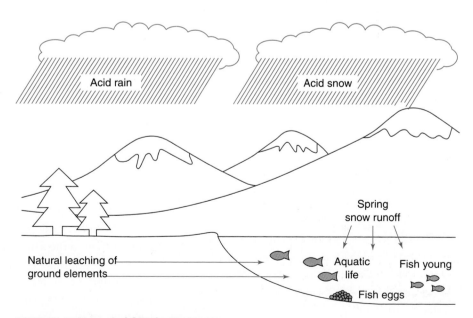

FIGURE 2 How Acid Rain Destroys

snow sits with little melting, so that by spring thaw, the acid released is greatly concentrated. Aluminum and other heavy metals normally present in soil are also released by acid rain and runoff. These toxic substances leach into waterways in heavy concentrations, affecting fish in all stages of development.

Summary

One complete cycle

Acid rain develops from nitrogen and sulfur oxides emitted by industrial and power plants burning fossil fuels. In the atmosphere, these oxides combine with ozone and water to form acid rain: precipitation with a lower-than-average pH. This acid precipitation returns to earth many miles from its source, severely damaging waterways that lack natural buffering agents. The northeastern United States and Canada are the most severely affected areas in North America.

Revision Checklist for Instructions

Use this checklist to evaluate the usability of instructions. (Numbers in parentheses refer to the first page of discussion.)

CONTENT

❑ Does the title promise exactly what the instructions deliver? (434)

❑ Is the background adequate for the intended audience? (438)

❑ Do explanations enable readers to understand what to do? (435)

❑ Do examples enable readers to see how to do it correctly? (439)

❑ Are the definition and purpose of each step given as needed? (438)

❑ Is all needless information omitted? (439)

❑ Are all obvious steps omitted? (439)

❑ Do notes, cautions, or warnings appear whenever needed, before the step? (441)

❑ Is the information rate appropriate for the reader's abilities and the difficulty of this procedure? (439)

❑ Are visuals adequate for clarifying the steps? (435)

❑ Do visuals repeat prose information whenever necessary? (439)

❑ Is everything accurate? (435)

ORGANIZATION

❑ Is the introduction adequate without being excessive? (449)

❑ Do the instructions follow the exact sequence of steps? (441)

❑ Is all the information for a particular step close together? (441)

❑ For a complex step, does each sentence begin on a new line? (443)

❑ Is the conclusion necessary and, if necessary, adequate? (452)

STYLE

❑ Do introductory sentences have enough variety to maintain interest? (442)

❑ Does the familiar material appear *first* in each sentence? (443)

❑ Do steps generally have short sentences? (443)

❑ Does each step begin with the action verb? (443)

❑ Are all steps in the active voice and imperative mood? (443)

❑ Do all steps have parallel phrasing? (444)

❑ Are transitions adequate for marking time and sequence? (442)

| Revision Checklist for Instructions *Continued* |

PAGE DESIGN
- ❏ Does each heading clearly tell readers what to expect? (444)
- ❏ On a typed page, are steps single-spaced within, and double-spaced between? (444)
- ❏ Do white space and highlights set off discussion from step? (444)

- ❏ Are notes, cautions, or warnings set off or highlighted? (445)
- ❏ Are visuals beside or near the step, and set off by white space? (445)

✔ EXERCISES

1. Improve the readability of these instructions by revising them for a more appropriate voice and design.

   What to Do Before Jacking Up Your Car
   Whenever the misfortune of a flat tire occurs, some basic procedures should be followed before the car is jacked up. If possible, your car should be positioned on as firm and level a surface as is available. The engine has to be turned off; the parking brake should be set; and the automatic transmission shift lever must be placed in "park," or the manual transmission lever in "reverse." The wheel diagonally opposite the one to be removed should have a piece of wood placed beneath it to prevent the wheel from rolling. The spare wheel, jack, and lug wrench should be removed from the luggage compartment.

2. Select part of a technical manual in your field or instructions for a general reader and make a copy of the material. Using the revision checklist, evaluate the sample's usability. In a memo to your instructor, discuss the strong and weak points of the instructions. Or be prepared to explain to the class why the sample is effective or ineffective.

3. Select a specialized process that you understand well and that has several distinct steps. Using the process analysis on pages 459–62 as a model, explain this process to classmates who are unfamiliar with it. Begin by completing an audience and use profile (page 459). Some possible topics: how the body metabolizes alcohol, how economic inflation occurs, how the federal deficit

affects our future, how a lake or pond becomes a swamp, how a volcanic eruption occurs.

4. Choose a topic from this list, your major, or an area of interest. Using the general outline in this chapter as a model, outline instructions for a procedure that requires at least three major steps. Address a general reader, and begin by completing an audience and use profile. Include (a) all necessary visuals or (b) a brief (page 333) and a rough diagram for each visual or (c) a "reference visual" (a copy of a visual published elsewhere) with instructions for adapting your visual from that one. (If you borrow visuals from other sources, provide full documentation.)

planting a tree	hitting a golf ball
hot-waxing skis	removing the rear
hanging wallpaper	wheel of a bicycle
filleting a fish	avoiding hypothermia

5. Assume you are assistant to the communications manager for a manufacturer of outdoor products. Among the company's best-selling items are its various models of gas grills. Because of fire and explosion hazards, all grills must be accompanied by detailed instructions for safe assembly, use, and maintenance.

   One of the first procedures in the manual is the "leak test," to ensure that the gas supply-and-transport apparatus is leak free. One of the engineers has prepared the instructions in Figure 19.4. Before being published in the manual, they must be approved by communications management. Your boss directs you to evaluate the instructions for accuracy, completeness, clarity, and appropriateness, and to report your findings in a memo. Because of the legal implications here, your evaluation must spell out all positive and negative details of content, organization, style, and format. (Use the page 462 Revision Checklist as a guide.) The boss is a busy and impatient reader, and expects your report to be no longer than two pages. Do the evaluation and write the memo.

6. Select any one of the instructional visuals in Figure 19.1 and write a prose version of those instructions—without using visual illustrations or special page design. Bring your version to class and be prepared to discuss the conclusions you've derived from this exercise.

7. Locate an example of five or more visuals from the following list.

A visual that shows:
- how to locate something
- how to operate something
- how to handle something
- how to assemble something
- how to position something
- how to avoid damage or injury
- how to diagnose and solve a problem
- how to identify safe or acceptable limits
- how to proceed systematically
- how to make the right decision
- why an action or procedure is important

Bring your examples to class for discussion, evaluation, and comparison.

☑ COLLABORATIVE PROJECTS

1. Select a specialized process you understand well (how gum disease develops, how an earthquake occurs, how steel is made, how a computer compiles and executes a program). Write a brief explanation of the process. Include visuals, as stipulated in Exercise 4 above. Exchange your analysis with a classmate in another major. Study your classmate's explanation for fifteen minutes and then write the same explanation *in your own words,* referring to your classmate's paper as needed. Now, evaluate your classmate's version of your original explanation for accuracy. Does it show that your explanation was understood? If not, why not? Discuss your conclusions in a memo to your instructor, submitted with all samples.

2. Draw a map of the route from your classroom to your dorm, apartment, or home—whichever is closest. Be sure to include identifying landmarks. When your map is completed, write instructions for a classmate who will try to duplicate your map from the information given in your written instructions. Be sure your classmate does not see your map! Exchange your instructions and try to duplicate your classmate's map. Compare your results with the original map. Discuss your conclusions about the usability of these instructions.

3. Divide into small groups and visit your computer center, library, or any place on campus or at work where you can find operating manuals for computers, lab or office equipment, or the like. (Or look through the documentation for your own computer hardware or software.) Locate fairly brief instructions that could use revision for improved content, organization, style, or format. Choose instructions for a procedure you are able to carry out. Make a copy of the instructions, test them for usability, and revise as needed. Submit all materials to your instructor, along with a memo explaining the improvements. Or be prepared to discuss your revision in class.

4. Visit your computer center and ask to borrow a software package that arrived without documentation or with very limited instructions. (Or perhaps you've received such programs as a member of a software club). Run the program, and then prepare instructions for the next users. Test the usability of your instructions by having a classmate use them to run the program. Revise as needed. Remember, you are writing for a user with no experience.

5. Using a common word processor or desktop publishing software, design a form to be used for advisee course scheduling, course evaluations, or some other school function. Conduct a usability study for this document and redesign it as needed. Then write a report analyzing and evaluating the form before and after the usability study.

Tank must be filled prior to this step (See "Propane Tank" section for details about filling this tank)

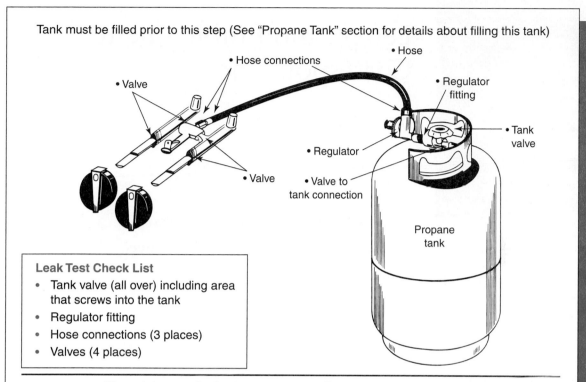

Leak Test Check List

• Tank valve (all over) including area
  that screws into the tank
• Regulator fitting
• Hose connections (3 places)
• Valves (4 places)

SAFETY! **All models must be Leak Tested! Take the hose. valve, and regulator assembly outdoors in a well ventilated area. Keep away from open flames or sparks. Do Not smoke during this test. Use only a soap and water solution to test for leaks.**

1. Have propane tank filled with propane gas only by a reputable propane gas dealer.
2. Attach regulator fitting to tank valve.
   ■ Regulator fitting has left-hand threads. Turn counterclockwise to attach.
3. Tighten regulator fitting securely with wrench.
4. Place the two matching control knobs onto the valve stems.
5. Turn the control knobs to the right, clockwise. This is the **"Off"** position.
6. Make a solution of half liquid detergent and half water.

7. Turn gas supply "ON" at the tank valve (counterclockwise).
8. Brush soapy mixture on all connections listed in the **Leak Test Check List.**
9. Observe each place for bubbles caused by leaks.
10. Tighten any leaking connections, if possible.
    ■ If leak cannot be stopped. **Do Not Use Those Parts!** Order new parts.
11. Turn gas supply **"Off"** at the tank valve.
12. Push in and turn control knobs to the left ("HI" position) to release pressure in hose.
13. Disconnect the regulator fitting from the tank valve.
    ■ Regulator fitting has left-hand threads. Turn fitting clockwise to disconnect.
14. Leave tank outdoors and return to the grill assembly with valve, hose, and regulator.

Figure 19.4   Instructions for Leak Testing a Grill. *Source: Instructions reprinted by permission of Thermos® Division.*

Letters and Employment Correspondence

A REPORT may be compiled by a team of writers for multiple readers, but a letter usually is written by one writer for one or more definite readers. Because a letter is more personal than a report and often has a persuasive purpose, proper tone is essential; you want the reader to be on your side. Because your signature indicates your approval—and responsibility—for the contents of the letter (which may serve as a legal document), precision is crucial.

This chapter covers three common letter types: inquiry letters, claim letters, and letters of application, along with résumés. Some other types are discussed in Chapters 16 and 22.

Standard Elements of Workplace Letters

Business and professional letters typically are laid out according to a conventional and familiar arangement that enables readers to identify specific parts immediately (Figure 20.1).

Basic Parts of Letters

A typical business letter has six parts, in order from top to bottom: heading, date, inside address, salutation, the letter text (introduction, body, conclusion), complimentary closing, and signature.

Heading and Date. If your stationery has a company letterhead, simply include the date two lines below the letterhead, at the right or left margin. On blank stationery, include your return address and the date (but not your name):

Street address	154 Sea Lane
City, state, zip code	Harwich, MA 02163
Month, Day, Year	July 15, 19xx

Use the Postal Service's two-letter state abbreviations (MA for Massachusetts, VT for Vermont) in your heading, inside address, and on the envelope. Depending on the letter's length, place the return address at least eight line spaces below the top of your page.

Inside Address. Two to six line spaces below the heading and abutting the left margin is the inside address.

Name and position	Dr. Ann Mello, Dean of Students
Company name	Western University
Street address	30 Mogul Hill Road
City, state, zip code	Stowe, VT 51350

Whenever possible, address your letter to a specifically named reader, and include your reader's job title, if any. Using a form of address such as "Mr." or "Ms." before the name is optional. (See page 469 for avoiding sexist usage in titles and salutations.)

Note: Depending on the length of your letter, adjust the horizontal placement of your return and inside address to achieve a page that appears balanced.

Heading

LEVERETT LAND & TIMBER COMPANY, INC.

creative land use
quality building materials
architectural construction

Date

January 17, 19XX

Inside
Address

Mr. Thomas E. Muffin
Clearwater Drive
Amherst, Massachusetts 01002

Salutation

Dear Mr. Muffin:

Letter
Text

I have examined the damage to your home caused by the ruptured water pipe
and consider the following repairs to be necessary and of immediate concern:

Exterior:
  Remove plywood soffit panels beneath overhangs
  Replace damaged insulation and plumbing
  Remove all built-up ice within floor framing
  Replace plywood panels and finish as required

Northeast Bedroom—Lower Level:
  Remove and replace all sheetrock, including closet
  Remove and replace all door casings and baseboards
  Remove and repair windowsill extensions and moldings
  Remove and reinstall electric heaters
  Respray ceilings and repaint all surfaces

This appraisal of damage repair does not include repairs and/or replacements
of carpets, tile work, or vinyl flooring. Also, this appraisal assumes that the
plywood subflooring on the main level has not been severely damaged.

Leverett Land & Lumber Company, Inc. proposes to furnish the necessary
materials and labor to perform the described damage repairs for the amount of
four thousand one hundred and eighty dollars ($4,180).

Complimentary
Closing
Signature

Sincerely,

P.A. Jackson

Title

Gerald A. Jackson, President

Typist's Initials

GAJ/ob

Enclosure
Notation

Enc. Itemized estimate

Figure 20.1   A Standard Design for a Workplace Letter

Salutation. The salutation, two line spaces below the inside address, begins with *Dear* and ends with a colon (*Dear Ms. Smith:*). If you don't know the person's name, use the position title (*Dear Manager*) or, preferably, an attention line (page 470). Address your reader by first name only if that is the way you would address this individual in person.

Salutations

> Dear Ms. Smith:
>
> Dear Managing Editor:
>
> Dear Professor Lexington-Trudeau:

No satisfactory guidelines exist for addressing several people within an organization. *Gentlemen* or *Dear Sirs* shows implied bias. *Ladies and Gentlemen* sounds too much like the beginning of a speech. *Dear Sir or Madam* is old-fashioned. *To Whom It May Concern* is vague and impersonal. Your best bet is to eliminate the salutation by using an attention line.

Letter Text. The text of your letter begins two line spaces below the salutation. Workplace letters typically include (1) a brief *introduction* paragraph (five lines or fewer) that identifies you and your purpose; (2) one or more *body* paragraphs that contain the details of your message; (3) a *conclusion* paragraph that sums up and encourages action.

Keep the paragraphs short (usually fewer than eight lines). If your body section is long, divide it into shorter paragraphs, as on page 477. Notice that the body section in that letter is broken down into four questions for easy answering.

Complimentary Closing. The closing, two line spaces below the last line of text, should coincide with the level of formality used in the salutation and should reflect your relationship to the addressee (polite but not overly intimate). These possibilities are ranked in decreasing order of formality:

Complimentary closings

> Respectfully,
>
> Sincerely, *(most often used)*
>
> Cordially,
>
> Best wishes,
>
> Warmest regards,
>
> Regards,
>
> Best,

In a modified block letter (Figure 20.2), align the closing with your heading or the date. In a full block letter (Figure 20.3), position the closing flush with the left margin.

Signature. Type your full name and title on the fourth and fifth lines below and aligned with the closing. Sign your name in the triple space between the two.

The signature block

> Sincerely yours,
>
> *Martha S. Jones*
>
> Martha S. Jones
> Personnel Manager

If you are writing as a representative of a company or group that bears legal responsibility for the correspondence, type the company's name in full caps two line spaces below your complimentary closing; place your typed name and title four line spaces below the company name, and sign in the triple space between.

Signature block representing the company

Yours truly,

HASBROUCK LABORATORIES

Lester Fong

L. H. Fong
Research Associate

Specialized Parts of Letters

Some letters require one or more of the following specialized parts. Examples appear in sample letters throughout the text.

Attention Line. Use an attention line when you write to an organization and do not know your reader's name but are directing the letter to a specific department or position.

An attention line can replace your salutation

Glaxol Industries, Inc.
232 Rogaline Circle
Missoula, MT 61347

ATTENTION: Director of Research and Development

Drop two line spaces below the inside address, and place the attention line either flush with the left margin or centered on the page.

Subject Line. Because it announces the topic of your letter, the subject line is a good device for attracting a busy person's attention.

A subject line can attract attention

SUBJECT: Placement of the Subject Line

Place the subject line below the inside address or attention line. Use one line space above and below. Underline the subject or highlight it typographically, for instance, by boldface.

Typist's Initials. If someone types your letter, your initials (in caps) and your typist's (in lowercase letters) should appear below the typed signature, flush with the left margin.

Typist's initials

JJ/pl

Some companies omit the writer's or dictator's initials because of the repetition from the signature block.

Enclosure Notation. When other documents accompany your letter, add an enclosure notation one line below the typist's initials, flush with the left margin. Indicate the number of enclosures.

Enclosure noted

> Enclosure
>
> Enclosures 2
>
> Encl. 3

If the enclosures are important documents such as legal certificates, checks, or specifications, name them in the notation.

Enclosure named

> Enclosures: 2 certified checks, 1 set of KBX plans.

An enclosure notation reminds whoever prepares your letter for mailing to enclose the appropriate material, and it alerts your reader.

Distribution Notation. If you distribute copies of your letter to other readers, so indicate by inserting the notation *CC, cc, Cc* (for *carbon copy*) or *Copy* one line space below any enclosure notation. Since most copies are now photocopies, some writers use the notation *pc* or *c.*

Distribution notations

> cc: office file
>
>   Melvin Blount
>
>
> c: S.Furlow
>
>   B.Smith

Most copies are distributed on an *FYI* (*For Your Information*) basis, but writers sometimes use the distribution notation to maintain a paper trail or to signal the primary reader that this information is being shared with others (e.g., superiors, legal authorities).

Postscript. A postscript (typed or handwritten) draws attention to a point you wish to emphasize or adds a personal note. Do not use a postscript if you forget to mention a point in the body of the letter. Rewrite the letter instead.

A postscript

> P. S. Because of its terminal position in your letter, a postscript can draw attention to a point that needs reemphasizing.

Place the postscript two line spaces below any other notation, and flush with the left margin. Because readers often regard postscripts as salesletter gimmicks, use them sparingly in your professional correspondence.

Design Factors

Although several formats are acceptable, and your company may have its own, the two most popular formats for workplace letters are *modified block* and *block.*

In the modified block form, the first line of a paragraph is not indented. Paragraphs are separated by a line space. The return address, complimentary closing, and signature align slightly to the right of page center (Figure 20.2).

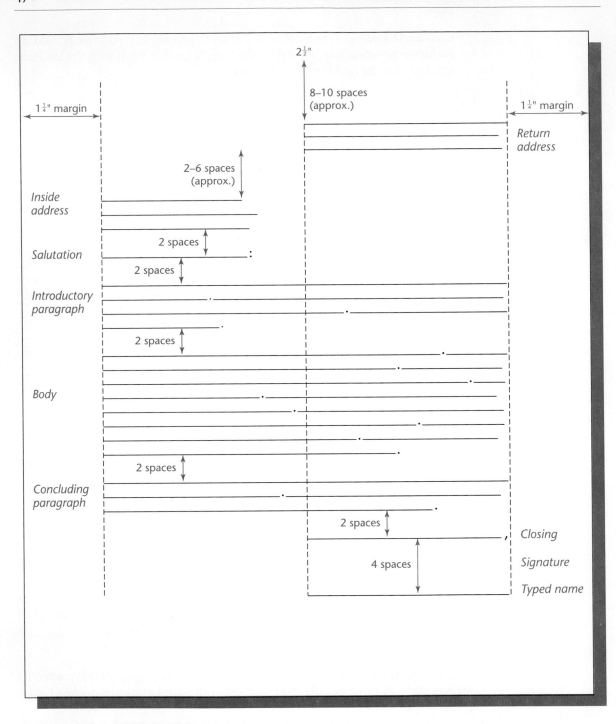

Figure 20.2   Modified Block Letter

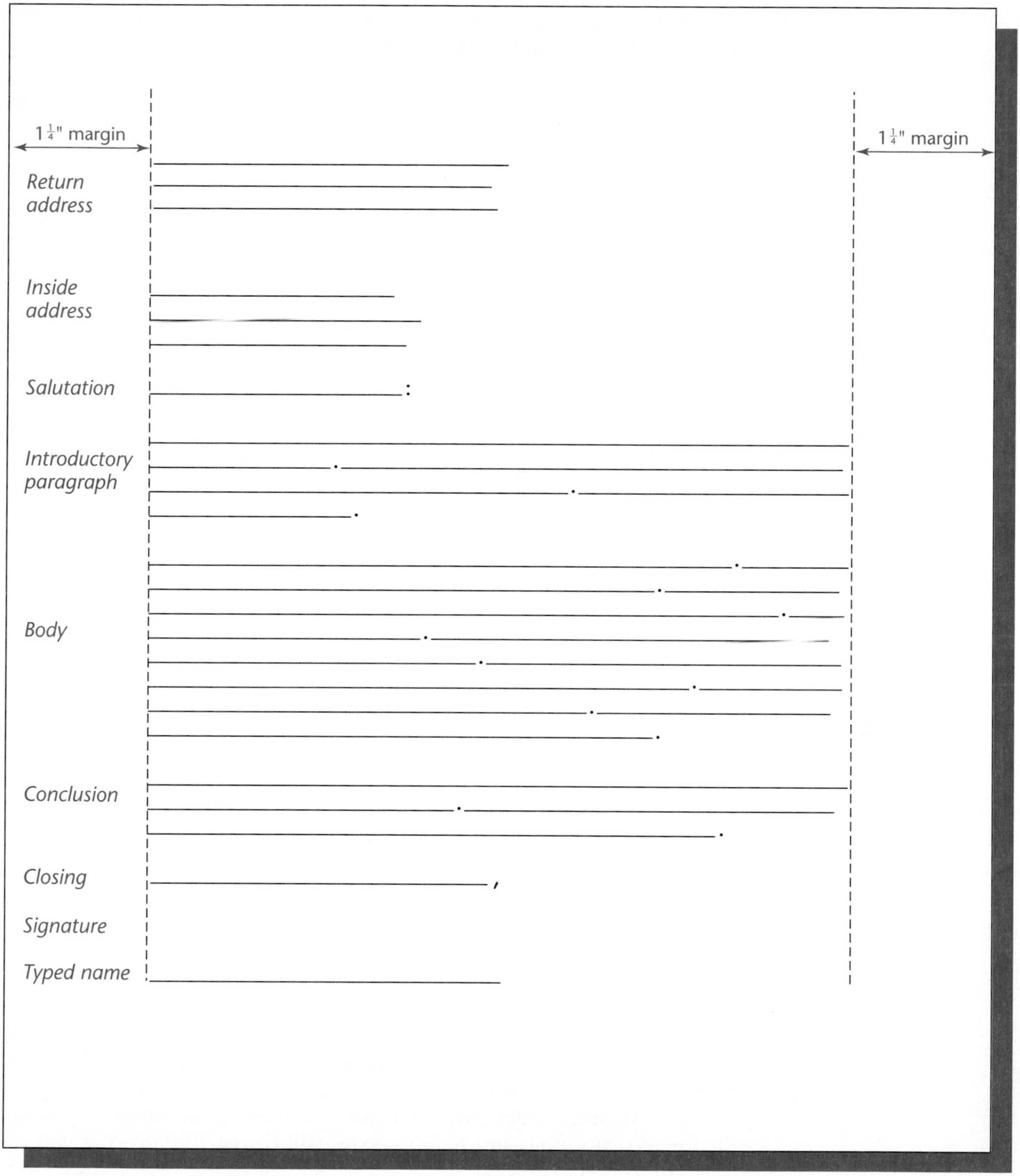

$1\frac{1}{4}$" margin

$1\frac{1}{4}$" margin

Return address

Inside address

Salutation

Introductory paragraph

Body

Conclusion

Closing

Signature

Typed name

Figure 20.3   Block Letter

In the block form, every line begins at the left margin (Figure 20.3). This form is popular because it looks businesslike and eliminates the need to tab and center.

Additional design factors enable workplace letters to appear inviting, accessible, and professional:

Quality Stationery. Whether you type or print your letters, use high-quality, 20-pound bond, $8^1/2$-by-11-inch, white stationery with a minimum fiber content of 25 percent. (Excellent stationery is available in continuous-feed form for letter-quality printers and as single sheets for laser printers.)

Uniform Margins and Spacing. When using stationery without a letterhead, frame your letter with $1^1/2$-inch top and side margins and bottom margins of 1 to $1^1/2$ inches. Use single spacing within paragraphs and double spacing between. Vary these guidelines to suit your needs, but strive for a balanced look.

Heading for Subsequent Pages. If your letter continues beyond a first page, begin each additional page with a notation identifying the addressee, date, and page number.

Subsequent-page heading

Adrianna Fonseca, June 25, 19xx, p. 2

Begin the text two spaces below this line. Never use an additional page solely for your complimentary closing and signature. Instead, reformat the letter so that at least two lines of final text appear above the closing.

The Envelope. Your $9^1/2$-by-$4^1/8$-inch envelope (also called a #10 envelope) should be of the same quality as your stationery. Place your reader's name and address block at a fairly central point on the envelope. Place your own name and address in the upper left corner. Single-space these elements.

Interpersonal Elements of Workplace Letters

In addition to presenting the reader with an accessible and inviting design, an effective letter enhances the relationship between writer and reader. Interpersonal elements are those that forge a *human* connection. Observe the following guidelines.

Focus on Your Reader's Interests: The "You" Perspective. In speaking face to face, you unconsciously modify your statements and expression as you read the listener's signals: a smile, a frown, a raised eyebrow, a nod. In a telephone conversation, a voice provides cues that signal approval, dismay, anger, or confusion. Writing a letter, however, carries a major disadvantage; you can easily forget that a flesh-and-blood person will be reacting to what you are saying—or seem to be saying.

A letter displaying a "you" perspective subordinates the writer's interests to those of the reader. In addition to focusing on what is important to the reader, the "you" perspective conveys respect for the reader's feelings and attitudes.

To achieve a "you" perspective, put yourself in the place of the person who will read your letter; ask yourself how the reader will react to what you have written. Even a single word, carelessly chosen, can offend. In writing to correct a billing error, for example, you might feel tempted to say this:

A needlessly offensive tone

> Our record keeping is very efficient and so this obviously is your error.

Such an accusatory tone might be appropriate after numerous failed attempts to achieve satisfaction on your part, but in your initial correspondence it will alienate the reader. Following is a more considerate version:

A tone that conveys the "you" perspective

> If my paperwork is wrong, please let me know and I will send you a corrected version immediately.

Instead of indicting the reader, this second version conveys respect for the reader's viewpoint.

Use Plain English. Workplace correspondence too often suffers from *letterese,* those tired, stuffy, and overblown phrases some writers think they need, to make their communications seem important. Here is a typically overwritten closing sentence:

Letterese

> Humbly thanking you in anticipation of your kind cooperation, I remain Faithfully yours,

Although no one *speaks* this way, some writers lean on such heavy prose instead of simply writing this:

Clear phrasing

> I will appreciate your cooperation.

Here are a few of the many old standards that are popular because they are easy to use but that make letters unimaginative and boring:

Letterese	*Plain-English*
As per your request	As you requested
Contingent upon receipt of	As soon as we receive
I am desirous of	I want, I would like
Please be advised that I	I
This writer	I
In the immediate future	Soon
In accordance with your request	As you requested
Due to the fact that	Because
I wish to express my gratitude.	Thank you.

Be natural. Write as you would speak in a classroom or office.

Anticipate Your Reader's Reaction. Like any effective writing, good letters do not just happen. Each is the product of a deliberate process. As you plan, write, and revise, answer these questions:

1. *What do I want the reader to do, think, or feel after reading this letter?* (offer me a job, give me advice or information, answer my inquiry, follow my instructions, grant me a favor, enjoy good news, accept bad news)

2. *What facts will my reader need?* (measurements, dates, costs, model numbers, enclosures, other details)

3. *To whom am I writing?* (Do you know the reader's name? When possible, write to a person, not a title.)

4. *What is my relationship to my reader?* (Is the reader a potential employer, an employee, a person doing a favor, a person whose products are disappointing, an acquaintance, an associate, a stranger?)

Answer those four questions *before* drafting the letter. After you have a draft, answer the following three questions, which pertain to the *effect* of your letter. Will readers be inclined to respond favorably?

5. *How will my reader react to what I've written?* (with anger, hostility, pleasure, confusion, fear, guilt, resistance, satisfaction)

6. *What impression of me will my reader get from this letter?* (intelligent, courteous, friendly, articulate, pretentious, illiterate, confident)

7. *Am I ready to sign my letter with confidence?* (Think about it.)

Mail the letter only when you have answered each question to your satisfaction.

Decide on a Direct or Indirect Plan. The reaction you anticipate from your reader should determine the organizational plan of your letter: either *direct* or *indirect*.

- Will the reader feel pleased or neutral about the message?
- Will the message cause resistance, resentment, or disappointment?

Each reaction calls for a different organizational plan. The direct plan puts the main point right in the first paragraph, followed by the explanation. Use the direct plan when you expect the reader to react with approval or when you want the reader to know immediately the point of your letter (e.g., in good-news, inquiry, or application letters—or other routine correspondence).

If you expect the reader to resist or to need persuading, consider an indirect plan. Give the explanation *before* the main point (as in refusing a request, admitting a mistake, or requesting a pay raise). An indirect plan might make readers more tolerant of bad news or more receptive to your argument.

Whenever you consider using an indirect plan, think carefully about its ethical implications. Never try to deceive the reader—and never create an impression that you have something to hide.

Inquiry Letters

Inquiry letters may be solicited or unsolicited. You often write the first type as a consumer requesting information about an advertised product. Such letters are welcomed because the reader stands to benefit. You can be brief: "Please send me your software price list."

Other inquiries will be unsolicited, that is, not in response to an ad, but requesting information for a report or project. Here, you are asking your reader to spend time reading your letter, considering your request, collecting the information, and writing a response. Always apologize for any imposition, express appreciation, and state a reasonable request clearly and briefly (long, involved inquiries are likely to go unanswered).

In order to ask specific questions, do your homework. Don't expect the respondent to read your mind. A general request ("Please send me all your data on . . .") is likely to be ignored.

Sample Situations

You are preparing an analytical report on the feasibility of harnessing solar energy for home heating in northern climates. You learn that a private, non-profit research group has been experimenting in solar energy systems. After deciding to write for details, you plan and compose your inquiry.

Tell the reader who you are and why you want information. Maintain the "you" perspective with opening statements that spark interest and goodwill.

In the body of your letter, write specific and clearly worded questions that are easy to understand and answer. If you have several questions, arrange them in a list. (Lists help readers organize their answers, increasing your chances of getting all the information you want.) Number each question, and separate it from the others, perhaps leaving space for responses right on the page. If you have more than five questions, consider placing them in an attached questionnaire.

Conclude by explaining how you plan to use the information and, if possible, how your reader might benefit. If you have not done so earlier, specify a date by which you need a response. Offer to send a copy of your finished report. Close with a statement of appreciation.

Your letter might resemble the one below.

Inquiry Letter

States the purpose

As a student at Evergreen College, I am preparing a report (April 15 deadline) on the feasibility of solar energy as a viable source of home heating in northern climates.

Makes a reasonable request
Presents a list of specific questions

While gathering data on home solar heating, I encountered references (in *Scientific American* and elsewhere) to your group's pioneering work in solar energy systems. Would you please allow me to benefit from your experience? I specifically would appreciate answers to these four questions:

1. At this stage of development, have you found active or passive solar heating more practical?

2. Do you hope to surpass the 60 percent limit of heating needs supplied by the active system? If so, what level of efficiency do you expect to achieve, and how soon?

3. What is the estimated cost of building materials for your active system, per cubic foot of living space?

4. What metal do you use in collectors to obtain the highest thermal conductivity at the lowest maintenance costs?

Tells how the material will be used, and offers to share findings

Your answers, along with any recent findings you can share, will enrich a learning experience I hope to put into practice after graduation by building my own solar-heated home. I would be glad to send you a copy of my report, along with the house plans I have designed. Thank you.

Sometimes your questions will be too numerous or involved to be answered by letter or questionnaire. You might then request an informative interview (if the respondent is nearby). Karen Granger, our next writer, sought a state representative's "opinions on the EPA's progress" in cleaning up local contamination. Anticipating a complex answer, she requested an interview.

Letter Requesting an Informative Interview

As a technical writing student at the University of Massachusetts, I am preparing a report evaluating the EPA's progress in cleaning up PCB contamination in New Bedford Harbor.

In my research, I have encountered your name repeatedly. Your dedicated work has had a definite influence on this situation, and I am hoping to benefit from your knowledge.

I was surprised to learn that, although this contamination is considered the most extensive anywhere, the EPA still has not moved beyond conducting studies. My own study questions the need for such extensive data gathering. Your opinion, as I can ascertain from *Standard Times* articles, is that the EPA definitely is moving too slowly.

The EPA refutes that argument by asserting they simply do not yet have the information necessary to begin a clean-up operation.

As both a writer and a New Bedford resident, I am very interested in your opinions on the EPA's progress. Could you find time in your busy schedule to grant me an interview? With your permission, I will phone in a few days to arrange an appointment. This interview would be an invaluable aid to me.

Your assistance would be deeply appreciated, and I would gladly send you a copy of my completed report.

Whenever you seek a written response to an unsolicited inquiry, include a stamped, self-addressed envelope for the reply.

Telephone and E-mail Inquiries

Unsolicited inquiries via telephone or E-mail can be efficient and productive but they also can be unwelcome and intrusive. A traditional letter implies greater respect for the respondent's privacy and provides a certain distance from which that person can contemplate a response—or even decide not to respond.

One alternative is to inquire by traditional letter and to invite a response by phone (collect) or E-mail, if your reader so prefers. Or, if you must inquire via telephone or E-mail, consider this suggestion: Establish a brief initial contact in which you apologize for any intrusion and then ask about the respondent's willingness to answer your questions at a convenient time in the near future. In this or any communication, we don't want to sacrifice goodwill in the interest of efficiency.

Claim Letters

Claim letters request adjustments for defective goods or poor services, or they complain about unfair treatment or the like. Claims can be routine or arguable. Routine claims follow the direct plan, because the claim is backed by a contract, guarantee, or the company's reputation. Arguable claims, on the other hand, are debatable, so they call for an indirect plan.

Routine Claims

In a routine claim, make your request or state the problem in your introductory paragraph; then explain in the body section. Close courteously, restating the action you request.

Make the tone courteous and reasonable. Your goal is not to express dissatisfaction but to achieve results: a refund, replacement, apology. Press your claim objectively yet firmly by explaining it clearly and by stipulating the *reasonable* action that will satisfy you.

Explain the problem in enough detail for a reader to understand the basis for your claim. Explain that your new alarm clock never rings, instead of merely saying it's defective. Identify the faulty item clearly, giving serial and model numbers, and date and place of purchase. Then propose a fair adjustment. Conclude by expressing goodwill and confidence in the reader's integrity.

The next writer does not ask whether the ski manufacturer will honor his claim; he assumes it will. He asks directly how to return the skis for repair. He uses an attention line to direct his claim to the right department. The subject line, and its re-emphasis in the first sentence, makes clear the nature of the claim.

Routine Claim Letter
Attention: <u>Consumer Affairs Department</u>

Subject: <u>Delaminated Skis</u>

States problem and action desired

This winter, my Tornado skis began to delaminate. I want to take advantage of your lifetime guarantee to have them relaminated.

Provides details	I bought the skis from the Ski House in Erving, Massachusetts, in November 1973. Although I no longer have the sales slip, I did register them with you. The registration number is P9965.
Explains basis for claim	I'm aware that you no longer make metal skis, but as I recall, your lifetime guarantee on the skis I bought was a major selling point. Only your company and one other were backing their skis so strongly.
Courteously states desired action	Would you please let me know how to go about returning my delaminated skis for repair?

Arguable Claims

When your request is in some way unusual, you must *persuade* the firm to grant your claim. Say your parked car is wrecked by a drunk driver, and the insurance company appraises the car at $1,500. But two months earlier, you had the engine rebuilt. By settling for the $1,500, you would lose $1,000, so you write a claim letter, explaining the circumstances and requesting a fair adjustment.

Use the indirect plan for an arguable claim. Readers are more likely to respond favorably *after* reading your explanation. Begin with a neutral statement both parties can agree to—but which also serves as the basis for your request (e.g., "Customer goodwill is often an intangible quality, but a quality that brings tangible benefits").

Once you've established agreement, explain and support your claim. Include enough information for a fair evaluation: date and place of purchase, order or policy number, dates of previous letters or calls, and background.

Conclude by requesting a *specific action* (a credit to your account, a replacement, a rebate). Ask confidently.

Our next writer employs a tactful, reasonable tone and an indirect plan.

	Arguable Claim Letter
Establishes early agreement	Your company has an established reputation as a reliable wholesaler of office supplies. For eight years we have counted on that reliability, but a recent episode has left us annoyed and disappointed.
Presents facts to support claim	On January 29, 1996, we ordered (#675198) five cartons of Maxell 3.5″ diskettes (#A74–866), eight cartons of Scotch 3.5″ diskettes (#A74–892), and three cartons of Epson MX 70/80 ribbon cartridges (#A19–556).
Offers more support	On February 5, the order arrived. But instead of double-sided, double-density Maxells ordered, we received single-sided Verbatim diskettes. Instead of Scotch 3.5″ diskettes, we received 8″ diskettes. And the Epson ribbons were blue, not the black we had ordered. We returned the order the same day.
Includes all relevant information	Also on the 5th, we called John Fitzsimmons at your company to explain our problem. He promised delivery of a corrected order by the 12th. Finally, on the 22nd we did receive an order—the original incorrect one—with a note claiming that the packages had been water damaged while in our possession.

Sticks to the facts—accuses no one

Our warehouse manager insists the packages were in perfect condition when he released them to the shipper. Because we had the packages only five hours, and had no rain on the 5th, we are certain the damage did not occur here.

Requests a specific adjustment

Responsibility for damages therefore rests with either the shipper or your warehouse staff. What bothers us is our outstanding bill from Hightone ($1,049.50) for the faulty shipment. We insist that the bill be canceled and that we receive a corrected statement. Until this misunderstanding, our transactions with your company were excellent. We hope they can be again.

Résumés and Job Applications

In today's job market, many applicants compete for few openings. Some large companies receive thousands of résumés yearly for a mere handful of job openings. Whether you are applying for your first professional job or changing careers, you must market your skills effectively. Your résumé and letter of application must stand out among the competition.

Job Prospecting

Begin your employment search by studying the job market to identify the careers and jobs for which you best qualify.

Evaluating Your Skills, Aptitudes, and Preferences. Focus on openings that suit your skills, aptitudes, and goals:

Identify your assets

- What skills have you acquired in school, on the job, through hobbies or other interests?
- Do you have skills in leadership or in group projects (as demonstrated in employment, social organizations, or extracurricular activities)?
- Do you speak a foreign language? Have musical or artistic talent?
- Do you have communication skills? Are you a good listener?
- What do you seek in a career: security, excitement, money, travel, power, prestige, or something else?

Besides helping focus your search, answers to these questions will be handy when you write your résumé and prepare for interviews.

Learning About the Job Market. Launch your search early. Don't wait for the job to come to you. Take the initiative by observing these suggestions:

Explore all resources

- Scan the Help Wanted section in major Sunday newspapers for job descriptions, salaries, and qualifications. The World Wide Web offers electronic access to job listings in major newspapers nationwide via CareerPath.com (http://www.careerpath.com). For more on electronic job hunting see *In Brief* (page 487).
- Ask a reference librarian to point out occupational handbooks, government publications, newsletters, and magazines or journals in your field.

- Visit your college placement service. Openings are posted there, interviews are scheduled, and counselors can provide good advice about job hunting.

- Sign up at your placement office for interviews with company representatives who visit the campus.

- Speak with someone in your field to get an inside view and some practical advice.

- Seek the advice of faculty in your major who do outside consulting or who have worked in business, industry, or government.

- Look for a summer job in your field; this experience may count as much as your education.

- Establish a network of contacts; don't be afraid to ask for advice. Make a list of names, addresses, and phone numbers of people willing to help.

- Many professional organizations invite student memberships (at reduced fees). Such affiliations can generate excellent contacts, and they look good on the résumé. If you do join a professional organization, try to attend meetings of the local chapter.

- Do whatever additional *networking* you can. Notice that several job-search methods in Figure 20.4 include talking with people and exploring useful contacts.

Increased use of E-mail and online job listings and résumé postings likely will have a major impact on job-search methods. (See page 487.)

By launching your search before your senior year, you may learn that certain courses make you more marketable. If you are changing jobs or careers, employers will be more interested in what you have accomplished *since* college. Be prepared to show how your experience is relevant to the new job. Capitalize on the network you have established.

You can also learn about career opportunities and the job market by consulting resources such as these at your library or newsstand:

Research potential employers

- *Fortune, Business Week, Forbes,* the *Wall Street Journal,* or trade publications in your field—for the latest developments and the big picture on business, economy, technology, and the like. Articles in these publications about specific topics and companies can be searched in the *Business Index.*

- *Occupational Outlook Handbook* and its quarterly update, *Occupational Outlook Quarterly,* published by the U.S. Department of Labor—for detailed descriptions of occupations, qualifications, employment prospects, salaries, and so on.

- *Almanac of Jobs and Salaries, Dun's® Employment Opportunities Directory,* or *Federal Career Opportunities*—for government and private-sector organizations that hire college graduates.

**Figure 20.4
How People
Usually Find Jobs**
Source: Tips for Finding the
Right Job. *U.S. Department
of Labor.*

Most Commonly Used Job-search Methods

Percent of Total Job-seekers Using the Method	Method	Effectiveness Rate*
66.0%	Applied directly to employer.....................................	47.7%
50.8	Asked friends about jobs where they work....................	22.1
41.8	Asked friends about jobs elsewhere...........................	11.9
28.4	Asked relatives about jobs where they work.................	19.3
27.3	Asked relatives about jobs elsewhere	7.4
45.9	Answered local newspaper ads................................	23.9
21.0	Private employment agency	24.2
12.5	School placement office...	21.4
15.3	Civil Service test	12.5
10.4	Asked teacher or professor.....................................	12.1
1.6	Placed ad In local newspaper	12.9
6.0	Union hiring hall...	22.2

* A percentage obtained by dividing the number of jobseekers who actually found work using the method, by the total number of jobseekers who tried to use that method, whether successfully or not.

- *Moody's Industrial Index* or Standard and Poor's *Register of Corporations*—for data on plant locations, major subsidiaries, products, executive officers, and corporate assets.
- Annual reports, for data on a company's assets, innovations, recent performance, and prospects. (Many libraries collect annual reports in a separate file cabinet. Ask your librarian.)

The previous listing offers a mere sampling of the resources—increasingly available in electronic versions—for motivated job-seekers.

Whether you are a beginner or a veteran, you might register with an employment agency. Of course, a fee is payable after you are hired, but employers often pay this fee. Ask about fee arrangements *before* you sign up.

Once you have a clear picture of where you fit into the job market, you will set out to answer the big question asked by all employers: *What do you have to offer?* Your answer must be a highly polished presentation of yourself, your education, work history, interests, and skills—in short, your résumé.

Preparing Your Résumé

The résumé is a summary of your experience and qualifications. Written before your application letter, the résumé provides background information to support your letter. This information supplies an employer with a one- or two-page reference. In turn, the letter will emphasize specific parts of your résumé and will discuss how your background is suited to a particular job.

Employers generally spend fifteen to forty-five seconds scanning a résumé initially. They look for an obvious and persuasive answer to this question: *What can you do for us?* Employers are impressed by a résumé that:

What employers expect in a résumé

1. looks good (conservative, tasteful, uncluttered, on quality paper)
2. reads easily (headings, typeface, spacing, and punctuation that provide clear orientation)
3. provides information the employer needs for making an interviewing decision

Employers generally discard résumés that are mechanically flawed, cluttered, sketchy, or hard to follow. Don't leave readers guessing or annoyed; make your résumé perfect.

Organize your information within these categories:

- name and address
- job and career objectives
- educational background
- work experience
- personal data
- interests, activities, awards, and skills
- references

Select and organize material to emphasize what you can offer. Don't just list *everything;* be selective. (We're talking about *communicating* instead of merely delivering information.) Don't abbreviate, because some readers may not know the referent. Use punctuation to clarify and emphasize, not to be "artsy." Try to limit your résumé to a single page, as most employers prefer. (Of course, if you are changing jobs or careers, or if your résumé looks cramped, you might need a second page.)

Begin your résumé well before your job search. You will need that much time to do a first-class job. Your final version can be duplicated for various similar targets—but each new type of job requires a new résumé that is tailored to fit the advertised demands of that job.

Caution: Never *invent* credentials. Your résumé should make you look as good as the facts allow. Distorting the facts is unethical and counterproductive. Companies routinely investigate claims made in a résumé, and people who have lied are fired.

Name and Address. As a heading, include your full name, mailing address, and phone number (many interview invitations and job offers are made by phone). If your school and summer address differ, include both, indicating dates when you can be reached at each.

Job and Career Objectives. From the information collected earlier (page 481), you should have a clear idea of the *specific* jobs for which you *realistically* qualify. Resist the impulse to be all things to all people. Be prepared to have different statements of objectives to meet the requirements of different job descriptions.

The key to a successful résumé is the image of *you* it projects—disciplined and purposeful, yet flexible. State your immediate and long-range goals, including any plans to continue your education:

Statement of career objective

> Intensive-care nursing in a teaching hospital, with the eventual goal of supervising and instructing.

Do not borrow a trite statement from some placement brochure. If applying to a company with various branches, express your willingness to relocate.

To save space, you can omit your statement of career objectives from the résumé and include it in your letter instead.

Educational Background. If your education is more substantial than your work experience, place it first for emphasis. Begin with your most recent school and work backward, listing degrees, diplomas, and schools attended *beyond* high school (unless the high school's prestige, its program, or your achievements warrant its inclusion). List the courses that have directly prepared you for the job you seek. If your class rank and grade point average are in the upper 30 percent, list them. Include any schools attended or courses completed while in military service. If you finance part or all of your education by working, say so, indicating the percentage of your contribution.

Work Experience. If you have solid experience that relates to the job you seek, list it before your education. Begin by listing your most recent job and work backward, listing each significant job you've held. Provide dates and names of employers. Indicate whether the job was full-time, part-time (hours weekly), or seasonal. Describe your exact duties in each job, indicating promotions. If the job was major (and related to this one), describe it in detail, and give your reason for leaving. Include military experience. If you have no real experience, show that you have potential by emphasizing your preparation and by writing an enthusiastic letter.

Complete sentences are unnecessary in résumés. They take up room best left for emphasis on your other qualifications. Use action verbs (*supervised, developed, built, taught, installed, managed, trained, solved, planned, directed*) to stress qualifications. If your résumé is likely to be scanned electronically, list key words as nouns (*leadership skills, software development, data processing, editing*) immediately below your name and address. (See *In Brief,* page 487.)

Personal Data. An employer cannot legally discriminate on the basis of sex, religion, race, age, national origin, disability, or marital status. Therefore, you aren't required to provide this information or a photograph. But if you believe that any of this information could advance your prospects, by all means include it.

Personal Interests, Activities, Awards, and Skills. List hobbies, sports, and other pastimes; memberships in teams and organizations; offices held; and any recognition you have received. Include dates and types of volunteer work.

Employers know that people with a well-rounded lifestyle are likely to take an active interest in their jobs. Be selective in this section. List only items that reflect the qualities employers seek.

References. Your list of references names at least three (five maximum) people *who have agreed* to write strong, positive assessments of your qualifications and personal qualities. Often a reference letter is the key to getting an employer to want to meet you; choose your references carefully.

Select references who can speak with authority about your ability and character. Avoid members of your family and close friends not in your field. Instead choose among previous employers, professors, and respected community figures who know you well enough to write *concretely* on your behalf.

A mediocre letter of reference is more damaging than no letter at all. Don't simply ask, "Could you please act as one of my references?" This question leaves the person little chance to refuse, but this person might not know you well or might be unimpressed by your work and therefore might write a watery letter that does more harm than good. Instead, make an explicit request: "Do you feel you know me and my work well enough to write a strong letter or reference? If so, would you act as one of my references?" This second version gives your respondent the option of declining gracefully or it elicits a firm commitment to a positive recommendation.

Detailed recommendations are time-consuming to write. Your references have no time to write individual letters to every prospective employer. Ask for only one letter, with no salutation. Your reference keeps a copy for her or his files; you keep the original for your personal dossier (so you can reproduce it as necessary); and a copy is sent to the placement office for inclusion in your placement dossier. Because the law permits you to read all material in your dossier, this arrangement provides you with your own copy of your credentials. (The dossier is discussed later in this chapter.)

Under some circumstances you may—if you wish to—waive the right to examine your recommendations. Some applicants, especially those applying to professional schools, as in medicine and law, waive this right in concession to a general feeling that a letter writer who is assured of confidentiality is more likely to provide a balanced, objective, and reliable assessment of a candidate. In your own case, seek the advice of your major adviser or a career counselor.

If people you select as references live elsewhere, you might make your request by letter, like this one:

Letter Requesting a Reference

From September 19xx to August 19xx, I worked at Teo's Restaurant as waiter, cashier, and then assistant manager. Because I enjoyed my work, I decided to study for a career in the hospitality field.

In three months I will graduate from San Jose City College with an A.A. degree in Hotel and Restaurant Management. Next month I begin my job search. Do you remember me and my work enough to write me a strong letter of recommendation? If so, would you kindly serve as one of my references?

IN BRIEF

Electronic Job Hunting

On-line Employment Resources

A 20-minute Web search via *Yahoo*™ or *Netscape*™ of the linked categories "Business and Economy: Employment: Jobs" yields resources like these: job listings in specific fields or locations, company profiles, employers who hire in particular specialties, résumé posting sites.

A sampling of Web sites: *Boston Job Bank, College Grad Job Hunter, Eurojobs on-line, HiTechCareers, Hospitality Industry Job Exchange, Jobs in Atomic and Plasma Physics, Positions in Bioscience and Medicine.*

The *Interactive Employment Network* lists specific help-wanted ads whenever the user types in a desired location or job category (Welz 52). At *CareerPath.com,* users scan employment listings from major newspapers (Stone and Gegax 16). Electronic recruiting centers such as *CareerWEB* match applicants with employers in engineering, marketing, data processing, and technical fields worldwide. These services provide guidelines for submitting electronic résumés and related information online.

Electronic Résumés

Today's electronic résumés usually are online versions of hardcopy résumés (with differences explained below). Increasingly these will appear as hypertext documents with links to the applicant's placement dossier, publications, and other support material (Welz 52). Multimedia résumés will incorporate sound, visuals, and animation.

Electronic Scanning of Résumés

Although present job hunting largely continues to depend on hardcopy résumés, these often are scanned electronically. Electronic storage of online or hardcopy résumés offers employers an efficient way to screen countless applicants; to compile a database of applicants, for later openings; and to evaluate all applicants fairly.

How Scanning Works. The computer captures and searches the printed image for key words (nouns instead of traditional "action verbs"). Those résumés containing the most key words ("hits") make the final cut (Pender 120).

How to Prepare a Scannable Résumé

Using nouns as key words, list all your skills, credentials, and job titles. (Help-wanted ads are a useful source for key words.)

- List specialized skills: *marketing, C++ programming, database management, user documentation, Internet collaboration, software development, graphic design, hydraulics, fluid mechanics, editing.*

- List general skills: *teamwork coordination, conflict management, oral communication, report and proposal writing, problem solving, troubleshooting, Spanish speaker.*

- List credentials: *student member, Institute of Electrical and Electronic Engineers, board-certified, B.S. Electrical Engineering, top 5 percent of class.*

- List job titles: *manager, director, supervisor, intern, coordinator, project leader.*

If you lack skills or experience, emphasize your personal qualifications: *analytical skills, energy, efficiency, flexibility, imagination, motivation.* In any résumé for the global marketplace, including *willingness to relocate and travel*—if indeed this is your sentiment.

In preparing a scannable résumé, use plain typeface, 10 to 14 point type, and white paper. Use **boldface** or FULL CAPS for emphasis. Avoid fancy fonts, italics, underlining, bullets, slashes, dashes, parentheses, or ruled lines (Pender 120). Avoid a two-column format, which often is jumbled by scanners that read across the page.

Continues on page 488

IN BRIEF

Submit the résumé in a large envelope so it will not be folded.

You may want to submit two versions of your résumé: one traditionally designed and one scannable—or include a keyword section, as in Figure 20.6.

Figure 20.7 shows a scannable version of Figure 20.6.

> To save your time, please omit the salutation from your letter. If you could send me the original, I will forward a copy to my college placement office.
>
> To update you on my recent activities, I've enclosed my résumé. If I can answer any questions, please contact me by phone (collect) at 123–555–4321 or by E-mail at JPURDY@AOL.com.
>
> Thank you for your help and support.

Opinion is divided about whether names and addresses of references should appear in a résumé. If saving space is important, simply state, "References available on request," keeping your résumé only one page long, but if your résumé already takes up more than one page, you probably should include names and addresses of references. (A prospective employer might recognize a name, and thus notice *your* name among the crowd of applicants.) If you are changing careers, a full listing of references is extremely important.

Organizing Your Résumé

Emphasize your assets

Organize your résumé in the order that conveys the strongest impression of your qualifications, skills, and experience. Depending on your background, you can arrange your material in reverse chronological order, functional order, or a combination of both.

Reverse Chronological Organization. In a reverse chronological résumé you list your most recent experience first, moving backward through your earlier experiences. Use this arrangement to show a pattern of job experiences or progress along a specific career track (as in Figure 20.5).

Functional Organization. In a functional résumé, you emphasize skills, abilities, and achievements that relate specifically to the job for which you are applying. Use this arrangement if you have limited job experience, gaps in your job record, or if you are changing careers.

Combined Organization. Most employers prefer résumés that are chronologically organized because these are easier to scan. However, electronic scanning of résumés (page 487) will call for a more functional pattern. One alternative

James David Purdy
203 Elmwood Avenue
San Jose, CA 95139
Tel.: 214-316-2419

Career Objective	Customer relations for a hospitality chain, leading to market management.

Education
1994–1997   *San Jose College, San Jose, CA*
Associate of Arts Degree in Hotel/Restaurant Management, June 1997.
Grade point average: 3.25 of a possible 4.00. All college expenses
financed by scholarship and part-time job (20 hours weekly).

Employment
1994–1997   *Peek-a-Boo Lodge, San Jose, CA*
Began as desk clerk and am now desk manager (part-time) of this 200-unit
resort. Responsible for scheduling custodial and room service staff,
convention planning, and customer relations.

1992–1994   *Teo's Restaurant, Pensacola, FL*
Beginning as waiter, advanced to cashier and finally to assistant manager.
Responsible for weekly payroll, banquet arrangements, and supervising
dining room and lounge staff.

1991–1992   *Encyclopaedia Britannica, Inc., San Jose, CA*
Sales representative (part-time). Received top bonus twice.

1990–1991   *White's Family Inn, San Luis Obispo, CA*
Worked as bus person, then waiter (part-time).

Personal   *Awards*
Captain of basketball team, 1992; Lion's Club Scholarship, 1994.

Special Skills
Speak French fluently: expert skier.

Activities
High school basketball and track teams (3 years); college student senate
(2 years); Innkeepers' Club—prepared and served monthly dinners at the
college (2 years).

Interests
Skiing, cooking, sailing, oil painting, and backpacking.

References   Placement Office, San Jose City College, San Jose, CA 90462

Figure 20.5   Résumé for an Entry-level Candidate (Reverse-Chronological Arrangement)

Karen P. Granger
82 Mountain Street
New Bedford, MA 02740
Telephone (617) 864-9318

Objective   A summer internship documenting microcomputer software.

Qualifications Software and hardware documentation. Editing. Desktop publishing. Usability testing. Computer science. Internet research. World Wide Web collaboration. Networking technology. Instructor-led training. IBM PC, Macintosh, DEC 20 mainframe and VAX 11/780 systems. Pagemaker, Netscape, Powerpoint, Excel, and Lotus Notes software. Logo, Pascal, HTML, and C++ program languages.

Education   Attending University of Massachusetts at Dartmouth; B.A. expected January 1998. Major: English/Communication. Minor: Computer Science. Dean's List, all semesters. GPA: 3.54. Class rank 110 of 1792.

Experience
Intern   **Conway Communications, Inc.,** 39 Wall Street, Marlboro, MA 02864.
Technical Writer   Learned local area network (LAN) technology and Conway's product line. Wrote, designed, and tested five hardware upgrade manuals. Produced a hardware installation/maintenance manual from another writer's work. Specified and approved all illustrations. Designed home page and hypertext links. Summers 1995, 1996.

Writing   **Writing/Reading Center,** UMD. Tutored writing and word processing
Tutor   for individuals and groups. Edited WRC student newsletter. Trained new tutors. Co-wrote and acted in a video about the WRC. Designed WRC home page for World Wide Web. Fall 1994–present.

Managing   **The Torch,** UMD Weekly Newspaper. Organized staff meetings, generated
Editor   story ideas, edited articles, and supervised page layout and pasteup. Fall 1995–present.

Achievements   Two writing samples published in Dr. John M. Lannon's *Technical Writing,* 7th ed. (Longman, 1997); Massachusetts State Honors Scholarship, 1994–1997.

Activities   Student member, Society for Technical Communication and American Soc. for Training and Development; student representative, College Curriculum Committee; UMD Literary Soc.

References   Placement Office, University of Massachusetts, No. Dartmouth, MA 02747

Figure 20.6   Résumé for a Summer-internship Candidate (Modified-Functional Arrangement)

KAREN P. GRANGER
82 Mountain Street
New Bedford, MA 02720
Telephone (617) 864-9318

OBJECTIVE
A summer internship in software documentation.

QUALIFICATIONS
Software and hardware documentation. Editing. Desktop publishing. Usability testing.
Computer science. Internet research. World Wide Web collaboration. Networking technology.
Instructor-led training. IBM PC, Macintosh, DEC 20 mainframe and VAX 11/780 systems.
Pagemaker, Netscape, Powerpoint, Excel, and Lotus Notes software. Logo, Pascal, HTML, and
C++ program languages.

EDUCATION
UNIVERSITY OF MASSACHUSETTS AT DARTMOUTH: B.A. expected January 1998. English and
Communications major. Computer Science minor. GPA: 3.54. Class rank: top 7 percent.

EXPERIENCE
CONWAY COMMUNICATIONS, INC., 39 WALL STREET, MARLBORO, MA 02864: Intern
Technical Writer. LAN technology. Writing, designing, and testing hardware upgrade manuals.
Desktop publishing of installation and maintenance manual. Specifying art and illustrations.
Designing home page and hypertext links. Summers 1995 and 1996.

WRITING AND READING CENTER, UMD: Tutor. Individual and group instruction in writing and
word processing. Training new tutors. Newsletter editing. Scriptwriting and acting in a training
video. Designing home page. Fall 1994–present.

THE TORCH, UMD WEEKLY NEWSPAPER: Managing Editor. Responsible for conducting staff
meetings, generating story ideas, supervising page layout, paste-up, and copyediting. Fall
1995–present.

ACHIEVEMENTS AND AWARDS
Writing samples published in Dr. John M. Lannon's TECHNICAL WRITING, 7th ed., Longman,
1997. Massachusetts State Honors Scholarship, 1994–1997. Dean's list, each semester.

ACTIVITIES
Student member, Society for Technical Communication and American Society for Training and
Development. Student representative, College Curriculum Committee. UMD Literary Society.

REFERENCES
Placement Office, University of Massachusetts, No. Dartmouth, MA 02747.

Figure 20.7   A Computer-Scannable Résumé

is a modified-functional résumé that preserves the logical progression that employers prefer but that also highlights your abilities and job skills (as in Figure 20.6).

A Sample Situation

Let's imagine you are a twenty-four-year-old student about to graduate from a community college with an A.A. degree in Hotel and Restaurant Management. Before college, you worked at related jobs for more than three years. You now seek a junior management position while you continue your education part time.

You have spent two weeks compiling information for your résumé and obtaining commitments from four references. Figure 20.5 shows your résumé. Notice that this résumé mentions nothing about salary. Wait until this matter comes up in your interview, or later. The information, specific but concise and accessible, describes what you have to offer. (For a résumé composed in search of summer employment, see Figure 20.6.)

Your résumé should be attractive, neat, and free of mechanical or grammar errors. A perfect copy is your responsibility. When fully satisfied with your résumé, make a number of high-quality copies. (Never send out carbon, thermofax, or mimeographed copies.) Consider using a word processor and laser printer, which enables you to make changes to suit various jobs and have a perfect document for each version.

The Job Application Letter

Your Image. Although it elaborates on your résumé, your application letter must emphasize personal qualities and qualifications in a convincing way. In your résumé you present raw facts. In your application letter you relate these facts to the company to which you are applying. The tone and insight you bring to your discussion suggest a good deal about who you are. The letter is your chance to explain how you see yourself fitting into the organization, to interpret your résumé, and to show how you can be an asset to this employer. Your letter's immediate purpose is to secure an interview.

Targets. Unlike the résumé, the letter never should be photocopied. Although you can base letters to different employers on the same model—with appropriate changes—each letter should be prepared anew.

Sometimes you will apply for positions advertised in print or by word of mouth (solicited applications). At other times you will write prospecting letters to organizations that have not advertised but might need someone like you (unsolicited applications). In either case tailor your letter to the situation.

The Solicited Letter. Imagine you are James Purdy (Figure 20.5). In *Innkeeper's Monthly,* you read the following advertisement and decide to apply:

Resort Management Openings

Liberty International, Inc., is accepting applications for several junior management positions at our new Lake Geneva resort. Applicants must have three years of practical experience, along with formal training in all areas of hotel/restaurant management. Please apply by June 1, 19xx, to

Sara Costanza
Personnel Director
Liberty International, Inc.
Lansdowne, PA 24135

Now plan and compose your letter.

Introduction. Create a confident tone by stating directly your reason for writing. Name the job, and remember you are talking *to* someone; use the pronoun "you" instead of awkward or impersonal constructions such as "One can see from the enclosed résumé." If you can, establish a connection by mentioning a mutual acquaintance—but only with that person's permission. Finally, after referring to your enclosed résumé, discuss your qualifications.

Body. Concentrate on the experience, skills, and aptitudes you can bring to this specific job. Follow these suggestions:

Make your letter dynamic

- Don't come across as a jack-of-all-trades. Relate your qualifications specifically to the job for which you're applying.
- Avoid flattery ("I am greatly impressed by your remarkable company").
- Be specific. Replace "much experience," "many courses," or "increased sales" with "three years of experience," "five courses," or "a 35 percent increase in sales between June and October 1996."
- Support your claims with *evidence*, to show how your qualifications will benefit this employer. Instead of saying, "I have leadership skills," say, "I was student senate president during my senior year and captain of the lacrosse team."
- Create a dynamic tone by using *active* voice and action verbs:

Weak	Management responsibilities were steadily given to me.
Strong	I steadily assumed management responsibilities.

- Trim the fat from your sentences:

Flabby	I have always been a person who enjoys a challenge.
Lean	I enjoy a challenge.

- Express self-confidence:

Unsure	It is my opinion that I have the potential to become a successful manager because ....
Confident	I will be a successful manager because ....

■ Never be vague:

Vague   I am familiar with the 1022 interactive database management system, and RUNOFF, the text processing system.

Definite   As a lab grader for one semester, I kept grading records on the 1022 database management system, and composed lab procedures on the RUNOFF text processing system.

■ Avoid letterese. Write in plain English.
■ Be enthusiastic. An enthusiastic attitude can be as important as your background, in some instances.

Conclusion. Restate your interest and emphasize your willingness to retrain or relocate (if necessary). If your reader is nearby, request an interview; otherwise, request a phone call, stating times you can be reached. Your conclusion should leave your reader with the impression that you are someone worth knowing.

Revision. *Never* settle for a first draft—or a second or third! This letter is your model for letters serving in various circumstances. Make it perfect.

After several revisions, James Purdy finally signed the letter shown below.

Job Application Letter

203 Elmwood Avenue
San Jose, CA 10462
April 22, 1997

Sara Costanza
Personnel Director
Liberty International, Inc.
Lansdowne, PA 24135

Dear Ms. Costanza:

Writer identifies self and purpose
Establishes a connection

Please consider my application for a junior management position at your Lake Geneva resort. I will graduate from San Jose City College on May 30 with an Associate of Arts degree in Hotel/Restaurant Management. Dr. H. V. Garlid, my nutrition professor, described his experience as a consultant for Liberty International and encouraged me to apply.

Relates specific qualifications to the job opening

For two years I worked as a part-time desk clerk, and I am now the desk manager at a 200-unit resort. This experience, combined with earlier customer relations work in a variety of situations, has given me a clear and practical understanding of customers' needs and expectations.

As an amateur chef, I know of the effort, attention, and patience required to prepare fine food. Moreover, my skiing and sailing background might be assets to your resort's recreation program.

I have confidence in my hospitality management skills. My experience and education have prepared me to work well with others and to respond creatively to changes, crises, and added responsibilities.

Should my background meet your needs, please phone me any weekday after 4 p.m. at 214–316–2419.

Sincerely,

James D. Purdy

James D. Purdy

Enclosure

Purdy wisely emphasizes practical experience because his background is varied and impressive. An applicant with less practical experience would emphasize education instead, discussing related courses, extracurricular activities, and aptitudes.

As an additional example, here is the letter composed by Karen Granger in her quest for a summer internship.

Internship Application Letter

Dear. Mr. White:

I read in Internships 1997 that your company offers a summer documentation internship. Because of my education and previous technical writing employment, I am higly interested in such a position.

In January 1998, I will graduate from the University of Massachusetts with a B.A. in English/Writing. I have prepared specifically for a computer documentation career by taking computer science, mathematics, and technical writing courses.

In one writing course, the Computer Documentation Seminar, I wrote three software manuals. One manual uses a tutorial to introduce beginners to the Apple Macintosh and MacWrite. The other manuals describe two IBM PC applications that arrived at the university's computer center with no documentation.

The enclosed résumé describes my work as the intern technical writer with Conway Communications, Inc., for two summers. I learned local area networking (LAN) by documenting Conway's LAN hardware and software. I was responsible for several projects simultaneously and spent much of my time talking with engineers and testing procedures. If you would like samples of my writing, please let me know.

Although Conway has invited me to return next summer and to work full time after graduation, I would like more varied experience before committing myself to permanent employment. I know I could make a positive contribution to Birchwood Group, Inc. May I telephone you next week to arrange a meeting?

Sincerely,

Margin notes (left column):

Expresses confidence and enthusiasm throughout

Makes follow-up easy for the reader

Begins by stating purpose

Identifies herself and college background

Expands on background

Describes work experience

Explains interest in this job

The Unsolicited Letter. Ambitious job seekers do not limit their search to advertised openings. (Studies indicate that fewer than 20 percent of all job openings are advertised.) The unsolicited, or prospecting, letter is a good way to uncover possibilities beyond the Help Wanted section. Such letters have advantages and disadvantages.

Disadvantages. The unsolicited approach has two drawbacks: (1) you might waste time writing to organizations that have no openings; and (2) you cannot tailor your letter to specific requirements.

Advantages. For an advertised opening you compete with legions of applicants, whereas your unsolicited letter might arrive just when an opening has materialized. Even when no opening exists, most companies welcome unsolicited letters and keep impressive applications on file, or pass them along to a company that has an opening. Unsolicited letters can be a sound investment if your targets are well chosen and your expectations are realistic.

Reader Interest. Because your unsolicited letter is unexpected, attract attention immediately. Don't begin: "I am writing to inquire about the possibility of obtaining a position with your company." By now, your reader is asleep. If you can't establish a connection through a mutual acquaintance, use a forceful opening:

Opens forcefully

> Does your hotel chain have a place for junior manager with a college degree in hospitality management, proven commitment to quality service, and customer relations experience that extends far beyond textbooks. If so, please consider my application for a position.

Address your letter to the person most likely in charge of hiring. (Consult the business directories listed on pages 482–483 for names of company officers.)

Employers will regard the quality of your application as an indication of the quality of work you will do. Businesses spend much money and time projecting favorable images. The image you project, in turn, must meet their standards.

Support for the Application

Your Dossier

Your dossier contains your credentials: college transcript, recommendation letters, and any other items (such as a notice of scholarship award or commendation letter) that document your achievements. An employer impressed by what you say about yourself will want to read what others think and will request your dossier. By collecting recommendations in one folder, you spare your references from writing the same letter over and over.

Your college placement office will keep your dossier on file and send copies to employers who request them. Always keep your own copy as well. Then, if an employer requests your dossier, you can make a photocopy and mail it, advising your reader that the placement copy is on the way. This is not needless repetition! Most employers establish a specific timetable for (1) advertising an opening, (2) reading letters and résumés, (3) requesting and reviewing dossiers, (4) holding interviews, and (5) making an offer. Timing is crucial. Too often, dossier requests from employers sit and gather dust in a busy placement office. In some cases, dossier requests have been lost permanently. Weeks can pass before your dossier is mailed. In these situations the only loser is you.

Employment Interviews

An employer impressed by your credentials will arrange an interview. The interview's purpose is to confirm the employer's impressions from your application. You are a finalist. But now you must be as impressive in person as you seem on paper—and the best person for the job.

You might meet with one interviewer, a group, or several groups in succession. You might be interviewed alone or with several candidates at once. Interviews can last one hour or less, a full day, or several days. The interview can range from a pleasant chat to grueling quiz sessions. Some interviewers may antagonize you deliberately to observe your reaction.

Unprepared interviewees make mistakes like the following (Dumont and Lannon 620):

How people fail job interviews

- know little or nothing about the company
- have inflated ideas about their worth
- have little idea of how their education prepares them for work
- dress inappropriately
- exhibit no self-confidence
- have only vague ideas of how they could benefit the employer
- inquire only about salary and benefits
- speak negatively of former employers or coworkers

Careful preparation is the key to a productive interview.

Prepare for the interview by learning about the company (its products or services, history, prospects, branch locations) in trade journals and industrial indexes. Request company literature and annual reports. Speak with people who know about the company. Prepare specific answers to the obvious questions:

- *Why do you wish to work here?*
- *What do you know about our company?*
- *What do you see as your biggest weakness? Your biggest strength?*
- *What would you like to be doing in ten years?*

Plan informative and direct answers to questions about your background, training, experience, and salary requirements. Prepare your own list of questions about the job and the organization; you will be invited to ask questions, and the questions you ask can be as revealing as the answers you give.

How to survive a job interview

Verify the interview's exact time and location. Come dressed as if you already work for the company. Maintain eye contact most of the time; if you stare at your shoes, the inteviewer will not be impressed. Relax but do not slouch. Do not smoke, even if invited. Do not pretend to know more than you do; if you cannot answer a question, say so. Avoid abrupt yes or no answers, as well as life stories. Make your answers concrete but to the point. Allow the interviewer to guide the conversation.

Don't be afraid to allow silence. An interviewer simply may *stop* talking, just to observe your reaction to silence. If you really have nothing more to say or ask, don't feel compelled to speak; caught off-guard, many of us are likely to say something stupid. Let the interviewer make the next move.

When your interviewer hints that the meeting is ending (perhaps by checking a wristwatch), don't wear out your welcome. Restate your interest, ask when you might expect further word, thank the interviewer, and leave.

If the interviewer invites you to lunch, display good manners and restraint. Don't order the most expensive dish on the menu; don't order an alcoholic beverage; don't salt your food before tasting it; don't eat too quickly; don't put elbows on the table; don't speak with your mouth full; don't smoke; and don't order a huge dessert. In short, remember this adage: "There's no such thing as a free lunch." And try to order last.

Take as many interviews as you can—as long as you are genuinely interested in the position. (To waste an interviewer's time just for practice would be unethical.) Expect your confidence and competence to increase with each interview.

The Follow-Up Letter

Within a few days after the interview, refresh the employer's memory with a letter restating your interest. James Purdy sent Sara Costanza this follow-up.

Refresh the employer's memory

> Thank you again for your hospitality during my visit to your Lansdowne offices. The tour of your resort facilities was informative and enjoyable.
>
> After meeting you and your colleagues and touring the resort, I believe I could be a productive member of your staff. I would welcome the opportunity to prove my abilities.

The Letter of Acceptance

You may receive a job offer by phone or letter. If by phone, request a written offer, and respond with a formal letter of acceptance. This letter may serve as part of your contract; spell out the terms you are accepting. Here is Purdy's letter of acceptance:

Accept an offer with enthusiasm

> I am delighted to accept your offer of a position as assistant recreation supervisor at Liberty International's Lake Geneva Resort, with a starting salary of $34,500.

As you requested, I will phone Bambi Druid in your personnel office for instructions on reporting date, physical, exam, and employee orientation.

I look forward to a long and satisfying career with Liberty International.

The Letter of Refusal

You may have to refuse offers. Even if you refuse by phone, write a prompt and cordial letter of refusal, explaining your reasons, and allowing for future possibilities. Purdy handled one refusal this way:

Decline an offer diplomatically

Although I was impressed by the challenge and efficiency of your company's operations, I am unable to accept your offer of a position as assistant desk manager of your London hotel.

I have accepted a position with Liberty International because Liberty has offered me the chance to participate in its manager-trainee program. Also, Liberty will pay tuition for the courses I take in completing my B.S. degree in hospitality management.

If any later openings should materialize at your Aspen resort, however, I would again appreciate your considering me as a candidate.

Thank you for your interest and courtesy.

IN BRIEF

Evaluating a Job Offer

Fortunately, most organizations will not expect you to accept or reject an offer on the spot. You probably will be given at least a week to make up your mind. Although there is no way to remove all risks from this career decision, you will increase your chances of making the right choice by thoroughly evaluating each offer—weighing all the advantages against all the disadvantages of taking the job.

The Organization. Background information on the organization—be it a company, government agency, or nonprofit concern—can help you decide whether it is a good place for you to work.

Is the organization's business or activity in keeping with your own interests and beliefs? It will be easier to apply yourself to the work if you are enthusiastic about what the organization does.

How will the size of the organization affect you? Large firms generally offer a greater variety of training programs and career paths, more managerial levels for advancement, and better employee benefits than small firms. Large employers also have more advanced technologies in their laboratories, offices, and factories. However, jobs in large firms tend to be highly specialized—workers are assigned relatively narrow responsibilities.

Continued on page 500

IN BRIEF

On the other hand, jobs in small firms may offer broader authority and responsibility, a closer working relationship with top management, and a chance to clearly see your contribution to the success of the organization.

Should you work for a fledgling organization or one that is well established? New businesses have a high failure rate, but for many people, the excitement of helping create a company and the potential for sharing in its success more than offset the risk of job loss. It may be almost as exciting and rewarding, however, to work for a young firm which already has a foothold on success.

Does it make any difference to you whether the company is private or public? A private company may be controlled by an individual or a family, which can mean that key jobs are reserved for relatives and friends. A public company is controlled by a board of directors responsible to the stockholders. Key jobs are open to anyone with talent.

Is the organization in an industry with favorable long-term prospects? The most successful firms tend to be in industries that are growing rapidly.

Where is the job located? If it is in another city, you need to consider the cost of living, the availability of housing and transportation, and the quality of educational and recreational facilities in the new location. Even if the place of work is in your area, consider the time and expense of commuting and whether it can be done by public transportation.

Where are the firm's headquarters and branches located? Although a move may not be required now, future opportunities could depend on your willingness to move to these places.

It frequently is easy to get background information on an organization simply by telephoning its public relations office. A public company's annual report to the stockholders tells about its corporate philosophy, history, products or services, goals, and financial status. Most govern-

ment agencies can furnish reports that describe their programs and missions. Press releases, company newsletters or magazines, and recruitment brochures also can be useful. Ask the organization for any other items that might interest a prospective employee.

Background information on the organization also may be available at your public or school library. If you cannot get an annual report, check the library for reference directories that provide basic facts about the company, such as earnings, products and services, and number of employees.

Stories about an organization in magazines and newspapers can tell a great deal about its successes, failures, and plans for the future. You can identify articles on a company by looking under its name in periodical or computerized indexes—such as the *Business Periodicals Index, Reader's Guide to Periodical Literature, Newspaper Index, Wall Street Journal Index, and New York Times Index.* It probably will not be useful to look back more than 2 or 3 years.

The library also may have government publications that present projections of growth for the industry in which the organization is classified. Long-term projections of employment and output for more than 200 industries, covering the entire economy, are developed by the Bureau of Labor Statistics and revised every other year—see the November 1995 Monthly Labor Review for the most recent projections. The *U.S. Industrial Outlook,* published annually by the U.S. Department of Commerce, presents detailed analysis of growth prospects for a large number of industries. Trade magazines also have frequent articles on the trends for specific industries.

Career centers at colleges and universities often have information on employers that is not available in libraries. Ask the career center librarian how to find out about a particular organization. The career center may have an entire file of information on the company.

Continued on page 501

IN BRIEF

The Nature of the Work. Even if everything else about the job is good, you will be unhappy if you dislike the day-to-day work. Determining in advance whether you will like the work may be difficult. However, the more you find out about it before accepting or rejecting the job offer, the more likely you are to make the right choice. Ask yourself questions like the following.

Does the work match your interests and make good use of your skills? The duties and responsibilities of the job should be explained in enough detail to answer this question.

How important is the job in this company? An explanation of where you fit in the organization and how you are supposed to contribute to its overall objectives should give an idea of the job's importance.

Are you comfortable with the supervisor?
Do employees seem friendly and cooperative?
Does the work require travel?
Does the job call for irregular hours?
How long do most people who enter this job stay with the company? High turnover can mean dissatisfaction with the nature of the work or something else about the job.

The Opportunities. A good job offers you opportunities to grow and move up. It gives you chances to learn new skills, increase your earnings, and rise to positions of greater authority, responsibility, and prestige.

The company should have a training plan for you. You know what your abilities are now. What valuable new skills does the company plan to teach you?

The employer should give you some idea of promotion possibilities within the organization. What is the next step on the career ladder? If you have to wait for a job to become vacant before you can be promoted, how long does this usually take? Employers differ on their policies regarding promotion from within the organiza-

tion. When opportunities for advancement do arise, will you compete with applicants from outside the company? Can you apply for jobs for which you qualify elsewhere within the organization or is mobility within the firm limited?

The Salary and Benefits. Wait for the employer to introduce these subjects. Most companies will not talk about pay until they have decided to hire you. In order to know if their offer is reasonable, you need a rough estimate of what the job should pay. You may have to go to several sources for this information. Talk to friends who recently were hired in similar jobs. Ask your teachers and the staff in the college placement office about starting pay for graduates with your qualifications. Scan the help-wanted ads in newspapers. Check the library or your school's career center for salary surveys, such as the College Placement Council Salary Survey and Bureau of Labor Statistics occupational wage surveys. If you are considering the salary and benefits for a job in another geographic area, make allowances for differences in the cost of living, which may be significantly higher in a large metropolitan area than in a smaller city, town, or rural area. Use the research to come up with a base salary range for yourself, the top being the best you can hope to get and the bottom being the least you will take. An employer cannot be specific about the amount of pay if it includes commissions and bonuses. The way the plan works, however, should be explained. The employer also should be able to tell you what most people in the job earn.

Also take into account that the starting salary is just that, the start. Your salary should be reviewed on a regular basis—many organizations do it every 12 months. If the employer is pleased with your performance, how much can you expect to earn after 1, 2, or 3 or more years?

Continued on page 502

IN BRIEF

Don't think of your salary as the only compensation you will receive—consider benefits. Benefits can add a lot to your base pay. Health insurance and pension plans are among the most important benefits. Other common benefits include life insurance, paid vacations and holidays, and sick leave. Benefits vary widely among smaller and larger firms, among full-time and part-time workers, and between the public and private sectors. Find out exactly what the benefit package includes and how much of the costs you must bear.

Asking yourself these kinds of questions won't guarantee that you make the best career decision—only hindsight could do that—but you probably will make a better choice than if you act on impulse.

☆   ☆   ☆

Source: Excerpted from U.S. Department of Labor. Tomorrow's Jobs. *Washington, D.C.: GPO, 1995.*

Revision Checklist for Letters

Use this checklist to refine the content, arrangement, and style of your letters. (Numbers in parentheses refer to the first page of discussion.)

CONTENT
❑ Is the letter addressed to a specifically named person? (467)

❑ Does the letter contain all of the standard parts? (467)

❑ Does the letter have all needed specialized parts? (470)

❑ Have you given the reader all necessary information? (476)

❑ Have you identified the name and position of your reader? (476)

ARRANGEMENT
❑ Does the introduction immediately engage the reader and lead naturally to the body? (493)

❑ Are transitions between letter parts clear and logical? (252)

❑ Does the conclusion encourage the reader to act? (494)

❑ Is the format correct? (471)

❑ Is the design acceptable? (471)

STYLE
❑ Is the letter in conversational language (free of letterese)? (475)

❑ Does the letter reflect a "you" perspective throughout? (474)

❑ Does the tone reflect your relationship with your reader? (476)

❑ Is the reader likely to react favorably to this letter? (476)

❑ Is the style throughout clear, concise, and fluent? (261)

❑ Is the letter grammatical? (Appendix)

❑ Does the letter's appearance enhance your image? (492)

✓ EXERCISES

1. Bring to class a copy of a business letter addressed to you or a friend. Compare letters. Choose the most and least effective.

2. Write and mail an unsolicited letter of inquiry about the topic you are investigating for an analytical report or research assignment. In your letter you might request brochures, pamphlets, or other informative literature, or you might ask specific questions. Submit to your instructor a copy of your letter and the response.

3. a. As a student in a state college, you learn that your governor and legislature have cut next year's operating budget for all state colleges by 20 percent. This cut will cause the firing of young and popular faculty members, drastically reduce admissions, financial aid, and new programs, and wreck college morale. Write a claim letter to your governor or representative, expressing your strong disapproval and justifying a major adjustment in the proposed budget.

   b. Write a claim letter to a politician about some issue affecting your school or community.

   c. Write a claim letter to an appropriate school official to recommend action on a campus problem.

4. Write a 500- to 700-word essay applying to a college for transfer or for graduate or professional school admission. Cover two areas: (1) what you can bring to this school by way of attitude, background, and talent; and (2) what you expect to gain in personal and professional growth.

5. Write a letter applying for a part-time or summer job, in response to a specific ad. Choose an organization related to your career goal. Identify the exact hours and calendar period during which you are free to work.

6. a. Identify a job you hope to have. Using newspaper ads, library (see your reference librarian for assistance), placement office, and personal sources, write your own description of the job: duties, responsibilities, work hours, salary range, requirements for promotion, highest promotion possible, unemployment rate in the field, employment outlook during the next decade, need for further education (advanced degrees, special training),

employment rate in terms of geography, optimum age bracket (as in football, is one considered "over the hill" after thirty-five?).

   b. Using these sources, construct a profile of the ideal employee for this job. If you were the personnel director screening applicants, what specific qualifications would you require (education, experience, age, physical ability, appearance, special skills, personality traits, attitude, outside interests, and so on)? Try to locate an actual newspaper ad describing job responsibilities and qualifications. Or assume you're a personnel officer, and compose an ad for the job.

   c. Assess your own credentials against the ideal employee profile. Review the plans you have made to prepare for this job: specific courses, special training, work experience. Assume you have completed your preparation. How do you measure up to the requirements in b? Are your goals realistic? If not, why not? What alternative plan should you create?

   d. Using your list from part c, compose a perfect résumé.

   e. Write a letter applying for the job described in part a.

   f. Write a follow-up letter thanking your fictional employer for your interview and restating your interest.

   g. Write a letter accepting a job offer from this same employer.

   h. Write a letter graciously declining this job offer.

   i. Submit each of these items, in order, to your instructor.

   Note: Use the sample letters in this chapter for guidance, but don't borrow specific expressions.

7. Most of these sentences need to be overhauled before being included in a letter. Identify the weakness in each statement, and revise as needed.

   a. Pursuant to your ad, I am writing to apply for the internship.

   b. I need all the information you have about methane-powered engines.

   c. You idiots have sent me a faulty disk drive!

   d. It is imperative that you let me know of your decision by January 15.

e. You are bound to be impressed by my credentials.

f. I could do wonders for your company.

g. I humbly request your kind consideration of my application for the position of junior engineer.

h. If you are looking for a winner, your search is over!

i. I have become cognizant of your experiments and wish to ask your advice about the following procedure.

j. You will find the following instructions easy enough for an ape to follow.

k. I would love to work for your wonderful company.

l. As per your request I am sending the country map.

m. I am in hopes that you will call soon.

n. We beg to differ with your interpretation of this leasing clause.

o. I am impressed by the high salaries paid by your company.

8. Write a complaint letter about a problem you've had with goods or service. State your case clearly and objectively, and request a specific adjustment.

9. *For Class Discussion:* Under what circumstances might it be acceptable to contact a potential inquiry respondent by E-mail? When should you just leave the person alone?

✓ COLLABORATIVE PROJECTS

1. Assume this advertisement has appeared in your school newspaper.

STUDENT CONSULTANT WANTED

The Dean of Students invites applications for the position of student consultant to the Dean for the next academic year. Duties: (1) meeting with students as individuals and groups to discuss issues, opinions, questions, complaints, and recommendations about all areas of college policy, (2) presenting oral and written reports to the Dean of Students, and (3) attending college planning sessions as student spokesperson. Time commitment: 15 hours weekly. Salary: $4,000.

Candidates should be full-time students with at least one year of student experience at this college. The ideal applicant will be skilled in report writing and oral communication, will work well in groups, and will demonstrate a firm commitment to our college. Application deadline: May 15.

a. Compose a résumé and a letter of application for this position.

b. Divide your class into screening, interview, and hiring committees.

c. Exchange your group's letters and résumés with those of another group.

d. As an individual committee member, read and evaluate each of the applications you have received. Rank each application privately on paper, according to the criteria in this chapter, before discussing them with your group. *Note:* While screening applicants, you will be competing for selection by another committee, which is reviewing your own application.

e. As a committee, select the three strongest applications, and interview each finalist for ten minutes, after you have prepared a list of standardized questions.

f. On the basis of these interviews, rank your preferences privately, on paper, giving specific reasons for your final choice.

g. Compare your conclusions with those of your colleagues and choose the winning candidate.

h. As a committee, compose a memo to your instructor, justifying your final recommendations. (See pages 509–11 for more on justification reports.

2. Form groups according to college majors. Prepare a set of instructions for entry-level job-seekers in your major, telling them how to launch their search. Base at least part of your advice on your analysis of Figure 20.4. Limit your document to one double-sided page, using an inviting and accessible design and any visuals you consider appropriate. Appoint one group member to present your final document to the class.

Memos and Short Reports

R EPORTS present ideas and facts to decision makers. In the professional world, superiors and colleagues rely on short reports as a basis for making *informed* decisions on matters as diverse as the most comfortable office chairs to buy or the best recruit to hire for management training.

Purpose of Memoranda and Short Reports

People on the job must communicate rapidly and precisely. Here are some of the kinds of reports you might write on any workday:

- a request for assistance on a project
- a requisition for parts and equipment
- a proposal for a new project
- a report of your progress on a specific assignment
- an hourly or daily account of your work activities
- a report on your inspection of a site, item, or process
- a record of a meeting
- a report of your survey to select the best prices or products

Many of these reports take the form of a memorandum.

The major form for internal written communication in organizations, memos leave a paper trail of directives, inquiries, instructions, requests, recommendations, and so on. Electronic-mail memos may leave a trail as well. Although E-mail messages tend to be less formal and more hastily written than paper documents, they are saved in both hard copy and online—and can inadvertently be forwarded to someone never intended to receive or read them.

Organizations rely heavily on memos to trace decisions, track progress, recheck data, or identify the person responsible. Therefore, any memo you write can have far-reaching ethical and legal implications. Be sure your memo includes the date, your initials for verification, and that your information is specific, accurate, and unambiguous.

The standard memo has a heading that names the organization and identifies the sender, recipient, subject (often in caps or underlined for emphasis), and date. (Placement of these items may differ among firms.) Other memo elements are shown in Figure 21.1.

Memo reports cover every conceivable topic. Common types include recommendation, justification, progress, periodic, and survey reports.

Recommendation Reports

Recommendation reports interpret data, draw conclusions, and make recommendations, often in response to a specific reader request. The recommendation report on page 508 is addressed to the writer's boss. This sample typifies the problem solving carried on daily by professionals and documented in memo form.

NAME OF ORGANIZATION

MEMORANDUM
Date: (also serves as a chronological record for future reference)
To: Name and title (the title also serves as a record for reference)
From: Your name and title (your initials for verification)
Subject: GUIDELINES FOR FORMATTING MEMOS

Subject Line
Announce the memo's purpose and contents, to orient readers to the subject and help them assess its importance. An explicit title also makes filing by subject easier.

Introductory Paragraph
Unless you have reason for being indirect, state your main point immediately.

Topic Headings
When discussing a number of subtopics, include headings (as we do here). Headings help you organize and they help readers locate information quickly.

Paragraph Spacing
Do not indent the first line of paragraphs. Single space within paragraphs and double space between them.

Second-page Headings
When the memo exceeds one page, begin the second and subsequent pages with recipient's name, date, and page number. For example: Ms. Baxter, June 12, 19xx, page 2. Place this information three lines from the page top and begin your text three lines below.

Memo Verification
Do not sign your memos. Initial the "From" line, after your name.

Copy Nototion
When sending copies to people not listed on the "To" line, include a copy notation two spaces below the last line, and list, by rank, the names and titles of those receiving copies. For example,

Copies: J. Spring, V.P., Production
            H. Baxter, General Manager, Production

Figure 21.1   Guidelines for Memo Formatting

A Problem-Solving Recommendation

Assume you are the writer, Bruce Doakes, in the situation that generated the Trans Globe memo below:

You are assistant manager of occupational health and safety for a major airline that employs over two hundred reservation and booking agents. Each agent spends eight hours daily seated at a workstation that has a computer, telephone, and other electronic equipment. Many agents have complained of chronic discomfort from their work: headache, eyestrain and irritation, blurred or double vision, backache, and stiff neck and joints.

Your boss asks you to study the problem and recommend improvements in the work environment. You survey employees and consult ophthalmologists, chiropractors, orthopedic physicians, and the latest publications on ergonomics (tailoring work environments for employees' physical and psychological well-being). After completing the study, you compose the report. ■

Bruce's report needs to be persuasive as well as informative.

Recommendation Memo

TRANS GLOBE AIRLINES

MEMORANDUM

To: R. Ames, Vice President, Personnel
From: B. Doakes, Health and Safety
Date: August 15, 19xx
Subject: **RECOMMENDATIONS FOR REDUCING
             AGENTS' DISCOMFORT**

*Provides immediate
orientation by giving
brief background and
main point*

In our July 20 staff meeting, we discussed physical discomfort among reservation and booking agents, who spend eight hours daily at automated workstations. Our agents complain of headaches, eyestrain and irritation, blurred or double vision, backaches, and stiff joints. This report outlines the apparent causes and recommends ways of reducing discomfort.

Causes of Agents' Discomfort

*Interprets findings and
draws conclusions*

For the time being, I have ruled out the computer display screens as a cause of headaches and eye problems because of the following reasons:

1.   Our new display screens have excellent contrast and no flicker.
2.   Research findings about the effects of low-level radiation from computer screens are inconclusive.

The headaches and eye problems seem to be caused by the excessive glare on display screens from background lighting.

Other discomforts, such as backaches and stiffness, apparently result from the agents' sitting in one position for up to two hours between breaks.

Recommended Changes

*Makes general
recommendations*

We can eliminate much discomfort by improving background lighting, workstation conditions, and work routines and habits.

Background Lighting. To reduce the glare on display screens, these are recommended changes in background lighting:

Expands on each recommendation

1. Decrease all overhead lighting by installing lower-wattage bulbs.
2. Keep all curtains and adjustable blinds on the south and west windows at least half-drawn, to block direct sunlight.
3. Install shades to direct the overhead light straight downward, so that it is not reflected by the screens.

Workstation Conditions. These are recommended changes in the workstations:

1. Reposition all screens so light sources are neither at front nor back.
2. Wash the surface of each screen weekly.
3. Adjust each screen so the top is slightly below the operator's eye level.
4. Adjust all keyboards so they are 27 inches from the floor.
5. Replace all fixed chairs with adjustable, armless, secretarial chairs.

Work Routines and Habits. These are recommended changes in agents' work routines and habits:

1. Allow frequent rest periods (10 minutes each hour instead of 30 minutes twice daily).
2. Provide yearly eye exams for all terminal operators, as part of our routine health-care program.
3. Train employees to adjust screen contrast and brightness whenever the background lighting changes.
4. Offer workshops on improving posture.

Discusses benefits of following the recommendations

These changes will give us time to consider more complex options such as installing hoods and antiglare filters on terminal screens, replacing fluorescent lighting with incandescent, covering surfaces with nonglare paint, or other disruptive procedures.

cc. J. Bush, Medical Director
     M. White, Manager of Physical Plant

For advice on how to think critically as you formulate, evaluate, and refine your recommendations, see pages 556–58.

Justification Reports

Justification reports (a type of recommendation report) are in a unique class in that they often are initiated by the writer rather than requested by the readers. As the name implies, such reports *justify* the writer's position on some issue. Justification reports therefore typically begin rather than end with the request or recommendation. Such reports answer this key question for readers: *Why should we?*

Typically, justification reports follow a version of this arrangement:

1. State the problem and your recommendations for solving it.
2. Point out the cost, savings, and benefits of your plan.
3. If needed, explain how your suggestion can be implemented.
4. Conclude by encouraging the reader to act.

Our next writer uses a version of the preceding arrangement: she begins with the problem and recommended solution, spells out costs and benefits, and concludes by re-emphasizing the major benefit. The tone is confident yet diplomatic—appropriate for an unsolicited recommendation to a superior.

Justification Memo

GREENTREE BIONOMICS, INC.

MEMORANDUM
To: D. Spring, Personnel Director
           Greentree Bionomics, Inc. (GBI)
From: M. Noll, Biology Division
Date: April 18, 19xx
Subject: The Need to Hire Additional Personnel

Introduction and Recommendation

Opens with the problem

With twenty-six active employees, GBI has been unable to keep up with scheduled contracts. As a result, we have a contract backlog of roughly $500,000. This backlog is caused by understaffing in the biology and chemistry divisions.

Recommends a solution

To increase production and ease the work load, I recommend that GBI hire three general laboratory assistants.

Expands on the recommendation

The lab assistants would be responsible for cleaning glassware and general equipment; feeding and monitoring fish stocks; preparing yeast, algae, and shrimp cultures; preparing stock solutions; and assisting scientists in various tests and procedures.

Cost and Benefits

Shows how benefits would offset costs

While costing $37,440 yearly (at $6.00/hour), three full-time lab assistants would have a positive effect on overall productivity:

1. Uncleaned glassware no longer would pile up, and the fish-holding tanks could be cleaned daily (as they should be) instead of weekly.
2. Because other employees no longer would need to work more than forty hours weekly, morale would improve.
3. Research scientists would be freed from general maintenance work (cleaning glassware, feeding and monitoring the fish stock, etc.). With more time to perform client tests, the researchers could eliminate our backlog.
4. With our backlog eliminated, clients no longer would have cause for impatience.

*Encourages acceptance
of recommendation*

Conclusion

Increased production at GBI is essential to maintaining good client relations. These additional personnel would allow us to continue a reputation of prompt and efficient service, thus ensuring our steady growth and development.

Progress Reports

Large organizations depend on progress (or status) reports to keep track of activities, problems, and progress on various projects. Daily progress reports are vital in a business that assigns crews to many projects. Management uses progress reports to evaluate the project and its supervisor, and to decide how to allocate funds.

Often, a progress report is one of a series. Together, the project proposal (Chapter 22), progress reports (the number varies with the scope and length of the project), and the final report (Chapter 23) provide a record and history of the project.

To give management the answers it needs, progress reports must, at a minimum, answer these questions:

1. *How much has been accomplished since the last report?*
2. *Is the project on schedule?*
3. *If not, what went wrong?*
   a. *How was the problem corrected?*
   b. *How long will it take to get back on schedule?*
4. *What else needs to be done?*
5. *What is the next step?*
6. *Any unexpected developments?*
7. *When do you anticipate completion? Or, on a long project, when do you anticipate completion of the next phase?*

If the report is part of a series, you might also refer to prior problems or developments.

Many organizations have forms for organizing progress reports, and so no one format is best. But each report in a series should follow the same organization. The following memorandum illustrates how one writer organized her report.

Progress Report (On the Job)

Subject: Progress Report: Equipment for New Operations Building

*Summarizes
achievements to date*

Work Completed

Our training group has met twice since our May 12 report in order to answer the questions you posed in your May 16 memo. In our first meeting, we identified the types of training we anticipate.

Details the achievements

Types of Training Anticipated

• Divisional Surveys
• Loan Officer Work Experience
• Divisional Systems Training
• Divisional Clerical Training (Continuing)
• Divisional Clerical Training (New Employees)
• Divisional Management Training (Seminars)
• Special/New Equipment Training

In our second meeting, we considered various areas for the training room.

Training Room

The frequency of training necessitates our having a training room available daily. The large training room in the Corporate Education area (10th floor) would be ideal. Before submitting our next report, we need your confirmation that this room can be assigned to us.

To support the training programs, we purchased this equipment:

• Audioviewer
• 16mm projector
• Videocassette recorder and monitor
• CRT
• Mini/micro computer, for computer-assisted instruction
• Slide projector
• Tape recorder

This equipment will allow us to administer training in a variety of modes, ranging from programmed and learner-controlled instruction to group seminars and workshops.

Describes what remains to be done

Work Remaining

To support the training, we need to furnish the room appropriately. Because the types of training will vary, the furniture should provide a flexible environment. Outlined here are our anticipated furnishing needs.

• Tables and chairs that can be set up in many configurations. These would allow for individual or group training and large seminars.
• Portable room dividers. These would provide study space for training with programmed instruction, as well as allow for simultaneous training.
• Built-in storage space for audiovisual equipment and training supplies. Ideally, this storage space should be multipurpose, providing work or display surfaces.
• A flexible lighting system, important for audiovisual presentations and individualized study.
• Independent temperature control, to ensure that the training room remains comfortable regardless of group size and equipment used.

Gives a rough timetable

The project is on schedule. As soon as we receive your approval of these specifications, we will proceed to the next step: sending out bids for room dividers, and having plans drawn for the built-in storage space.

cc: R. S. Pike, SVP
      G. T. Bailey, SVP

As you work on a longer report or term project, your instructor might require a progress report. In this next memo, Karen Granger documents her progress on her term project: an evaluation of the Environmental Protection Agency's effectiveness in cleaning a heavily contaminated harbor.

Progress Report on Term Project

PROGRESS REPORT

To: Dr. John Lannon
From: Karen P. Granger
Date: April 17, 19xx
Subject: Evaluation of the EPA's *Remedial Action Master Plan*

Project Overview
As my term project, I have been evaluating the issues of politics, scheduling, and safety surrounding the EPA's published plan to remove PCB contaminants from New Bedford Harbor.

Work Completed

February 23: Began general research on the PCB contamination of the New Bedford Harbor.

Summarizes achievements to date

March 8: Decided to analyze the *Remedial Action Master Plan* (RAMP) in order to determine whether residents are being "studied to death" by the EPA.

March 9–19: Drew a map of the harbor to show areas of contamination. Obtained the RAMP from Pat Shay of the EPA.

Interviewed Representative Grimes briefly by phone; made an appointment to interview Grimes and Sharon Dean on April 13.

Interviewed Patricia Chase, President of the New England Sierra Club, briefly, by phone.

March 24: Obtained *Public Comments on the New Bedford RAMP,* a collection of reactions to the plan.

April 13: Interviewed Grimes and Dean; searched Grimes' files for information. Also searched the files of Raymond Soares, New Bedford Coordinator, EPA.

Work in Progress
Contacting by telephone the people who commented on the RAMP.

Describes work remaining, with timetable

Work to Be Completed

April 25: Finish contacting commentators on the RAMP.

April 26: Interview an EPA representative about the complaints that the commentators raised on the RAMP.

Date for Completion: May 3, 19xx.

Describes the problems encountered

Complications

The issue of PCB contamination is complicated and emotional. The more I uncovered, the more difficult I found it to remain impartial in my research and analysis. As a New Bedford resident, I expected to find that we are indeed being studied to death; because my research seems to support my initial impression, I am not sure I have remained impartial.

Lastly, the people I talk to do not always have the time to find answers for my questions. Everyone, however, has been interested and encouraging, if not always informative.

Progress reports serve as a paper trail on a project.

Periodic Activity Reports

The periodic activity report is similar to the progress report in that it summarizes activities over a specified period. But unlike progress reports, which summarize specific accomplishments on a given project, periodic reports summarize general activities during a given period. Manufacturers requiring periodic reports often have prepared forms, because most of their tasks are quantifiable (e.g., Units produced). But most white-collar jobs do not lend themselves to prepared-form reports. You will probably have to develop your own format, as our next writer does.

DeWitt's report answers her boss's primary question: *What did you accomplish last month?* Her response has to be detailed amd informative.

Periodic Activity Report

Date: 6/15/xx
To: N. Morgan, Assistant Vice President
From: F. C. DeWitt
Subject: Coordinating Meetings for Training

Gives overview of recent activities, and their purpose

For the past month, I've been working on a cooperative project with the Banking Administration Institute, Computron Corporation, and several banks. My purpose has been to develop training programs, specific to banking, appropriate for computer-assisted instruction (CAI).

We have focused on three major areas: Proof/Encoding Training, Productivity Skills for Management, and Banking Principles.

Gives details

I hosted two meetings for this task force. On June 16, we will discuss Proof/Encoding Training, and on June 17, Productivity Training. The objective for the Proof/Encoding meeting was to compare ideas, information, and current training packages available on this topic. We are now designing a training course.

The objective of the meeting on Productivity was to discuss skills that increase productivity in banking (specifically Banking Operations). Discussion included instances in which computer-assisted instruction is appropriate for teaching productivity skills. Computron also discussed computer applications used to teach productivity.

On June 10, I attended a meeting in Washington, D.C., to design a course in basic banking principles for high-level clerical/supervisory-level employees. We also discussed the feasibility of adapting this course to CAI. This type of training, not currently available through Corporate Education, would meet a definite supervisor/management need in the division.

Explains the benefits of these activities

My involvement in these meetings has two benefits. First, structured discussions with trainers in the banking industry provide an exchange of ideas, methods, and experiences. This involvement expedites development of our training programs because it saves me time on research. Second, microcomputers will continue to affect future training. With a working knowledge of these systems and their applications, I now am able to assist my group in designing programs specific to our needs.

Periodic reports are used by employees to inform management of what they are doing and how well they are doing it. Therefore, accuracy, clarity, and appropriate level of detail are important—as is the persuasive (and ethical) dimension. Try to ensure that your readers have all the necessary facts—and that they understand these facts as clearly as you do.

Survey Reports

Brief survey reports often are used to examine conditions that affect an organization (consumer preferences, available markets). The following memo, from the research director for a midwestern grain distributor, gives clear and specific information directly. To simplify interpretation, the data are arranged in a table.

Survey Report
Subject: Food-Grain Consumption, U.S., 1993–96

Announces purpose of memo

Here are the data you requested on April 7, as part of your division's annual marketing survey.

Presents data in tabular form, for easy comparison

U.S. Per Capita Consumption of Food Grains in Pounds, 1993–96				
	1993	1994	1995	1996
Corn Products				
Cornmeal and other	16.4	16.4	16.6	16.7
Corn syrup and sugar	22.3	22.7	23.3	23.8
Oat Products	3.8	4.1	4.9	4.9
Barley Products	1.2	1.2	1.2	1.2
Wheat				
Flour	116.2	116.4	117.1	117.8
Breakfast cereals	2.9	2.9	2.9	2.9
Rye, Flour	1.2	1.2	1.2	1.2
Rice, Milled	8.3	6.7	7.7	7.0

Names the source

Source: U.S. Dept. of Agriculture, Economic Research Service

Discusses the conclusions to be drawn from the table

Food-grain consumption during the period remained stable or increased by less than 10 percent, with the notable exception of oat products (21 percent increase).

Of greatest concern is the impact on our long-term marketing objectives of the 17 percent decrease in milled-rice consumption.

cc: C. B. Schulz, Vice President, Marketing

If the above data had originated from *primary* sources (interviews, questionnaire, company records) the report also would describe how these data were collected.

Minutes of Meetings

Many team or project meetings on the job require someone to record the proceedings. Minutes are the records of such meetings. Copies of minutes are distributed to all members and interested parties, to keep track of proceedings. The appointed secretary records the minutes. When you record minutes, answer these questions:

- What group held the meeting? When, where, and why?
- Who chaired the meeting? Who else was present?
- Were the minutes of the last meeting approved (or disapproved)?

- Who said that?
- Was anything resolved?
- Who made which motions and what was the vote?

Minutes are filed as part of an official record, and so must be precise, clear, highly informative, and free of the writer's personal commentary ("As usual, Ms. Jones disagreed with the committee") or judgmental words ("good," "poor," "irrelevant").

Minutes of a Meeting

Subject: Minutes of Managers' Meeting, October 5, 19xx

Members Present

Tells who came

Harold Tweeksbury, Jeannine Boisvert, Sheila DaCruz, Ted Washington, Denise Walsh, Cora Parks, Cliff Walsh, Joyce Capizolo.

Agenda

Summarizes discussion of each item

1.   The meeting was called to order on Wednesday, October 5, at 10 A.M. by Cora Parks.

2.   The minutes of the September meeting were approved unanimously.

3.   The first order of new business was to approve the following policies for the Christmas season:

    a. Temporary employees should list their ID numbers in the upper left-corner of their receipt envelopes in order to help verification. Discount Clerical assistant managers will be responsible for seeing that this procedure is followed.

    b. When temporary employees turn in their envelopes, personnel from Discount Clerical should spot-check them for completeness and legibility. Incomplete or illegible envelopes should be corrected, completed, or rewritten. Envelopes should not be sealed.

Tells who said what
Tells what was voted

4.   Jeannine Boisvert moved that we also hold one-day training workshops for temporary employees in order to teach them our policies and procedures. The motion was seconded. Joyce Capizolo disagreed, saying that on-the-job training (OJT) was enough. The motion for the training session carried 6–3. The first workshop, which Jeannine agreed to arrange, will be held October 25.

5.   Joyce Capizolo requested that temporary employees be sent a memo explaining the temporary employee discount procedure. The request was converted to a motion and seconded by Cliff Walsh. The motion passed by a 7–2 vote.

6.   Cora Parks adjourned the meeting at 11:55 A.M.

Revision Checklist for Short Reports

Use this checklist as a guide to refining and revising your short reports. (Numbers in parentheses refer to the first page of discussion.)

❏ Have you chosen the best report form for your purpose? (506)

❏ Does the memorandum have a complete heading? (507)

❏ Does the subject line forecast the memo's contents? (507)

❏ Are readers given enough information for an *informed* decision? (506)

❏ Are the conclusions and recommendations clear? (509)

❏ Are paragraphs single-spaced within and double-spaced between? (506)

❏ Do headings, charts, or tables appear whenever needed? (507)

❏ If more than one reader is receiving copies, does the memo include a distribution notation (cc:) to identify other readers? (507)

❏ Is the writing style clear, concise, exact, fluent, and appropriate? (261)

❏ Does the document's appearance create a favorable impression? (507)

❏ Is the document ethically acceptable? (506)

☑ EXERCISES

1. We all would like to see changes in our schools' policies or procedures, whether they are changes in our majors, school regulations, social activities, grading policies, registration procedures, or the like. Find some area of your school that needs obvious changes, and write a justification report to the person who might initiate that change. Explain why the change is necessary and describe the benefits. Follow the format on page 510.

2. Think of an idea you would like to see implemented in your job (e.g., a way to increase productivity, improve service, increase business, or improve working conditions). Write a justification memo, persuading your audience that your idea is worthwhile.

3. Write a memo to your employer, justifying reimbursement for this course. *Note:* You might have written another version of this assignment for Exercise 5 in Chapter 1. If so, compare early and recent versions for content, organization, style, and format.

4. Identify a dangerous or inconvenient area or situation on campus or in your community (endless cafeteria lines, a poorly lit intersection, slippery stairs, a poorly adjusted traffic light).

Observe the problem for several hours during a peak-use period. Write a justification report to a *specifically identified* decision maker, describing the problem, listing your observations, making recommendations, and encouraging reader support or action.

5. Assume you have received a $10,000 scholarship, $2,500 yearly. The only stipulation for receiving installments is that you send the scholarship committee a yearly progress report on your education, including courses, grades, school activities, and cumulative average. Write the report.

6. In a memo to your instructor, outline your progress on your term project. Describe your accomplishments, plans for further work, and any problems or setbacks. Conclude your memo with a specific completion date.

7. Keep accurate minutes for one class session (preferably one with debate or discussion). Submit the minutes in memo form to your instructor.

8. Conduct a brief survey (e.g., of comparative interest rates from various banks on a car loan, comparative tax and property evaluation rates in three local towns, or comparative prices among local retailers for an item). Arrange your data and report your findings to your instructor in a memo

that closes with specific recommendations for the most economical choice.

9. Although the campus store (or cafeteria) is a convenient place to buy small items (or food), you believe students are overcharged for the convenience. To prove your point, you decide to do a comparative analysis of the cost of five items at the campus store (or cafeteria) and two local stores (or restaurants). Be sure to compare like sizes, weights, brands. Conduct your survey, analyze your findings, and submit the results in memo form to the student senate. Your instructor might want you to include recommendations.

10. Recommendation Report (choose one)

a. You are legal consultant to the leadership of a large auto-workers' union. Before negotiating its next contract, the unions needs to anticipate effects of robotics technology on assembly-line auto workers within ten years. Do the research and write a report recommending a course of action.

b. You are a consulting engineer to an island community of 200 families suffering a severe shortage of fresh water. Some islanders have raised the possibility of producing drinking water from salt water (desalination). Write a report for the town council, summarizing the process and describing instances in which desalination has been used successfully or unsuccessfully. Would desalination be economically feasible for a community this size? Recommend a course of action.

c. You are a health officer in a town less than one mile from a massive radar installation. Citizens are disturbed about the effects of microwave radiation. Do they need to worry? Should any precautions be taken? Find the facts and write your report.

d. You are an investment broker for a major firm. A longtime client calls to ask your opinion. She is thinking of investing in a company that is fast becoming a leader in fiber optics communication links. "Should I invest in this technology?" your client wants to know. Find out, and give her your recommendations in a short report.

e. The buildings in the condominium complex you manage have been invaded by carpenter ants. Can the ants be eliminated by any insecticide *proven* nontoxic to humans or pets? (Many dwellers have small children and pets.) Find out, and write a report making recommendations to the maintenance supervisor.

f. The "coffee generation" wants to know about the properties of caffeine and the chemicals used on coffee beans. What are the effects of these substances on the body? Write your report, making specific recommendations about precautions that coffee drinkers can take.

g. As a consulting dietitian to the school cafeteria in Blandville, you've been asked by the school board to report on the most dangerous chemical additives in foods. Parents want to be sure that foods containing these additives are eliminated from school menus, insofar as possible. Write your report, making general recommendations about modifying school menus.

h. Dream up a scenario of your own in which information and recommendations would make a real difference. (Perhaps the question could be one you've always wanted answered.)

☑ COLLABORATIVE PROJECT

Organize into groups of four or five and choose a topic upon which all group members can take the same position. Here are some possibilities:

- Should your college abolish core requirements?
- Should every student in your school pass a writing proficiency exam before graduating?
- Should courses outside one's major be graded pass/fail at the student's request?
- Should your school drop or institute student evaluation of teachers?
- Should all students be required to be computer literate before graduating?
- Should campus police carry guns?

- Should dorm security be improved?
- Should students with meal tickets be charged according to the type and amount of food they eat, instead of paying a flat fee?

As a group, decide your position on the issue, and brainstorm collectively to justify your recommendation to a stipulated primary audience in addition to your classmates and instructor. Complete an audience and use profile (page 65), and compose a justification report. Appoint one member to present the report in class.

Proposals

A PROPOSAL is an offer to do something or a recommendation that something be done. A proposal's general purpose is to *persuade* readers to improve conditions, authorize work on a project, accept a service or product (for payment), or otherwise support a plan for solving a problem or doing a job.

Your own proposal may be a letter to your school board to suggest changes in the English curriculum; it may be a memo to your firm's vice president to request funding for a training program for new employees; or it may be a 1,000-page document to the Defense Department to bid for a missile contract (competing with proposals from other firms). You might write the proposal alone or as part of a team. It might take hours or months.

Purpose of Proposals

Whether in science, business, industry, government, or education, proposals are written for decision makers: managers, executives, directors, clients, trustees, board members, community leaders, and the like. Inside or outside your organization, these are the people who decide whether your suggestions are worthwhile, whether your project ever will materialize, whether your service or product is useful. If your job depends on funding from outside sources, proposals might be the most important documents you produce.

The Proposal Process

The basic proposal process can be summarized simply: someone offers a plan for something that needs to be done. In business and government, this process has three stages:

1. Client *X* needs a service or product.
2. Firms *A, B,* and *C* propose a plan for meeting the need.
3. Client *X* awards the job to the firm offering the best proposal.

The complexity of each phase will of course depend on the situation. Here is a typical situation:

Assume that you manage a mining engineering firm in Tulsa, Oklahoma. You regularly read the *Commerce Business Daily,* an essential reference tool for anyone whose firm seeks government contracts. This daily publication lists the government's latest needs for *services* (salvage, engineering, maintenance) and for *supplies, equipment, and materials* (guided missiles, engine parts, building materials, and so on). On Wednesday, February 19, you spot this announcement in the *Commerce Business Daily:*

A solicitation for proposals

> **Development of Alternative Solutions to Acid Mine Water Contamination from Abandoned Lead and Zinc Mines** near Tar Creek, Neosho river, Ground Lake, and the Boone and Roubidoux aquifers in northeastern Oklahoma. This will include assessment of environmental impacts of mine drainage followed by

development and evaluation of alternate solutions to alleviate acid mine drainage in receiving streams. An optional portion of the contract, to be bid upon as an add-on and awarded at the discretion of the OWRB, will be to prepare an Environmental Impact Assessment for each of three alternative solutions as selected by the OWRB. The project is expected to take 6 months to accomplish, with anticipated completion date of September 30, 19XX. The projected effort for the required task is 30 person-months. Requests for proposals will be issued and copies provided to interested sources upon receipt of a written request. Proposals are due March 1. (044)

Oklahoma Water Resources Board
P.O. Box 53585
1000 Northeast 10th Street
Oklahoma City, OK 73151
(405) 555-2541

Your firm has the personnel, experience, and time to do the job, so you decide to compete for the contract. Because the March 1 deadline is fast approaching, you write immediately for a copy of the request for proposal (RFP). The RFP will give you the guidelines for developing the proposal—guidelines for spelling out your plan to solve the problem (methods, timetables, costs).

You receive a copy of the RFP on February 21—only one week before deadline. You get right to work with the two staff engineers you have appointed to your proposal team. Because the credentials of your staff could affect the client's acceptance of the proposal, you ask team members to update their résumés for inclusion in an appendix to the proposal.

Several other firms will be competing for this project. The client will award it to the firm submitting the best proposal, based on the following criteria (and perhaps others):

Criteria by which clients evaluate proposals

- understanding of the client's needs, as described in the RFP
- soundness of the firm's technical approach
- quality of the project's organization and management
- ability to complete the job by the deadline
- ability to control costs
- specialized experience of the firm in this type of work
- qualifications of staff to be assigned to the project
- the firm's record for similar projects

A client's specific evaluation criteria often are listed (in order of importance or on a point scale) in the RFP. Although these criteria may vary, every client expects a proposal that is *clear, informative,* and *realistic.*

Some clients hold a preproposal conference for the competing firms. During this briefing, the firms are informed of the client's needs, expectations, specific start-up and completion dates, criteria for evaluation, and any other details that might help guide proposal development.

In most cases, the merits of your proposed plan are judged *solely* by what you have *on paper.* This reality is summed up in the advice below from a government publication that solicits proposals.

> [Proposal] reviewers will base their conclusions only on information contained in the proposal. It cannot be assumed that readers are acquainted with the firm or key individuals or any referred-to experiments.[1]

Make any proposal able to stand alone in meaning and persuasiveness.

A helpful source of information about proposals is *The Grantsmanship Center News,* published six times yearly, which contains a wealth of information about federal grants for nonprofit organizations in areas such as rural development, media, science, energy and environment, and education. This publication also lists notices of training programs in proposal writing, and offers books on proposal writing and evaluation, along with bibliographies of publications about the proposal process.

Proposal Types

Despite their variety, proposals are classified in three ways: according to *origin, audience,* or *intention.* Based on its origin, a proposal is either *solicited* or *unsolicited*—that is, requested by an employer or potential client or initiated on your own because you have recognized a need. Business and government proposals usually are solicited, and they originate from a customer's request (as shown in the sample situation on pages 525–26).

Based on its audience, a proposal may be *internal* or *external*—written for members of your organization or for clients and funding agencies. (The situation on pages 539–40 calls for an external proposal.)

Based on its intention, a proposal may be a *planning, research,* or *sales* proposal. These last categories by no means account for all variations among proposals. Some proposals may in fact fall within all three categories, but these are the types you most likely will have to write. A discussion of each type follows.

The Planning Proposal

A planning proposal suggests ways of solving a problem or bringing about improvement. It might be a request for funding to expand the campus newspaper, an architectural plan for new facilities at a ski area, or a plan to develop energy alternatives to fossil fuels. The successful planning proposal always answers this main question for readers:

- *What are the benefits of following your suggestions?*

The planning proposal that follows is external and solicited. The XYZ Corporation has contracted a team of communication consultants to design

---

1. *Small Business Innovation Research Program* (Washington: U.S. Department of Defense, 1993): 9.

in-house writing workshops. The consultants need to persuade the reader that their methods are likely to succeed. In their proposal, addressed to the company's education officer, the consultants offer concrete and specific solutions to clearly identified problems.

After a brief introduction summarizing the problem, our writers develop their proposal under two headings, "Assessment of Needs" and "Proposed Plan."

Under "Proposed Plan," subheadings offer an even more specific forecast. The "Limitations" section shows that our writers are careful to promise no more than they can deliver.

Because this proposal is external, it is cast in letter form. Notice, however, that the word choice ("thanks," "what we're doing," "Jack and Terry") creates an informal, familiar tone. Such a tone is appropriate in this external document because the writers and reader have spent many hours in conferences, luncheons, and phone conversations.

Planning Proposal

Dear Mary:

States purpose

Thanks for sending the writing samples from your technical support staff. Here is what we're doing to design a realistic approach.

Assessment of Needs

Identifies problem

After conferring with technicians in both Jack's and Terry's groups, and analyzing their writing samples, we identified this hierarchy of needs:

- improving readability
- achieving precise diction
- summarizing information
- organizing a set of procedures
- formulating various memo reports
- analyzing audiences for upward communication
- writing persuasive bids for transfer or promotion
- writing persuasive suggestions

Proposed Plan

Proposes solution

Based on the needs listed above, we have limited our instruction package to eight carefully selected and readily achievable goals.

Course Outline. Our eight 2-hour sessions are structured as follows:

Details what will be done

1. achieving sentence clarity
2. achieving sentence conciseness
3. achieving fluency and precise diction
4. writing summaries and abstracts
5. outlining manuals and procedures
6. editing manuals and procedures

7. designing various reports for various purposes

8. analyzing the audience and writing persuasively

Details how it will be done

Classroom Format. The first three meetings will be lecture-intensive with weekly exercises to be done at home and edited collectively in class. The remaining five weeks will combine lecture and exercises with group editing of work-related documents. We plan to remain flexible so we can respond to needs that arise.

Limitations

Sets realistic expectations

Given our limited contact time, we cannot realistically expect to turn out a batch of polished communicators. By the end of the course, however, our students will have begun to appreciate writing as a deliberate process.

Encourages reader response

If you have any suggestions for refining this plan, please let us know.

The Research Proposal

Research (or grant) proposals request approval (and often funding) for a research project. A chemistry professor might address a research proposal to the Environmental Protection Agency for funds to identify toxic contaminants in local groundwater. Research proposals are solicited by many government and private agencies: National Science Foundation, National Institutes of Health, and others. Each granting agency has its own requirements for proposal format and content, but any successful research proposal answers these questions:

- *Why is this project worthwhile?*
- *What qualifies you to undertake the project?*
- *What are its chances of succeeding?*

In college, you might submit proposals for independent study, field study, or a thesis project. Here is the title of a research proposal submitted to the thesis committee in a geology department:

PROPOSAL FOR A MASTER'S THESIS PROJECT

TO INVESTIGATE THE TERTIARY GEOLOGY

OF THE ST. MARIES RIVER DRAINAGE

FROM ST. MARIES TO CLARKIA, IDAHO

A technical writing student might submit an informal proposal requesting the instructor's approval for a term project (which, in turn, may be a formal proposal). The introduction of the next proposal describes the problem and justifies the need for the study. The body outlines the scope, method, and sources for the proposed investigation. The conclusion describes the goal of the investigation and encourages reader support. This proposal is convincing because it answers questions about *what, why, how, when,* and *where*. Because this proposal is internal, it is cast informally as a memo.

Research Proposal

To: Dr. John Lannon
From: T. Sorrells Dewoody
Date: March 16, 19XX
Subject: Proposal for Determining the Feasibility of Marketing Dead Western
           White Pine

Introduction

Opens with background and causes of the problem

Over the past four decades, huge losses of western white pine have occurred in the northern Rockies, primarily attributable to white pine blister rust and the attack of the mountain pine beetle. Estimated annual mortality is 318 million board feet. Because of the low natural resistance of white pine to blister rust, this high mortality rate is expected to continue indefinitely.

If white pine is not harvested while the tree is dying or soon after death, the wood begins to dry and check (warp and crack). The sapwood is discolored by blue stain, a fungus carried by the mountain pine beetle. If the white pine continues to stand after death, heart cracks develop. These factors work together to cause degradation of the lumber and consequent loss in value.

Statement of Problem

Describes problem

White pine mortality reduces the value of white pine stumpage because the commercial lumber market will not accept it. The major implications of this problem are two: first, in the face of rising demand for wood, vast amounts of timber lie unused; second, dead trees are left to accumulate in the woods, where they are rapidly becoming a major fire hazard here in northern Idaho and elsewhere.

Proposed Solution

Describes one possible solution

One possible solution to the problem of white pine mortality and waste is to search for markets other than the conventional lumber market. The last few years have seen a burst of popularity and growing demand for weathered barn boards and wormy pine for interior paneling. Some firms around the country are marketing defective wood as specialty products. (These firms call the wood from which their products come "distressed," a term I will use hereafter to refer to dead and defective white pine.) Distressed white pine quite possibly will find a place in such a market.

Scope

Defines scope of the proposed study

To assess the feasibility of developing a market for distressed white pine, I plan to pursue six areas of inquiry:

1. What products presently are being produced from dead wood, and what are the approximate costs of production?
2. How large is the demand for distressed-wood products?
3. Can distressed white pine meet this demand as well as other species meet it?
4. Does the market contain room for distressed white pine?

5. What are the costs of retrieving and milling distressed white pine?

6. What prices for the products can the market bear?

Methods

Describes how study will be done

My primary data sources will include consultations with Dr. James Hill, Professor of Wood Utilization, and Dr. Sven Bergman, Forest Economist—both members of the College of Forestry, Wildlife, and Range. I will also inspect decks of dead white pine at several locations, and visit a processing mill to evaluate it as a possible base of operations. I will round out my primary research with a letter and telephone survey of processors and wholesalers of distressed material.

Secondary sources will include publications on the uses of dead timber, and a review of a study by Dr. Hill on the uses of dead white pine.

My Qualifications

Gives the writer's qualifications for this project

I have been following Dr. Hill's study on dead white pine for two years. In June of this year I will receive my B.S. in forest management. I am familiar with wood milling processes and have firsthand experience at logging. My association with Drs. Hill and Bergman gives me the opportunity for an in-depth feasibility study.

Conclusion

Encourages reader acceptance

Clearly, action is needed to reduce the vast accumulations of dead white pine in our forests. The land on which they stand is among the most productive forests in northern Idaho. By addressing the six areas of inquiry mentioned earlier, I can determine the feasibility of directing capital and labor to the production of distressed white pine products. With your approval I will begin research at once.

The Sales Proposal

A sales proposal offers a service or product. Sales proposals may be solicited or unsolicited. If they are solicited, several firms may compete with proposals of their own. Because sales proposals are addressed to readers outside your organization, they are cast as letters if they are brief. Long sales proposals, like long reports, are formal documents with supplements (cover letter, title page, table of contents).

The sales proposal, a major marketing tool in business and industry, will be successful if it answers this question:

- *How will you serve our needs better than your competitors:*

or

- *Why should we hire you instead of someone else?*

The following solicited proposal offers a service. Because the writer is competing with other firms, he explains specifically *why* his machinery is best for the job, *how* the job can best be completed, *what* his qualifications are for getting the job done, and *how much* the job will cost. He will be legally bound by his estimate. To protect himself, he points out possible causes of increased costs. *Never underestimate costs by failing to account for all variables*—a sure way to lose money or clients.

The introduction describes the subject and purpose of the proposal. The conclusion reinforces the confident tone throughout and encourages readers' acceptance by ending with—and thus emphasizing—two vital words: "economically" and "efficiently."

Sales Proposal

Subject: Proposal to Dig a Trench and Move Boulders at Bliss Site

Dear Mr. Haver:

Offers to undertake the job

I've inspected your property and would be happy to undertake the landscaping project necessary for the development of your farm.

Gives the writer's qualifications

The backhoe I use cuts a span 3 feet wide and can dig as deep as 18 feet—more than an adequate depth for the mainline pipe you wish to lay. Because this backhoe is on tracks rather than tires, and is hydraulically operated, it is particularly efficient in moving rocks. I have more than twelve years of experience with backhoe work and have completed many jobs similar to this one.

Explains how the job will be done

After examining the huge boulders that block access to your property, I am convinced they can be moved only if I dig out underneath and exert upward pressure with the hydraulic ram while you push forward on the boulders with your D–9 Caterpillar. With this method, we can move enough rock to enable you to farm that now inaccessible tract. Because of its power, my larger backhoe will save you both time and money in the long run.

Gives a qualified cost estimate

This job should take 12 to 15 hours, unless we encounter subsurface ledge formations. My fee is $100 an hour. The fact that I provide my own dynamiting crew at no extra charge should be an advantage to you because you have so much rock to be moved.

Encourages reader acceptance

Please phone me anytime for more information. I'm sure we can do the job economically and efficiently.

The proposal categories (planning, research, and sales) discussed in this section are neither exhaustive nor mutually exclusive. A research proposal, for example, may request funds for a study that will lead to a planning proposal. The Vista proposal partially shown on pages 530–33 is a combined planning and sales proposal; if clients accept the writer's preliminary plan, they will hire the firm to install the automated system.

Proposal Guidelines

Readers will evaluate your proposal according to how clearly, informatively, and realistically you answer these questions:

- *What are you proposing?*
- *What problem will you solve?*
- *Why is your plan worthwhile?*
- *What is unique about your plan?*

- *What are your (or your firm's) credentials?*
- *How will the plan be implemented?*
- *How long will the project take to complete?*
- *How much will it cost?*
- *How will we benefit if we accept your plan?*

In addition to answering the questions above, successful proposal writers adhere to the following guidelines.

Design an Accessible and Appealing Format. Format is the *look* of a document, including such features as

- the layout of words and graphics
- typeface, type size, and white space
- highlights and lists
- headings

A poorly designed proposal suggests to readers the writer's careless attitude toward the project.

Signal Your Intent with a Clear Title. Decision makers are busy people who have no time for guessing-games. The title should clearly signal the proposal's purpose and content.

     Unclear   PROPOSED OFFICE PROCEDURES FOR VISTA FREIGHT, INC.

What kinds of office procedures are being proposed? This title is too broad.

     Revised   A PROPOSAL FOR AUTOMATING VISTA'S FREIGHT BILLING SYSTEM

Don't write "Recommended Improvements" when you mean "Recommended Wastewater Treatment." A specific and comprehensive title signals the proposal's intent.

Include Supporting Material and Appropriate Supplements. Both short and long proposals may include supporting materials (maps, blueprints, specifications, calculations, and so forth). Place supporting material in an appendix to avoid interrupting the discussion.

    Depending on your readers, appropriate supplements for a long proposal might include a title page, cover letter, table of contents, summary, abstract, and appendixes. Readers with various responsibilities will be interested in different parts of your proposal. Some know about the problem and will read only your plan; some look only at the summary; others will study recommendations or costs; still others need all the details. If you're unsure as to which supplements to include in an internal proposal, ask the intended reader or study other proposals. For a solicited proposal (one written for an outside agency) follow the agency's instructions *exactly.*

Focus on the Problem and the Objective. Readers want specific suggestions for filling specific needs. Their biggest question is "What's in this for me?" Show them you understand their problem and offer a plan for improving their products, sales, or services.

Notice how proposal writer Gerald Beaulieu focuses on Vista's inefficient office procedures, and then outlines specific solutions.

Gives background

Statement of the Problem

Vista provides two services. (1) It locates freight carriers for its clients. The carriers, in turn, pay vista a 6 percent commission for each referral. (2) Vista handles all shipping paperwork for its clients. For this auditing service, clients pay Vista a monthly retainer.

Describes problem and its effects

Although Vista's business has increased steadily for the past three years, record-keeping, accounting, and other paperwork still are done *manually*. These inefficient procedures have caused a number of problems, including late billings, lost commissions, and poor account maintenance. Unless its office procedures are updated, Vista stands to lose clients.

Enables readers to visualize results

Objective

This proposal offers a realistic and efficient plan for Vista to streamline office procedures. We first identify the burden imposed on your staff by the current system, and then we show how to reduce inefficiency, eliminate client complaints, and improve your cash flow by automating most office procedures.

Treat Contingencies and Limitations Realistically. Do not underestimate the project's complexity. Identify contingencies (occurrences subject to chance) readers might not anticipate, and propose realistic methods for dealing with the unexpected. Here is how the Vista proposal qualifies its promises:

Assesses contingencies realistically

As outlined below, Vista can realize tangible benefits by automating office procedures. But, as countless firms have learned, *imposing* automated procedures on a staff can create severe morale problems—particularly among senior staff who feel coerced. To diminish employee resistance, invite your staff to comment on this proposal. To help avoid hardware and software problems once the system is operational, we have included recommendations and a budget for staff training. (Firms have learned that inadequate training is counterproductive to the automation process.)

If the best available solutions have limitations, let readers know. Otherwise, you and your firm could be liable in the case of project failure. Avoid overstatement. Notice how the above solutions are qualified ("diminish" and "help avoid" instead of "eliminate") so as not to promise more than the writer can deliver. Ethical communication is essential.

Explain the Benefits of Implementing Your Plan. A persuasive proposal shows readers how they (or their organization) will benefit by adopting your plan.

The Vista proposal specifies the following benefits. (Each benefit will be described at length in the body section.)

Relates benefits directly to reader's needs

Once your automated system is operational, you will be able to
- identify cost-effective carriers
- coordinate shipments (which will ensure substantial discounts for clients)
- print commission bills
- track shipments by weight, miles, fuel costs, and destination
- send clients weekly audit reports on their shipments
- bill clients on a 25-day cycle
- produce weekly or monthly reports

Additional benefits include eliminating repetitive tasks, improving cash flow, and increasing productivity.

Provide Concrete and Specific Information. Vagueness is a fatal flaw in a proposal. Before you can persuade readers, you must inform them; therefore, you need to *show* as well as *tell.* Instead of writing, "We will install state-of-the-art equipment," write,

Spells out details

To meet your automation requirements, we will install twelve Power Macintosh 8500 computers with 2-Gigabyte hard drives. The system will be networked for rapid file transfer between offices. The plan also includes interconnection with four Hewlett-Packard 5 MP printers, and one HP Desk Jet 1600 CM color printer.

To avoid any misunderstanding and to reflect your ethical commitment, a proposal must elicit *one* interpretation only.

Include Effective Visuals. If they enhance your proposal, use visuals, properly introduced and discussed. Page 533 shows one visual from the Vista proposal.

Gives a framework for interpreting the visual

As the flowchart illustrates (Figure 1), your routing and billing system creates a good deal of redundant work for your staff. The routing sheet alone is handled at least six times. Such extensive handling leads to errors, misplaced sheets, and late billing.

Use a Tone That Connects with Readers. Your tone is the voice and personality that appear between the lines. Make the tone confident, encouraging, and diplomatic. Show readers you believe in your plan; urge them to act, and anticipate how they will react to your suggestions. But do not come across with a bossy or insulting tone, as in this example:

Had Vista's managers been better trained, they could have streamlined office procedures by eliminating some of the more obvious paper-shuffling problems. Moreover, the clerical staff could have avoided errors *had they been more vigilant.*

Here is a more diplomatic version:

Avoids blaming anyone

Vista's manual office procedures have resulted in an inefficient flow of information. The problems include repetitive tasks, late billings, lost documents, and poor account maintenance.

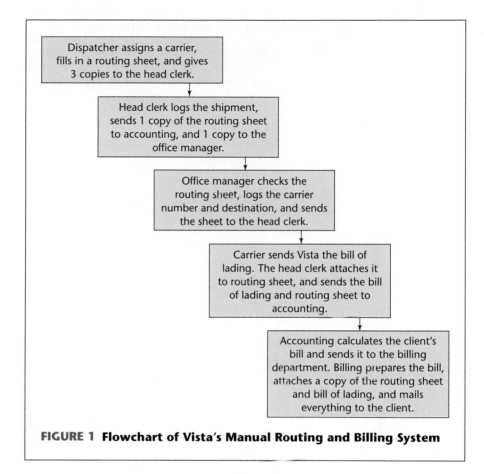

FIGURE 1 Flowchart of Vista's Manual Routing and Billing System

Analyze Audience Needs. Proposals address diverse audiences. A research proposal might be read by experts, who would then advise the granting agency whether to accept or reject it. Planning and sales proposals might be read by colleagues, superiors, and clients (often laypersons). Informed and expert readers will be most interested in the technical details. Nontechnical readers will be interested in the expected results, but will need an explanation of technical details as well. Learn all you can about the needs, interests, and biases of your audience.

If the primary audience is expert or informed, keep the proposal itself technical. For uninformed secondary readers (if any), provide an informative abstract, a glossary, and appendixes explaining specialized information. If the primary audience has no expertise and the secondary audience does, follow this pattern: Write the proposal itself for laypersons, and provide appendixes with the technical details (formulas, specifications, calculations) that the informed readers will use to evaluate your plan.

A General Model for Proposals

Like all informative writing, proposals are organized into these main sections: introduction, body, and conclusion. Depending on project complexity, each section contains some or all of the subsections listed in the following general outline:

I. INTRODUCTION
    A. Statement of Problem and Objective
    B. Background
    C. Need
    D. Benefits
    E. Qualifications of Personnel
    F. Data Sources
    G. Limitations and Contingencies
    H. Scope

II. BODY
    A. Methods
    B. Timetable
    C. Materials and Equipment
    D. Personnel
    E. Available Facilities
    F. Needed Facilities
    G. Cost
    H. Expected Results
    I. Feasibility

III. CONCLUSION
    A. Summary of Key Points
    B. Request for Action

Headings can be rearranged, combined, divided, or deleted as needed. Not every proposal has all subsections, but each major section must answer specific readers' questions, as illustrated next.

Introduction

The introduction answers all these questions—or all those that apply to the situation:

- *What problem do you propose to solve?*
- *In general, what solution are you proposing?*
- *Why are you proposing it?*
- *What are the benefits?*
- *What are your qualifications for this project?*

From the beginning, your goal is to sell your idea, to persuade readers the job needs doing and you are the one to do it. If your introduction is long-winded, evasive, or vague, readers might stop reading. Make it concise, specific, and clear.

Spell out the problem to make it clear to the audience—and to show you understand it fully. Explain the benefits of solving the problem or undertaking the project. Identify any sources of data. In a research or sales proposal, state your qualifications for doing the job. If your plan has limitations or contingencies, explain them. Finally, give the scope of your plan by listing the subsections to be discussed in the body section.

Following is the introduction for a planning proposal titled "A Proposal for Solving the Noise Problem in the University Library." Jill Sanders, a library work-study student, addresses her proposal to the chief librarian and the administrative staff. Because this proposal is unsolicited, it must first make the problem vivid through details that arouse concern and interest. This introduction is longer than it would be in a solicited proposal, whose readers already would agree on the severity of the problem.

Introduction

Statement of Problem

Concise descriptions of problem and objective immediately alert the readers

During the October 19XX Convocation at Margate University, students and faculty members complained about noise in the library. Soon afterward, areas were designated for "quiet study," but complaints about noise continue. To create a scholarly atmosphere, the library should take immediate action to decrease noise.

Objective

This proposal examines the noise problem from the viewpoint of students, faculty, and library staff. It then offers a plan to make areas of the library quiet enough for serious study and research.

Sources

This section comes early because it is referred to in next section

My data come from a university-wide questionnaire, interviews with students, faculty, and library staff, inquiry letters to other college libraries, and my own observations for three years on the library staff.

Details of the Problem

Details enable readers to understand the problem

This subsection examines the severity and causes of the noise.

Severity. Since the 19XX Convocation, the library's fourth and fifth floors have been reserved for quiet study, but students hold group-study sessions at the large tables and disturb others working alone. The constant use of computer terminals on both floors adds to the noise, especially when students converse. Moreover, people often chat as they enter or leave study areas.

On the second and third floors, designed for reference, staff help patrons locate materials, causing constant shuffling of people and books, as well as loud conversation. At the computer service desk on the third floor, conferences between students and instructors create more noise.

<table>
<tr><td>

Shows how campus feels about problem

</td><td>

The most frequently voiced complaint from the faculty members interviewed was about the second floor, where people using the Reference and Government Documents services converse loudly. Students complain about the lack of a quiet spot to study, especially in the evening, when even the "quiet" floors are as noisy as the dorms.

</td></tr>
</table>

Shows how campus feels about problem

Shows concern is widespread and pervasive

The most frequently voiced complaint from the faculty members interviewed was about the second floor, where people using the Reference and Government Documents services converse loudly. Students complain about the lack of a quiet spot to study, especially in the evening, when even the "quiet" floors are as noisy as the dorms.

More than 80 percent of respondents (530 undergraduates, 30 faculty, 22 graduate students) to a university-wide questionnaire (Appendix A) insisted that excessive noise discourages them from using the library as often as they would prefer. Of the student respondents, 430 cited quiet study as their primary reason for wishing to use the library.

The library staff recognizes the problem but has insufficient personnel. Because all staff members have assigned tasks, they have no time to monitor noise in their sections.

Causes. Respondents complained specifically about these causes of noise (in descending order of frequency):

Identifies specific causes

1. Loud study groups that often lapse into social discussions.
2. General disrespect for the library, with some students' attitudes characterized as "rude," "inconsiderate," or "immature."
3. The constant clicking of computer terminals on all five floors, and of typewriters on the first three.
4. Vacuuming by the evening custodians.

All complaints converged on lack of enforcement by library staff.

Because the day staff works on the first three floors, quiet-study rules are not enforced on the fourth and fifth floors. Work-study students on these floors have no authority to enforce rules not enforced by the regular staff. Small, black-and-white "Quiet Please" signs posted on all floors go unnoticed, and the evening security guard provides no deterrent.

Needs

This statement of need evolves logically and persuasively from earlier evidence

Excessive noise in the library is keeping patrons away. By addressing this problem immediately, we can help restore the library's credibility and utility as a campus resource. We must reduce noise on the lower floors and eliminate it from the quiet-study floors.

Scope

Previews the plan

The proposed plan includes a detailed assessment of methods, costs and materials, personnel requirements, feasibility, and expected results.

Body

The body (or plan) section of your proposal will receive most attention from readers. It answers all these questions that are applicable:

- *How will it be done?*
- *When will it be done?*

- *What materials, methods, and personnel will it take?*
- *What facilities are available?*
- *How long will it take?*
- *How much will it cost, and why?*
- *What results can we expect?*
- *How do we know it will work?*
- *Who will do it?*

Here you spell out your plan in enough detail for readers to evaluate its soundness. If this section is vague, your proposal stands no chance of being accepted. Be sure your plan is realistic and promises no more than you can deliver. The main goal of this section is to prove that your plan will work.

Proposed Plan

This plan takes into account the needs and wishes of our campus community, as well as the available facilities in our library.

Methods

Tells how plan will be implemented

Noise in the library can be reduced in three complementary phases: (1) improving publicity, (2) shutting down and modifying our facilities, and (3) enforcing the quiet rules.

Describes first phase

Improving Publicity. First, the library must publicize the noise problem. This assertive move will demonstrate the staff's interest. Publicity could include articles by staff members in the campus newspaper, leaflets distributed on campus, and a freshman library orientation acknowledging the noise problem and asking cooperation from new students. All forms of publicity should detail the steps being taken by the library to solve the problem.

Describes second phase

Shutting Down and Modifying Facilities. After notifying campus and local newspapers, you should close the library for one week. To minimize disruption, the shutdown should occur between the end of summer school and the beginning of the fall term.

During this period, you can convert the fixed tables on the fourth and fifth floors to cubicles with temporary partitions (six cubicles per table). You could later convert the cubicles to shelves as the need increases.

Then you can take all unfixed tables from the upper floors to the first floor, and set up a space for group study. Plans already are under way for removing the computer terminals from the fourth and fifth floors.

Describes third phase

Enforcing the Quiet Rules. Enforcement is the essential, long-term element in this plan. No one of any age is likely to follow all the rules all the time—unless the rules are enforced.

First, you can make new "Quiet" posters to replace the present, innocuous notices. A visual-design student can be hired to draw up large, colorful posters that attract attention. Either the design student or the university print shop can take charge of poster production.

Next, through publicity, library patrons can be encouraged to demand quiet from noisy people. To support such patron demands, the library staff can begin monitoring the fourth and fifth floors, asking study groups to move to the first floor, and revoking library privileges of those who refuse. Patrons on the second and third floors can be asked to speak in whispers. Staff members should set an example by regulating their own voices.

Costs and Materials

Estimates costs and materials needed

- The major cost would be for salaries of new staff members who would help monitor. Next year's library budget, however, will include an allocation for four new staff members.
- A design student has offered to make up four different posters for $200. The university printing office can reproduce as many posters as needed at no additional cost.
- Prefabricated cubicles for 26 tables sell for $150 apiece, for a total cost of $3,900.
- Rearrangement on various floors can be handled by the library's custodians.

The Student Fee Allocations Committee and the Student Senate routinely reserve funds for improving student facilities. A request to these organizations presumably would yield at least partial funding for the plan.

Personnel

Describes personnel needed

The success of this plan ultimately depends on the willingness of the library administration to implement it. You can run the program itself by committees made up of students, staff, and faculty. This is yet another area where publicity is essential to persuade people that the problem is severe and that you need their help. To recruit committee members from among students, you can offer Contract Learning credits.

The proposed committees include an Antinoise Committee overseeing the program, a Public Relations Committee, a Poster Committee, and an Enforcement Committee.

Feasibility

Assesses probability of success

On March 15, 19XX, I mailed survey letters to 25 New England colleges, inquiring about their methods for coping with noise in the library. Among the respondents, 16 stated that publicity and the administration's attitude toward enforcement were main elements in their success.

Improved publicity and enforcement could work for us as well. And slight modifications in our facilities, to concentrate group study on the busiest floors, would automatically lighten the burden of enforcement.

Benefits

Offers a realistic and persuasive forecast of benefits

Publicity will improve communication between the library and the campus. An assertive approach will show that the library is aware of its patrons' needs and willing to meet those needs. Offering the program for public inspection will draw the entire community into improvement efforts. Publicity, begun now, will pave the way for the formation of committees.

The library shutdown will have a dual effect: it will dramatize the problem to the community, and also provide time for the physical changes. (An antinoise program begun with carpentry noise in the quiet areas would hardly be effective.) The shutdown will be both a symbolic and a concrete measure, leading to reopening of the library with a new philosophy and a new image.

Continued strict enforcement will be the backbone of the program. It will prove that staff members care enough about the atmosphere to jeopardize their friendly image in the eyes of some users, and that the library is not afraid to enforce its rules.

Conclusion

The conclusion reaffirms the need for the project and persuades readers to act. It answers the questions readers can be expected to ask:

- *How badly do we need this change?*
- *Why should we accept your proposal?*
- *How do we know this is the best plan?*

End on a strong note, with a conclusion that is assertive, confident, and encouraging—and keep it short.

Re-emphasizes need and feasibility and encourages action

Conclusion and Recommendation

The noise in Margate University library has become embarrassing and annoying to the whole campus. Forceful steps are needed to restore the academic atmosphere.

Aside from the intangible question of image, close inspection of the proposed plan will show that it will work if you take the recommended steps and—most important—if daily enforcement of quiet rules becomes a part of the library's services.

In long proposals, especially those beginning with a comprehensive abstract, the conclusion can be omitted.

A Sample Situation

Audience and Use Profile for a Formal Proposal

Audience Identity and Needs

This proposal is aimed at one audience only: the citizens of Amherst and surrounding towns, who will be voting whether to grant the funding for the school renovation. Most readers are broadly familiar with this issue, having seen it debated for nearly two years. But they still need an item-by-item explanation of the problems created by ever-increasing enrollment. Since the project would be funded partially by increased property taxes, voters are asking tough questions such as these:

- Just how big a problem is this?
- Says who?
- Why, exactly, do we need this project?
- What will it cost?
- Why can't we wait?
- Suppose we do nothing?
- What will we get for our money?

ATTITUDE AND PERSONALITY

Nearly everyone (parents, taxpayers without school-age children, people on fixed incomes) has a real stake in this topic. Many voters tend to resist any request for more money by arguing that everyone has to economize during these difficult economic times. In short, audience attitude ranges from supportive to receptive to hesitant to defiant. This proposal is aimed at the hesitant population, since there is little chance of influencing those who reject the project outright.

To have a positive impact on readers, this proposal needs to make a case for the project and spell out a realistic plan, showing that the various committees are sincere in their intention to eliminate nonessential building costs.

EXPECTATIONS ABOUT THE DOCUMENT

Voters will scrutinize and evaluate this proposal for its soundness. Especially in a budget request, the audience expects no shortcuts. Each element of the project will have to be justified. All the relevant numbers (enrollment figures, projected tax increase, costs of delaying the project, and so on) will have to be given and explained in ways readers can understand.

The proposal will be designed as an 8 1/2 X 11-inch, fold-out brochure, to be mailed to each resident in the regional district. The document will be organized in a sequence that (1) identifies the problem, (2) lists costs to taxpayers, (3) justifies immediate need, (4) spells out relevant basic data and project background, (6) explains consequences of inaction, and (7) gives a detailed view of the additions and renovations. ■

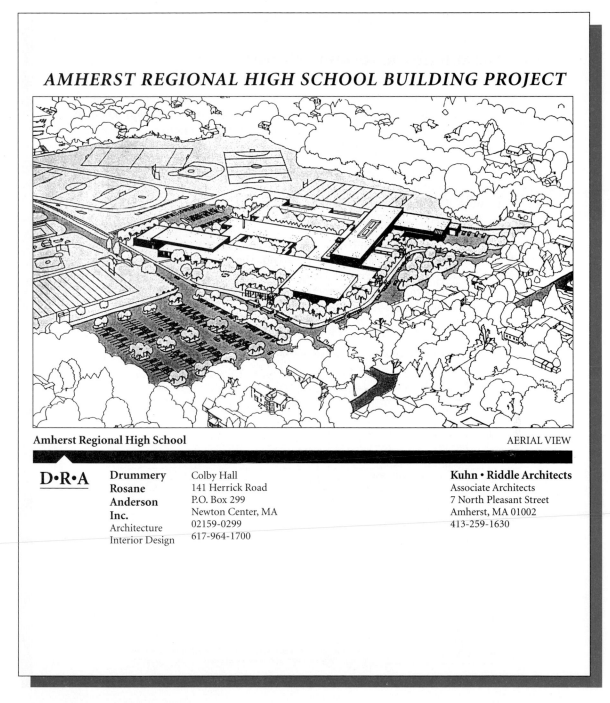

AMHERST REGIONAL HIGH SCHOOL BUILDING PROJECT

Amherst Regional High School   AERIAL VIEW

D•R•A   **Drummery**   Colby Hall   **Kuhn • Riddle Architects**
             Rosane   141 Herrick Road   Associate Architects
             Anderson   P.O. Box 299   7 North Pleasant Street
             Inc.   Newton Center, MA   Amherst, MA 01002
             Architecture   02159-0299   413-259-1630
             Interior Design   617-964-1700

Figure 22.1   A Planning Proposal.   *Reprinted with permission. Copyright © 1994 Amherst/Pelham Regional School Board; Drummery, Rosane, and Anderson, Inc; Kuhn Riddle Architects.*

1

WHY IS THE ADDITION/RENOVATION NEEDED?

Within one year both the junior and senior high schools will be above capacity and enrollments will continue to grow (Figure 1). Both buildings will require additions if the 7–9 and 10–12 split is maintained. By moving the 9th grade from the junior high school to the high school, we will eliminate the need for extensive additions to both buildings and create better opportunities for early adolescent and high school programming.

The students who will attend grades 9–12 over the next nine years are already enrolled in the elementary school of the four regional towns or in grades 7 and 8 in the present junior high. The number of students expected in the high school by the year 2000–1 nearly doubles from the current 759 (10–12) to 1,450 (9–12).

The high school was built in 1956 with an addition in 1964. All its systems are now between 30 and 40 years old and in need of upgrading. The state will only reimburse a school district for renovation if it is part of a large addition or expansion project. This project gives us the opportunity to renovate the existing building and bring it into compliance with current codes.

FIGURE 1. NO GROWTH vs. PROJECTED GROWTH IN GRADES 9–12 FOR 1993–2004

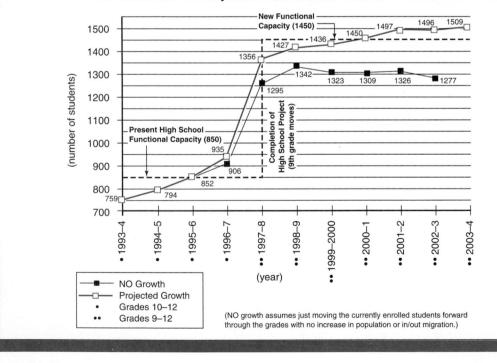

Figure 22.1   A Planning Proposal   *Continued*

2

WHAT WILL THE PROJECT COST?

The total construction project cost will not exceed $23,100,000, which includes the cost for construction, equipment, furnishings, and all fees. With the state reimbursing 67% of the project cost, the towns will be responsible for 33%, or approximately $7,700,000. This $7.7 million will be divided as follows:

Amherst $5,953,600 (77.32%)
Leverett $646,800 (8.40%)
Pelham $495,800 (6.44%)
Shutesbury $603,800 (7.84%)

The construction cost includes $13,475,000 for construction of the new addition and related site work plus $5,000,000 for renovation of the existing building. An additional $4,625,000 is budgeted for other project related costs, including furnishings, equipment, surveys, testing, architectural/engineering fees and contingencies.

WHY NOW?

Enrollments in the 1994–95 school year (Table 1) will effectively fill our secondary schools. Approval at this time by the four communities will allow ample time for preparation of construction documents necessary for state approval of the expansion. The project can then be bid in Spring 1995, with construction beginning in summer. Even under these circumstances, construction will not be complete until September 1997.

The School Building Committee is seeking approval at this time in order to:

1. allow construction to begin as soon as possible, providing additional space at the earliest possible date
2. provide ample time for design and final construction documents
3. minimize cost of temporary space arrangements
4. take advantage of favorable interest rates
5. avoid potential changes to the reimbursement formula which could increase the cost to the towns. Under the current formula the Amherst-Pelham Regional School District is eligible for 67% reimbursement from the state.

TABLE 1 OCTOBER 1, 1993 ACTUAL ENROLLMENTS

GRADE	K	1	2	3	4	5	6	7	8	9	10	11	12
AMHERST	222	265	252	239	252	253	272	273	231	216	227	198	183
PELHAM	24	17	15	17	22	17	16	12	11	20	11	11	11
LEVERETT	20	28	35	31	28	23	26	23	28	22	18	12	29
SHUTESBURY	26	27	29	30	39	27	31	28	24	18	26	15	18
TOTALS	292	337	331	317	341	320	345	336	294	276	282	236	241

Figure 22.1   A Planning Proposal   *Continued*

3

THE BASICS:

School Growth: General increase in the school population plus the addition of ninth grade to the high school will require housing 1,450 students by the year 2000. Current high school enrollment is 759.

The Scope: New construction of 118,500 square feet to include:
* Enlarged cafeteria
* Enlarged library
* New labs for science and technology education
* 22 multi-purpose classrooms
* Large group instruction room
* 2 exercise rooms for physical education
* Expanded space for:
>special education
>T.B.E. (transitional bilingual education)
>E.S.L. (English as a second language)
>art instruction
>music instruction
>teacher resource areas
* Improved parking and bus access

Refurbishing/renovation of 118,750 of existing square feet to include:

* Heating	* Lighting
* Ventilating	* Electrical
* Plumbing	* Fire protection
* Reroofing	* Energy conservation

The Cost: $23.1 million for the total construction project including furnishings, equipment and fees, of which 67% ($15.4 million) is reimbursable by the state. The state will also reimburse 67% of the interest costs.

The Tax Impact: The tax impact will be spread over the 20-year bonding period. In general it will be higher in the earliest years and will decline over the remaining time, with the state reimbursement being constant over this period. The tax impact will vary among the four towns.

The Time Table: With a positive vote in all four towns this Spring, ground could be broken by June 1995. Phased construction will allow the school to remain open during construction, with completion and full occupancy by September 1997.

Figure 22.1   A Planning Proposal   *Continued*

4

BACKGROUND:

* Regional School Building Needs Committee, over 3 years, studied:
  a. Student population growth
  b. Space alternatives
  c. Possibility of third secondary school
  d. Feasibility of building additions/renovations to both junior and senior high school.

* Regional School Committee voted to move 9th grade from junior high to high school because the change:
  a. Unites the 9th grade with 9-12 curriculum in one location
  b. Requires only one building project.

* Specifications Committee, using State Department of Education criteria, determined:
  a. Use of existing space
  b. Total space requirements for a school with an enrollment of 1,450, based on continuation of existing school programs
  c. Flexibility needed for the future.

* School Committee, Building Committee, and State Department of Education have approved the Educational Specifications which reflect the current and projected needs for space by departments, programs and other building functions.

* Regional School Building Committee:
  a. Selected an architect for the project
  b. Selected a project design from several alternatives
  c. Reviewed thoroughly the conditions of the existing building
  d. Consulted with the state's School Building Assistance Bureau
  e. Reviewed and revised the scope and costs of the project.

WHAT WILL HAPPEN IF NO ACTION IS TAKEN?

If the high school is not expanded, both the junior and senior high schools will experience crippling overcrowding in the very near future. Both schools would likely require staggered sessions, and/or temporary modular classrooms, with early and late arrival/dismissal times. This will lead to a general erosion of the quality of education because of severely limited program and extra curricular opportunities for students. School operating expenses and transportation costs will increase. The cost of temporary classrooms is not reimbursable by the state. Additionally, the cost for ADA compliance and future repairs to major building systems would have to be borne 100% by local taxpayers.

Figure 22.1   A Planning Proposal   *Continued*

5

WHAT IS THE PROPOSED DESIGN?

The Building Committee established a program based on the educational specifications and a desire to:

* Avoid a sprawling floor plan
* Produce a functional and economical design
* Centralize locations of core facilities, such as the library
* Consolidate and centralize circulation of students.

Working together, the architects and Building Committee reviewed many options for adding onto the existing school. These options included adding over the existing building, filling in courtyards, adding many small additions and consolidating all the space in a one, two, or three-story addition. The final design, as described below, is a functional, compact, and economic response to the program requirements (Figures 2a and 2b).

The recommended addition/renovation project has two components:

(1) It adds 118,500 square feet of new construction to house the additional square footage needed due to increased enrollments and,

(2) It renovates 118,750 square feet of the existing building to upgrade the 40-year old structure and bring it into the 21st century.

The central, organizing element of the project will be a new, wide, ramped corridor. or "Main Street". that will run the length of the school between the existing building on the west and the 3-story addition on the east. The west side of the corridor will tie into all three of the main corridors of the existing school. Along the east side of the corridor will be a new library, new cafeteria and new entry into the school from the east side of the addition. At the north end of the "Street" on the first floor will be the technical education wing. The second floor will house the new science labs, the art classrooms and several multi-purpose classrooms. Finally, the third floor will be devoted to multi-purpose classroom space. To accomplish the meshing of old and new, and to achieve an efficient and compact plan, several existing science labs and shops will be razed.

The addition will house those core facilities which need to be expanded due to increased enrollments (i.e., library and cafeteria) and those spaces where substantial technological improvements were needed (i.e., art, tech. education and science labs). Finally, the third floor's multi-purpose classrooms give the entire building flexibility to adapt to a variety of educational philosophies, such as a "house" concept, should this approach be used in the future.

Figure 22.1   A Planning Proposal   *Continued*

6

The addition will be clad in masonry and massed in a way so that it blends in with the existing school. An elevator and three stairways will link the three floors in an efficient, functional manner. The addition will be energy-efficient, handicapped accessible, technologically up-to-date, and sprinklered for fire safety.

In order for the addition and existing school to blend together as a cohesive unit, (and to obtain state approval for our reimbursement), the existing building must be updated and renovated. The existing cafeteria will be converted into a large group instruction area, expanded bathrooms for the auditorium, and special education spaces. The existing library will be converted into small classroom spaces. The present art classroom will be converted to an additional music room. The auditorium will be updated with new lighting and sound systems. For the most part, the existing classroom layout will remain the same and will be upgraded.

A concerted effort has been made to use existing systems where possible. All HVAC, electrical, plumbing, and fire safety systems will be upgraded as needed for energy efficiency and code compliance. A new roof, new lockers, insulation and windows will be installed and all interior spaces will be updated for compliance with ADA and state accessibility guidelines.

A small addition will be constructed on the west side of the gym to house two, new, 2,500-square-foot exercise rooms to augment the need for additional physical education teaching stations. It will also house a training room and additional storage space.

Finally, the site will be improved in several ways (Figure 3). The existing parking lot will be expanded north for an additional 54 parking spaces, and an additional 58-car parking lot will be added on the west side of the gym. The road and intersection immediately south of the school will be rearranged to allow for safer and more efficient traffic patterns. A new entrance canopy on the south side will help focus attention to the existing main entrance. The Junior High connector road will be altered to allow for an automobile drop-off area adjacent to the new east entrance to the addition.

In summary, all the new square footage (except for the exercise rooms), is consolidated into one new addition, linked to the existing school by a new major circulation corridor. Every effort has been made to tie together an updated 40-year-old school with a new addition so they will work as a whole and will serve our four regional communities well into the 21st Century.

Figure 22.1   A Planning Proposal   *Continued*

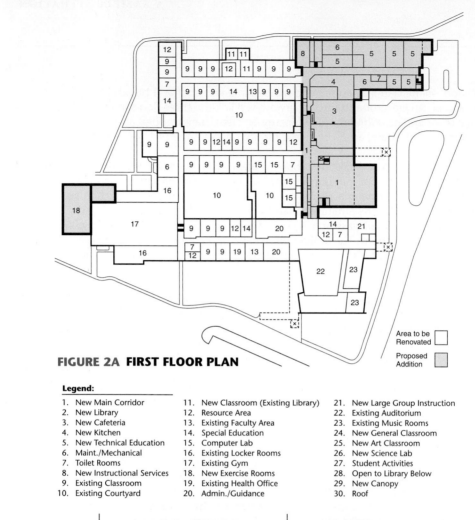

FIGURE 2A FIRST FLOOR PLAN

Legend:

1. New Main Corridor	11. New Classroom (Existing Library)	21. New Large Group Instruction
2. New Library	12. Resource Area	22. Existing Auditorium
3. New Cafeteria	13. Existing Faculty Area	23. Existing Music Rooms
4. New Kitchen	14. Special Education	24. New General Classroom
5. New Technical Education	15. Computer Lab	25. New Art Classroom
6. Maint./Mechanical	16. Existing Locker Rooms	26. New Science Lab
7. Toilet Rooms	17. Existing Gym	27. Student Activities
8. New Instructional Services	18. New Exercise Rooms	28. Open to Library Below
9. Existing Classroom	19. Existing Health Office	29. New Canopy
10. Existing Courtyard	20. Admin./Guidance	30. Roof

FIGURE 2B SECOND FLOOR PLAN AND THIRD FLOOR PLAN

Figure 22.1   A Planning Proposal *Continued*

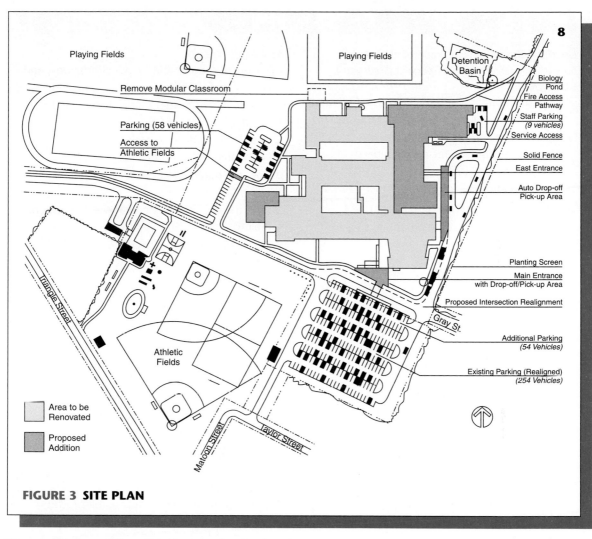

Playing Fields

Playing Fields

Detention
Basin

8

Biology
Pond

Fire Access
Pathway

Staff Parking
(9 vehicles)

Service Access

Solid Fence

East Entrance

Auto Drop-off
Pick-up Area

Remove Modular Classroom

Parking (58 vehicles)

Access to
Athletic Fields

Planting Screen

Main Entrance
with Drop-off/Pick-up Area

Proposed Intersection Realignment

Gray St.

Additional Parking
(54 Vehicles)

Existing Parking (Realigned)
(254 Vehicles)

Triangle Street

Athletic
Fields

Matoon Street

Taylor Street

Area to be
Renovated

Proposed
Addition

FIGURE 3 SITE PLAN

Figure 22.1 A Planning Proposal *Continued*

Revision Checklist for Proposals

Use this list to revise your proposal. (Numbers in parentheses refer to the first page of discussion.)

FORMAT

❑ Is the short internal proposal in memo form and the short external program in letter form? (525)

❑ Does the long proposal have adequate supplements to serve the different needs of different readers? (530)

❑ Is the format professional in appearance? (530)

❑ Are headings logical and adequate? (534)

❑ Does the title forecast the proposal's subject and purpose? (530)

CONTENT

❑ Is the problem clearly identified? (530)

❑ Is the objective clearly identified? (530)

❑ Does everything in the proposal support its objective? (533)

❑ Does the proposal *show* as well as *tell?* (532)

❑ Is the proposed plan, service, or product beneficial? (531)

❑ Are the proposed methods practical and realistic? (537)

❑ Are all foreseeable limitations and contingencies identified? (531)

❑ Is the proposal free of overstatement? (289)

❑ Are visuals used effectively and whenever appropriate? (532)

❑ Is the proposal's length appropriate to the subject? (522)

ARRANGEMENT

❑ Is there a recognizable introduction, body, and conclusion? (534)

❑ Are all *relevant* headings from the general outline included? (534)

❑ Does the introduction provide sufficient orientation to the problem and the plan? (534)

❑ Does the body explain *how, where,* and *how much?* (536)

❑ Does the conclusion encourage acceptance of the proposal? (539)

❑ Are there clear transitions between related ideas? (Appendix)

STYLE

❑ Is the level of technicality appropriate for primary readers? (533)

❑ Do supplements follow the appropriate style guidelines? (376)

❑ Will the informative abstract be understood by all readers? (382)

❑ Does the tone connect with readers? (532)

❑ Is the writing style throughout clear, concise, and fluent? (261)

❑ Is the language convincing and precise? (261)

❑ Is the proposal written in grammatical English? (Appendix)

❑ Is the proposal ethically acceptable? (531)

☑ EXERCISES

1. Assume the head of your high school English Department has asked you, as a recent graduate, for suggestions about revising the English curriculum to prepare students for writing. Write a proposal, based on your experience since high school. (Primary audience: the English Department head and faculty; secondary audience: school committee.) In your external, solicited proposal, identify problems, needs, and benefits and spell out a realistic plan. Review the outline on page 534 before selecting specific headings.

2. After identifying your primary and secondary audience, compose a short planning proposal for improving an unsatisfactory situation in the classroom, on the job, or in your dorm or

apartment (e.g., poor lighting, drab atmosphere, health hazards, poor seating arrangements). Choose a problem or situation whose resolution is more a matter of common sense and lucid observation than of intensive research. Be sure to (a) identify the problem clearly, give brief background, and stimulate the readers' interest; (b) state clearly the methods proposed to solve the problem; and (c) conclude with a statement designed to gain readers' support for your proposal.

3. Write a research proposal to your instructor (or an interested third party) requesting approval for the final term project (an analytical report or formal proposal). Identify the subject, background, purpose, and benefits of your planned inquiry, as well as the intended audience, scope of inquiry, data sources, methods of inquiry, and a task timetable. Be certain that adequate primary and secondary sources are available. Convince your reader of the soundness and usefulness of the project.

4. As an alternate term project to the formal analytical report (Chapter 23), develop a long proposal for solving a problem, improving a situation, or satisfying a need in your school, community, or job. Choose a subject sufficiently complex to justify a formal proposal, a topic requiring research (mostly primary). Identify an audience (other than your instructor) who will use your proposal for a specific purpose. Compose an audience and use profile, using the sample on pages 539–40 as a model. Here are possible subjects for your proposal:

- improving living conditions in your dorm or fraternity/sorority
- creating a student-oriented advertising agency on campus

- creating a day care center on campus
- creating a new business or expanding a business
- saving labor, materials, or money on the job
- improving working conditions
- improving campus facilities for the disabled
- supplying a product or service to clients or customers
- increasing tourism in your town
- eliminating traffic hazards in your neighborhood
- reducing energy expenditures on the job
- improving security in dorms or in the college library
- improving in-house training or job orientation programs
- creating a one-credit course in job hunting or stress management for students
- improving tutoring in the learning center
- making the course content in your major more relevant to student needs
- Creating a new student government organization
- finding ways for an organization to raise money
- improving faculty advising for students
- purchasing new equipment
- improving food service on campus
- easing first-year students through the transition to college
- changing the grading system at your school
- establishing more equitable computer terminal use

Analytical Reports

CHAPTER

23

THE formal analytical report, like the short recommendation reports in Chapter 21, leads to recommendations. The formal report replaces the memo when the topic requires lengthy discussion.

A form of critical thinking in which we separate a whole into its parts, analysis is essential to workplace problem solving.

Imagine, for example, you receive this assignment from your supervisor:

> Recommend the best method for removing the heavy-metal contamination from our company dump site.

To make the best recommendation, you will have to analyze the situation to learn everything you can about the nature and extent of the problem and the feasibility of various solutions. This is where you apply critical thinking skills from Chapter 2 along with the research process discussed in Chapters 8 and 9. As part of your research, you might do a literature search to discover whether anyone has been able to solve a problem like yours, or to learn about the newest technologies for toxic cleanup. You then have to decide how much, if any, of what others have done can apply to your situation.

Purpose of Analysis

As an employee you may be asked to evaluate a new assembly technique on the production line, or to locate and purchase the best equipment at the best price. You might have to identify the cause behind a monthly drop in sales, the reasons for low employee morale, the causes of an accident, or the reasons for equipment failure. You might need to assess the feasibility of a proposal for a company's expansion or investment. This list is endless but the procedure remains the same: (1) asking the right questions, (2) searching for information, (3) evaluating and interpreting your findings, and (4) drawing conclusions and making recommendations.

Typical Analytical Problems

Far more than an encyclopedia presentation of information, the analytical report shows how you arrived at your conclusions and recommendations. Here are some typical analytical problems.

Will X Work for a Specific Purpose? Analysis can answer practical questions. Say your employer is concerned about the effects of stress on employees. She asks you to investigate the claim that low-impact aerobics has therapeutic benefits—with an eye toward such a program for employees. You design your analysis to answer this question: *Do long-impact aerobics programs significantly reduce stress?* The analysis follows a *questions-answers-conclusions* sequence. Because the report could lead to action, you include recommendations based on your conclusions.

Is X or Y Better for a Specific Purpose? Analysis is essential in comparing machines, processes, business locations, computer systems, or the like. Assume that you manage a ski lodge and need to answer this question: *Which of the two most popular ski bindings is best for our rental skis?* In a comparative analysis of the Salomon 555 and the Americana bindings, you would assess the strengths and weaknesses of each binding on the basis of specific criteria (safety, cost, ease of repair, ease of adjustment, dependability, and so on) which you would rank in order of importance.

The comparative analysis follows a *questions-answers-conclusions* sequence and is designed to help the reader make a choice. Examples appear in magazines such as *Consumer Reports* and *Consumer's Digest.*

Why Does X Happen? The causal analysis is designed to answer questions like this: *Why do small businesses have a high failure rate?* This kind of analysis follows a variation of the questions-answers-conclusions structure: namely, *problem-causes-solution.* Such an analysis follows this sequence:

1. identifying the problem
2. examining possible and probable causes, and isolating definite ones
3. recommending solutions

An analysis of low employee morale would investigate causal relationships.

How Can X Be Improved or Avoided? Another form of problem solving focuses on desired results and recommends methods of achieving these results. This type of analysis answers questions like these: *How can we improve safety procedures in our plant? How can we operate our division more efficiently? How can we improve campus security?*

What Are the Effects of X? An analysis of the consequences of an event or action would answer questions like these: *How has air quality been affected by the local power plant's change from burning oil to coal? Does electromagnetic radiation pose a significant health risk?*

Another kind of problem-solving analysis is done to predict an effect: "What are the consequences of my changing majors?" Here, the sequence is *proposed action–probable effects–conclusions and recommendations.*

Is X Practical in This Situation? The feasibility analysis assesses the practicality of an idea or plan: *Will the consumer interests of Hicksville support a microcomputer store?* In a variation of the questions-answers-conclusions structure, a feasibility analysis presents *reasons for–reasons against,* with both sides supported by evidence. Business owners often use this type of analysis.

Combining Types of Analysis. Types of analytical problems overlap considerably. Any one study may in fact require answers to two or more of the previous questions. The sample report on pages 572–81 is both a feasibility analysis and a comparative analysis. It is designed to answer these questions: Is technical marketing the right career for me? If so, how do I enter the field?

Elements of Analysis

The analytical report incorporates many writing strategies from earlier assignments, along with the guidelines that follow.

Clearly Identified Problem or Question

Know what you're looking for. If your car's engine fails to turn over when you switch on the ignition, you would wisely check battery and electrical connections before dismantling parts of the engine. Apply a similar focus to your report.

On page 553, a hypothetical employer posed this question: *Will a low-impact aerobics program significantly reduce stress among my employees?* The aerobics question obviously requires answers to three other questions: *What are the therapeutic claims for aerobics? Are they valid? Will aerobics work in this situation?* How aerobic exercise got established, how widespread it is, who practices, and other such questions are not relevant to this problem (although some questions about background might be useful in the report's introduction). Always begin by defining the main questions and thinking through any subordinate questions they may imply. Only then can you determine the data or evidence you need.

With the main questions identified, the writer of the aerobics report can formulate her statement of purpose:

Definite

> The purpose of this report is to examine and evaluate claims about the therapeutic benefits of low-impact aerobic exercise.

The writer might have mistakenly begun instead with this statement:

Vague

> This report examines low-impact aerobic exercise.

Words such as *examine* and *evaluate* (or *compare, identify, determine, measure, describe,* and so on) enable readers to understand the specific analytical activity that is the subject of the report.

Notice how the first version sharpens the focus by expressing the precise subject of the analysis: not aerobics (a huge topic), but the alleged *therapeutic benefits* of aerobics.

Define your purpose by condensing your approach to a basic question: *Does low-impact aerobic exercise have therapeutic benefits?* or *Why have our sales dropped steadily for three months?* Then restate the question as a declarative sentence in your statement of purpose.

Subordination of Personal Bias

Interpret evidence impartially. Throughout your analysis, stick to your evidence. Do not force viewpoints on your material that are not substantiated by dependable evidence.

Accurate and Adequate Data

Do not distort the original data by excluding vital points. Say you are asked to recommend the best chainsaw for a logging company. Reviewing test reports, you come across this information.

> Of all six brands tested, the Bomarc chainsaw proved easiest to operate. It also had the fewest safety features, however.

If you cite these data, present *both* findings, not simply the first—even though you may prefer the Bomarc brand. *Then* argue for the feature you think should receive priority.

As space permits, include the full text of interviews or questionnaires in appendixes.

Fully Interpreted Data

Explain the significance of your data. Interpretation is the heart of the analytical report. You might interpret the chainsaw data in this way:

> Our cutting crews often work suspended by harness, high above the ground. Also, much work is in remote areas. Safety features therefore should be our first requirement in a chainsaw. Despite its ease of operation, the Bomarc saw does not meet our safety needs.

By saying "therefore" you engage in analysis—not mere information sharing. *Merely listing your findings is not enough.* Tell readers what your findings mean.

Clear and Careful Reasoning

Each stage of your analysis requires decisions about what to record, what to exclude, and where to go next. As you evaluate your data *(Is this reliable and important?)*, interpret your evidence *(What does it mean?)*, and make recommendations based on your conclusions *(What action is needed?)*, you might have to alter your original plan. Remain flexible enough to revise your thinking if contradictory new evidence appears.

Appropriate Visuals

Use visuals generously (Chapter 14). Graphs are especially useful in an analysis of trends (rising or falling sales, radiation levels). Tables, charts, photographs, and diagrams work well in comparative analyses.

Valid Conclusions and Recommendations

Along with the informative abstract, conclusions and recommendations are the sections of a long report that receive most attention from readers. The goal of analysis is to reach a valid *conclusion*—an overall judgment about what all the material means (that X is better than Y, that B failed because of C, that A is a good plan of action). Here is the conclusion of a report on the feasibility of installing an active solar heating system in a large building:

Offer a final judgment

1. Active solar space heating for our new research building is technically feasible because the site orientation will allow for a sloping roof facing due south, with plenty of unshaded space.
2. It is legally feasible because we are able to obtain an access easement on the adjoining property, to ensure that no buildings or trees will be permitted to shade the solar collectors once they are installed.
3. It is economically feasible because our sunny, cold climate means high fuel savings and faster payback (fifteen years maximum) with solar heating. The long-term fuel savings justify our short-term installation costs (already minimal because the solar system can be incorporated during the building's construction—without renovations).

Conclusions are valid when they are logically derived from accurate interpretation.

Having explained *what it all means*, you then recommend *what should be done*. Taking into account all possible alternatives, your recommendations urge specific action (to invest in *A* instead of *B*, to replace *C* immediately, to follow plan *A*, or the like). Here are the recommendations based on the previous interpretations:

Tell what should be done

1. I recommend we install an active solar heating system in our new research building.
2. We should arrange an immediate meeting with our architect, building contractor, and solar-heating contractor. In this way, we can make all necessary design changes before construction begins in two weeks.
3. We should instruct our legal department to obtain the appropriate permits and easements immediately.

Recommendations are valid when they propose an appropriate response to the problem or question.

Because they culminate your research and analysis, recommendations challenge your imagination, your creativity, and—above all—your critical thinking skills. Having reached a valid conclusion about *what is,* you now must decide *what ought to be done.* But what strikes one person as a brilliant idea might be seen by others as idiotic or offensive. Depending on whether recommendations are carefully thought out or off the wall, writers earn an audience's respect or its scorn. (Figure 23.1 depicts the kinds of decisions writers encounter in formulating, evaluating, and refining their recommendations.)

When you do achieve definite conclusions and recommendations, express them with assurance and authority. Unless you have reason to be unsure, avoid noncommittal statements ("It would seem that" or "It looks as if"). Be direct and assertive ("The earthquake danger at the reactor site is acute," or "I recommend an immediate investment"). Let readers know where you stand.

If, however, your analysis yields nothing definite, do not force a simplistic conclusion on your material. Instead, explain your position ("The contradictory responses to our consumer survey prevent me from reaching a definite conclusion. Before we make any decision about this product, I recommend a full-scale market analysis"). The wrong recommendation is far worse than no recommendation at all.

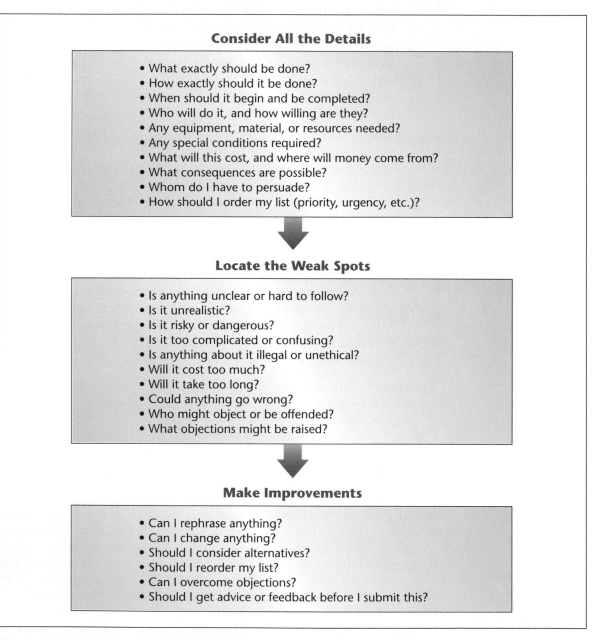

Consider All the Details

- What exactly should be done?
- How exactly should it be done?
- When should it begin and be completed?
- Who will do it, and how willing are they?
- Any equipment, material, or resources needed?
- Any special conditions required?
- What will this cost, and where will money come from?
- What consequences are possible?
- Whom do I have to persuade?
- How should I order my list (priority, urgency, etc.)?

Locate the Weak Spots

- Is anything unclear or hard to follow?
- Is it unrealistic?
- Is it risky or dangerous?
- Is it too complicated or confusing?
- Is anything about it illegal or unethical?
- Will it cost too much?
- Will it take too long?
- Could anything go wrong?
- Who might object or be offended?
- What objections might be raised?

Make Improvements

- Can I rephrase anything?
- Can I change anything?
- Should I consider alternatives?
- Should I reorder my list?
- Can I overcome objections?
- Should I get advice or feedback before I submit this?

Figure 23.1   How to Think Critically About Your Recommendations
Source: Questions adapted from Ruggiero, Vincent R. The Art of Thinking, *3rd ed. New York: Harper, 1991: 162–65.*

A General Model for Analytical Reports

Whether you outline earlier or later, the finished report depends on a good outline. This model outline can be adapted to most analytical reports:

I. INTRODUCTION
    A. Definition, Description, and Background
    B. Purpose of the Report, and Intended Audience
    C. Sources of Information
    D. Working Definitions (here or in a glossary)
    E. Limitations of the Study
    F. Scope of the Inquiry (topics listed in order of importance)

II. COLLECTED DATA
    A. First Topic for Investigation
        1. Definition
        2. Findings
        3. Interpretation of findings
    B. Second Topic for Investigation
        1. First subtopic
            a. Definition
            b. Findings
            c. Interpretation of findings
        2. Second subtopic (and so on)

III. CONCLUSION
    A. Summary of Findings
    B. Overall Interpretation of Findings (as needed)
    C. Recommendations (as needed and appropriate)

This outline is only tentative. Modify if necessary.

Two sample reports in this text follow our model outline. The first report, "Children Exposed to Electromagnetic Radiation: A Risk Assessment" (minus the document supplements that ordinarily accompany a long report), begins on page 560. The second report, "The Feasibility of a Technical Marketing Career," appears in Figure 23.2. (Supplements for the latter report appear in Chapter 16.)

Each report responds to slightly different questions. The first tackles these questions: *What are the effects of X and what should we do about them?* The second tackles two questions: *Is X feasible, and which version of X is better for our purposes?* At least one of these reports should serve as a model for your own analysis.

Introduction

The introduction engages and orients readers, and provides background—as briefly as possible. Identify the origin and significance of the topic, define or describe the problem or issue, and explain the report's purpose. Identify your audience only in a report for your instructor, and only if you don't attach an audience and use profile. Identify briefly your data sources (which you will later document fully), and explain any omissions of data (e.g., a person unavailable for an interview, or research still in progress). List working definitions, unless you have so many that a glossary is needed. If you do include a glossary and appendixes, notify readers early. Finally, preview the scope of your analysis by listing the major topics you will discuss in the body section.

Not all reports require every element. Give readers only what they need and expect.

As you read the following introduction, think about the elements designed to engage and orient readers (i.e., local citizens), and evaluate their effectiveness. (Review pages 124–29 for the situation that gave rise to this report.)

Children Exposed to Electromagnetic Radiation: A Risk Assessment

Laurie A. Simoneau

Introduction

Definition and background

Wherever electricity flows—through the largest transmission line or the smallest appliance—it emits varying intensities of charged waves known as an elctromagnetic field (EMF). Some medical studies have linked human exposure to EMFs with definite physiological changes and with possible illness, including cancer, miscarriage, and depression.

Some experts disagree over the extent of health risk, if any, from EMFs. Others question whether EMF risk is greater from high-voltage transmission lines, the smaller distribution lines strung on utility poles, or household appliances. Conclusive research may take years; meanwhile, concerned citizens worry about avoiding potential risks.

Description of problem

In Bocaville, four sets of transmission lines—two at 115 Kilovolts (kV) and two at 500 kV—cross residential neighborhoods and public property. The Adams elementary school is less than 100 feet from this power-line corridor. EMF risks, whatever they may be, are thought to increase with proximity of exposure.

Purpose and data sources

On the basis of examination of recent research and interviews with local authorities, this report assesses whether potential health risks from EMF exposure seem significant enough for Bocaville to (a) increase public awareness, (b) divert the transmission lines that run adjacent to the elementary school, and (c) implement widespread precautions in the transmission and distribution of electrical power throughout Bocaville.

Scope of inquiry

Five major topics are covered in this report: what we know about the various sources of EMFs, what research indicates about the physiological and health effects from EMF exposure, how experts differ in their interpretations of the research, what the power industry has to say, and what is being done locally and nationwide to avoid risk.

Body

The body section describes and explains your findings. Present a clear and detailed picture of the evidence, interpretations, and reasoning on which you will base your conclusion. Divide topics into subtopics, and use clear, informative headings that show readers exactly where they are at any point in the discussion. Remember your ethical responsibility for presenting a *fair* and *balanced* treatment of the material, instead of loading the report with only those findings that support your viewpoint.

As you read the following body section, evaluate how effectively it informs readers, keeps them on track, reveals a clear line of reasoning, and presents an impartial analysis.

Data Section

Sources of EMF Exposure

First topic

Definition

Electromagnetic intensity is measured in *milligauss* (mG), a unit of electrical measurement. The higher the milligauss reading, the stronger the electromagnetic field. Consistent exposure above 1–2 mG is thought to increase cancer risk significantly (White 15), but no scientific evidence concludes that exposure even below 2.5 mG is safe.

Findings

Table 1 gives the EMF intensities generated by electric power lines during average and peak usage. These intensities are measured within the power line right-of-way and at varying distances from the lines.

Table 1 EMF Emissions from Power Lines (in milligauss)

Types of Transmission Lines	Maximum on Right-of-Way	Distance from lines			
		50'	100'	200'	300'
115 Kilovolts (kV)					
Average usage	30	7	2	0.4	0.2
Peak usage	63	14	4	0.9	0.4
230 Kilovolts (kV)					
Average usage	58	20	7	1.8	0.8
Peak usage	118	40	15	3.6	1.6
500 Kilovolts (kV)					
Average usage	87	29	13	3.2	1.4
Peak usage	183	62	27	6.7	3.0

Source: United States Environmental Protection Agency, EMF in Your Environment. *Washington: GPO, 1992. Data from Bonneville Power Administration.*

Interpretation

As Table 1 indicates, EMF intensity drops substantially as distance from the power lines increases.

Although the 2 million miles of power lines crisscrossing the country have been the focus of the EMF controversy, potentially harmful electromagnetic waves also are emitted by household wiring, appliances, electric blankets, and computer terminals—and even from the earth's natural magnetic field. The background magnetic field (at a safe distance from any electrical appliances) in the average American home varies from 0.5 to 4.0 mG (U.S. EPA 10). Table 2 compares approximate intensities of various sources inside the home and from the earth's magnetic field.

Table 2 EMF Emissions from Selected Sources (in milligauss)

Source	Range[a,b]
Earth's magnetic field	0.1–2.5
Blow-dryer	60–1400
Four in. from TV screen	40–100
Four ft. from TV screen	0.7–9
Flourescent lights	10–12
Electric razor	1200–1600
Electric blanket	2–25
Computer terminal (12 inches away)	3–15
Toaster	10–60

[a]Data from Paul Brodeur, "ANNALS OF RADIATION: The Cancer at Slater School," *The New Yorker* 7 Dec. 1992: 188; Michael Castleman, "Electromagnetic Fields," *Sierra* Jan./Feb. 1992: 21–22; John Miltane, Interview, 5 Apr. 1993. "What Utilities Can Do about the EMF Threat," *Electrical World* Nov. 1988: 37; M. Yakutchik, "A New Jolt of Concern," *USA WEEKEND* 1 Jan. 1993: 5.

[b]Readings are done with a gaussmeter, and tend to vary greatly, depending on technique, the proximity of the gaussmeter to the source, its direction of aim, and other random factors.

EMF intensity from certain appliances tends to be higher than from transmission lines because of the amount of current involved.

Definitions

Finding

Voltage measures the speed and pressure of electricity in wires, but *current* measures the volume of electricity passing through wires. Current is what produces electromagnetic fields, and current is measured in amperage. The current flowing through a transmission line typically ranges from 200 to 400 amps. Most homes have a 200-amp service. This means that if every electrical item in the house were turned on at the same time, the house could run about 200 amps—almost the same as the transmission line. Consumers then have the ability to put 200 amps of current-flow into their homes, while transmission lines carrying 200 to 400 amps are at least 50 feet away (Miltane).

Proximity and duration of exposure, however, are other factors in determining risk. People are exposed to EMFs from home appliances at close proximity, but appliances run only periodically. Exposure is therefore sporadic, and intensity diminishes sharply within a few feet (Figure 1).

As Figure 1 indicates, EMF intensity drops dramatically over very short distances from the typical appliance.

Finding

Power line exposure, on the other hand, is at a greater distance (usually 50 feet or more), but is constant. Moreover, intensity can remain strong well beyond 100 feet (Castleman 21; Miltane). In fact, buildings as far as 150 feet from high-current lines routinely show constant readings of 2 to 3 milligauss (Brodeur 87).

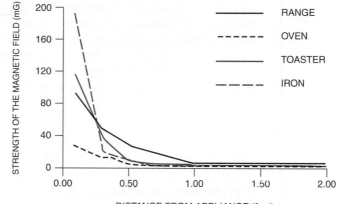

FIGURE 1 EMF Strengths of Typical Electric Appliances. *Source: United States Environmental Protection Agency.* EMF In Your Environment. *Washington: GPO, 1992.*

Interpretation

Research has not yet determined which type of exposure might be more harmful: to higher intensities for short intervals, or to lower intensities constantly. But in either case, proximity seems most significant because EMF intensity diminishes rapidly with distance.

Physiological Effects and Health Risks from EMF Exposure

Second topic

Research on EMF exposure falls into two categories: epidemiologic studies and laboratory studies. The findings are sometimes controversial and inconclusive, but nonetheless disturbing.

First subtopic
General findings

Epidemiologic Studies

Epidemiologic studies look for statistical correlations between EMF exposure and certain diseases and physical disorders. Of 77 such studies between 1979 and 1992, slightly over 70 percent suggest that EMF exposure increases the incidence of the following conditions (Lee et al. 12–28; Pinsky 155–215):

- cancer, especially leukemia and brain tumors
- miscarriage
- stress and depression
- learning disabilities

In one study, for example, pregnant women working at computer terminals at least 20 hours weekly had a risk of miscarriage twice that of other pregnant clerical workers (Pinsky 212–13).

Following are summaries of four noted epidemiologic studies implicating EMFs in childhood cancer:

Detailed findings

A Landmark Study of the EMF/Cancer Connection. A Denver, Colorado study by Wertheimer and Leeper in 1979 was the first to implicate EMFs as a cause of cancer. These researchers compared hundreds of homes in which children had developed cancer with similar homes in which children were cancer free. Children with cancer were found to be two to three times as likely to live in "high-current homes" (within 130 feet of a transmission line or within 50 feet of a distribution line).

This study, however, has been criticized because (1) it was not "blind" (researchers knew which homes cancer victims were living in), and (2) they never took gaussmeter readings to verify their designation of "high-current" homes (Pinsky 160–62; Taubes 96).

Detailed findings

Follow-up Studies. Several major studies of the EMF/cancer connection have confirmed Wertheimer's earlier findings:

• In 1988, Savitz studied hundreds of Denver houses and found that children with cancer were 1.7 times as likely to live in high-current homes. Unlike his predecessors, Savitz did not know whether a cancer victim lived in the home he was measuring, and he took gaussmeter readers to verify that houses could be designated "high-current" (Pinsky 162–63).

• In 1990, London and Peters found that Los Angeles children had 2.5 times more risk of developing leukemia if they lived near power lines (Brodeur 115).

• In 1992, a massive Swedish study found that children in houses with average intensities of greater than one milligauss had twice the normal risk of developing leukemia; at greater than two milligauss, the risk nearly tripled; at greater than three milligauss, it nearly quadrupled (Brodeur 115).

• In 1995, University of Colorado researchers evaluated and synthesized findings from eleven separate studies of power line EMF effects (a *meta-analysis*). Among their conclusions: (1) prolonged human exposure to criss-crossed power lines significantly increases the risk for cancers in general; (2) prolonged exposure to any power line increases the risk for leukemia (Miller et al. 281–87).

Detailed findings

Workplace Studies. More than 80 percent of 51 studies from 1981 to 1993— most notably a 1992 Swedish study—concluded that electricians, electrical engineers, and power line workers constantly exposed to an average of 1.5 to 4.0 milligauss had a significantly elevated cancer risk (Brodeur, 115; Pinsky 177–209).

Notable Recent Studies. Two workplace studies published most recently seem to support or even amplify the above findings:

• A 1994 study by the Univerity of Southern California indicates that relatively high occupational exposure to EMFs increases the risk of Alzheimer's disease more than three times (DesMarteau 38; Maugh 3).

• A 1995 study of 138,905 electric utility workers by the University of North Carolina concluded that occupational EMF exposure increases the risk of brain cancer between 1.5 and 2.3 times. This study, however, found no increased risk for leukemia (Cavanaugh 8; Moore 16).

Interpretation

None of the above studies can be said to prove a direct cause-effect relationship, but their strikingly similar results suggest a conceivable link between prolonged EMF exposure and illness.

Second subtopic

Laboratory Studies
Laboratory studies assess the physiological effects of EMFs on humans and animals at cellular, metabolic, and behavioral levels. Such studies show that EMFs directly cause the following physiological changes (Brodeur 88; Pinsky 24–29):

General findings

- reduced heart rate
- altered brain waves
- impaired immune system
- interference with the synthesis of genetic material
- disrupted regulation of cell growth
- interaction with the biochemistry of cancer cells
- altered hormonal activity

These changes are documented in the following summaries of several significant laboratory studies.

Detailed findings

EMF Effects on Cell Growth and Division. A 1995 study at the School of Veterinary Medicine, Hanover, Germany, found marked increases of ornithine decarboxylase (a key enzyme that promotes tumor growth) in the tissues of rats exposed to electromagnetic fields for six weeks (Mevissen, Keitzmann, and Loscher 207–14).

Detailed findings

A 1995 study at the University of Wisconsin showed that cell metabolism is influenced by electromagnetic fields, with the extent of the effect depending on the age and health of the cell. While this type of cellular stress does not appear to initiate cancer, it might help promote the growth of an existing tumor (Goodman, Greenebaum, and Marron 278–338).

EMF Effects on Hormones. Several studies have found that EMF exposure (e.g., from an electric blanket) inhibits the production of melatonin, a hormone that fights cancer and depression, stimulates the immune system, and regulates natural bodily rhythms (Reiter 3–4). (*Note:* Only self-regulating electric blankets, those that switch on and off automatically, and therefore require more current, affected melatonin levels.)

Detailed findings

EMF Effects on Life Expectancy. A 1994 study at the University of the Orange Free State, Republic of South Africa, measured the life span of mice exposed to EMFs. Both first and second generations of exposed mice showed significantly shortened life-expectancy (de Jager and deBruyn 221–24).

Detailed findings

EMF Effects on Behavior. Recent studies indicate that people who live adjacent to power lines have roughly twice the normal rate of depression, and that rats exposed to EMFs exhibit a roughly 20 percent slower rate of learning (Pinsky 31–32).

Interpretation

Although laboratory studies seem more conclusive than the epidemiologic studies, what these findings *mean* is debatable.

Third Topic

Conflicting Interpretations of EMF Research
Experts differ over the meaning of EMF research findings largely because of the following limitations attributed to various studies:

Findings

- Epidemiologic studies are criticized for overstating the evidence. For example, the American Physical Society, a group of 45,000 physicists, has reviewed the EMF research and seen "no consistent link between cancer and power line fields" (Broad 19). Some critics claim that reported EMF-cancer links are the product of "data dredging" (i.e., making multiple comparisons between types of cancers and EMF sources until certain random correlations appear) (Palfreman 31; Taubes 99). Some studies also are accused of mistaking *coincidence* for *correlation*, without exploring other "confounding factors" (toxic substances or adverse conditions to which subjects might have been exposed)—including the earth's natural electromagnetic field (Moore 16).

    Supporters of EMF research argue that the sheer volume of evidence seems overwhelming (Kirkpatrick 81, 83; White 15). Moreover, the recent Swedish studies cited earlier seem to invalidate the above criticisms (Brodeur 115).

- Laboratory studies are criticized, even by the scientists who conduct them, because effects on an isolated culture of cells are not always equal to effects on the total human or animal organism. Also, effects in experimental animals are not always analogous to equivalent effects in humans (Jauchem 190–94). Finally, no scientist thus far has offered a reasonable hypothesis to explain the possible health effects (Palfreman 26).

- Some critics claim that the research and resulting publicity about EMFs has become a profit venture, spawning "a new growth industry among researchers, as well as marketers of EMF monitors" ("Electrophobia" 1).

Interpretation

Although most scientists agree that EMFs have effects on the human body, they disagree about whether a real hazard exists. Another point of agreement is the need for more and better research (Schneider 31).

Fourth topic

Views from the Electric Power Industry

What position on the EMF issue is taken by the Electrical Power Research Institute (EPRI), the research arm for the nation's electric utilities? EPRI claims that recent EMF studies have provided a body of valuable but inconclusive information that warrants further study (Moore 17).

What does our local power company know and think about the alleged EMF risk? Marianne Halloran-Barney, Energy Service Advisor for County Electric, expressed this view in an E-mail correspondence:

Findings

> There are definitely some links, but we don't know, really, what the effects are or what to do about them. . . . There are so many variables in EMF research that it's a question of whether the studies were even done correctly. . . . Maybe in a few years there will be really definite answers.

Echoing Halloran-Barney's views, John Miltane, Chief Engineer for County Electric, added this political perspective:

> The public needs and demands electricity, but in regard to the negative effects of generation and transmission, the pervasive attitude seems to be

"not in my back yard!" Utilities in general are scared to death of the EMF issue, but at County Electric we're trying to do the best job we can while providing reliable electricity to 24,000 customers.

Interpretation

Industry views seem to parallel the national perspective among the broader population: Informed people genuinely are concerned, but remain unsure about what level of anxiety is warranted or what exactly should be done.

Fifth topic

Steps Being Taken Toward Risk Avoidance

Although conclusive answers about EMFs may require another decade of research, concerned citizens already are taking action about potential hazards.

First subtopic

Risk Avoidance Nationwide

Communities around the country have taken steps, such as the following, to protect schoolchildren from EMF exposure:

Findings

- Hundreds of individuals and community groups have taken legal action to block proposed construction of new power lines. A single Washington law firm has defended roughly 140 utilities in cases related to EMFs (Dana and Turner 32).
- Houston schools "forced a utility company to remove a transmission line that ran within 300 feet of three schools. Cost: $8 million" (Kirkpatrick 85).
- California parents and teachers are pressuring reluctant school and public health officials to investigate cancer rates in the roughly 1,000 schools located within 300 feet of transmission lines, and to close at least one school (within 100 feet) in which cancer rates far exceed normal (Brodeur 118).

Given the expense of reducing EMFs from power lines (an estimated $1 billion to $3 billion yearly), critics argue that the questionable risks and the minimal benefits do not justify the costs of measures like those above (Broad 19; Palfreman 33). Nonetheless, widespread expressions of concern about EMF exposure continue to grow.

Second subtopic
Findings

Risk Avoidance Locally

Local awareness of the EMF issue seems very low. The major public concern seems to be with property values of homes near power lines. According to Energy Service Advisor Halloran-Barney, County Electric receives one or two calls monthly from concerned customers, including people buying homes near power lines. The lack of public awareness adds another dimension to the EMF problem: People can't avoid a health threat they don't know exists.

According to Chief engineer John Miltane, County Electric takes the EMF issue very seriously. Whenever possible, new distribution lines are run underground and configured to diminish EMF intensity:

Although EMFs are impossible to eliminate altogether, we design anything we build to emit minimal intensities. . . . Also, we are considering underground cable (at $1,200 per foot) to replace 8,000 feet of transmission lines, but the bill would ring up to nearly $10 million, for the cable alone,

Interpretation

without labor. You don't get environmental stuff for free, which is one of the problems.

Before risk avoidance can be implemented on a broader community level, the public must first be informed about EMFs and the associated risks of exposure.

Conclusion

The conclusion of an analytical report is likely to be of most interest to readers because it answers the questions that sparked the analysis in the first place. Many workplace reports, therefore, are submitted with the conclusion *preceding* the introduction and body sections.

In the conclusion, you summarize, interpret, and recommend. Although you have interpreted evidence at each step in your analysis, your conclusion pulls the strands together in a broader interpretation and suggests a course of action, where appropriate. This final section must be consistent in three ways:

1. Your summary must reflect accurately the body of your report.
2. Your overall interpretation must be consistent with the findings in your summary.
3. Your recommendations must be consistent with the purpose of the report, the evidence presented, and the interpretations given.

The summary and interpretations should lead logically to your recommendations.

As you read the following conclusion, evaluate how effectively it provides a clear and consistent perspective on the whole document.

Conclusion

Summary and Overall Interpretation of Findings

Review of major points

Electromagnetic fields exist wherever electricity flows; the stronger the current the higher the EMF intensity. While no "safe" EMF level has been identified, long-term exposure to intensities greater than 2.5 milligauss is considered dangerous. Although home appliances can generate high EMFs during use, power lines can generate constant EMFs, typically causing readings of 2 to 3 milligauss in buildings within 150 feet. Our elementary school is less than 100 feet from a high-voltage line corridor.

Notable epidemiologic studies implicate EMFs with increased rates of medical disorders such as cancer, miscarriage, stress, depression, and learning disabilities—all directly related to intensity and duration of exposure. Laboratory studies show that EMFs cause the kinds of cellular, metabolic, and behavioral changes that could produce the above medical disorders.

An overall judgment about what the findings mean

Though still controversial and inconclusive, the various findings are strikingly similar and underscore the need for more research and for risk avoidance, especially as far as children are concerned.

Concerned citizens nationwide are beginning to prevail over resistant school and health officials and utility companies in reducing EMF risk to schoolchildren. Even though our local power company is taking reasonable risk avoidance steps, our community can do more to learn about the issues and diminish potential risk.

Recommendations

Given the limitations of our present knowledge, drastic and enormously expensive actions (such as burying all the town's power lines, increasing the height of utility towers, or using metal shields to deflect EMFs) seem inadvisable, as pointed out by John Moulder of the Medical College of Wisconsin. Moreover, these might turn out to be the wrong actions. However, the following inexpensive steps can immediately address EMF risk:

A reasonable and realistic course of action

- A version of this report should be distributed to all Bocaville residents.
- The school board should hire a qualified contractor to take milligauss readings throughout the elementary school, to determine the extent of the problem and to suggest reasonable corrective measures.
- The Town Council should meet with County Electric Company representatives, to explore options and costs for rerouting or burying that segment of the power lines near the school.
- A town meeting should then be held, to answer citizen's questions and to solicit opinions.
- A committee (consisting of at least one physician, one engineer, and other experts) should be appointed to review emerging research as it relates specifically to our school and town.

A closing call to action

As we await conclusive answers, we need to learn all we can about the EMF issue, and to do all we can to diminish this potentially significant health risk.

The Works Cited section for the preceding report appears on pages 204 and 206. Our author employs MLA documentation style.

Supplements

Submit your completed report with these supporting documents, in order:

- title page
- letter of transmittal
- table of contents
- list of tables and figures
- abstract
- **report text (introduction-body-conclusion)**
- glossary (as needed)
- appendixes (as needed)
- works cited page (or alphabetical or numbered list of references)

A Sample Situation

The report in Figure 23.2, patterned after our model outline, combines a feasibility analysis with a comparative analysis.

The situation: Richard Larkin, author of the following report, has a work-study job fifteen hours weekly in his school's placement office. His boss, John Fitton (placement director), likes to keep abreast of trends in various fields. Larkin, an engineering major, has become interested in technical marketing and sales. In need of a report topic for his writing course, Larkin offers to analyze the feasibility of a technical marketing and sales career, both for himself and for technical and science graduates in general. Fitton accepts Larkin's offer, looking forward to having the final report in his reference file for use by students choosing careers. Larkin wants his report to be useful in three ways: to satisfy a course requirement, to help him in choosing his own career, and to help other students with their career choices.

With his topic approved, Larkin begins gathering his primary data, using interviews, letters of inquiry, telephone inquiries, and lecture notes. He supplements these primary sources with articles in recent publications. He will document his findings in APA (author-date) style.

As a guide for designing his final report, Larkin completes the following audience and use profile (based on the worksheet, page 65).

Audience and Use Profile for a Formal Report

AUDIENCE IDENTITY AND NEEDS

My primary audience consists of John Fitton, Placement Director, and the students who will be referring to my report as they choose careers. The secondary audience is my writing instructor. The data I've uncovered will help me make my own career choice.

Fitton is highly interested in this project, and he has promised to study my document carefully and to make copies available to interested students. Because he already knows something about the technical marketing field, Fitton will need very little background to understand my report. Many student readers, however, may know little or nothing about technical marketing, and so will need background, definitions, and detailed explanations. Here are the questions I can anticipate from my collective audience:

- What, exactly, is technical marketing and sales?
- What are the requirements for this career?
- What are pros and cons of this career?
- Could this be the right career for me?
- How do I enter the field?
- Is there more than one option for entering the field? If so, which option would be best for me?

ATTITUDE AND PERSONALITY

Readers likely to be most affected by my document are students who will be making career choices. I would expect my readers' attitudes to vary widely:

1. Some readers will approach my document with great interest, especially those seeking a career that is more people oriented than technology oriented.

2. Some readers will be only casually interested in my document as they investigate a range of possible careers.

3. Some readers are likely to feel overwhelmed by the variety and importance of the career choices they face. Members of this group might be looking for easy answers to the problem of choosing a career.

4. Other readers may approach my document skeptically, perhaps unwilling to reexamine their earlier decisions about a more traditional career.

To connect with this array of readers, I will need to persuade them that my conclusions are based on dependable data and careful reasoning.

EXPECTATIONS ABOUT THE DOCUMENT

Although I initiated this document, Fitton has been greatly supportive, and eager to read the final version. All my readers will expect me to spell things out. But I know that my readers are busy and impatient, so I'll want to make this report concise enough to be read in no more than fifteen or twenty minutes.

Essential information will include an expanded definition of technical marketing and sales, the skills and attitudes needed for success, the career's advantages and drawbacks, and a description of various paths for entering the career. Throughout, I'll relate my material to many technical and science majors, not just engineers.

The body of this report combines a feasibility analysis with a comparative analysis. Therefore, I'll use a reasons-for and reasons-against structure in the feasibility section. In the comparison section, I'll use a block structure followed by a table that presents a point-by-point comparison of the four entry paths. Because I want this report to lead to informed decisions, I will include concrete recommendations that are based solidly on my conclusions.

To address various readers who may not wish to read the entire report, I will include an informative abstract.

My tone throughout should be conversational. Because I am writing for a mixed audience (placement director, students, and writing instructor), I will use a third-person point of view. ■

This report's front matter (title page and so on) and end matter are shown and discussed in Chapter 16, pages 377-86.

1

Feasibility Analysis of a Career in Technical Marketing

INTRODUCTION

The escalating cutbacks in aerospace, defense-related, and other industries have narrowed career opportunities for many of today's science and engineering graduates. A study by the Massachusetts Institute of Technology, for example, found that leading industries hired 80 to 90 percent fewer engineers in the mid 1990s than in the mid 1980s (Solomon, 1996, p. 24). This trend is expected to continue--if not worsen--in the foreseeable future.

Employment opportunities in all engineering specialties (except for computer engineering) are projected to grow at rates that range from average to far below average to near-static through the year 2005. In some specialties (such as petroleum engineering) employment actually will decline (Gradler & Schrammel, 1994, pp. 14-16).

Given such bleak employment prospects, recent graduates would do well to consider careers that are more promising. One especially attractive alternative is technical marketing, a career that involves identifying, reaching, and selling to customers a technical product or service.

Customer orientation is an ever-growing part of today's business and manufacturing climate. As early as the mid 1980s, U. S. industry ceased to be "manufacturing driven" (where customers would buy any products available). Instead, companies became "service driven" by customers demanding products that were designed to exact specifications and that could be serviced efficiently (Basta, 1988, p. 84). These customer demands account for the fact that top technical managers typically have sales and marketing experience in addition to their technical background.

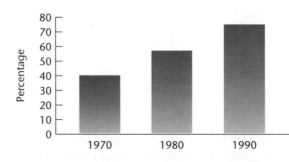

FIGURE 1 Top Managers with Sales/Marketing Backgrounds. *Source: Data from Campbell (1993).*

Figure 23.2   An Analytical Report

2

Nationwide, roughly 75 percent of top managers have worked their way up through sales and marketing ranks (Campbell, 1993, p. 28). Figure 1 shows how marketing background has become a fast track to top management.

In the product-oriented industries of 1970, technical marketing accounted for only 39 percent of top-management backgrounds. But that number nearly doubled by 1990 because of growth in customer orientation. Also, in 1995, employment listings for recent graduates showed more than one hundred major companies offering positions in technical marketing and sales (College Placement Council, 1995, p. 393).

Any undergraduates interested in this career need answers to these basic questions:

- Is this the right career for me?
- How do I enter the field?

To help answer these questions, this report analyzes information gathered from professionals as well as from the literature.

The following analysis includes these areas: definition of technical marketing, employment outlook, technical skills required, other skills and qualities required, career advantages and drawbacks, and a comparison of four entry options.

COLLECTED DATA

Key Factors in Technical Marketing as a Career

Anyone considering technical marketing needs to assess whether this career is practical, considering the individual's interests, abilities, and expectations.

THE TECHNICAL MARKETING PROCESS. Although the terms *marketing* and *sales* usually occur interchangeably, technical marketing involves far more than sales work. The process itself (identifying, reaching, and selling to customers) entails six major activities (Cornelius & Lewis, 1993, p. 44):

1. *Market research:* gathering information about the size and character of the market for a product or service.
2. *Product development and management:* producing the goods to fill a specific market need.
3. *Cost determination and pricing:* measuring every expense in the production, distribution, advertising, and sales of the product, to determine its price.

Figure 23.2   An Analytical Report *Continued*

3

4. *Advertising and promotion:* developing and implementing all strategies for reaching customers.
5. *Product distribution:* coordinating all elements of a technical product or service, from its conception through its final delivery to the customer.
6. *Sales and technical support:* creating and maintaining customer accounts, and servicing and upgrading products.

Fully engaged in all these activities, the marketing professional gains a good understanding of the industry, the product, and the customer's needs.

EMPLOYMENT OUTLOOK. For graduates with the right combination of technical and personal qualifications, the employment outlook for technical marketing appears excellent, as depicted in Figure 2. Compared with jobs in engineering, which will increase at roughly the same rate as overall employment, jobs in marketing management will increase at nearly a 50 percent higher rate.

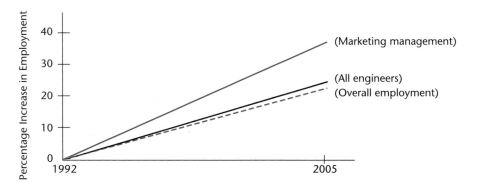

FIGURE 2 The Employment Outlook for Technical Marketing.
Source: Data from Gradler & Schrammel (1994).

Although highly competitive, these marketing positions will call for the very kinds of technical, analytical, and problem-solving skills that engineers can offer (Solomon, 1996, p. 24)—especially in an increasingly automated environment.

Figure 23.2   An Analytical Report *Continued*

4

TECHNICAL SKILLS REQUIRED. Computer networks, interactive media, and multimedia will exert an increasing impact on the way products are advertised and sold. Admittedly, initial sales volume on the net has been disappointing because of problems with navigation, security, and overall buyer hesitation (Resnick, 1995, pp. 71-72). However, despite these short-term deficiencies, online marketing specialist Graham Albatro is among many experts who predict unlimited long-term potential for marketing via the Internet and World Wide Web (Internet discussion list, April 4, 1996).

To enhance the level of direct customer contact, marketing representatives increasingly work out of a "virtual" office. Using laptop computers, fax networks, and personal digital assistants, representatives in the field have real-time access to electronic catalogs of product lines, multimedia presentations, pricing for customized products, inventory data, product distribution, and customized sales contracts (Young, 1995, pp. 86-93).

With their rich background in problem-solving, computer, and technical skills, engineering graduates are ideally suited for (a) working in an automated environment and (b) implementing and troubleshooting these complex and often sensitive electronic systems.

OTHER SKILLS AND QUALITIES REQUIRED. In marketing and sales, not even the most sophisticated automation can substitute for the "human factor": the ability to connect with customers on a person-to-person level (Young, 1995, p. 95). One senior sales engineer praises the efficiency of her automated sales system, but thinks that automation "will get in the way" of direct customer contact. Other technical marketing professionals express similar views about the continued importance of human interaction (94).

Besides a strong technical background, marketing requires a generous blend of those traits summarized in Figure 3.

Motivation is essential for marketing work. Professionals must be energetic and able to function with minimal supervision. Ideal candidates are creative people who can plan and program their own tasks, who can manage their time, and who have no fear of hard work (personal interview, February 22, 1996). Leadership potential, as demonstrated by extracurricular activities, is an asset.

Motivation, however, is no guarantee of success. Marketing professionals are paid to communicate the virtues of their products or services to customers. This career therefore calls for skill in communication, both written and oral. Writing for readers outside the organization includes advertising copy, product descriptions, sales proposals, sales letters, and operating instructions for customers. In-house writing includes recommendation reports, feasibility studies, progress reports, and assorted memos (U.S. Department of Labor, 1994, p. 7)

Figure 23.2   An Analytical Report *Continued*

5

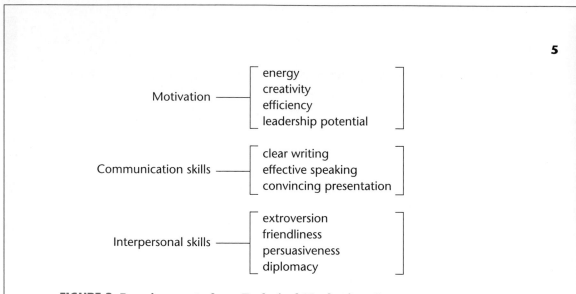

FIGURE 3 Requirements for a Technical Marketing Career

Skilled presentation is vital to any sales effort. Technical marketing professionals need to speak with a level of confidence that will enable them to be persuasive--to represent their products and services in the best possible light (personal interview, February 8, 1996). Sales presentations require public speaking at conventions, trade shows, and similar forums.

Besides motivation and communication skills, interpersonal skills are the ultimate requirement for success in marketing. (Solomon, 1996, pp. 28-29). Consumers are more likely to buy a product or service if they like the person selling it. Marketing professionals are extroverted and friendly; they enjoy meeting people. Because they understand diplomacy, they are able to motivate customers without alienating them.

ADVANTAGES OF THE CAREER. A technical marketing career offers diverse experience in every phase of a company's operation, from the design to the sale of a product. Many companies encourage employees to rotate periodically between marketing and their technical specialties (Basta, 1988, p. 84). Because of its broad exposure, a marketing position can provide a direct path to upper-management jobs. In fact, sales engineers with solid experience often open their own firms as "manufacturers' agents" representing a variety of companies ("Engineering careers," 1995, p. 247).

Figure 23.2   An Analytical Report *Continued*

6

Another benefit is the possibility of a high salary. Most marketing professionals receive a base pay plus commissions. According to John Turnbow, managing recruiter of National Electric's Technical Marketing Program, some of NE's new marketing engineers earn over $60,000 in their first year. In fact, many salaries in this field reach six figures--sometimes higher than a company's executive salaries (personal communication, April 5,1996).

This career is particularly attractive for its geographic and job mobility. Companies nationwide seek recent graduates, but especially in the Southeast and on east and west coasts ("Electronic sales," 1996, pp. 1134-37). In addition, one develops and refines the kinds of interpersonal and communication skills that are highly portable. This is especially important at a time in which more than one out of seven jobs created since 1991 has been temporary and in which job security is disappearing (Alderman, 1995, p. 37).

DRAWBACKS OF THE CAREER. Technical marketing is not a career for everyone. Personnel might spend 50 to 75 percent of work time traveling. Success requires hard work for long hours and occasional weekends (Campbell, 1993, p. 24). Above all, the job is stressful because of constant pressure to sell and to meet quotas. Anyone considering this career should be willing and able to work and thrive in a highly competitive environment.

A Comparison of Entry Options

Engineers and other technical graduates enter technical marketing through one of four options. Some graduates join small companies and begin marketing work immediately. Others join companies that offer formal training programs. Some begin by acquiring experience in their specialty. Others earn a graduate degree beforehand. Each option is described below and then evaluated on the basis of specific criteria.

OPTION 1: ENTRY-LEVEL MARKETING WITHOUT TRAINING PROGRAMS: Direct positions in marketing are available with smaller companies and with firms that represent an array of manufacturers. Because this option offers no formal training, candidates must be motivated and enterprising.

Elaine Carto, president of Abco Electronics, believes small companies can offer a unique opportunity; entry-level salespersons learn about all facets of an organization, and have a good possibility for rapid advancement (personal interview, February 10, 1996). Career counselor Phil Hawkins says, "It is all a matter of whether you want to be a big fish in a small pond or a small fish in a big pond" (personal interview, February 12, 1996).

Manufacturers' representatives constitute another entry-level position. These professionals represent products for manufacturers who have no marketing staff of

Figure 23.2   An Analytical Report *Continued*

7

their own. Manufacturers' representatives are, in effect, their own bosses, choosing from among many offers the products they wish to represent (Tolland, 1996).

Entry-level marketing offers immediate income and a chance for early promotion. A disadvantage, however, might be the gradual loss of any technical edge that one might have acquired in college.

OPTION 2: A MARKETING AND SALES TRAINING PROGRAM. Formal training programs offer the most popular route into sales and marketing. Large to mid-size companies typically offer two formats: (a) a product-specific program, in which trainees learn about a particular product or line of products, or (b) a rotational program, in which trainees learn about various products and work in various positions. Programs last from several weeks to several months, covering products, markets, competition, and customer relations.

Former trainees speak of the diversity and satisfaction offered by such training programs. Ralph Lang, of Allied Products, enjoyed the constant interaction with company personnel (phone interview, April 10, 1996). Bill Collins, sales engineer with Intrex Computers, values his broad knowledge of Intrex's product line, instead of being narrowly focused on one technical area. Sarah Watts, also with Intrex, enjoys applying her training to a variety of sales challenges (Campbell, 1993, p. 29). Like direct entry , this option offers immediate income. A possible disadvantage, however, is that trainees might find their technical expertise compromised because they have had no chance to practice in their specialty.

OPTION 3: PRACTICAL EXPERIENCE IN ONE'S SPECIALTY. Instead of directly entering marketing, some candidates first gain practical experience in their specialty. This option provides direct exposure to the workplace environment and a chance to sharpen technical ability in practical applications. Moreover, some companies will offer marketing and sales positions to their outstanding engineers as a first step toward upper-management ("Engineering careers," 1995, p.183).

Many industry experts consider the practical-experience option the wisest choice. Jane Doser, recruitment manager for Trans Electric, has this view of on-the-job experience: "Until you've done some work in your specialty, you really don't know yourself as a person or know what kind of person you're going to be" (Schranke, 1985, p. 15).

This option ensures technical expertise, but delays a candidate's entry to technical marketing.

Figure 23.2   An Analytical Report *Continued*

8

OPTION 4: GRADUATE PROGRAM. Instead of direct entry, some people choose to pursue either a Master's of Science (MS) degree in their specialty, a Master's of Business Administration (MBA) degree, or both. According to engineering professor Mary McClane, MS degrees usually are unnecessary for technical marketing unless a particular sale is highly complex (1996). The MBA carries a real advantage in terms of general business and marketing skills. In general, jobseekers with an MBA have a distinct competitive advantage (Gradler & Schrammel, 1994, p. 3). More significantly, new MBAs with a technical bachelor's degree and 1 to 2 years of experience command salaries more than 30 percent higher than MBAs who have neither work experience nor a technical bachelor's degree (Baxter, 1994, p. 7).

A motivated student might combine degrees. Dora Anson, president of Susimo Cosmic Systems, sees the MS and MBA as the ideal combination for technical marketing (lecture, January 15, 1996). To employers, the advanced degree means greater technical competence, stronger motivation, and higher performance levels.

One disadvantage of a full-time graduate program is lost salary, compounded by school expenses. These costs must be weighed against the prospect of promotion and monetary rewards later in one's career.

AN OVERALL COMPARISON BY RELATIVE ADVANTAGES. Table 1 compares the four technical marketing entry options on the basis of three criteria: immediate income, rate of advancement through marketing ranks, and long-term potential.

Table 1 Relative Advantages Among Four Technical-Marketing Entry Options

	Relative Advantages		
Option	Early Immediate income	Rapid advancement in marketing	long-term potential
Entry level, no experience	yes	yes	no
Training program	yes	yes	no
Practical experience	yes	no	yes
Graduate program	no	no	yes

As Table 1 shows, the choice of any particular entry option can have important career implications.

Figure 23.2   An Analytical Report *Continued*

9

CONCLUSION

Summary of Findings

Technical marketing and sales involves identifying, reaching, and selling the customer a technical product or service. Besides a solid technical background, anyone entering the field will need motivation, communication skills, and interpersonal skills. This career offers job diversity and excellent income potential, balanced against hard work and relentless pressure to perform.

College graduates who seek a career in technical marketing and sales have four entry options:

1. Direct entry with no formal training
2. A formal training program
3. Practical experience in a technical specialty prior to entry
4. Graduate programs

Each option has advantages and disadvantages in immediacy of income, rate of advancement, and long-term potential.

Interpretation of Findings

For graduates with a strong technical background and the right skills and motivation, technical marketing offers attractive prospects for income and advancement. Anyone contemplating this field, however, needs to enjoy customer contact and be able to thrive in a highly competitive environment.

Those who decide that technical marketing is for them can choose among four entry options:

- If immediate income is unimportant, graduate school is an attractive option.
- For hands-on experience, an entry-level job is the logical option.
- For sharpening technical skills, prior work in one's specialty is invaluable.
- For sophisticated sales training, a formal program with a large company is best.

Figure 23.2   An Analytical Report *Continued*

10

Recommendations

If your interests and abilities match the requirements outlined earlier, follow these suggestions in planning for a technical marketing career.

1. To get a firsthand view, seek the opinions and advice of people in the field.

2. Before settling on an entry option, consider all its advantages and disadvantages and decide whether this particular option best coincides with your personal goals.

3. Remember that you are never committed to one entry option only. For a fuller understanding of technical marketing, you might pursue any combination of options during your professional life.

4. When making any vital career decision, consider career counselor Warren Polgar's advice: "Listen to your brain and your heart" (lecture, November 19, 1995). Choose an option or options that offer not only professional advancement, but personal satisfaction as well.

REFERENCES

[The complete list of references is shown and discussed in Chapter 10, page 218.]

Figure 23.2   An Analytical Report *Continued*

Revision Checklist for Analytical Reports

Use this list to refine the content, arrangement, and style of your report. (Numbers in parentheses refer to the first page of discussion.) For evaluating your research methods and reasoning, refer also to the checklist on page 187.

CONTENT

❑ Does the report grow from a clear statement of purpose? (555)

❑ Is the report's length adequate and appropriate for the subject? (553)

❑ Are all limitations of the analysis clearly acknowledged? (560)

❑ Are visuals used whenever possible to aid communication? (556)

❑ Are all data accurate? (556)

❑ Are all data unbiased? (555)

❑ Are all data complete? (556)

❑ Are all data fully interpreted? (556)

❑ Is the documentation adequate, correct, and consistent? (191)

❑ Are the conclusions logically derived from accurate interpretation? (556)

❑ Do the recommendations constitute an appropriate response to the question or problem? (556)

ARRANGEMENT

❑ Is there a distinct introduction, body, and conclusion? (559)

❑ Are headings appropriate and adequate? (368)

❑ Are there enough transitions between related ideas? (Appendix)

❑ Is the report accompanied by all needed front matter? (569)

❑ Is the report accompanied by all needed end matter? (569)

STYLE AND PAGE DESIGN

❑ Is the level of technicality appropriate for the stated audience? (571)

❑ Is the writing style throughout clear, concise, and fluent? (261)

❑ Is the language convincing and precise? (261)

❑ Is the report written in grammatical English? (Appendix)

❑ Is the page design inviting and accessible? (355)

✅ EXERCISE

Prepare an analytical report, using some sequence of these guidelines:

a. Choose a subject for analysis from the list your instructor provides, from your major, or from a subject of interest.

b. Identify the problem or question so that you will know exactly what you are looking for.

c. Restate the main question as a declarative sentence in your statement of purpose.

d. Identify an audience—other than your instructor—who will use your information for a specific purpose.

e. Hold a private brainstorming session to generate major topics and subtopics.

f. Use the topics to make an outline based on the model outline in this chapter. Divide as far as necessary to identify all points of discussion.

g. Make a tentative list of all sources (primary and secondary) that you will investigate. Verify that adequate sources are available.

h. Write your instructor a proposal memo, describing the problem or question and your plan for analysis. Attach a tentative bibliography.

i. Use your working outline as a guide to research and observation. Evaluate sources and evidence, and interpret all evidence fully. Modify your outline as needed.

j. Submit a progress report to your instructor describing work completed, problems encountered, and work remaining.

k. Compose an audience and use profile. (Use the sample on pages 570 and 571 as models, along with the profile worksheet on page 65).

l. Write the report for your stated audience. Work from a clear statement of purpose, and be sure that your reasoning is shown clearly. Verify that your evidence, conclusions, and recommendations are consistent. Be especially careful that your recommendations observe the critical-thinking guidelines in Figure 23.1.

m. After writing your first draft, make any needed changes in the outline and revise your report according to the revision checklist. Include all necessary supplements.

n. Exchange reports with a classmate for further suggestions for revision.

o. Prepare an oral report of your findings for the class as a whole.

☑ COLLABORATIVE PROJECTS

1. Divide into small groups. Choose a subject for group analysis—preferably, a campus issue—and partition the topic by group brainstorming. Next, select major topics from your list and classify as many items as possible under each major topic. Finally, draw up a working outline that could be used for an analytical report on this subject.

2. Prepare a questionnaire based on your work above, and administer it to members of your campus community. List the findings of your questionnaire and your conclusions in clear and logical form. (Review pages 155–62, on questionnaires and surveys.)

Oral Presentations

Avoiding Presentation Pitfalls

Planning Your Presentation

Work from an Explicit Purpose Statement

Analyze Your Listeners

Analyze Your Speaking Situation

Select an Appropriate Delivery Method

Preparing Your Presentation

Research Your Topic

Aim for Simplicity and Conciseness

Anticipate Audience Questions

Outline Your Presentation

Plan Your Visuals

Prepare Your Visuals

Rehearse Your Delivery

Delivering Your Presentation

Cultivate the Human Landscape

Keep Your Listeners Oriented

Manage Your Visuals

Manage Your Presentation Style

Manage Your Speaking Situation

ORAL presentations vary in style, range, complexity, and formality. They may include convention speeches, reports at national meetings, reports via teleconferencing networks, technical briefings for colleagues, and speeches to community groups. These talks may be designed to inform (to describe new government safety requirements), to persuade (to induce company officers to vote a pay raise), or to do both. The higher your status, on the job or in the community, the more you will give oral presentations.

Avoiding Presentation Pitfalls

An oral presentation is only the tip of a pyramid built from many earlier labors. But such presentations often serve as the concrete measure of your overall job performance. In short, your audience's only basis for judgment may be the brief moments during which you stand before them.

The podium or lectern can be a lonely place. In fact, people usually consider speaking before a group as the most intimidating communication task. In the words of two experts, "most persons in most presentational settings do not perform well" (Goodall and Waagen 14–15). Despite the fact that they can help make or break one's career, oral presentations often are boring, confusing, unconvincing, or too long. Many are poorly delivered, with the presenter losing her or his place, fumbling through notes, apologizing for forgetting something, or generally seeming unprofessional. Table 24.1 lists some of the pitfalls.

Given such difficulties, how can any presenter display skill and confidence? By proceeding systematically through careful analysis, planning, and preparation.

Table 24.1 ■ Common Pitfalls in Oral Presentations

Speaker . . .	Visuals * * *	Setting ■ ■ ■
• looks no one in the eye	* are nonexistent	■ is too noisy
• seems like a robot	* are hard to see	■ is too hot or cold
• hides behind the lectern	* are hard to interpret	■ is too large or small
• speaks too softly/loudly	* are out of sequence	■ is too bright for visuals
• sways, fidgets, paces	* are shown too rapidly	■ is too dark for notes
• rambles	* are shown too slowly	■ has equipment missing
• never gets to the point	* have typos/errors	■ has broken equipment
• loses her/his place	* are word-filled	
• fumbles with notes or visuals		

Planning Your Presentation

A successful presentation involves more than just getting up there and talking. We have all sat through enough lectures and presentations during our student careers to know how to separate the excellent from the awful. The successful presenter knows how to forge a relationship with the listeners, how to establish rapport and persuade listeners their time has been well spent.

Work from an Explicit Purpose Statement

Formulate, on paper, a statement of purpose in two or three sentences. Why, exactly, are you speaking on this subject? Who are your listeners? What do you want the listeners to think, know, or do? (A solid purpose statement also can serve as the introduction to your presentation.)

Assume, for example, that you represent an environmental engineering firm that has completed a study of groundwater quality in your area. The organization that sponsored your study has asked for an oral version of your written report, at a town meeting. Report title: "Pollution Threats to Local Groundwater."

After careful thought, you settle on this purpose statement:

Purpose statement

> *Purpose:* By informing Cape Cod residents about the dangers to the Cape's freshwater supply posed by rapid population growth, this report is intended to increase local interest in the problem.

Now you are prepared to focus on the listeners and the speaking situation.

Analyze Your Listeners

Assess your listeners' needs, knowledge, concerns, level of involvement, and possible objections (Goodall 16).

Many audiences include people with varied technical backgrounds. Unless you essentially know each person's background, speak to a general, heterogeneous audience, as in a classroom of mixed majors.

Questions for Analyzing Your Listeners

- Who are my listeners (strangers, peers, superiors, clients)?
- What is their attitude toward me or the topic (hostile, indifferent, needy, friendly)?
- Why are they here (they want to be here, they are forced to be here, they are curious)?
- What kind of presentation do they expect (brief, informal; long, detailed; lecture)?
- What do these listeners already know (nothing, a little, a lot)?

- What do they need or want to know (overview, bottom line, nitty gritty)?
- How large is their stake in this topic (about layoffs, new policies, pay raises)?
- Do I want to motivate, mollify, inform, instruct, or warn my listeners?
- What are their biggest concerns or objections about this topic?
- What do I want them to think, know, or do?

Analyze Your Speaking Situation

The more you can discover about the circumstances, the setting, and the constraints for your presentation, the more deliberately you will be able to prepare.

Questions for Analyzing Your Speaking Situation

- How much time will I have to speak?
- Will other people be speaking before or after?
- How formal or informal is the setting?
- How large is the audience?

- How large is the room?
- How bright and adjustable is the lighting?
- What equipment is available?
- How much time do I have to prepare?

Later parts of this chapter explain how to incorporate your answers to the preceding questions in your preparation.

Select an Appropriate Delivery Method

Your presentation's effectiveness will depend largely on *how* it connects with listeners. Different types of delivery create different connections.

The Memorized Delivery. A memorized delivery offers little prospect of connecting with listeners because the speaker is too busy reciting the lines and trying to avoid disaster. This type of delivery takes forever to prepare, offers no chance for revision in mid-presentation, and spells disaster if you lose your place. Use only when absolutely necessary.

The Impromptu Delivery. An impromptu (off the cuff) delivery can be a natural way of connecting with listeners—but only when you really know your material, feel comfortable with your listeners, and have an informal speaking situation (group brainstorming, or a response to a question: "Tell us about your team's progress on the automation project"). Avoid impromptu deliveries for complex information—no matter how well you know the material.

Too many things can go wrong in an unplanned, spontaneous presentation: You might say something offensive or irrelevant; you might seem disorganized; you might not make sense. If you anticipate being called on, never assume "It's all in my head." Get your plan down on paper. If you have little warning, at least jot a few notes about what you want to say.

The Scripted Delivery. For a complex technical presentation, a conference paper, or a formal speech, you may want to read your material verbatim from a prepared script. Scripted presentations work well if you have many details to present, are long-winded, or have a strict time limit (e.g., at a conference), or if this audience makes you very nervous. Consider a scripted delivery when you want the content, organization, and style of your presentation to be as near perfect as possible.

Although a scripted delivery helps you control your material, it offers little chance for listener contact or for revision in mid-presentation, and it can be boring and mechanical.

If you *do* plan to read aloud, allow ample preparation time. Leave plenty of white space between lines and paragraphs. Rehearse until you are able to glance up from the script periodically without losing your place. Plan on roughly two minutes per double-spaced page.

The Extemporaneous Delivery. An extemporaneous delivery is carefully planned, practiced, and based on notes that keep you on track. In this natural way of addressing an audience, you glance at your material and speak in a conversational style. Extemporaneous delivery is based on key ideas in sentence or topic outline form, rather than fully developed paragraphs to be read or memorized.

The dangers in extemporaneous delivery are that you might lose track of your material, forget something important, say something unclearly, or exceed your time limit. Careful preparation is the key.

Table 24.2 summarizes the various uses and drawbacks among the most common types of delivery. Keep in mind that these need not be fixed, exclusive categories. In many instances, some combination of methods can be very effective. For example, in an orientation for new employees, you might prefer the flexibility of an extemporaneous format but also read a brief passage aloud from time to time (e.g., excerpts from the company's formal code of ethics).

Preparing Your Presentation

Plan the presentation systematically, to stay in control and build confidence. For our limited purposes here, we will assume your presentation is extemporaneous.

Research Your Topic

Do your homework. Be prepared to support each assertion, opinion, conclusion, and recommendation with evidence and reason. Check your facts for accuracy. Your audience expects to hear a knowledgeable speaker. Don't disappoint them.

Begin gathering material well ahead of time. Use summarizing techniques from Chapter 11 to identify and organize major points.

If your presentation is simply a spoken version of a written report, you need far less preparation. Simply expand your outline for the written report into a sentence outline.

Aim for Simplicity and Conciseness

Keep your presentation short and simple. Boil the material down to a few main points. Listeners' normal attention span is about twenty minutes. After that, they begin tuning out. Time yourself in practice sessions and trim as

Table 24.2 ■ A Comparison of Oral Presentation Methods

Delivery Method	* Main Uses *	• Main Drawbacks •
IMPROMPTU *(Inventing as you speak)*	* in-house meetings * small, intimate groups * simple topics	• offers no chance to prepare • speaker might ramble • speaker might lose track
SCRIPTED *(reading verbatim from a written work)*	* formal speeches * large, unfamiliar groups * highly complex topics * strict time limit * cross-cultural audiences * highly nervous speaker	• takes a long time to prepare • speaker can't move around • limits human contact • can appear stiff and unnatural • might bore listeners • makes working with visuals difficult
EXTEMPORANEOUS *(speaking from an outline of key points)*	* face-to-face presentations * medium-sized, familiar groups * moderately complex topics * somewhat flexible time limit * visually based presentations	• speaker might lose track • speaker might leave something out • speaker might get tongue-tied • speaker might exceed time limit • speaker might fumble with notes, visuals, or equipment

needed. (If the material requires a lengthy presentation, plan a short break, with refreshments if possible, about half way.)

Anticipate Audience Questions

Consider those parts of your presentation that listeners might question or challenge. You might need to clarify or justify information that is new, controversial, disappointing, or surprising.

Outline Your Presentation

Review Chapter 12 for organizing and outlining strategies. Each sentence in the following presentation outline is a topic sentence for a paragraph that a well-prepared speaker can develop in detail.

Presentation outline

Pollution Threats to Local Groundwater

Arnold Borthwick

I. Introduction to the Problem
A. Do you know what you are drinking when you turn on the tap and fill a glass?
B. The quality of our water is good, but not guaranteed to last forever.
C. Cape Cod's rapid population growth poses a serious threat to our freshwater supply. (Visual # 1)
D. Measurable pollution in some town water supplies already has occurred. (Visual # 2)
E. What are the major causes and consequences of this problem and what can we do about it? (Visual # 3)

II. Description of the Aquifer
A. The groundwater is collected and held in an aquifer.
   1. This porous rock formation forms a broad, continuous arch beneath the entire Cape. (Visual # 4)
   2. The lighter freshwater flows on top of the heavier salt water.
B. This type of natural storage facility, combined with rapid population growth, creates potential for disaster.

III. Hazards from Sewage and Landfills
A. With increasing population, sewage and solid waste from landfill dumps increasingly invade the aquifer.
B. The Cape's sandy soil promotes rapid seepage of wastes into the groundwater.
C. As wastes flow naturally toward the sea, they can invade the drawing radii of town wells. (Visual # 5)

IV. Hazards from Saltwater Intrusion
A. Increased population also causes overdraw on some town wells, resulting in saltwater intrusion. (Visual # 6)
B. Salt and calcium used in snow removal add to the problem by seeping into the aquifer from surface runoff.

V. Long-term Environmental and Economic Consequences
A. The environmental effects of continuing pollution of our water table will be far-reaching. (Visual # 7)
   1. Drinking water will have to be piped in more than 100 miles from Quabbin Reservoir.
   2. The Cape's beautiful freshwater ponds will be unfit for swimming.
   3. Aquatic and aviary marsh life will be threatened.
   4. The Cape's sensitive ecology might well be damaged beyond repair.
B. Such environmental damage would, in turn, spell economic disaster for Cape Cod's major industry—tourism.

VI. Conclusion and Recommendations
A. This problem is becoming more real than theoretical.
B. The conclusion is obvious: If the Cape is to survive ecologically and

financially, we must take immediate action to preserve our *only* water supply.

   C. These recommendations offer a starting point for action. (Visual # 8)

     1. Restrict population density in all Cape towns by creating larger building-lot requirements.

     2. Keep strict watch on proposed high-density apartment and condominium projects.

     3. Create a committee in each town to educate residents about water conservation.

     4. Prohibit salt, calcium, and other additives in sand spread on snow-covered roads.

     5. Explore alternatives to landfills for solid-waste disposal.

   D. Given its potential impact on our quality of life, such a crucial issue deserves the active involvement of every Cape resident.

Before practicing the delivery, transfer your presentation outline to three-by-five notecards (one side only, each card numbered), which you can hold in one hand and shuffle as needed. Or insert the outline pages in a looseleaf binder for easy flipping. Type or print clearly, leaving enough white space to locate material at a glance.

Plan Your Visuals

Visuals increase listeners' interest, focus, understanding, and memory. Select visuals that will clarify and enhance your talk—without making you fade into the background.

Decide Where Visuals Will Work Best. Use visuals to emphasize a point and enhance your presentation whenever *showing* would be more effective than merely *telling*.

Decide Which Visuals Will Work Best. Will you need numerical or prose tables, graphs, charts, graphic illustrations, computer graphics? How fancy should these visuals be? Should they impress or merely inform? Use the visual planning sheet in Chapter 14 to guide your decisions.

Decide How Many Visuals Are Appropriate. Prefer an array of lean and simple visuals that present material in digestible amounts instead of one or two overstuffed visuals that people end up staring at endlessly.

Create a Storyboard. A storyboard is a double-column format in which your discussion is outlined in the left column, aligned with the specific supporting visuals in the right column (Figure 24.1).

Decide Which Visuals Are Achievable. Fit each visual to the situation. The visuals you select will depend on the room, the equipment, and the production resources available.

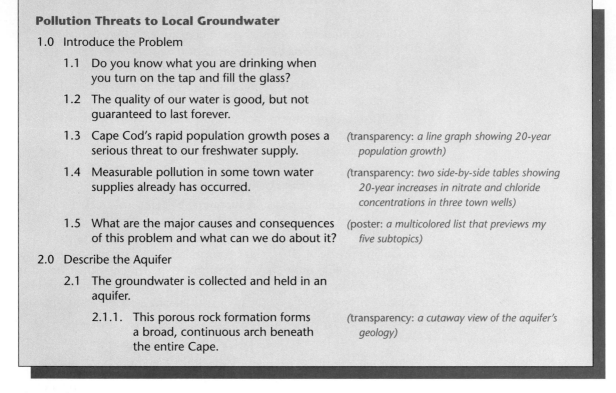

Figure 24.1   A Partial Storyboard

Fit each visual to the situation

How large is the room and how is it arranged? Some visuals work well in small rooms, but not large ones, and vice versa. How well can the room be darkened? Which lights can be left on? Can the lighting be adjusted selectively? What size should visuals be, to be seen clearly by the whole room? (A smaller, intimate room usually is better than a room that is too big and cavernous.)

What hardware is available (slide projector, opaque projector, overhead projector, film projector, videotape player, terminal with large-screen monitor)? What graphics programs are available? Which is best for your purpose and listeners? How far in advance does this equipment have to be requested?

What resources are available for producing the visuals? Can drawings, charts, graphs, or maps be created as needed? Can transparencies (for overhead projection) be made or slides collected? Can handouts be typed and reproduced? Can multimedia displays be created?

Select Your Media. Fit the medium to the situation. Which medium or combination is best for the topic, setting, and listeners? How fancy do listeners expect this to be? Which media are appropriate for this occasion? Examples appear on page 593.

PREPARING YOUR PRESENTATION 593

Fit the medium to the situation

- For a weekly meeting with colleagues in your department, scribbling on a blank transparency, chalkboard, or dry-erase markerboard might suffice.
- For interacting with listeners, you might use a chalkboard to record audience responses to your questions.
- For immediate orientation, you might begin with a poster listing key visuals/ideas/themes to which you will refer repeatedly.
- For helping listeners take notes, absorb technical data, or remember complex material, you might distribute a presentation outline as a preview or provide handouts.
- For displaying and discussing written samples listeners bring in, you might use an opaque projector.
- For a presentation to investors, clients, or upper management, you might require polished and professionally prepared visuals, including computer graphics, multimedia, and state-of-the-art technology.

Figure 24.2 presents the various common media in approximate order of availability and ease of preparation.

Prepare Your Visuals

As you prepare visuals, focus on economy, clarity, and simplicity.

Be Selective. Use a visual only when it truly serves a purpose. Use restraint in choosing what to highlight with visuals. Try not to begin or end the presentation with a visual. At those times, listeners' attention should be focused on the presenter instead of the visual.

Make Visuals Easy to Read and Understand. Think of each visual as an image that flashes before your listeners. They will not have the luxury of studying the visual at leisure. Listeners need to know at a glance what they are looking at and what it means. Following are guidelines for achieving readability.

Guidelines for Readable Visuals

- Make visuals large enough to be read anywhere in the room.
- Don't cram too many words, ideas, designs, or typestyles.
- Keep wording and image simple.
- Boil your message down to fewest words.
- Break things into small amounts.
- Summarize with key words, phrases, or short sentences.
- Use 18–24 point typesize and sans serif typeface (White, *Great Pages* 80).

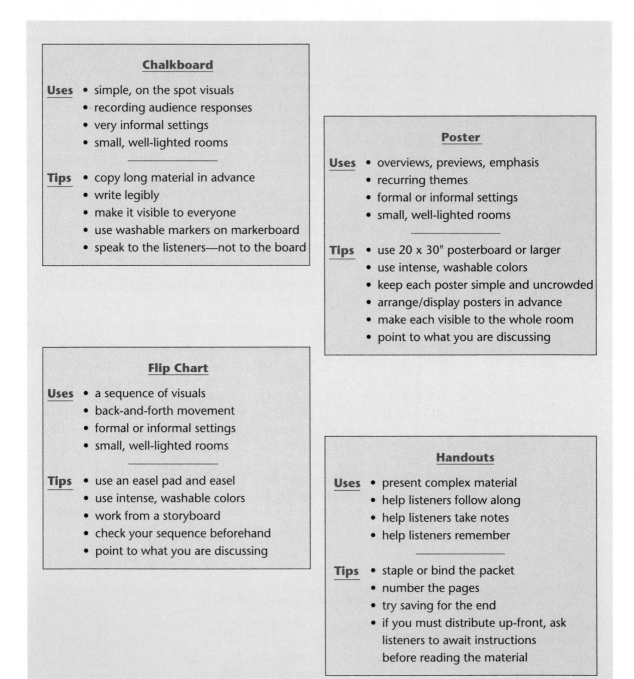

Chalkboard

Uses • simple, on the spot visuals
     • recording audience responses
     • very informal settings
     • small, well-lighted rooms

Tips • copy long material in advance
     • write legibly
     • make it visible to everyone
     • use washable markers on markerboard
     • speak to the listeners—not to the board

Poster

Uses • overviews, previews, emphasis
     • recurring themes
     • formal or informal settings
     • small, well-lighted rooms

Tips • use 20 x 30" posterboard or larger
     • use intense, washable colors
     • keep each poster simple and uncrowded
     • arrange/display posters in advance
     • make each visible to the whole room
     • point to what you are discussing

Flip Chart

Uses • a sequence of visuals
     • back-and-forth movement
     • formal or informal settings
     • small, well-lighted rooms

Tips • use an easel pad and easel
     • use intense, washable colors
     • work from a storyboard
     • check your sequence beforehand
     • point to what you are discussing

Handouts

Uses • present complex material
     • help listeners follow along
     • help listeners take notes
     • help listeners remember

Tips • staple or bind the packet
     • number the pages
     • try saving for the end
     • if you must distribute up-front, ask listeners to await instructions before reading the material

Figure 24.2   Selecting Media for Visual Presentations

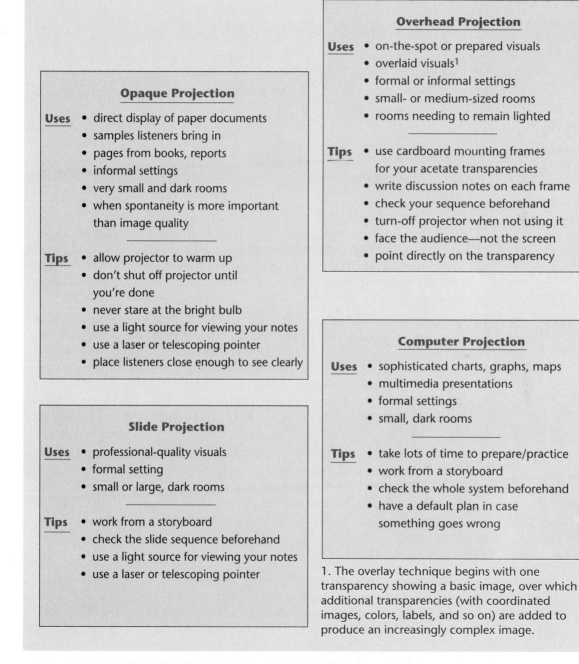

Opaque Projection

Uses
- direct display of paper documents
- samples listeners bring in
- pages from books, reports
- informal settings
- very small and dark rooms
- when spontaneity is more important than image quality

Tips
- allow projector to warm up
- don't shut off projector until you're done
- never stare at the bright bulb
- use a light source for viewing your notes
- use a laser or telescoping pointer
- place listeners close enough to see clearly

Slide Projection

Uses
- professional-quality visuals
- formal setting
- small or large, dark rooms

Tips
- work from a storyboard
- check the slide sequence beforehand
- use a light source for viewing your notes
- use a laser or telescoping pointer

Overhead Projection

Uses
- on-the-spot or prepared visuals
- overlaid visuals[1]
- formal or informal settings
- small- or medium-sized rooms
- rooms needing to remain lighted

Tips
- use cardboard mounting frames for your acetate transparencies
- write discussion notes on each frame
- check your sequence beforehand
- turn-off projector when not using it
- face the audience—not the screen
- point directly on the transparency

Computer Projection

Uses
- sophisticated charts, graphs, maps
- multimedia presentations
- formal settings
- small, dark rooms

Tips
- take lots of time to prepare/practice
- work from a storyboard
- check the whole system beforehand
- have a default plan in case something goes wrong

1. The overlay technique begins with one transparency showing a basic image, over which additional transparencies (with coordinated images, colors, labels, and so on) are added to produce an increasingly complex image.

Figure 24.2   Selecting Media for Visual Presentations *Continued*

In addition to being able to *read* the visual, listeners need to *understand* it. Following are guidelines for achieving clarity.

Guidelines for Understandable Visuals

- Display only one point per visual—unless previewing or reviewing (White 79).
- Give each visual a title that announces the topic.
- Use color, sparingly, to highlight key word, facts, or the bottom line.
- Use brightest color for what is most important (White, *Great Pages* 78–79).
- Label each part of a diagram or illustration.
- Proofread each visual carefully.

When your material is extremely detailed or complex, distribute handouts to each listener.

Look for Alternatives to Word-Filled Visuals. Instead of presenting mere overhead versions of printed pages, explore the full *visual* possibilities of your media. For example, anyone who tries to write a verbal equivalent of the visual message in Figure 24.3 soon will appreciate the power of images in relation to words alone. Whenever possible, use drawings, graphs, charts, photographs, and other visual representation discussed in Chapter 14.

Use all Available Technology. Using desktop publishing systems and presentation software such as Powerpoint™ or Inspiration™ you can create professional-quality visuals and display technical concepts. Using hypertext and multimedia systems, you can create dynamic presentations that appeal to the listener's multiple senses. Using an automatic, remote-controlled transparency feeder and a laser pointer (pencil-sized), you can deliver a smooth and elegant presentation. These are just a few of the possibilities inherent in the technology.

Check the Room and Setting Beforehand. Make sure you have enough space, electrical outlets, and tables for your equipment. If you will be addressing a large audience by microphone and plan to point to features on your visuals, be sure the microphone is movable. Pay careful attention to lighting, especially for chalkboards, flipcharts, and posters. Don't forget a pointer if you need one.

Rehearse Your Delivery

Hold ample practice sessions to learn the geography of your report. Try to rehearse at least once in front of friends. Or use a full-length mirror and a tape recorder. Assess your delivery from listeners' comments or from your taped voice (which will sound high to you). Use the evaluation sheet on page 602 as a guide.

If at all possible, rehearse using the actual equipment (overhead projector and so on) in the actual setting, to ensure that you have all you need and that everything works.

**Figure 24.3
Images More
Powerful Than
Words**

*Source: "Distribution of the
World's Water," as
appeared in* WWF Atlas of
the Environment *by
Geoffrey Lean and Don
Hinrichsen. Copyright
©1994 Banson Marketing
Ltd. Reprinted with
permission.*

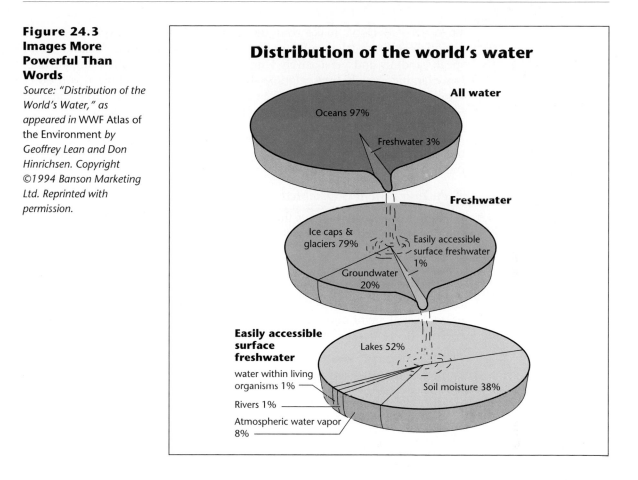

Delivering Your Presentation

You have planned and prepared carefully. Now consider the following simple steps to make your actual presentation enjoyable instead of terrifying.

Cultivate the Human Landscape

A successful presentation involves relationship building between presenter and audience.

Get to Know Your Audience. Try to meet some audience members before your presentation. We all feel more comfortable with people we know. Don't be afraid to smile.

Display Enthusiasm and Confidence. Nobody likes a speaker who seems half dead. Clean up verbal tics ("er," "ah," "uuh"). Overcome your shyness; research indicates that shy people are seen as less credible, trustworthy, likable, attractive, and knowledgeable.

Be Reasonable. Don't make your point at someone else's expense. If your topic is controversial (layoffs, policy changes, downsizing), decide how to speak candidly and persuasively with the least chance of offending anyone. For example, in your presentation about groundwater pollution, you don't want to attack the developers, since the building trade is a major producer of jobs, second only to tourism, on Cape Cod. Avoid personal attacks.

Don't Preach. Speak like a person talking—not someone giving a sermon or the Gettysburg Address. Use *we, you, your, our,* to establish commonality with the audience. Avoid jokes or wisecracks.

Keep Your Listeners Oriented

Enable your listeners to focus their attention and organize their understanding. Give them a map, some guidance, and highlights.

Introduce Your Topic Clearly. Open with a preview of your discussion:

A presentation preview

> Today, I want to discuss A, B, and C.

Use visuals to highlight your main points and reveal your organization. For example, you might outline main points on a poster, a chalkboard, or a transparency, or hand out a presentation outline.

Focus on Listeners' Concerns. Say something to establish immediate common ground. Show how your presentation has meaning for them, personally.

An appeal to listener's concerns

> Do you know what you will be drinking when you turn on the tap?

Listeners who have a definite stake in the issues will be far more attentive and receptive.

Provide Explicit Transitions. Alert your listeners whenever you are preparing to switch gears:

Explicit transitions

> For my next point. . . .
>
> Turning now to my second point. . . .
>
> The third point I want to emphasize. . . .

Repeat key points or terms to keep them fresh in listeners' minds.

Give Concrete Examples. Mobilize the informative and persuasive force of good examples.

A concrete example

> Overdraw from town wells in Maloket and Tanford (two of our most rapidly growing towns) has resulted in measurable salt infusion at a yearly rate of 0.1 mg per liter.

Use examples that focus on listener concerns.

Review and Interpret. Last things are best remembered. Help listeners remember the main points:

A review of main points

> To summarize the dangers to our groundwater. . . .

Also, be clear about what this material means. Be emphatic about what listeners should be doing, thinking, or feeling:

An emphatic conclusion   ▌   The conclusion is obvious: If the Cape is to survive, we must. . . .

Try to conclude with a forceful answer to this implied question from each listener: "What does this all mean to me?"

Manage Your Visuals

Presenting visuals effectively is a matter of good timing and careful management.

Prepare Everything Beforehand. If you plan to draw on a chalkboard or poster, do the drawings beforehand (in multicolors). Otherwise, listeners will be sitting idly while you draw away.

Prepare handouts if you want listeners to remember or study certain material. Distribute these *after* your talk. (You want the audience to be looking at and listening to you, instead of reading the handout.) Distribute handouts before or during the talk only if you want listeners to take notes—or if your equipment breaks down. When you do distribute handouts beforehand, ask listeners to await your instructions before they turn to a particular page.

Use transparency mounting frames for easy handling (the white cardboard frames enable you to number the transparencies and prepare notes for yourself on the frame).

Increasingly available are automatic transparency feeders with a remote control, enabling your transparency presentation to work like a slide show. This device attaches easily to your overhead projector and enables you to reveal each point line by line.

Arrange Everything Beforehand. Make sure you organize your media materials and the physical layout beforehand, to avoid fumbling during the presentation. Check your visual sequence against your storyboard.

Follow a Few Simple Guidelines. Make your visuals part of a seamless presentation. Avoid listener distraction, confusion, and frustration by observing the following guidelines:

Guidelines for Presenting Visuals

- Try not to begin with a visual.
- Try not to display a visual until you are ready to discuss it.
- Tell viewers what they should be looking for in the visual.
- Point to what is important.

- Stand aside when discussing a visual, so everyone can see it.
- Don't turn your back on the audience.
- After discussing the visual, remove it promptly.
- Switch off equipment that is not in use.
- Try not to end with a visual.

Manage Your Presentation Style

Think about how you are moving, how you are speaking, where you are looking. These all are elements of your personal style.

Use Natural Movements and Reasonable Postures. Move and gesture as you normally would in conversation, and maintain reasonable postures. Avoid foot shuffling, pencil tapping, swaying, slumping, or fidgeting.

Adjust Volume, Pronunciation, and Rate. With a microphone, don't speak too loudly. Without one, don't speak too softly. Be sure you can be heard clearly without shattering eardrums. Ask your audience about the sound and speed of your delivery after a few sentences.

Nervousness causes speakers to gallop along and mispronounce things. Slow down and pronounce clearly. Usually, a rate that seems a bit slow to you will be just right for listeners.

Maintain Eye Contact. Look directly into listeners' eyes. With a small audience, eye contact is one of your best connectors. As you speak, establish eye contact with as many listeners as possible. With a large group, maintain eye contact with those in the first rows. Establish eye contact immediately—before you even begin to speak—by looking around.

Manage Your Speaking Situation

Do everything you can to keep things running smoothly.

Be Responsive to Listener Feedback. Assess listener feedback continually and make adjustments as needed. If you are laboring through a long list of facts or figures and people begin to doze or fidget, you might summarize. Likewise, if frowns, raised eyebrows, or questioning looks indicate confusion, skepticism, or indignation, you can backtrack with a specific example or explanation. By tuning in to your audience's reactions, you can avoid leaving them confused, hostile, or simply bored.

Stick to Your Plan. Say what you came to say, then summarize and close—politely and on time. Don't punctuate your speech with digressions that pop into your head. Unless a specific anecdote was part of your original plan to clarify a point or increase interest, avoid excursions. We often tend to be more interested in what we have to say than our listeners are!

Leave Listeners with Something to Remember. Before ending, take a moment to summarize the major points and reemphasize anything of special importance. Are listeners supposed to remember something, have a different attitude, take a specific action? Let them know! As you conclude, thank your listeners.

Allow Time for Questions and Answers. At the very beginning, tell your listeners that a question and answer period will follow. Observe the following guidelines on page 601 for managing listener questions diplomatically and efficiently.

Guidelines for Managing Listener Questions

- Announce a specific time limit, to avoid prolonged debates.
- Listen carefully to each question.
- If you can't understand a question, ask that it be rephrased.
- Repeat every question, to ensure that everyone hears it.
- Be brief in your answers.

- If you need extra time, arrange for it after the presentation.
- If anyone attempts lengthy debate, offer to continue *after* the presentation.
- If you can't answer a question, say so and move on.
- End the session with, "We have time for one more question," or some such signal.

IN BRIEF

Oral Presentations for Cross-Cultural Audiences

Imagine you've been assigned to represent your company at an international conference or before international clients (e.g., of passenger aircraft or mainframe computers). As you plan and prepare your presentation, remain sensitive to various cultural expectations.

For example, some cultures might be offended by a presentation that gets right to the point without first observing formalities of politeness, well wishes, and the like.

Certain communication styles are welcomed in some cultures, but considered offensive in others. In southern Europe and the Middle East, people expect direct and prolonged eye contact as a way of showing honesty and respect. In Southeast Asia, this may be taken as a sign of aggression or disrespect (Gesteland 24). A sampling of the questions to consider:

- *Should I smile a lot or look serious? (Hulbert 42)*
- *Should I rely on expressive gestures and facial expressions?*
- *How loudly or softly, rapidly or slowly should I speak?*

- *Should I come out from behind the podium and approach the audience or keep my distance?*
- *Should I get right to the point or take plenty of time to lead into and discuss the matter thoroughly?*
- *Should I focus on only the key facts or on all the details and various interpretations?*
- *Should I be assertive in offering interpretations and conclusions, or should I allow listeners to reach conclusions on their own?*
- *Which types of visuals and which media might work or not work?*
- *Should I invite questions from this audience, or would this be offensive?*

To account for language differences, prepare a handout of your entire script for distribution after the presentation, along with a copy of your visuals. This way, your audience will be able to study your material at leisure.

Peer Evaluation Sheet for Oral Presentations

Presentation Evaluation for (name/topic) _____

Comments

Content
☐ Began with a clear purpose.   _____
☐ Showed command of the material.   _____
☐ Supported assertions with evidence.   _____
☐ Used adequate and appropriate visuals.   _____
☐ Used material suited to this audience's
   needs, knowledge, concerns, and interests.   _____
☐ Acknowledged opposing views.   _____
☐ Gave the right amount of information.   _____

Organization
☐ Presented a clear line of reasoning.   _____
☐ Used transitions effectively.   _____
☐ Avoided needless digressions.   _____
☐ Summarized before concluding.   _____
☐ Was clear about what the listeners
   should think or do.   _____

Style
☐ Seemed confident, relaxed, and likable.   _____
☐ Seemed in control of the speaking situation.   _____
☐ Showed appropriate enthusiasm.   _____
☐ Pronounced, enunciated, and spoke well.   _____
☐ Used appropriate gestures, tone,
   volume, and delivery rate.   _____
☐ Had good posture and eye contact.   _____
☐ Answered questions concisely and convincingly.   _____

Overall professionalism: Superior _____ **Acceptable** _____ **Needs work** _____

Evaluator's signature: _____

✓ EXERCISES

1. In a memo to your instructor, identify and discuss the kinds of oral reporting duties you expect to encounter in your career.

2. Design an oral presentation for your class. (Base it on a written report.) Make a sentence outline, and a storyboard that includes at least three visuals. Practice with a tape recorder or a friend. Use the oral report checklist to evaluate your delivery.

3. Observe a lecture or speech, and evaluate it according to the oral report checklist. Write a memo to your instructor (without naming the speaker), identifying strong and weak areas and suggesting improvements.

4. In an oral presentation to the class, present your findings, conclusions, and recommendations from the analytical report assignment in Chapter 23.

Review of Grammar, Usage, and Mechanics

Common Sentence Errors

Effective Punctuation

Transitions and Other Connectors

Effective Mechanics

N O MATTER how vital and informative a message may be, its credibility is damaged by basic errors. Any of these errors—an illogical, fragmented, or run-on sentence; faulty punctuation; or a poorly chosen word—stands out and mars otherwise good writing. Not only do such errors annoy the reader, but they speak badly for the writer's attention to detail. Your career will make the same demands for good writing that your English classes do. The difference is that evaluation (grades) in professional situations usually shows in promotions, reputation, and salary.

The Correction Symbol Table, on the last page of this book, lists correction symbols and page numbers.

Common Sentence Errors

Any piece of writing is only as good as each of its sentences. Here are common sentence errors, with suggestions for easy repairs.

frag

Sentence Fragment

A sentence expresses a logically complete idea. Any complete idea must have a subject and a verb and must not depend on another complete idea to make sense. Your sentence *might* contain several complete ideas, but it *must* contain at least one!

Although Mary was nervous, she grabbed the line, and she saved the sailboat.
    (incomplete idea)   *(complete idea)*   *(complete idea)*

If the idea is not complete—if your reader is left wondering what you mean—you probably have omitted some essential element (the subject, the verb, or another complete idea). Such a piece of a sentence is a *fragment.*

Grabbed the line. *(a fragment because it lacks a subject)*
Although Mary was nervous. *(a fragment because, although it has a subject and a verb, it needs to be joined with a complete idea)*

The only exception to the sentence rule applies when we give a command (Run!), in which the subject (you) is understood. Logically complete, this statement is properly called a sentence. So is this one:

Sam is an electronics technician.

Readers cannot miss your meaning: Somewhere is a person; the person's name is Sam; the person is an electronics technician. Suppose instead we write:

Sam an electronics technician.

This statement is not logically complete, therefore not a sentence. The reader is left asking, "What about Sam the electronics technician?" The verb—the word that makes things happen—is missing. By adding a verb we can easily change this fragment to a complete sentence.

Simple verb	Sam **is** an electronics technician.
Verb and adverb	Sam, an electronics technician, **works hard.**
Dependent clause, verb, and subjective complement	**Although he is well paid,** Sam, an electronics technician, **is not happy.**

Do not, however, mistake the following statement—which seems to have a verb—for a complete sentence:

Sam being an electronics technician.

Such "ing" forms do not function as verbs unless accompanied by such other verbs as **is, was,** and **will be.** Again, readers are confused unless you complete your idea with an independent clause.

Sam, being an electronics technician, checked all **circuitry.**

Likewise, remember that the "to + verb" form does not function as a verb.

To become an electronics technician.

The meaning is unclear unless you complete the thought.

To become an electronics technician, **Sam had to pass an exam.**

Sometimes we inadvertently create fragments by adding certain words (**because, since, if, although, while, unless, until, when, where,** and others) to an already complete sentence, transforming our independent clause (complete sentence) into a dependent clause.

Although Sam is an electronics technician.

Such words subordinate the words that follow them so that an additional idea is needed to make the first statement complete. That is, they make the statement dependent on an additional idea, which must itself have a subject and a verb and be a complete sentence. (See "Faulty Subordination.") Now we have to round off the statement with a complete idea (independent clause).

Although Sam is an electronics technician, **he hopes to be an engineer.**

Note: Be careful not to use a semicolon or a period, instead of a comma, to separate elements in the preceding sentence. Because the dependent clause depends on the independent clause for its meaning, you need only a *pause* (symbolized by a comma), not a *break* (symbolized by a semicolon), between these ideas. In fact, many fragments are created when too strong a mark of punctuation (period or semicolon) severs the connection between a dependent and an independent clause. (See the later discussion of punctuation.)

Here are some fragments from technical documents. Each is repaired in more than one way. Can you think of other ways of making these statements complete?

Fragment	She spent her first week on the job as a researcher. **Compiling information from digests and journals.**
Correct	She spent her first week on the job as a researcher, compiling information from digests and journals.
	She spent her first week on the job as a researcher. She compiled information from digests and journals.
Fragment	**Because the operator was careless.** The new computer was damaged.
Correct	Because the operator was careless, the new computer was damaged.
	The operator's carelessness resulted in damage to the new computer.

Comma Splice

In a comma splice, two complete ideas (independent clauses), which should be *separated* by a period or a semicolon, are incorrectly *joined* by a comma:

Comma splice	Sarah did a great job, she was promoted.

You can choose among several possibilities for correcting this error:

1. Substitute a period followed by a capital letter:
   Sarah did a great job. She was promoted.
2. Substitute a semicolon to signal the relationship:
   Sarah did a great job; she was promoted.
3. Use a semicolon with a connecting adverb (a transitional word):
   Sarah did a great job; **consequently,** she was promoted.
4. Use a subordinating word to make the less important sentence incomplete, thereby dependent on the other:
   Because Sarah did a great job, she was promoted.
5. Add a connecting word after the comma:
   Sarah did a great job, **and** she was promoted.

Your choice of construction will depend, of course, on the exact meaning or tone you wish to convey. The following comma splice can be repaired in the ways described above.

Comma splice	This is a new technique, therefore, some people mistrust it.
Correct	This is a new technique. Some people mistrust it.
	This is a new technique; therefore, some people mistrust it.
	Because this is a new technique, some people mistrust it.
	This is a new technique, **so** some people mistrust it.

ro

Run-on Sentence

The run-on sentence, a cousin to the comma splice, crams too many ideas without needed breaks or pauses.

> Run-on   The hourglass is more accurate than the water clock because water in a water clock must always be at the same temperature to flow at the same speed since water evaporates and must be replenished at regular intervals, thus being less effective than the hourglass for measuring time.

Like a runaway train, this statement is out of control. Here is a corrected version:

> Revised   The hourglass is more accurate than the water clock because water in a water clock must always be at the same temperature to flow at the same speed. Also, water evaporates and must be replenished at regular intervals. These temperatures and volume problems make the water clock less effective than the hourglass for measuring time.

coord

Faulty Coordination

Give equal emphasis to ideas of equal importance by joining them, within simple or compound sentences, with coordinating conjunctions: **and, but, or, nor, for, so,** and **yet.**

> This course is difficult **but** worthwhile.
>
> My horse is old **and** gray.
>
> We must decide to support **or** reject the dean's proposal.

But do not confound your meaning by coordinating excessively.

> Excessive coordination   The climax in jogging comes after a few miles **and** I can no longer feel stride after stride **and** it seems as if I am floating **and** jogging becomes almost a reflex **and** my arms **and** legs continue to move **and** my mind no longer has to control their actions.
>
> Revised   The climax in jogging comes after a few miles when I can no longer feel stride after stride. By then I am jogging almost by reflex, nearly floating, my arms and legs still moving, my mind no longer having to control their actions.

Notice how the meaning becomes clear when the less important ideas (**nearly floating, arms and legs still moving, my mind no longer having**) are shown as dependent on, rather than equal to, the most important idea (**jogging almost by reflex**)—the idea that contains the lesser ones.

Avoid coordinating ideas that cannot be sensibly connected:

> Faulty   John had a drinking problem **and** he dropped out of school.
>
> Revised   John's drinking problem depressed him so much that he couldn't study, so he quit school.
>
> Faulty   I was late for work **and** wrecked my car.

Revised	Late for work, I backed out of the driveway too quickly, hit a truck, and wrecked my car.

Instead of *try and,* use *try to.*

Faulty	I will try and help you.
Revised	I will try to help you.

sub

Faulty Subordination

Proper subordination shows that a less important idea is dependent on a more important idea. By using subordination you can combine simple sentences into complex sentences and emphasize the most important idea. Consider these complete ideas:

> Joe studies hard. He has severe math anxiety.

Because these ideas are expressed as simple sentences, they appear to be coordinate (equal in importance). But if you wanted to indicate your opinion of Joe's chances of succeeding in math, you would need a third sentence: **His disability probably will prevent him from succeeding,** or **His willpower will help him succeed.** To communicate the intended meaning concisely, combine ideas and subordinate the one that deserves less emphasis:

> Despite his severe math anxiety *(subordinate idea),* Joe studies hard *(independent idea).*

This first version suggests that Joe will succeed. Below, subordination is used to suggest the opposite meaning:

> Despite his diligent studying *(subordinate idea),* Joe has severe math anxiety *(independent idea).*

A dependent (or subordinate) clause in a sentence is signalled by a subordinating conjunction: **because, so, if, unless, after, until, since, while, as,** and **although,** among others. Be sure to place the idea you want emphasized in the independent clause; do not write

> Although Mary is receiving excellent medical treatment, she is seriously ill.

if you mean to suggest that Mary has a good chance of recovering.

Do not coordinate when you should subordinate:

Weak	Television viewers can relate to an athlete they idolize and they feel obliged to buy the product endorsed by their hero.

Of the two ideas in the sentence above, one is the cause, the other the effect. Emphasize this relationship through subordination:

Revised	Because television viewers can relate to an athlete they idolize, they feel obliged to buy the product endorsed by their hero.

When combining several ideas within a sentence, decide which is most important, and subordinate the other ideas to it—do not merely coordinate:

Faulty   This employee is often late for work, and he writes illogical reports, and he is a poor manager, and he should be fired.

Revised   Because this employee is often late for work, writes illogical reports, and has poor management skills, **he should be fired.** *(The last clause is independent.)*

Do not overstuff sentences by subordinating excessively:

Overstuffed   This job, which I took when I graduated from college, while I waited for a better one to come along, which is boring, where I've gained no useful experience, makes me anxious to quit.

Revised   Upon college graduation, I took this job while waiting for a better one to come along. Because I find it boring and have gained no useful experience, I am eager to quit.

agr sv

Faulty Agreement—Subject and Verb

The subject should agree in number with the verb. We are not likely to use faulty agreement in short sentences, where subject and verb are not far apart. Thus, we are not likely to say "Jack eat too much" instead of "Jack eats too much," but in more complicated sentences—those in which the subject is separated from its verb by other words—we sometimes lose track of the subject-verb relationship.

Faulty   The lion's **share** of diesels **are** sold in Europe.

Although **diesels** is closest to the verb, the subject is **share,** a singular subject that must agree with a singular verb.

Correct   The lion's **share** of diesels **is** sold in Europe.

Agreement errors are easy to correct when subject and verb are identified.

Faulty   A **system** of lines **extend** horizontally to form a grid.

Correct   A **system** of lines **extends** horizontally to form a grid.

A second problem with subject-verb agreement occurs when we use indefinite pronouns such as **each, everyone, anybody,** and **somebody.** They function as subjects and usually take a singular verb.

Faulty   **Each** of the crew members **were** injured.

Correct   **Each** of the crew members **was** injured.

Faulty   **Everyone** in the group **have** practiced long hours.

Correct   **Everyone** in the group **has** practiced long hours.

Agreement problems can be caused by collective nouns such as **herd, family, union, group, army, team, committee,** and **board.** They can call for a singular or plural verb—depending on your intended meaning. When denoting the group as a whole, use a singular verb.

Correct   The **committee meets** weekly to discuss new business.

The editorial **board** of this magazine **has** high standards.

To denote individual members of the group, however, use a plural verb.

> **Correct**   Not all members of the editorial **board are** published authors.

Yet another problem occurs when two subjects are joined by **either . . . or** or **neither . . . nor.** Here, the verb is singular if both subject are singular, and plural if both subjects are plural. If one subject is plural and one is singular, the verb agrees with the subject closer to the verb.

> **Correct**   Neither **John** nor **Bill works** regularly.
>
> **Either apples** or **oranges are** good vitamin sources.
>
> Either Felix or his **friends are** crazy.
>
> Neither the boys nor their **father likes** the home team.

If, on the other hand, two subjects (singular, plural, or mixed) are joined by **both . . . and,** the verb will be plural. Whereas **or** suggests "one or the other," **and** announces a combination of the two subjects, thereby requiring a plural verb.

> **Correct**   **Both** Joe and Bill **are** resigning.
>
> The **book and** the **briefcase appear** expensive.

A single *and* between singular subjects makes for a plural subject.

agr p

Faulty Agreement—Pronoun and Referent

A pronoun can make sense only if it refers to a specific noun (its referent or antecedent), with which it must agree in gender and number.

> **Correct**   **Joe** lost **his** blueprints.
>
> The **workers** complained that **they** were treated unfairly.

Some instances, however, are not so obvious. When an indefinite pronoun such as **each, everyone, anybody, someone,** and **none** serves as the pronoun referent, the pronoun itself is singular.

> **Correct**   **Anyone** can get **his or her** degree from that college.
>
> **Each** candidate described **her** plans in detail.

ref

Faulty or Vague Pronoun Reference

Whenever a pronoun is used, it must refer to one clearly identified referent; otherwise, your message will be confusing.

> **Ambiguous**   **Sally** told **Sarah** that **she** was obsessed with her job.
> **Correct**   Sally told Sarah, "I'm obsessed with my job."
>
> Sally told Sarah, "I'm obsessed with your job."
>
> Sally told Sarah, "You're obsessed with [your, my] job."
>
> Sally told Sarah, "She's obsessed with [her, my, your] job."

Avoid using **this, that,** or **it**—especially to begin a sentence—unless the pronoun refers to a specific antecedent (referent).

Vague	He drove away from his menial **job,** boring **lifestyle,** and damp **apartment,** happy to be leaving **it** behind.
Correct	He drove away, happy to be leaving behind his menial job, boring lifestyle, and damp apartment.
Vague	The problem with our **defective machinery** is compounded by the **operator's incompetence. That** annoys me!
Correct	I am annoyed by the problem with our defective machinery as well as by the new operator's incompetence.

ca

Faulty Pronoun Case

A pronoun's case (nominative, objective, or possessive) is determined by its role in a sentence: as subject, object, or indicator of possession.

If the pronoun serves as the subject of a sentence (**I, we, you, she, he, it, they, who**), its case is *nominative.*

> **She** completed her graduate program in record time.
>
> **Who** broke the chair?

When a pronoun follows a version of **to be** (a linking verb), it explains (complements) the subject, and so its case is nominative.

> It was **she.**
>
> The chemist who perfected our new distillation process is **he.**

If the pronoun serves as the object of a verb or a preposition (**me, us, you, her, him, it, them, whom**), its case is *objective.*

| Object of the verb | The employees gave **her** a parting gift. |
| Object of the preposition | Several colleagues left with **him.**
To **whom** do you wish to complain? |

If a pronoun indicates possession (**my, mine, ours, your, yours, his, her, hers, its, their, theirs, whose**), its case is *possessive.*

> The brown briefcase is **mine.**
>
> **Her** offer was accepted.
>
> **Whose** opinion do you value most?

Here are some frequent errors in pronoun case:

Faulty	**Whom** is responsible to **who**? *(The subject should be nominative and the object should be objective.)*
Correct	**Who** is responsible to **whom**?
Faulty	The debate was between Marsha and **I.** *(As object of the preposition, the pronoun should be objective.)*
Correct	The debate was between Marsha and **me.**

Faulty	**Us** board members are accountable for our decisions. *(The pronoun accompanies the subject, "board members," and thus should be nominative.)*
Correct	**We** board members are accountable for our decisions.
Faulty	A group of **we** managers will fly to the convention. *(The pronoun accompanies the object of the preposition, "managers," and thus should be objective.)*
Correct	A group of **us** managers will fly to the convention.

Hint: By deleting the accompanying noun from the two latter examples, we can easily identify the correct pronoun case ("We . . . are accountable"; "A group of us . . . will fly").

`mod`

Faulty Modification

A sentence's word order (syntax) helps determine its effectiveness and meaning. Words or groups of words are modified by adjectives, adverbs, phrases, or clauses. Modifiers explain, define, or add detail to other words or ideas. Prepositional phrases, for example, usually define or limit adjacent words:

the foundation **with the cracked wall**

the repair job **on the old Ford**

the journey **to the moon**

As do phrases with "-ing" verb forms:

the student **painting the portrait**

Opening the door, we entered quietly.

Phrases with "to + verb" form limit:

To succeed, one must work hard.

Some clauses limit:

the person **who came to dinner**

Problems with word order occur when a modifying phrase begins a sentence, and has no word to modify.

Dangling modifier	**Dialing the phone,** the cat ran out the open door.

The cat obviously did not dial the phone, but because the modifier **Dialing the phone** has no word to modify, the word order suggests that the noun beginning the main clause *(cat)* names the one who dialed the phone. Without any word to join itself to, the modifier *dangles.* By inserting a subject, we can repair this absurd message.

`dgl`

Correct	As Joe dialed the phone, the cat ran out the open door.

A dangling modifier also can obscure your meaning.

| Dangling modifier | **After completing the student financial aid application form,** the Financial Aid Office will forward it to the appropriate state agency. |

Who completes the form—the student or the financial aid office?

Here are some other dangling modifiers that make the message confusing, inaccurate, or downright absurd:

| Dangling modifier | **While walking,** a cold chill ran through my body. |
| Correct | While **I** walked, a cold chill ran through my body. |

| Dangling modifier | Impurities have entered our bodies **by eating chemically processed foods.** |
| Correct | Impurities have entered our bodies by **our** eating chemically processed foods. |

| Dangling modifier | **By planting different varieties of crops,** the pests were unable to adapt. |
| Correct | By planting different varieties of crops, **farmers** prevented the pests from adapting. |

The order of adjectives and adverbs in a sentence is as important as the order of modifying phrases and clauses. Notice how changing word order affects the meaning of these sentences:

I **often** remind myself of the need to balance my checkbook.

I remind myself of the need to balance my checkbook **often.**

Be sure that modifiers and the words they modify follow an order that reflects your meaning.

| Misplaced modifier | Joe typed another memo on our computer **that was useless.** *(Was the typewriter or the memo useless?)* |
| Correct | Joe typed another useless memo on our computer. |

or

Joe typed another memo on our useless computer.

| Misplaced modifier | He read a report on the use of nonchemical pesticides **in our conference room.** *(Are the pesticides to be used in the conference room?)* |
| Correct | In our conference room he read a report on the use of nonchemical pesticides. |

| Misplaced modifier | She volunteered **immediately** to deliver the radioactive shipment. *(Volunteering immediately, or delivering immediately?)* |
| Correct | She immediately volunteered to deliver . . . |

or

She volunteered to deliver immediately . . .

<table>
<tr><td>par</td></tr>
</table>

Faulty Parallelism

To reflect relationships among items of equal importance, express them in identical grammatical form:

Correct   We here highly resolve . . . that government **of the people, by the people, for the people** shall not perish from the earth.

The statement above describes the government with three modifiers of equal importance. Because the first modifier is a prepositional phrase, the others must be also. Otherwise, the message would be garbled, like this:

Faulty   We here highly resolve . . . that government **of the people, which the people created and maintain, serving the people** shall not perish from the earth.

If you begin the series with a noun, use nouns throughout the series; likewise for adjectives, adverbs, and specific types of clauses and phrases.

Faulty   The new apprentice is **enthusiastic, skilled,** and **you can depend on her.**

Correct   The new apprentice is **enthusiastic, skilled,** and **dependable.** *(all subjective complements)*

Faulty   In his new job, he felt **lonely** and **without a friend.**

Correct   In his new job, he felt **lonely** and **friendless.** *(both adjectives)*

Faulty   She plans **to study** all this month and **on scoring well** in her licensing examination.

Correct   She plans **to study** all this month and **to score well** in her licensing examination. *(both infinitive phrases)*

Faulty   She **sleeps** well and **jogs** daily, **as well as eating** high-protein foods.

Correct   She **sleeps** well, **jogs** daily, and **eats** high-protein foods. *(all verbs)*

To improve coherence in long sentences, repeat words that introduce parallel expressions:

Faulty   Before buying this property, you should decide whether you will settle down and raise a family, travel for a few years, or pursue a graduate degree.

Correct   Before buying this property, you should decide whether **to settle** down and raise a family, **to travel** for a few years, or **to pursue** a graduate degree.

<table>
<tr><td>shift</td></tr>
</table>

Sentence Shifts

Shifts in point of view damage coherence. If you begin a sentence or paragraph with one subject or person, do not shift to another.

Shift in person	When **you** finish the job, **one** will have a sense of pride.
Correct	When **you** finish the job, **you** will have a sense of pride.
Shift in number	**One** should sift the flour before **they** make the pie.
Correct	**One** should sift the flour before **one** makes the pie. (Or better: Shift the flour before making the pie.)

Do not begin a sentence in the active voice and then shift to passive.

Shift in voice	**He delivered** the plans for the apartment complex, and the building site **was also inspected by him.**
Correct	**He delivered** the plans for the apartment complex and also **inspected** the building site.

Do not shift tenses without good reason.

Shift in tense	She **delivered** the blueprints, **inspected** the foundation, **wrote** her report, and **takes** the afternoon off.
Correct	She **delivered** the blueprints, **inspected** the foundation, **wrote** her report, and **took** the afternoon off.

Do not shift from one mood to another (as from imperative to indicative mood in a set of instructions).

Shift in mood	**Unscrew** the valve and then steel wool **should be used** to clean the fitting.
Correct	**Unscrew** the valve and then **use** steel wool to clean the fitting.

Do not shift from indirect to direct discourse within a sentence.

Shift in Discourse	Jim wonders **if he will get the job** and **will he like it?**
Correct	Jim wonders **if he will get the job** and **if he will like it.** Will Jim get the job, and will he like it?

Effective Punctuation

pct

Punctuation marks are like road signs and traffic signals. They govern reading speed and provide clues for navigation through a network of ideas; they mark intersections, detours, and road repairs; they draw attention to points of interest along the route; and they mark geographic boundaries. In short, punctuation marks give us a simple way of making ourselves understood.

End Punctuation

The three marks of end punctuation—period, question mark, and exclamation point—work like a red traffic light by signaling a complete stop.

./

Period. A period ends a sentence. Periods end some abbreviations.

Ms.	Assn.	Dr.
M.D.	Inc.	B.A.

Periods serve as decimal points for figures.

$15.95

2.14%

?/

Question Mark. A question mark follows a direct question.

Where is the balance sheet?

Do not use a question mark to end an indirect question.

Faulty	He asked if all students had failed the test?
Correct	He asked if all students had failed the test.

or

He asked, "Did all students fail the test?"

!/

Exclamation Point. Because exclamation points symbolize strong feeling, don't overuse them. Otherwise you might seem hysterical or insincere.

Correct	Oh, no!
	Pay up!

Use an exclamation point only when expression of strong feeling is appropriate.

;/

Semicolon

A semicolon usually works like a blinking red traffic light at an intersection by signaling a brief but definite stop.

Semicolon Separating Independent Clauses. Semicolons separate independent clauses (logically complete ideas), whose contents are closely related and are not connected by a coordinating conjunction.

The project was finally completed; we had done a good week's work.

The semicolon can replace the conjunction-comma combination that joins two independent ideas.

The project was finally completed, and we were elated.

The project was finally completed; we were elated.

The second version emphasizes the sense of elation.

Semicolons Used with Adverbs as Conjunctions and Other Transitional Expressions. Semicolons accompany conjunctive adverbs, and other expressions that connect related independent ideas (**besides, otherwise, still, however, furthermore, consequently, therefore, in contrast, in fact,** or the like).

The job is filled; however, we will keep your résumé on file.

Your background is impressive; in fact, it is the best among our applicants.

Semicolons Separating Items in a Series. When items in a series contain internal commas, semicolons provide clear separation between items.

We are opening branch offices in the following cities: Santa Fe, New Mexico; Albany, New York; Montgomery, Alabama; and Moscow, Idaho.

Members of the survey crew were John Jones, a geologist; Hector Lightweight, a draftsman; and Mary Shelley, a graduate student.

Colon

`: /`

Like a flare in the road, a colon signals you to stop and then proceed, paying attention to the situation ahead, the details of which will be revealed as you move along. Usually, a colon follows an introductory statement that requires a follow-up explanation.

We need the following equipment immediately: a voltmeter, a portable generator, and three pairs of insulated gloves.

She is an ideal colleague: honest, reliable, and competent.

Two candidates clearly are superior: John and Marsha.

With the exception of **Dear Sir:** and other salutations in formal correspondence, colons follow independent (logically and grammatically complete) statements. Because colons, like end punctuation and semicolons, signal a full stop, they never are used to fragment a complete statement.

Faulty   My plans include: finishing college, traveling for two years, and settling down in Boston.

No punctuation should follow "include."

Colons can introduce quotations.

The supervisor's message was clear enough: "You're fired."

A colon normally replaces a semicolon in separating two related, complete statements when the second statement explains or amplifies the first.

His reason for accepting the lowest-paying job offer was simple: he had always wanted to live in the Northwest.

The statement following the colon explains the "reason" mentioned in the statement preceding the colon.

Comma

`, /`

The comma is the most frequently used—and abused—punctuation mark. Unlike the period, semicolon, and colon, which signal a full stop, the comma signals a *brief pause*. The comma works like a blinking yellow traffic light, for which you slow down without stopping. Never use a comma to signal a *break* between independent ideas; it is not strong enough.

Comma as a Pause Between Complete Ideas. In a compound sentence where a coordinating conjunction (**and, or, nor, for, but**) connects equal (independent) statements, a comma usually precedes the conjunction.

> This is a high-paying job, but the stress is high.
>
> This vacant shop is just large enough for our boutique, and the location is excellent for walk-in customer traffic.

Without the conjunction, these statements would suffer from a comma splice, unless the comma were replaced by a semicolon or period.

Comma as a Pause Between an Incomplete and a Complete Idea. A comma usually appears between a complete and an incomplete statement in a complex sentence to show that the incomplete statement depends for its meaning on the complete statement. (The incomplete statement cannot stand alone, separated by a break such as a semicolon, colon, or period.)

> **Because he is a fat cat,** Jack diets often.
>
> **When he eats too much,** Jack gains weight.

Above, the first idea is made incomplete by a subordinating conjunction (**since, when, because, although, where, while, if, until**), which here connects a dependent with an independent statement. The first (incomplete) idea depends on the second (complete) for wholeness. When the order is reversed (complete idea followed by incomplete), the comma usually is omitted.

> Jack diets often **because he is a fat cat.**
>
> Jack gains weight **when he eats too much.**

Because commas take the place of speech signals, reading a sentence aloud should tell you whether or not to pause (and use a comma).

Commas Separating Items (Words, Phrases, or Clauses) in a Series. Use a comma to separate items in a series.

> **Jane, Joe, Marsha,** and **John** are joining us on the hydroelectric project.
>
> The office was **yellow, orange,** and **red.**
>
> She works hard **at home, on the job,** and even **during her vacation.**
>
> The employee claimed **that the hours were long, that the pay was low, that the work was boring,** and **that the supervisor was paranoid.**

Use no commas when *or* or *and* appears between all items in a series.

> She is willing to work in San Francisco or Seattle or even in Anchorage.

Add a comma when *or* or *and* is used only before the final item in the series.

> Our luncheon special for Thursday will be rolls, steak, beans, and ice cream.

Without the comma, the sentence might cause readers to conclude that beans and ice cream is an exotic new dessert.

Comma Setting off Introductory Phrases. Infinitive, prepositional, or verbal phrases introducing a sentence usually are set off by commas.

Infinitive phrase	**To be or not to be,** that is the question.
Prepositional phrase	**In Rome,** do as the Romans do.
Participial phrase	**Moving quickly,** the army surrounded the enemy.

When an interjection introduces a sentence, it is set off by a comma.

Oh, is that the final verdict?

When a noun in direct address introduces a sentence, it is set off by a comma.

Mary, you've done a great job.

Commas Setting off Nonrestrictive Elements. A restrictive phrase or clause limits or defines the subject in such a way that deleting the modifier would change the meaning of the sentence.

All candidates **who have work experience** will receive preference.

The clause, **who have work experience,** defines **candidates** and is essential to the meaning of the sentence. Without this clause, the meaning would be entirely different.

All candidates will receive preference.

This next sentence also contains a restriction:

All candidates **with work experience** will receive preference.

The phrase, **with work experience,** defines **candidates** and thus specifies the meaning of the sentence. Because this phrase *restricts* the subject by limiting the category, **candidates,** it is essential to the sentence's meaning and so is not separated from the sentence by commas.

A nonrestrictive phrase or clause does not limit or define the subject; a nonrestrictive element could be deleted without changing the basic meaning of the sentence.

Our draftsperson, **who has little experience,** is highly competent.
This house, **riddled with carpenter ants,** is falling apart.

In each of those sentences, the modifying phrase or clause does not restrict the subject; each could be deleted:

Our new draftsperson is highly competent.
This house is falling apart.

A nonrestrictive clause or phrase is set off from the sentence by commas.

EFFECTIVE PUNCTUATION   621

Commas Setting off Parenthetical Elements. Items that interrupt sentence flow are called parenthetical and are enclosed by commas. Expressions such as **of course, as a result, as I recall,** and **however** are parenthetical and may denote emphasis, afterthought, clarification, or transition.

Emphasis	This deluxe model, **of course,** is more expensive.
Afterthought	Your report format, **by the way,** was impeccable.
Clarification	The loss of my job was, **in a way,** a blessing.
Transition	Our warranty, **however,** does not cover tire damage.

Direct address is parenthetical.

Listen, **my children,** and you shall hear . . .

A parenthetical expression at the beginning or the end of a sentence is set off by a comma.

Naturally, we will expect a full guarantee.

My friends, I think we have a problem.

You've done a good job, **Jim.**

Yes, you may use my name in your advertisement.

Commas Setting off Quoted Material. Quoted items included within a sentence are often set off by commas.

The customer said, **"I'll take it,"** as soon as he laid eyes on our new model.

Commas Setting off Appositives. An appositive, a word or words explaining a noun and placed immediately after it, is set off by commas.

Martha Jones, **our new president,** is overhauling all personnel policies.

Alpha waves, **the most prominent of the brain waves,** are typically recorded in a waking subject whose eyes are closed.

Please make all checks payable to Sam Sawbuck, **company treasurer.**

Commas Used in Common Practice. Commas set off the day of the month from the year, in a date.

May 10, 1984

They set off numbers in three-digit intervals.

11,215

6,463,657

They set off street, city, and state in an address.

The bill was sent to John Smith, 184 Sea Street, Albany, NY 01642.

When the address is written vertically, however, the omitted commas are those which would otherwise occur at the end of each address line.

John Smith
184 Sea Street
Albany, NY 01642

Commas set off an address or date in a sentence.

Room 3C, Margate Complex, is the site of our newest office.
December 15, 1997 is my retirement date.

They set off degrees and titles from proper names.

Roger P. Cayer, M.D.
Gordon Browne, Jr.
Sandra Mello, Ph.D.

Commas Used Erroneously. Avoid needless or inappropriate commas. You are probably safer using too few commas than too many. Reading sentences aloud is one way to identify inappropriate pauses.

Faulty   As I opened the door, he told me, that I was late. *(separates the indirect from the direct object)*

The universal symptom of the suicide impulse, is depression. *(separates the subject from its verb)*

This has been a long, difficult, project. *(separates the final adjective from its noun)*

John, Bill, and Sally, are joining us on the design phase of this project. *(separates the final subject from its verb)*

An employee, who expects rapid promotion, must quickly prove her or his worth. *(separates a modifier that should be restrictive)*

I spoke in a conference call with John, and Marsha. *(separates two words linked by a coordinating conjunction)*

The room was, eighteen feet long. *(separates the linking verb from the subjective complement)*

We painted the room, red. *(separates the object from its complement)*

---

ap/

Apostrophe

Apostrophes serve three purposes: to indicate the possessive, a contraction, and the plural of numbers, letters, and figures.

Apostrophe Indicating the Possessive. At the end of a singular word, or of a plural word that does not end in *s*, add an apostrophe plus *s* to indicate the possessive.

The people's candidate won.
The chain saw was Bill's.
The men's locker room burned.
I borrowed Doris's book.
Have you heard Ray Charles's song?

Do not use an apostrophe to indicate the possessive form of either singular or plural pronouns:

> The book was hers.
> Ours is the best sales record.
> The fault was theirs.

At the end of a plural word that ends in *s*, add an apostrophe only.

> the cows' water supply
> the Jacksons' wine cellar

At the end of a compound noun, add an apostrophe plus *s*.

> my father-in-law's false teeth

At the end of the last word in nouns of joint possession, add an apostrophe plus *s* if both own one item.

> Mary and Sam's lakefront cottage

Add an apostrophe plus *s* to both nouns if each owns specific items.

> Mary's and Sam's passports

Apostrophe Indicating a Contraction. An apostrophe shows that you have omitted one or more letters in a phrase that is usually a combination of a pronoun and a verb.

I'm	they're
he's	we're
you're	who's

Don't confuse *they're* with *their* or *there.*

Faulty	there books
	Their now leaving.
	living their
Correct	their books
	They're now leaving.
	living there

Remember the distinction in this way:

> Their boss knows they're there.

Don't confuse **it's** and **its. It's** means "it is." **Its** is the possessive.

> It's watching its reflection in the pond.

Don't confuse **who's** and **whose. Who's** means "who is," whereas **whose** indicates the possessive.

> Who's interrupting whose work?

Other contractions are formed from the verb and the negative.

isn't	can't
don't	haven't
won't	wasn't

Apostrophe Indicating the Plural of Numbers, Letters, and Figures.

The 6's on this new typewriter look like smudged G's, the 9's are illegible, and the %'s are unclear.

Use an apostrophes only when its absence would create ambiguity.

"/"

Quotation Marks

Quotation marks set off exact words borrowed from another speaker or writer. At the end of a quotation the period or comma is placed within quotation marks.

"Hurry up," he whispered.
She told me, "I'm depressed."

The colon or semicolon is always placed outside quotation marks:

Our contract clearly defines "middle-management personnel"; however, it does not state salary range.
You know what to expect when Honest John offers you a "bargain": a piece of junk.

Sometimes a question mark is used within a quotation that is part of a larger sentence. (Do not follow a question mark with a comma.)

"Can we stop the flooding?" inquired the supervisor.

When the question mark or exclamation point is part of the quotation, it belongs within quotation marks, replacing the comma or period.

"Help!" he screamed.
She asked John, "Can't we agree about anything?"

When the question mark or exclamation point denotes the attitude of the person quoting instead of the person quoted, it belongs outside the quotation mark.

Why did he wink and tell me, "It's a big secret"?
He actually accused me of being an "elitist"!

When quoting a passage of fifty words or longer, indent the entire passage five spaces and single-space between its lines to set it off from the text. Do not enclose the indented passage in quotation marks.

Use quotation marks around titles of articles, book chapters, poems, and unpublished reports.

The enclosed article, "The Job Market for College Graduates," should provide some helpful insights.

The title of a published work—book, journal, newspaper, brochure, or pamphlet—should be underlined or italicized.

Finally, use quotation marks (with restraint) to indicate your ironic use of a word.

He is some "friend"!

Ellipses

... /

Three dots in a row (...) indicate that you have omitted some material from a quotation. If the omitted words come at the end of the original sentence, a fourth dot indicates the period. Use several dots centered in a line to indicate that a paragraph or more has been left out. Ellipses help you save time and zero in on the important material within a quotation.

"Three dots. . . . indicate that you have omitted some material . . . . A fourth dot indicates the period. . . . Several dots centered in a line · · · indicate a paragraph or more. . . . Ellipses help you . . . zero in. . . .

Italics

ital

In typing or longhand writing, indicate italics by *underlining*. On a word processor, use italic print for titles of books, periodicals, films, newspapers, and plays; for the names of ships; for foreign words or scientific names; for emphasizing a word (use sparingly); for indicating the special use of a word.

The Oxford English Dictionary is a handy reference tool.

The *Lusitania* sank rapidly.

She reads *The Boston Globe* often.

My only advice is *caveat emptor.*

Bacillus anthracis is a highly virulent organism.

Do *not* inhale these spores, under any circumstances!

Our contract defines a *full-time employee* as one who works a minimum of thirty-five hours weekly.

Parentheses

( ) /

Use commas normally to set off parenthetical elements, dashes to give some emphasis to the material that is set off, and parentheses to enclose material that defines or explains the statement that precedes it.

This organism requires an anaerobic (oxygenless) environment.

The cost of manufacturing our Beta II transistors has increased by 10 percent in one year. (See Appendix A for full cost breakdown.)

This new three-color model (made by Ilco Corporation) is selling well.

Notice that material within parentheses, like all other parenthetical material discussed earlier, can be deleted without harming the logical and grammatical structure of the sentence.

Also, use parentheses to enclose numbers or letters that segment items of information in a series.

The three basic steps in this procedure are (1) ..., (2) ..., and (3) ....

Brackets

> [ ] /

Use brackets within a quotation to add material that was not in the original quotation but is needed for clarification. Sometimes a bracketed word will provide an antecedent (or referent) for a pronoun.

"She [Jones] was the outstanding candidate for the job."

Brackets can enclose information from some other location within the context of the quotation.

"It was in early spring [April 2, to be exact] that the tornado hit."

Use brackets to correct a quotation.

"His report was [full] of mistakes."

Use *sic* ("thus" or "so") when quoting a mistake in spelling, usage, or logic.

Her secretary's comment was clear: "She don't [sic] want any of these."

Dashes

> – – /

Dashes can be effective indicators of meaning—as long as they are not overused. Make dashes on your typewriter by placing two hyphens side by side. Parentheses de-emphasize enclosed material; dashes emphasize it.

Used selectively, dashes can provide emphasis, but they are not a substitute for all other punctuation. When in doubt, do not use a dash!

Dashes can denote an afterthought.

Have a good vacation—but don't get sunstroke.

They can enclose an interruption in the middle of a sentence.

This building's designer—I think it was Wright—was, above all, an artist.

Our new team—Jones, Smith, and Brown—already is compiling outstanding statistics.

Although they can often be used interchangeably with commas, dashes dramatize a parenthetical statement more than commas do.

Mary, a true friend, spent hours helping me rehearse for my interview.

Mary—a true friend—spent hours helping me rehearse for my interview.

Notice the added emphasis in the second version.

| – / |

Hyphen

Use a hyphen to join compound modifiers (two or more words preceding the noun as an adjective), but not compound nouns.

> the rough-hewn wood
> the well-written report
> the all-too-human error
> a three-part report

Do not hyphenate these same words if they *follow* the noun

> The wood was rough hewn.
> The report is well written.
> The error was all too human.

Hyphenate an adverb-participle compound preceding a noun.

> the high-flying glider

Do not hyphenate compound modifiers if the adverb ends in **-ly.**

> the finely tuned engine

Hyphenate most words that begin with the prefix **self-.** (Check your dictionary.

> self-reliance
> self-discipline
> self-actualizing

Hyphenate to avoid ambiguity.

> re-creation *(a new creation)*
> recreation *(leisure activity)*

Hyphenate words that begin with **ex-** only if **ex-** means "past."

> ex-employee
> expectant

Hyphenate all fractions, along with ratios that are used as adjectives and that precede the noun.

> a two-thirds majority
> In a four-to-one ratio they defeated the proposal.

Do not hyphenate ratios if they do not immediately precede the noun.

> The proposal was voted down four to one.

Hyphenate compound numbers from twenty-one through ninety-nine.

> Thirty-eight windows were broken.

Hyphenate a series of compound adjectives preceding a noun.

> The subjects for the motivation experiment were fourteen-, fifteen-, and six-teen-year-old students.

Use a hyphen to divide a word at the right-hand margin. Consult your dictionary for the correct syllable breakdown:

> com-puter
>
> comput-er

Actually, it is best to avoid altogether this practice of dividing words at the ends of lines in a typewritten text.

Transitions and Other Connectors

trans

Transitions help make your meaning clear, signaling readers that you are in a specific time or place, that you are giving an example, showing a contrast, shifting gears, or concluding your discussion. Here are some common transitions and the relations they indicate:

Addition
I am majoring in naval architecture; **furthermore,** I spent three years crewing on a racing yawl.

moreover	and
in addition	again
also	as well as

Place
Here is the switch that turns on the stage lights. **To the right** is the switch that dims them.

beyond	to the left
over	nearby
under	adjacent to
opposite to	next to
beneath	where

Time
The crew will mow the ball field this morning; **immediately afterward,** we will clean the dugouts.

first	the next day
next	in the meantime
second	in turn
then	subsequently
meanwhile	while
at length	since
later	before
now	after

Comparison
Our reservoir is drying up because of the drought; **similarly,** water supplies in neighboring towns are low.

	likewise	
	in the same way	
	in comparison	
Contrast	Felix worked hard; **however,** he received poor grades.	
	however	but
	nevertheless	on the other hand
	yet	to the contrary
	still	notwithstanding
	in contrast	conversely
Results	Jack fooled around; **consequently,** he was fired.	
	thus	thereupon
	hence	as a result
	therefore	so
	accordingly	as a consequence
Example	Competition for part-time jobs is fierce; **for example,** eighty students applied for the clerk's job at Sears.	
	for instance	namely
	to illustrate	specifically
Explanation	She had a terrible semester; **that is,** she flunked four courses.	
	in other words	in fact
	simply stated	put another way
Summary or conclusion	Our credit is destroyed, our bank account is overdrawn, and our debts are piling up; **in short,** we are bankrupt.	
	in closing	to sum up
	to conclude	all in all
	to summarize	on the whole
	in brief	in retrospect
	in summary	in conclusion

Pronouns serve as connectors, because a pronoun refers back to a noun in a preceding clause or sentence.

As the **crew** neared the end of the project, **they** were all willing to work overtime to get the job done.

Low employee morale is damaging our productivity. **This** problem needs immediate attention.

Repetition of key words or phrases is another good connecting device— as long as it is not overdone. Here is another example of effective repetition:

Overuse and drought have depleted our water supply critically. Because of our **depleted supply,** we need to enforce strict **water**-conservation measures.

Here, the repetition also emphasizes a critical problem.

Because this next paragraph lacks transitions, sentences seem choppy and awkward:

Choppy   Technical writing is a difficult but important skill to master. It requires long hours of work and concentration. This time and effort are well spent. Writing is indispensable for success. Good writers derive pride and satisfaction from their effort. A highly disciplined writing course should be part of every student's curriculum.

Here is the same paragraph rewritten to improve coherence:

Revised   Technical writing is a difficult but important skill to master. It requires long hours of work and concentration. This time and effort, **however,** are well spent **because** writing is an indispensable tool for success. **Moreover,** good writers derive pride and satisfaction from their effort. A highly disciplined writing course, **therefore,** should be part of every student's curriculum.

Besides increasing coherence *within* a paragraph, transitions and connectors emphasize relationships **between** paragraphs by linking related groups of ideas. Here are two transitional sentences that could serve as conclusions for some paragraphs or as topic sentences for paragraphs that would follow; or they could stand alone for emphasis as single-sentence paragraphs:

Because the A-12 filter has decreased overall engine wear by 15 percent, it should be included as a standard item in all our new models.

With the camera activated and the watertight cover sealed, the diving bell is ready to be submerged.

Topic headings, like those in this book, are another connecting device. A topic heading is both a link and a separation between related yet distinct groups of ideas.

Sometimes a whole paragraph can serve as a connector between major sections of your report. Assume that you have just completed a section in a report on the advantages of a new oil filter and are now moving to a section on selling the idea to the buying public. Here is a paragraph you might write to link the two sections:

Because the A-12 filter has decreased overall engine wear by 15 percent, it should be included as a standard item in all our new models. However, tooling and installation adjustments will add roughly $100 to the list price of each model. Therefore, we have to explain to customers the filter's long-range advantages. Let's look at ways of explaining these advantages.

Notice that this transitional paragraph *contains* transitional expressions as well.

Effective Mechanics

Correctness in abbreviation, capitalization, use of numbers, and spelling is an important sign of your attention to detail.

ab

Abbreviations

Whenever you abbreviate, consider your audience; never use an abbreviation that might confuse your reader. Abbreviations are often inappropriate in formal writing. When in doubt, write the word out.

Abbreviate some words and titles when they precede or immediately follow a proper name.

Correct	Mr. Jones	Raymond Dumont, Jr.
	Dr. Jekyll	Warren Weary, Ph.D.
	St. Simeon	

Do not however, write abbreviations such as these:

Faulty	Mary is a Dr.
	Pray hard and you might become a St.

In general, do not abbreviate military, religious, and political titles.

Correct	Reverend Ormsby
	Captain Hook
	President Lincoln

Abbreviate time designations only when they are used with actual times.

Correct	A.D. 576
	400 B.C.
	5:15 A.M.

Do not abbreviate these designations when they are used alone.

Faulty	Plato lived sometime in the B.C. period.
	She arrived in the A.M.

In formal writing, do not abbreviate days of the week, months, words such as **street** and **road**, or names of disciplines such as **English.** Avoid abbreviating states, such as **Me.** for **Maine;** countries, such as **U.S.** for **United States;** and book parts such as **Chap.** for **Chapter, pp.** for **page,** and **fig.** for **figure.**

Use **no.** for **number** only when the actual number is given.

Correct	Check switch No. 3.

Abbreviate a unit of measurement only when it appears often in your report and is written out in full on first use. Use only abbreviations you are sure readers will understand. Abbreviate items in a visual aid only if you need to save space.

Here are common abbreviations for units of measurement:

AC	alternating current	kw	kilowatt
amp	ampere	kwh	kilowatt hour
Å	angstrom	l	liter
az	azimuth	lat	latitude
bbl	barrel	lb	pound
BTU	British Thermal Unit	lin	linear
C	Celsius	long	longitude
cal	calorie	log	logarithm
cc	cubic centimeter	m	meter
circ	circumference	max	maximum
cm	centimeter	mg	milligram
CPS	cycles per second	min	minute
cu ft	cubic foot	ml	milliliter
dB	decibel	mm	millimeter
DC	direct current	mo	month
dm	decimeter	mph	miles per hour
doz	dozen	oct	octane
DP	dewpoint	oz	ounce
F	Fahrenheit	psf	pounds per square foot
F	farad	psi	pounds per square inch
fbm	foot board measure	qt	quart
fl oz	fluid ounce	r	roentgen
FM	frequency modulation	rpm	revolutions per minute
freq	frequency	sec	second
ft	foot	sp gr	specific gravity
ft lb	foot pound	sq	square
gal	gallon	t	ton
GPM	gallons per minute	temp	temperature
gr	gram	tol	tolerance
hp	horsepower	ts	tensile strength
hr	hour	V	volt
in	inch	VA	volt ampere
IU	international unit	W	watt
J	joule	wk	week
ke	kinetic energy	WL	wavelength
kg	kilogram	yd	yard
km	kilometer	yr	year

Here are some common abbreviations for reference in manuscripts:

anon.	anonymous	fig.	figure
app.	appendix	i.e.	that is
b.	born	illus.	illustrated
©	copyright	jour.	journal
c., ca.	about (c. 1988)	l., ll.	line(s)
cf.	compare	ms., mss.	manuscript(s)
ch.	chapter	no.	number
col.	column	p., pp.	page(s)
d.	died	pt., pts.	part(s)
ed.	editor	rev.	revised or review
e.g.	for example	sec.	section
esp.	especially	sic	thus, so (to cite an error in the quotation)
et al.	and others		
etc.	and so on	trans.	translation
ex.	example	vol.	volume
f. or ff.	the following page or pages		

For abbreviations of other words, consult your dictionary. Most dictionaries lists abbreviations at the front or back or alphabetically with the word entry.

Capitalization

cap

Capitalize these items: proper nouns, titles of people, books and chapters, languages, days of the week, the months, holidays, names of organizations or groups, races and nationalities, historical events, important documents, and names of structures or vehicles. In titles of books, films, and so on, capitalize first and following words, except articles or short prepositions.

A Tale of Two Cities	the Chevrolet Corvette
Protestant	Russian
Wednesday	Labor Day
the *Queen Elizabeth II*	DuPont Chemical Company
the Statue of Liberty	Senator John Pasteur
April	France
Chicago	the War of 1812
the Bill of Rights	the Emancipation Proclamation

Do not capitalize the seasons, names of college classes (**senior, junior**), or general groups (**the younger generation,** or **the leisure class**).

Capitalize adjectives that are derived from proper nouns.

Chaucerian English

Capitalize titles preceding a proper noun but not those following:

> State Senator Marsha Smith
> Marsha Smith, state senator

Capitalize words such as **street, road, corporation, college** only when they accompany a proper noun.

> Bob Jones University
> High Street
> the Rand Corporation

Capitalize **north, south, east,** and **west** when they denote specific locations, not when they are simple directions.

> the South
> the Northwest
> Turn east at the next set of lights.

Begin all sentences with capitals.

> [#]

Numbers

If numbers can be expressed in one or two words, you can write them out or you can use the numerals.

fourteen	14
eighty-one	81
ninety-nine	99

For larger numbers, use numerals.

4,364	2,800,357
543	200 million
3¼	

Use numerals to express decimals, precise technical figures, or any other exact measurements. Numerals are more easily read and better remembered than numbers that are spelled out.

50 kilowatts	15 pounds of pressure
14.3 milligrams	4,000 rpm

Express these in numerals: dates, census figures, addresses, page numbers, exact units of measurement, percentages, ages, times with A.M. or P.M. designations, and monetary and mileage figures.

page 14	1:15 P.M.
18.4 pounds	9 feet

115 miles   12 gallons
the 9-year-old tractors $15
15.1 percent

Do not begin a sentence with a numeral.

Six hundred students applied for the 102 available jobs.

Do not use numerals to express approximate figures, time not designated as A.M. or P.M., or streets named by numbers less than 100.

about seven hundred fifty
four fifteen
108 East Forty-second Street

If one number immediately precedes another, spell out the first and use a numeral for the second:

Please deliver twelve 18-foot rafters.

In contracts and other documents in which precision is vital, a number can be stated both in numerals and in words:

The tenant agrees to pay a rental fee of three hundred seventy-five dollars ($375.00) monthly.

Works Cited

Adams, Gerald R., and Jay D. Schvaneveldt. *Understanding Research Methods.* New York: Longman, 1985.

The Aldus Guide to Basic Design. Aldus Corporation, 1988.

American Psychological Association. *Publication Manual of the American Psychological Association.* 4th ed. Washington: Author, 1994.

"Are We in the Middle of a Cancer Epidemic?" *University of California at Berkeley Wellness Letter* 10.9 (1994): 4–5.

Bailey, Edward P. *Writing Clearly: A Contemporary Approach.* Columbus, OH: Merrill, 1984.

Barbour, Ian. *Ethics in an Age of Technology.* New York: Harper, 1993.

Barfield, Woodrow, Mark Haselkorn, and Catherine Weatbrook. "Information Retrieval with a Printed User's Manual and with Online Hypercard Help." *Technical Communication* 37.1 (1990): 22–27.

Barnett, Arnold. "How Numbers Can Trick You." *Technology Review* Oct. 1994: 38–45.

Barnum, Carol, and Robert Fisher, "Engineering Technologists as Writers: Results of a Survey." *Technical Communication* 31.2 (1984): 9–11.

Baumann, K. E., et al. "Three Mass Media Campaigns to Prevent Adolescent Cigarette Smoking." *Preventive Medicine* 17 (1988): 510–30.

Beamer, Linda. "Learning Intercultural Communication Competence." *Journal of Business Communication* 29.3 (1992): 285–303.

Bedford, Marilyn S., and F. Cole Stearns. "The Technical Writer's Responsibility for Safety." *IEEE Transactions on Professional Communication* 30.3 (1987): 127–32.

Begley, Sharon. "Is Science Censored?" *Newsweek* 14 Sept. 1992.

Benson, Phillipa J. "Visual Design Considerations in Technical Publications." *Technical Communication* 32.4 (1985): 35–39.

Bernstein, Mark. "Deeply Intertwingled Hypertext: The Navigation Problem Reconsidered." *Technical Communication* 38.1 (1991): 41–47.

Bjerklie, David. "E-Mail: The Boss Is Watching." *Technology Review* 14 Apr. 1993: 14–15.

———. "Telecommuting: Preparing for Round Two." *Technology Review* 20 July 1995: 20–21.

Bogert, Judith, and David Butt. "Opportunities Lost, Challenges Met: Understanding and Applying Group Dynamics in Writing Projects." *Bulletin of the Association for Business Communication* 53.2 (1990): 51–53.

Boiarsky, Carolyn. "Using Usability Testing to Teach Reader Response." *Technical Communication* 39.1 (1992): 100–02.

Boucher, Norman. "Back to the Everglades." *Technology Review* Aug./Sept. 1995: 24–35.

Brittan, David. "Being There: The Promise of Multimedia Communications." *Technology Review* May/June 1992: 43–50.

Brody, Herb. "Great Expectations: Why Technology Predictions Sometimes Go Awry." *Technology Review* July 1991: 38–44.

Brownell, Judi, and Michael Fitzgerald. "Teaching Ethics in Business Communication: The Effective/Ethical Balancing Scale." *Bulletin of the Association for Business Communication* 55.3 (1992): 15–18.

Bruhn, Mark J. "E-Mail's Conversational Value." *Business Communication Quarterly* 58.3 (1995): 43–44.

Bryan, John. "Down the Slippery Slope: Ethics and the Technical Writer as Marketer." *Technical Communication Quarterly* 1.1 (1992): 73–88.

Burghardt, M. David. *Introduction to the Engineering Profession.* New York: Harper, 1991.

Burnett, Rebecca E. "Substantive Conflict in a Cooperative Context: A Way to Improve the Collaborative Planning of Workplace Documents." *Technical Communication* 38.4 (1991): 532–39.

Callahan, Sean. "Eye Tech." *Forbes ASAP* 7 June 1993.

Caswell-Coward, Nancy. "Cross-Cultural Communication: Is It Greek to You?" *Technical Communication* 39.2 (1992): 264–66.

Chauncey, Caroline. "The Art of Typography in the Information Age." *Technology Review* Feb./Mar. (1986): 26+.

Christians, Clifford G., et al. *Media Ethics: Cases and Moral Reasoning.* 2nd ed. White Plains, NY: Longman, 1978.

Clark, Gregory. "Ethics in Technical Communication: A Rhetorical Perspective." *IEEE Transactions on Technical Communication* 30.3 (1987): 190–95.

Clement, David E. "Human Factors, Instructions, and Warnings, and Product Liability." *IEEE Transactions on Professional Communication* 30.3 (1987): 149–56.

Cochran, Jeffrey K., et al. "Guidelines for Evaluating Graphical Designs." *Technical Communication* 36.1 (1989): 25–32.

Coletta, W. John. "The Ideologically Biased Use of Language in Scientific and Technical Writing." *Technical Communication Quarterly* 1.1 (1992): 59–70.

Communication Concepts, Inc. "Electronic Media Poses New Copyright Issues." *Writing Concepts* ©. Reprinted in *INTERCOM* [Newsletter of the Society for Technical Communication] Nov. 1995: 13+.

Consumer Product Safety Commission. *Fact Sheet No. 65.* Washington: GPO, 1989.

Cotton, Robert, ed. *The New Guide to Graphic Design.* Secaucus, NJ: Chartwell, 1990.

Council of Biology Editors. *Scientific Style and Format: The CBE Manual for Authors, Editors, and Publishers.* 6th ed. Chicago: Cambridge UP, 1994.

Cross, Mary. "Aristotle and Business Writing: Why We Need to Teach Persuasion." *Bulletin of the Association for Business Communication* 54.1 (1991): 3–6.

Crossen, Cynthia. *Tainted Truth: The Manipulation of Fact in America.* New York: Simon, 1994.

Dana-Farber Cancer Institute. *Facts and Figures about Cancer.* Boston: Author, 1995.

Debs, Mary Beth, "Collaborative Writing in Industry." *Technical Writing: Theory and Practice.* Ed. Bertie E. Fearing and W. Keats Sparrow. New York: Modern Language Assn., 1989: 33–42.

———. "Recent Research on Collaborative Writing in Industry." *Technical Communication* 38.4 (1991): 476–85.

Dombrowski, Paul M. "Challenger and the Social Contingency of Meaning: Two Lessons for the Technical Communication Classroom." *Technical Communication Quarterly* 1.3 (1992): 73–86.

Dragga, Sam, and Gwendolyn Gong. *Editing: The Design of Rhetoric.* Amityville, NY: Baywood, 1989.

Dumont, R. A. "Writing, Research, and Computing." *Critical Thinking: An SMU Dialogue,* 1988.

Dumont, R. A., and Lannon, J. M. *Business Communications.* 3rd ed. Glenview, IL: Scott, 1990.

Dyson, Esther. "Intellectual Value." *Wired* July 1995: 134+.

Dyson, Esther, and Nicholas Negroponte. "How Smart Agents Will Change Selling." *Forbes ASAP* 28 Aug. 1995: 95.

"Earthquake Hazard Analysis for Nuclear Power Plants." *Energy and Technology Review* June 1984: 8.

Elbow, Peter. *Writing without Teachers.* New York: Oxford, 1973.

"Electronic Mentors." *The Futurist* May 1992: 56.

Facts and Figures about Cancer. Boston: Dana-Farber Cancer Institute, 1990.

Felker, Daniel B., et al. *Guidelines for Document Designers.* Washington: American Institutes for Research, 1981.

Figgins, Ross. "The Future of Business Communication Technology: Where Are We Headed, Captain?" *Communication and Technology: Today and Tomorrow.* Ed. Al Williams. Denton, TX: Assn. for Business Communication, 1994. 123–42.

Fineman, Howard, "The Power of Talk." *Newsweek* 8 Feb. 1993: 24–28.

Finkelstein, Leo, Jr. "The Social Implications of Computer Technology for the Technical Writer." *Technical Communication* 38.4 (1991): 466–73.

Florman, Samuel, quoting Donald D. Rikard. "Toward Liberal Learning for Engineers." *Technology Review* Mar. 1986: 18–25.

Franke, Earnest A. "The Value of the Retrievable Technical Memorandum System to the Engineering Company." *IEEE Transactions on Professional Communication* 32.1 (Mar. 1989): 12–16.

Garfield, Eugene, "What Scientific Journals Can Tell Us about Scientific Journals." *IEEE Transactions on Professional Communication* 16.4 (1973): 200–02.

Gartiganis, Arthur. "Lasers." *Occupational Outlook Quarterly* Winter 1984: 22–26.

Gesteland, Richard R. "Cross-Cultural Compromises." *Sky* May 1993: 20+.

Gibaldi, Joseph. *MLA Handbook for Writers of Research Papers.* 4th ed. New York: Modern Language Assn., 1995.

Gibaldi, Joseph, and Walter S. Achtert. *MLA Handbook for Writers of Research Papers.* 3rd ed. New York: Modern Language Assn., 1988.

Gilbert, Nick, "1-8000-ETHIC." *Financial World* 16 Aug. 1994: 20+.

Gilsdorf, Jeanette W. "Executives' and Academics' Perception of the Need for Instruction in Written Persuasion." *Journal of Business Communication* 23.4 (1986): 55–68.

———. "Write Me Your Best Case for . . ." *Bulletin of the Association for Business Communication* 54.1 (1991): 7–12.

Girill, T. R. "Technical Communication and Art. *Technical Communication* 31.2 (1984): 35.

———. "Technical Communication and Ethics." *Technical Communication* 34.3 (1987): 178–79.

———. "Technical Communication and Law." *Technical Communication* 32.3 (1985): 37.

Glidden, H. K. *Reports, Technical Writing, and Specifications.* New York: McGraw, 1964.

Golen, Steven, et al. "How to Teach Ethics in a Basic Business Communications Class." *Journal of Business Communication* 22.1 (1985): 75–84.

Goodall, H. Lloyd, Jr., and Christopher L. Waagen. *The Persuasive Presentation.* New York: Harper, 1986.

Goodman, Danny. *Living at Light Speed.* New York: Random, 1994.

Gouran, Dennis S., et al. "A Critical Analysis of Factors Related to Decisional Processes Involved in the Challenger Disaster." *Central States Speech Journal* 37.3 (1986): 119–35.

Gribbons, William M. "Organization by Design: Some Implications for Structuring Information." *Journal of Technical Writing and Communication* 22.1 (1992): 57–74.

Grice, Roger A. "Document Development in Industry." *Technical Writing: Theory and Practice.* Ed. Bertie E. Fearing and W. Keats Sparrow. New York: Modern Language Assn., 1989. 27–32.

———. "Focus on Usability: Shazam!" *Technical Communication* 42.1 (1995): 131–33.

Grice, Roger A., and Lenore S. Ridgway. "Presenting Technical Information in Hypermedia Format: Benefits and Pitfalls." *Technical Communication Quarterly* 4.1 (1995): 35–46.

Gruman, Galen. "The Paper Chase." *MACWORLD* Apr. 1995: 126–31.

Hafner, Kate, "Have Your Agent Call My Agent." *Newsweek* 27 Feb. 1995: 76–77.

Halpern, Jean W. "An Electronic Odyssey." *Writing in Nonacademic Settings.* Ed. Dixie Goswami and Lee Odell. New York: Guilford, 1985. 157–201.

Harcourt, Jules. "Teaching the Legal Aspects of Business Communication." *Bulletin of the Association for Business Communication* 53.3 (1990): 63–64.

Hartley, James. *Designing Instructional Text.* 2nd ed. London: Kogan Page, 1985.

Hauser, Gerald. *Introduction to Rhetorical Theory.* New York: Harper, 1986.

Haynes, Kathleen J. M., and Linda K. Robertson. "An Application of Usability Criteria in the Classroom." *Technical Writing Teacher* 18.3 (1991): 236–42.

Hays, Robert. "Political Realities in Reader/Situation Analysis." *Technical Communication* 31.1 (1984): 16–20.

Hein, Robert G. "Culture and Communication." *Technical Communication* 38.1 (1991): 125–26.

Hill-Duin, Ann. "Terms and Tools: A Theory and Research-Based Approach to Collaborative Writing." *Bulletin of the Association for Business Communication* 53.2 (1990): 45–50.

Holler, Paul F. "The Challenge of Writing for Multimedia." *INTERCOM* [Newsletter of the Society for Technical Communication] July/Aug. 1995: 25.

Horton, William. "Is Hypertext the Best Way to Document Your Product?" *Technical Communication* 38.1 (1991): 20–30.

———. "Mix Media, Not Metaphors." *Technical Communication* 41.4 (1994): 781–83.

Howard, Tharon. "Property Issue in E-Mail Research." *Bulletin of the Association for Business Communication* 56.2 (1993): 40–41.

Huff, Darrell. *How to Lie with Statistics.* New York: Norton, 1954.

Hulbert, Jack E. "Developing Collaborative Insights and Skills." *Bulletin of the Association for Business Communication* 57.2 (1994): 53–56.

———. "Overcoming Intercultural Communication Barriers." *Bulletin of the Association for Business Communication* 57.2 (1994): 41–44.

Humphreys, Donald S. "Making Your Hypertext Interface Usable." *Technical Communication* 40.4 (1993): 754–61.

Hutheesing, Nikhil. "Who Needs the Middleman?" *Forbes* 28 Aug. 1995.

Hyakawa, S. I. *Language in Thought and Action.* 3rd ed. New York: Harcourt, 1972.

James-Catalano, "The Virtual Library." *Internet World* June 1995: 26+.

Jameson, Daphne A. "Using a Simulation to Teach Intercultural Communication in Business Communication Courses." *Bulletin of the Association for Business Communication* 56.1 (1993): 3–11.

Janis, Irving L. *Victims of Groupthink: A Psychological Study of Foreign Policy Decisions and Fiascos.* Boston: Houghton, 1972.

Johannesen, Richard L. *Ethics in Human Communication.* 2nd ed. Prospect Heights, IL: Waveland, 1983.

Journet, Debra. Unpublished review of *Technical Writing*. 3rd ed.

Kawasaki, Guy. "The Rules of E-Mail." *MACWORLD* Oct. 1995: 286.

Kelley-Reardon, Kathleen. *They Don't Get It Do They?: Communication in the Workplace—Closing the Gap between Women and Men*. Boston: Little, 1995.

Kelman, Herbert C. "Compliance, Identification, and Internalization: Three Processes of Attitude Change." *Journal of Conflict Resolution* 2 (1958): 51–60.

Keyes, Elizabeth. "Typography, Color, and Information Structure." *Technical Communication* 40.4 (1993): 638–54.

Kiely, Thomas. "The Idea Makers." *Technology Review* Jan. 1993: 31–40.

Kipnis, David, and Stuart Schmidt. "The Language of Persuasion." *Psychology Today* Apr. 1985: 40–46. Rpt. in Raymond S. Ross, *Understanding Persuasion*. 3rd ed. Englewood Cliffs: Prentice, 1990.

Kirsh, Lawrence. "Take It from the Top." *MACWORLD* Apr. 1986: 112–15.

Kleimann, Susan D. "The Complexity of Workplace Review." *Technical Communication* 38.4 (1991): 520–26.

Kohl, John R., et al. "The Impact of Language and Culture on Technical Communication in Japan." *Technical Communication* 40.1 (1993): 62–72.

Kraft, Stephanie. "Whistleblower Bill's Holiday Adventures." *The Valley Advocate* [Northhampton, MA] 6 Jan. 1994: 5–6.

Kremers, Marshall. "Teaching Ethical Thinking in a Technical Writing Course." *IEEE Transactions on Professional Communication* 32.2 (1989): 58–61.

Lambert, Steve. *Presentation Graphics on the Apple® Macintosh*. Bellevue, WA: Microsoft, 1984.

LaPlante, Alice. "Brainstorming." *Forbes ASAP* 25 Oct. 1993: 45–61.

———. "Imaging Your Sea of Data." *Forbes ASAP* 29 Aug. 1994: 37–41.

Larson, Charles U. *Persuasion: Perception and Responsibility*. 7th ed. Belmont, CA: Wadsworth: 1995.

Lavin, Michael R. *Business Information: How to Find It, How to Use it*. 2nd ed. Phoenix, AZ: Oryx, 1992.

Lederman, Douglas. "Colleges Report Rise in Violent Crime," *Chronicle of Higher Education* 3 Feb. 1995, sec. A: 31–42.

Leki, Ilona. "The Technical Editor and the Non-native Speaker of English." *Technical Communication* 37.2 (1990): 148–52.

Levi, Stephen. "Optimism about the Net." *MACWORLD* July 1994: 179–80.

Lewis, Philip L., and N. L. Reinsch. "The Ethics of Business Communication." *Proceedings of the American Business Communication Conference*. Champaign, IL., 1981. In *Technical Communication and Ethics*. Ed. John R. Brockman and Fern Rook. Washington: Society for Technical Communication, 1989: 29–44.

Littlejohn, Stephen W. *Theories of Human Communication*. 2nd ed. Belmont, CA: Wadsworth, 1983.

Littlejohn, Stephen W., and David M. Jabusch. *Persuasive Transactions*. Glenview, IL: Scott, 1987.

McDonald, Kim A. "Some Physicists Criticize Research Purporting to Show Links between Low-Level Electromagnetic Fields and Cancer." *Chronicle of Higher Education* 3 May 1991, sec. A: 5+.

MacKenzie, Nancy, Unpublished review of *Technical Writing*. 5th ed.

Mackin, John. "Surmounting the Barrier between Japanese and English Technical Documents." *Technical Communication* 36.4 (1989): 346–51.

Maeglin, Thomas. Unpublished review of *Technical Writing*, 7th ed.

Martin, James A. "A Road Map to Graphics Web Sites." *MACWORLD* Oct. 1995: 135–36.

Martin, Jeanette S., and Lillian H. Chaney. "Determination of Content for a Collegiate Course in Intercultural Business Communication by Three Delphi Panels." *Journal of Business Communication* 29.3 (1992): 267–83.

Max, Robert R. "Wording It Correctly." *Training and Development Journal* Mar. 1985: 5–6.

McGuire, Gene. "Shared Minds: A Model of Collaboration." *Technical Communication* 39.3 (1992): 467–68.

Meyer, Benjamin D. "The ABCs of New-Look Publications." *Technical Communication* 33.1 (1986): 13–20.

Microsoft Word User's Guide: Word Processing Program for the Macintosh, Version 5.0. Redmond, WA: Microsoft Corporation, 1992.

Mirel, Barbara, Susan Feinberg, and Leif Allmendinger. "Designing Manuals for Active Learning Styles." *Technical Communication* 38.1 (1991): 75–87.

Mokhiber, Russell. "Crime in the Suites." *Greenpeace* May 1989: 14–16.

Monatersky, R. "Do Clouds Provide a Greenhouse Thermostat?" *Science News* 142 (1992): 69.

Morgan, Meg. "Patterns of Composing: Connections between Classroom and Workplace Collaborations." *Technical Communication* 38.4 (1991): 540–42.

Morgenson, Gretchen. "Would Uncle Sam Lie to You?" *Worth* Nov. 1994: 53+.

Morse, June. "Hypertext—What Can We Expect?" *STC Intercom* [Newsletter of the Society for Technical Communication] Feb. 1992: 6–7.

Nantz, Karen S., and Cynthia L. Drexel. "Incorporating Electronic Mail with the Business Communication Course." *Business Communication Quarterly* 58.3 (1995) 45–51.

Nelson, Sandra J., and Douglas C. Smith. "Maximizing Cohesion and Minimizing Conflict in Collaborative Writing Groups. *Bulletin of the Association for Business Communication* 53.2 (1990): 59–62.

Nickels-Shirk, Henrietta. "'Hyper' Rhetoric: Reflections on Teaching Hypertext." *Technical Writing Teacher* 18.3 (1991): 189–200.

"Notes." *Technology Review* July 1993: 72.

Nydell, Margaret K. *Understanding Arabs: A Guide for Westerners.* New York: Logan, 1987.

"On Line." *Chronicle of Higher Education* 21 Sept. 1992, sec. A: 1.

"Olive Oil and Breast Cancer: How Strong a Connection?" *University of California at Berkeley Wellness Letter* 11.7 (1995): 1–2.

Ornatowski, Cezar M. "Between Efficiency and Politics: Rhetoric and Ethics in Technical Writing." *Technical Communication Quarterly* 1.1 (1992): 91–103.

Pace, Roger C. "Technical Communication, Group Differentiation, and the Decision to Launch the Space Shuttle Challenger." *Journal of Technical Writing and Communication* 18.3 (1988): 207–20.

Pender, Kathleen. "Dear Computer, I Need a Job." *Worth* Mar. 1995: 120–21.

Perelman, Lewis, J. "How Hypermation Leaps the Learning Curve." *Forbes ASAP* 25 Oct. 1993: 78+.

Perloff, Richard M. *The Dynamics of Persuasion.* Hillsdale, NJ: Erlbaum, 1993.

Peyser, Marc, and Steve Rhodes. "When E-Mail Is Oops-Mail." *Newsweek* 16 Oct. 1995: 82.

Pinelli, Thomas E., et al. "A Survey of Typography, Graphic Design, and Physical Media in Technical Reports." *Technical Communication* 33.2 (1986): 75–80.

Plumb, Carolyn, and Jan H. Spyridakis. "Survey Research in Technical Communication: Designing and Administering Questionnaires." *Technical Communication* 39.4 (1992): 625–38.

Porter, James E. "Truth in Technical Advertising: A Case Study." *IEEE Transactions on Professional Communication* 30.3 (1987): 182–89.

Presidential Commission. *Report to the President on the Space Shuttle Challenger Accident.* Vol I. Washington: GPO, 1986.

Pugliano, Fiore. Unpublished review of *Technical Writing.* 5th ed.

Redish, Janice C., and David A. Schell. "Writing and Testing Instructions for Usability." *Technical Writing: Theory and Practice.* Ed. Bertie E. Fearing and W. Keats Sparrow. New York: Modern Language Assn., 1987: 61–71.

Redish, Janice C., et al. "Making Information Accessible to Readers." *Writing in Nonacademic Settings.* Ed. Lee Odell and Dixie Goswami: New York: Guilford, 1985.

Richards, Thomas O., and Ralph A. Richards, "Technical Writing." Paper presented at the University of Michigan, 11 July 1941. Published by the Society for Technical Communication.

Riney, Larry A. *Technical Writing for Industry.* Englewood Cliffs: Prentice, 1989.

Rokeach, Milton. *The Nature of Human Values.* New York: Free, 1973.

Ross, Philip E. "Lies, Damned Lies, and Medical Statistics." *Forbes* 14 Aug. 1995: 130–35.

Ross, Raymond S. *Understanding Persuasion.* 3rd ed. Englewood Cliffs: Prentice, 1990.

Rothschild, Michael. "When You're Gagging on E-Mail." *Forbes ASAP* 6 June 1994: 25–26.

Rottenberg, Annette, T. *Elements of Argument.* 3rd ed. New York: St. Martin's, 1991.

Rowland, D. *Japanese Business Etiquette: A Practical Guide to Success with the Japanese.* New York: Warner: 1985.

Rowland Robert C. "The Relationship between the Public and the Technical Spheres of Argument: A Case Study of the Challenger Seven Disaster." *Central States Speech Journal* 37.3 (1986): 136–46.

Rubens, Philip M. "Reinventing the Wheel?: Ethics for Technical Communicators." *Journal of Technical Writing and Communication* 11.4 (1981): 329–39.

Ruggiero, Vincent R. *The Art of Thinking.* 3rd ed. New York: Harper, 1991.

Ruhs, Michael A. "Usability Testing: A Definition Analyzed." *Boston Broadside* [Newsletter of the Society for Technical Communication] May/June 1992: 8+.

Samuelson, Robert. "Merchants of Mediocrity." *Newsweek* 1 Aug. 1994: 44.

Schenk, Margaret T., and James K. Webster. *Engineering Information Resources.* New York: Decker, 1984.

Schwartz, Marilyn, et al. *Guidelines for Bias-Free Writing.* Bloomington, IN: Indiana UP, 1995.

Scott, James C., and Diana J. Green. "British Perspectives on Organizing Bad-News Letters: Organizational Patterns Used by Major U.K. Companies." *Bulletin of the Association for Business Communication* 55.1 (1992): 17–19.

Selzer, Jack. "Composing Processes for Technical Discourse." *Technical Writing: Theory and Practice.* Ed. Bertie E. Fearing and W. Keats Sparrow. New York: Modern Language Assn., 1989: 43–50.

Sherblom, John C., Claire F. Sullivan, and Elizabeth C. Sherblom, "The What, the Whom, and the Hows of Survey Research," *Bulletin of the Association for Business Communication* 56:12 (1993): 58–64.

Sherif, Muzapher, et al. *Attitude and Attitude Change: The Social Judgment-Involvement Approach.* Philadelphia: Saunders, 1965.

Snyder, Joel. "Finding It on Your Own." *Internet World* June 1995: 89–90.

Spruell, Geraldine. "Teaching People Who Already Learned How to Write, to Write." *Training and Development Journal* Oct. 1986: 32–35.

Spyridakis, Jan H. "Conducting Research in Technical Communication: The Application of True Experimental Design." *Technical Communication* 39.4 (1992): 607–24.

Spyridakis, Jan H., and Michael J. Wenger. "Writing for Human Performance: Relating Reading Research to Document Design." *Technical Communication* 39.2 (1992): 202–15.

Stanton, Mike. "Fiber Optics." *Occupational Outlook Quarterly* Winter 1984: 27–30.

Staudennmaier, S. J. "Engineering with a Human Face." *Technology Review* July 1991: 66–67.

Steinberg, Stephen. "Travels on the Net." *Technology Review* July 1994: 20–31.

Stevenson, Richard W. "Workers Who Turn in Bosses Use Law to Seek Big Rewards." *New York Times* 10 July 1989, sec A: 1.

Stone, Brad, and T. Trent Gegax. "Your Favorite Sites." *Newsweek* 20 Nov. 1995: 16.

Stone, Peter H. "Forecast Cloudy: The Limits of Global Warming Models." *Technology Review* Feb./Mar. 1992: 32–40.

Stonecipher, Harry. *Editorial and Persuasive Writing.* New York: Hastings, 1979.

Strizich, Martha. "Companywide Online Services." *MACWORLD* Jan. 1995: 12–25.

Sturges, David L. "Internationalizing the Business Communication Curriculum." *Bulletin of the Association for Business Communication* 55.1 (1992): 30–39.

Teague, John H. "Marketing on the World Wide Web." *Technical Communication* 42.2 (1995): 236–42.

Templeton, Brad. "10 Big Myths about Copyright Explained." 29 Nov. 1994. Online posting. Listserve law/copyright/FAQ/myths/part 1 BITNET. 6 May 1995.

Thrush, Emily A. "Bridging the Gap: Technical Communication in an Intercultural and Multicultural Society." *Technical Communication Quarterly* 2.3 (1993): 271–83.

Unger, Stephen H. *Controlling Technology: Ethics and Responsible Engineer.* New York: Holt, 1982.

U.S. Air Force Academy. *Executive Writing Course.* Washington: GPO, 1981.

U.S. Department of Commerce. *Statistical Abstract of the United States* Washington: GPO, 1994.

U.S. Department of Labor. *Tips for Finding the Right Job.* Washington: GPO, 1993.

———. *Tomorrow's Jobs.* Washington: GPO 1995.

U.S. General Services Administration. *Your Rights to Federal Records.* Washington: GPO, 1995.

"Using Icons As Communication." *Simply Stated* [Newsletter of the Document Design Center, American Institutes for Research] 75 (Sept./Oct. 1987): 1+.

van der Meij, Hans, and John M. Carroll. "Principles and Heuristics for Designing Minimalist Instruction." *Technical Communication* 42.2 (1995): 243–61.

Van Pelt, William. Unpublished review of *Technical Writing.* 3rd ed.

Varner, Iris I., and Carson H. Varner. "Legal Issues in Business Communications." *Journal of the American Association for Business Communication* 46.3 (1983): 31–40.

Vaughan, David K. "Abstracts and Summaries: Some Clarifying Distinctions." *Technical Writing Teacher* 18.2 (1991): 132–41.

Velotta, Christopher. "How to Design and Implement a Questionnaire." *Technical Communication* 38.3 (1991): 387–92.

Victor, David A. *International Business Communication.* New York: Harper, 1992.

Walker, Janice R. "MLA-Style Citations of Electronic Sources." April 1995. The Alliance for Computers and Writing. http://prarie-island.ttu.edu/acw.html (10 Jan. 1996).

Walter, Charles, and Thomas F. Marsteller. "Liability for the Dissemination of Defective Information." *IEEE Transactions on Professional Communication* 30.3 (1987): 164–67.

Watkins, Beverly T. "Many Campuses Start Building Tomorrow's Electronic Library." *Chronicle of Higher Education* 2 Sept. 1992, sec. A: 1+.

Weinberg, Neil. "New Age Documents." *Forbes* 6 June 1994: 75.

Weinstein, Edith K. Unpublished review of *Technical Writing.* 5th ed.

Weiss, Edmond H. *How to Write a Usable User Manual.* Philadelphia: ISI, 1985.

Welz, Gary. "Job Seeking on the Net." *Internet World* May 1995: 52.

Weymouth, L. C. "Establishing Quality Standards and Trade Regulations for Technical Writing in World Trade." *Technical Communication* 37.2 (1990): 143–47.

White, Jan. *Color for the Electronic Age.* New York: Watson-Guptill, 1990.

———. *Editing by Design.* 2nd ed. New York: Bowker, 1982.

———. *Great Pages,* El Segundo, CA: Serif, 1990.

———. *Visual Design for the Electronic Age.* New York: Watson-Guptill, 1988.

Wicclair, Mark, R., and David K. Farkas. "Ethical Reasoning in Technical Communication: A Practical Framework." *Technical Communication* 31.2 (1984): 15–19.

Wickens, Christopher D. *Engineering Psychology and Human Performance.* 2nd ed. New York: Harper, 1992.

Wight, Eleanor, "How Creativity Turns Facts into Usable Information." *Technical Communication* 32.1 (1985): 9–12.

Williams, Joseph. *Style.* 4th ed. New York: Harper, 1994.

Williams, Robert I. "Playing with Format, Style, and Reader Assumptions." *Technical Communication* 30.3 (1983): 11–13.

Winsor, D. A. "Communication Failures Contributing to the Challenger Accident: An Example for Technical Communicators." *IEEE Transactions on Professional Communication* 31.3 (1988): 101–07.

Wojahn, Patricia G. "Computer-Mediated Communication: The Great Equalizer between Men and Women?" *Technical Communication* 41.4 (1994): 747–51.

Wriston, Walter. *The Twilight of Sovereignty.* New York: Scribner's, 1992.

Yoos, George. "A Revision of the Concept of Ethical Appeal." *Philosophy and Rhetoric* 12.4 (1979): 41–58.

Zinsser, William. *On Writing Well.* New York: Harper, 1980.

Index

Correction Symbols

Symbol	Meaning	Page	Symbol	Meaning	Page
ab	abbreviation	631	*– – /*	dashes	626
agr p	pronoun/referent		*. . . /*	ellipses	625
	agreement	611	*! /*	exclamation point	617
agr sv	subject/verb		*– /*	hyphen	627
	agreement	610	*ital*	italics	625
amb	ambiguity	262	*( ) /*	parentheses	625
appr	inappropriate diction	298	*. /*	period	617
bias	biased tone	298	*? /*	question mark	617
ca	pronoun case	612	*" / "*	quotation	624
cap	capitalization	633	*; /*	semicolon	617
chop	choppy sentences	279	*qual*	needless qualifier	278
cl	clutter word	278	*red*	redundancy	272
coh	paragraph coherence	252	*rep*	needless repetition	272
cont	contraction	296	*ref*	faulty reference	611
coord	coordination	608	*ro*	run-on sentence	608
cs	comma splice	607	*seq*	sequence of development	
dgl	dangling modifier	613		in a paragraph	254
euph	euphemism	288	*sexist*	sexist usage	300
exact	inexact word	284	*shift*	sentence shift	615
frag	sentence fragment	605	*st mod*	stacked modifiers	265
gen	generalization	289	*str*	paragraph structure	250
jarg	needless jargon	286	*sub*	subordination	609
len	paragraph length	253	*th op*	"th" sentence openers	273
lev	level of technicality	29	*trans*	transition	628
mng	meaning unclear	262	*trite*	triteness	288
mod	misplaced modifier	613	*un*	paragraph unity	252
noun ad	noun addiction	276	*v*	voice	266
om	omitted word	263	*var*	sentence variety	281
over	overstatement	289	*w*	wordiness	271
par	parallelism	615	*wo*	word order	265
pct	punctuation	616	*ww*	wrong word	290
ap/	apostrophe	622	*#*	numbers	634
[ ] /	brackets	626	¶	begin new paragraph	250
: /	colon	618	*ts*	topic sentence	251
, /	comma	618			